SCIENCE AND CIVIL SOCIETY

EDITED BY

Lynn K. Nyhart and Thomas H. Broman

OSIRIS | 17

A Research Journal Devoted to the History of Science and Its Cultural Influences

SCIENCE AND CIVIL SOCIETY

Acknowledgments

The papers gathered here were first discussed at a workshop on "Science and Civil Society," held in Madison, Wisconsin in April of 2000. We wish to thank the National Science Foundation (Grant No. SES-9910994), the University of Wisconsin's Anonymous Fund, and the University of Wisconsin's History of Science Department for providing material support to the workshop, and the workshop participants for their engaged and thoughtful participation. Our heartfelt thanks go as well to Kathryn Olesko for inviting us to edit this volume and for her attentive and tireless work in guiding it through the production process.

Lynn K. Nyhart
Thomas H. Broman
Madison, Wisconsin
February 2002

History and the History of Science *Redux*: A Preface

*By Kathryn Olesko**

ABSTRACT

This preface announces the shift in the intellectual orientation of *Osiris* toward the continuing mediation between history and the history of science, an issue first addressed by Thomas S. Kuhn. The background for this volume on science and civil society in the work of Jürgen Habermas is explained. Future issues of *Osiris* will mediate history of science and urban history, environmental history, comparative politics, and international affairs.

IN AN AGE of outspoken interdisciplinarity, it seems appropriate to return to how closely related disciplines communicate with one another. The history of science has since its inception embraced interdisciplinary perspectives and methods that other fields of knowledge have only more recently integrated into practice. The burgeoning field of science studies has multiplied the angles from which science as a human activity can be cast and understood. Whereas initially interdisciplinarity in the history of science remained confined to the broad but self-contained domains of history, philosophy, theology, sociology, and the natural sciences, by the end of the twentieth century anthropology, ethnology, political science, and literary and cultural studies had also treated science as an object of scholarly investigation, and had done so with great success. Yet the conceptual richness of science studies has also blurred distinctions between what is historical and what is not in the understanding of science as a human activity. Perhaps it is merely a sign of the pervasive "postmodern condition" that history as a mode of understanding is viewed as no different than any other. Nonetheless, as Friedrich Nietzsche argued over a century ago, we can no more do without the historical than the unhistorical.[1]

The history of science has always had an uncertain relationship with the larger

* BMW Center for German & European Studies, Georgetown University, Washington, DC 20057-1022; oleskok@georgetown.edu.

I extend my heartfelt thanks to those who assisted me in the preparation of this volume, especially Margaret Rossiter, Marc Rothenberg, Jarelle Stein, Jennifer Paxton, Teresa Mullen, Christopher Wiley, Eric Tesdall and members of the *Osiris* Editorial Board. It was a pleasure to work with Lynn Nyhart, Tom Broman and contributors to this volume. My gratitude is extended to the BMW Center for German & European Studies for its support of the Managing Editor's position.

[1] "On the Uses and Disadvantages of History for Life," reprinted in Friedrich Wilhelm Nietzsche, *Nietzsche: Untimely Meditations* (Cambridge: Cambridge University Press, 1997), pp. 57–124.

historical discipline of which it is a part. In 1971 Thomas S. Kuhn addressed the vexing issue of the relations between history and the history of science.[2] Since then the intra-disciplinary divide has narrowed but not entirely closed. We would no longer agree with the radical distinction that Kuhn drew between science and technology, nor would we place responsibility for the mediation entirely upon historians. And we certainly would not argue that the gap could only be closed by understanding the technical core of science. But we would still acknowledge, as he did, that a boundary between history and history of science still exists despite the central role of science in western civilization—to which we would quickly add that since Kuhn's essay, the role of science in world history has increased considerably with globalization, making the interweaving of the two fields even more essential. Now history of science doctoral candidates take more fields in history than in the past, but few history doctoral students take fields in the history of science. The integration of history of science into general history textbooks is certainly better, but there is still a tendency to approach science through great men, great sources, and the grand narrative of the Enlightenment promise of emancipation through reason. In the general education courses built on the synthesis of major historical problems, casting questions about science in the framework of larger issues—the nation-state and empire, social change, religious conflict, economic development, democracy and totalitarianism, and so on—is still all too rare. There are nonetheless hopeful signs for the future. More historians of science teach from within history departments now, a propitious opportunity for greater dialogue. Historians of science and technology have joined and are still joining other scholars in rewriting textbooks for the popular history courses in American History, World History and Western Civilization.[3] Postmodern debates on history are engaging historians of science, who in the process are attacking iconoclastic images of science.[4] Most encouraging for the mediation are the recent debates engaging historians and historians of science on the pages of the *American Historical Review*.[5]

To date, though, there has been no journal dedicated exclusively to the mediation of history and history of science. I took on the Editor's position with the intention of shifting *Osiris* decisively in that direction, and luckily for me, the History of Science Society Executive Committee, the Committee on Publications, and the *Osiris* Editorial Board posed no major opposition to my plan. Yet not much needed to be done at all, for in reorienting *Osiris*, I stood on the shoulders of giants. Since

[2] "The Relations between History and the History of Science," reprinted in Thomas S. Kuhn, *The Essential Tension: Selected Studies in Scientific Tradition and Change* (Chicago: University of Chicago Press, 1977), pp. 127–161.

[3] For example, Marvin Perry, Myrna Chase, James R. Jacob, Margaret C. Jacob and Theodore H. Von Laue, *Western Civilization: Ideas, Politics, and Society*, 6th ed. (New York: Houghton Mifflin, 2000); Peter Sterns, Michael Adas, Stuart Schwartz and Marc Jason Gilbert, *World Civilizations: The Global Experience*, 3rd ed. (NY: Longman, 2001).

[4] Joyce Oldham Appleby, Lynn Hunt and Margaret C. Jacob, *Telling the Truth about History* (NY: W. W. Norton, 1995). See also Margaret C. Jacob's integrative monographs, especially *Scientific Culture and the Making of the Industrial West* (Oxford: Oxford University Press, 1997).

[5] Anthony Grafton, Elizabeth Eisenstein, and Adrian Johns, "*AHR* Forum: How Revolutionary Was the Print Revolution?," *American Historical Review* 107 (2002): 84–128. On Alfred W. Crosby's provocative *The Measure of Reality: Quantification and Western Society, 1250–1600* (Cambridge: Cambridge University Press, 1997), see Roger Hart, Margaret C. Jacob, and Jack A. Goldstone, "Review Essays: Counting and Power," *American Historical Review* 105 (2000): 485–508.

its reestablishment in 1985, *Osiris* had gradually become a journal dedicated to broad issues mediating major themes in history and history of science, especially under Margaret Rossiter's editorship. There is no better example of this intellectual evolution than *Colonialism and Science*, guest edited by Roy MacLeod.[6] The essays in that volume so closely interweave science with colonial and postcolonial experiences that to separate the strands of their analyses would be to unravel the fabric of history. Like MacLeod's volume, the present one on *Science and Civil Society* seeks to open up new categories of analysis, new areas of investigation, and new ways of mediating history and the history of science. Future volumes under my editorship will concern the history of science in its relationship to urban history, environmental history, comparative politics, and international affairs.

Science and Civil Society builds on the approaches to historical scholarship for which *Osiris* has become known. A virtual point of origin for the essays in this volume, as many authors acknowledge, is Jürgen Habermas's pathbreaking work, *The Structural Transformation of the Public Sphere*.[7] Following the English translation of his work in 1989 scholars criticized Habermas for not including science as a part of the public sphere.[8] Yet science, technology, and the forms of rationality associated with them were in fact foundational in Habermas's understanding of how opinion formation, speech, and communication occurred in the public sphere. The rationality of science and technology was equally central to his conceptualization of civil society—as associational life based on voluntary agreements—for the behavioral norms characterizing civil society in the modern period drew so heavily upon rational principles in the evolution of forms of sociability (a symbolic form of communication), civility (including the capacity to listen), critical expression, and social trust. What Habermas found troublesome was that rationality had gone awry: rather than remaining the basis of a pristine, noise-free form of communication, scientific and technical rationality became bound, from the late nineteenth-century onward, to processes of production and so became contaminated by economic interests. Instead of promoting novel forms of critical expression, scientific and technical rationality contributed to the narrowing of choices in the public sphere and civil society; communication was stifled, opinion-formation controlled, and behavioral norms too rigidly standardized. Habermas sought to articulate the conditions and the qualities of "another reason," another form of rationality that would extend communication, keep it free from domination, and reawaken the emancipatory potential of rationality though the expansion of critical thinking. By the late 1960s he thought he had found the sources of "another reason" in the student revolts of that era, but that potential was never realized.[9]

Historians of science have already taken up elements of the Habermasian analysis of forms of rationality—associational life and social trust are two of the most

[6] Roy MacLeod, ed., *Colonialism and Science*, *Osiris* 15 (2000).

[7] Jürgen Habermas, *The Structural Transformation of the Public Sphere: An Inquiry into a Category of Bourgeois Society* (Cambridge, Mass.: MIT Press, 1989). Originally published as *Structurwandel der Öffentlichkeit: Untersuchungen zu einer Kategorie der bürgerlichen Gesellschaft* (Neuwied: Luchterhand, 1962).

[8] E.g., Paul Wood, "Science, the Universities, and the Public Sphere in Eighteenth-Century Scotland," *History of Universities* 8 (1994): 99–135.

[9] Jürgen Habermas, *Toward a Rational Society: Student Protest, Science, and Politics* (Boston: Beacon Press, 1970), especially the essay "Technology and Science as Ideology," pp. 81–122.

popular categories investigated[10] As Jan Golinski has admirably demonstrated in his examination of chemistry's role in Enlightenment public life, the new forms of behavior that characterized civil society had a profound impact not only upon the public presentation of scientific knowledge, but also upon the public roles of science and scientists. For some the rational sobriety of civil society clashed with the private interests of industrial development, circumscribing the sphere over which scientists could have influence. Golinski thus highlights the deeply symbiotic relationship between the development of chemistry and that of civil culture at the end of the eighteenth century.[11] *Science and Civil Society* seeks to define new areas of investigation that will further develop broad historical issues in which the civil and public roles of science are such important parts, including in the area that Celia Applegate identifies in her commentary as rational-critical debate. How did science and technology shape the forms of rational exchange in daily life? What cultural forms circumscribed the use of reason in daily life? How might expertise and professionalization, two categories used by historians of science, be reconceptualized in terms of the historical issues of the public sphere and civil society? If science and scientists shaped civil society, then how did scientists' participation in the public sphere and civil society shape science? These are some of the issues that continue the discussion found in this volume.

Marking the shift in the intellectual orientation of *Osiris* is a redesigned cover. There is no more appropriate image of civil society than the town square. Friedrich Weinbrenner's design for the Marktplatz in Karlsruhe, Germany (conceived 1797, built 1806–1826), symbolically captures the theme of science and civil society. Weinbrenner conceived of his public architectural projects as spaces designed for the social participation and intellectual exchange of an educated public willing to intermingle with people from all walks of life. Influenced by the panorama, whose popularity was then rising, Weinbrenner constructed his public projects so as to engage vision, the Enlightenment's key sense, in challenging ways. Whereas his theater interiors capitalized upon perspective's potential for creating uncanny optical illusions, his public architecture accentuated the role of reason in daily life. The imposing pyramidal structure at the center of Karlsruhe's town square, which was intended to draw the public's attention to geometrical regularities, is perhaps the best exemplar of Weinbrenner's socially conscious "architectonic reason." Weinbrenner's town square is a fitting metaphor for the themes contained in this volume.[12]

[10] E.g., Margaret C. Jacob, *Living the Enlightenment: Freemasonry and Politics in Eighteenth-Century Europe* (Oxford: Oxford University Press, 1991) and Steven Shapin, *A Social History of Truth: Civility and Science in Seventeenth-Century England* (Chicago: University of Chicago Press, 1994).

[11] Jan Golinski, *Science as Public Culture: Chemistry and Enlightenment in Britain, 1760–1820*

[12] See Friedrich Weinbrenner, *Über Theater in architektonischer Hinsicht mit Beziehung auf Plan und Ausführung des neuen Hoftheaters zu Carlsruhe* (Tübingen: J. G. Cotta, 1809) and his *Architektonisches Lehrbuch,* 3 vols. (Tübingen: J. G. Cotta, 1810–1825).

Introduction:
Some Preliminary Considerations on Science and Civil Society

*By Thomas H. Broman**

ABSTRACT

Until now, the interest displayed in civil society by historians and political scientists has not found much echo among historians of science. This is regrettable, because the doctrine of civil society offers historians of science a means of investigating the relationship between the forms of social life characteristic of civil society (voluntary associations, civic institutions, etc.) and engagement with science. The doctrine of civil society also permits historians to inquire into how science either legitimates or undermines political authority. Most importantly, civil society permits an analysis of the basis of science's public authority. As the articles in this collection show, the attempt to occupy the position of representing science as public knowledge has been a central concern of social and professional groups in a wide variety of settings.

DURING THE PAST FIFTEEN YEARS, the concept of civil society has become quite prominent in the thinking of political scientists and other diagnosticians of the modern condition.[1] The interest in civil society, taken here in its broadest meaning as describing a realm of social life positioned between the family and the state, is not all that difficult to trace. One source is provided by the political transformations seen in Eastern Europe in the 1980s, when independent political and social movements, such as Solidarity in Poland, contributed to the overthrow of totalitarian regimes and the restoration of democratic political institutions. Nor have only communist and totalitarian regimes been involved in this process, as evidenced by the overthrow of the autocratic Ferdinand Marcos from the presidency of the Philippines in 1986. More recently still, the election of Vicente Fox to the presidency of Mexico in 2000 signaled the end of more than seventy years of virtual one-party rule by the Institutional Revolutionary Party (PRI).

* Department of History of Science, University of Wisconsin–Madison, 7133 Social Science, 1180 Observatory Dr., Madison, WI 53706-1393; thbroman@facstaff.wisc.edu.

More even than is usually the case, the author wishes to thank those who have patiently discussed these matters with him, critiqued drafts, and otherwise attempted to deepen his understanding of science and civil society: Cathy Carson, John Carson, Peter Dear, Volker Hess, Clark Miller, Ted Porter, Mary Terrall, and Jessica Wang. Most especially, he thanks Lynn Nyhart, who has had to endure his recurrent bouts of conceptual incoherence and over-the-top prose for far too long.

[1] Ernest Gellner, *Conditions of Liberty: Civil Society and Its Rivals* (London: Allen Lane, 1994); John Keane, *Civil Society: Old Images, New Visions* (London: Polity, 1999); R. Fine and Shirin Rai, eds., *Civil Society: Democratic Perspectives* (London: Frank Cass, 1997); Ralf Dahrendorf, *After 1989: Morals, Revolution, and Civil Society* (New York: St. Martin's, 1997). Additional literature will be cited below.

 0369-7827/02/1701-0001$10.00

Another source of interest in civil society has been the systemic weakening of the welfare state in Europe and North America. These changes have been signaled by the growing inability of governments to provide high levels of social support for retirees and the physically disabled, for example, and the increasing pressure on countries such as Canada and Great Britain to maintain their systems of state-supported medical care in the face of constantly rising costs. In the United States, which did not have a well-developed welfare state to begin with, there has been a strong movement to cut long-term unemployment benefits and to privatize functions such as education (although this has failed to show much effect) and the operation of prisons. Indicative of this trend is the sight of so-called public universities now chasing after patents and private endowments with the fanatical zeal once shown only by schools such as Harvard and Princeton.

These developments have combined to direct the attention of historians and other scholars away from the state and toward those institutions that structure social life independently of government. The concept of civil society is attractive in these contexts because, according to Keith Tester, it is a resource for understanding what it is that has "made the social world social."[2] Moreover, it is not only the patterns of social life that are of interest here; just as significant are the interactions between social institutions and the state. Civil society is a tool for investigating these interactions as well.

To date, historians of science have paid scant attention to these developments, but, as this collection of essays hopes to demonstrate, the concept of civil society opens new and important questions for the history of science. By way of preliminary orientation, we can point to two general types of questions here and then refine them later on. The first concerns the voluntary associations, clubs, and other institutions characteristic of civil society. How has engagement with science figured in the composition and evolution of those institutions? Has engagement with science offered the members of civic associations a way of acquiring social status, and if so, why has this happened? Have the institutions of civil society favored the cultivation of certain sciences, to the relative exclusion of others? Here, too, we can ask why this would be the case.

A second general question about science and civil society can be asked with respect to the relationship between civil society and political authority. As we shall see below, one of the doctrine's central tenets makes the constitution of political authority dependent on the formation of society itself. By highlighting the mechanisms by which political authority is created, the doctrine of civil society brings the issue of legitimacy to center stage, offering ground for studying how political legitimacy is maintained. In light of these considerations, then, how does the cultivation of science in civil society either support or undermine the legitimacy of political authority? Does the invocation of certain kinds of scientific knowledge by members of civil society provide the foundation for criticizing the state's policies or for denying its legitimacy? Moreover, it is not just political legitimacy that is under consideration here. Insofar as all societies require widely accepted forms of legitimation to function effectively and avoid chaos—for example, in regulating matters as mundane as marriage and inheritance—how does science affect these other kinds of legitimation?

[2] Keith Tester, *Civil Society* (London and New York: Routledge, 1992), p. 5.

In alluding to the relationship between science and social legitimation, we bring the issues raised in this volume into territory that has already been opened up by the work on "gentlemanly science" and science and the court in early modern Europe.[3] Often inspired or informed by the work of the German sociologist Norbert Elias, this literature has shown how the practice and the legitimization of science were part of a broader pattern of emergent behavioral norms, cultural values, and social institutions in Europe between the sixteenth century and the French Revolution. At first glance, the study of science and civil society would appear to be quite similar to this work because it includes many of the same norms of civility characteristic of the process described by Elias and central to the historical literature on science and civility. But the difference is that civil society represents an entirely different conception of social structure. Therefore a clear distinction must be drawn between the hierarchically stratified and aristocratic forms of social organization in which civility has largely been discussed and the nonhierarchical vision of social formation that permeates the doctrine of civil society. "We hold these Truths to be self-evident," intones the American Declaration of Independence, "that all Men are created equal, that they are endowed by their Creator with certain unalienable Rights, that among these are Life, Liberty, and the Pursuit of Happiness." Along with women and African Americans in the cast of excluded thousands, the category of "all Men" in the Declaration of Independence left no room for hereditary aristocrats, either.

Although the doctrine of civil society has much to offer the history of science, it does have two significant drawbacks. First, it is a maddeningly imprecise and frankly contentious concept. Does the definition of civil society include economic activity? Montesquieu and Adam Smith certainly thought so, as did G. W. F. Hegel and Karl Marx. But contemporary political scientists tend to dissociate the two, preferring to focus on social intercourse that brings people into various kinds of associations, thereby reinforcing civic cohesiveness and becoming the basis for independent political action. Does civil society include the home and the family? John Locke specifically excluded these from civil society. Hegel, in a quintessentially Hegelian formulation, regarded the family as the primitive, absolute form of society in which "[some]one's frame of mind is to have self-consciousness of one's individuality within this unity as the absolute essence of oneself," a condition radically different from the atomized and selfish individualism that Hegel found in civil society.[4] Jürgen Habermas, too, drew a sharp boundary between civil society and the family. By contrast, Jean L. Cohen and Andrew Arato, whose 1992 volume *Civil Society and Political Theory* offers one of the most thorough and theoretically sophisticated

[3] For a sampling of this literature, see Steven Shapin, *A Social History of Truth* (Chicago: Univ. of Chicago Press, 1994); Bruce T. Moran, ed., *Patronage and Institutions: Science, Technology, and Medicine at the European Court, 1500–1750* (Rochester, N.Y.: Boydell Press, 1991); Paula Findlen, "Controlling the Experiment: Rhetoric, Court Patronage, and the Experimental Method of Francesco Redi (1626–1697)," *Hist. Sci.* 31 (1993): 35–64; Jay Tribby, "Dante's Restaurant: The Cultural Work of Experiment in Early Modern Tuscany," in *The Consumption of Culture, 1600–1800: Image, Object, Text*, ed. Ann Bermingham and John Brewer (London and New York: Routledge, 1995), pp. 319–37; Jay Tribby, "Club Medici: Natural Experiment and the Imagineering of 'Tuscany,'" *Configurations* 2 (1994): 215–35; and Mario Biagioli, *Galileo, Courtier* (Chicago: Univ. of Chicago Press, 1993). For discussion of the transition from this model of court-oriented science to a more bourgeois and domestic form in the eighteenth century, see Alice N. Walters, "Conversation Pieces: Science and Politeness in Eighteenth-Century England," *Hist. Sci.* 35 (1997): 121–54.

[4] G. W. F. Hegel, *The Philosophy of Right*, trans. T. M. Knox (Oxford: Oxford Univ. Press, 1979), p. 110. Compare this with the definition of "civil society" on pp. 122–3.

guides to the concept, include the family within civil society, in recognition of the evolving understanding of household work as a form of labor and of the distinctive situation faced by women as both mothers and members of the labor force.[5]

A more serious drawback to the concept of civil society is the tendency to treat it as a stable empirical object, amenable to social-scientific analysis just like any other institution. For example, political scientists have recently taken to bemoaning the "decline" of civil society in America and filling their writings with proposals for its "restoration," as if the object of their lamentations were as easily identifiable an empirical phenomenon as monthly church socials or bowling leagues.[6] Nor is this the only example of such usages. There is an enormous scholarly literature on economic development and political change in Eastern Europe and Africa that makes the presence or absence of a fully developed civil society a crucial marker of "progress" in these and other parts of the world.[7] Indeed, such ideas have been the basis of economists' and political scientists' discussion of "modernization" since it was first formulated theoretically in the 1950s and 1960s.[8] Even Cohen and Arato waver between treating civil society analytically as a theory of social formation and as an empirical term.[9]

The danger inherent in treating civil society as if it referred to a particular segment or location in society, in the manner of terms such as "trade unions" or "the inhabitants of St. Cecilia's parish," is that by doing so we run the risk of subscribing, unwittingly or deliberately, to civil society's own ideological underpinnings. With respect to empirical content, what the historian or sociologist actually identifies in practice are particular institutions, such as museums, professional associations, and sundry clubs and other voluntary groups. Yet when the next step is taken and those institutions are made to be representative of "civil society," this attribution has the effect of situating them in the context of a liberal-democratic political system and

[5] Jürgen Habermas, *The Structural Transformation of the Public Sphere*, trans. Thomas Burger and Frederick Lawrence (Cambridge, Mass.: MIT Press, 1989), p. 45; Jean L. Cohen and Andrew Arato, *Civil Society and Political Theory* (Cambridge, Mass.: MIT Press, 1992), pp. 532–48.

[6] See, e.g., Brian O'Connell, *Civil Society: The Underpinnings of American Democracy* (Hanover and London: Univ. Press of New England, 1999); Robert K. Fullinwider, ed., *Civil Society, Democracy, and Civic Renewal* (Lanham, Md.: Rowman & Littlefield, 1999); and Robert Putnam, "Bowling Alone: America's Declining Social Capital," *Journal of Democracy* 6 (1995): 65–78.

[7] See, e.g., Rasul Bakhsh Rais, ed., *State, Society, and Democratic Change in Pakistan* (Karachi: Oxford Univ. Press, 1997); *Africa's Second Wave of Freedom: Development, Democracy, and Rights* (Lanham, Md.: Univ. Press of America, 1998); Gordon White, Jude Howell, and Shang Xiaoyuan, *In Search of Civil Society: Market Reform and Social Change in Contemporary China* (Oxford: Clarendon Press, 1996); and Alberto Gasparini and Vladimir Yadov, eds., *Social Actors and Designing the Civil Society of Eastern Europe* (Greenwich, Conn.: JAI Press, 1995). A perusal of books cataloged under the Library of Congress designation HN380.7 will yield a flood of similar literature on Eastern Europe alone. For a refreshing antidote to this point of view, see Mahmood Mamdani, "The Politics of Civil Society and Ethnicity: Reflections on an African Dilemma," *Political Power and Social Theory* 12 (1998): 221–33. My thanks to my colleague Gay (not Ann) Seidman for this reference and discussions of the larger issues at work here.

[8] See C. E. Black, *The Dynamics of Modernization: A Study in Comparative History* (New York: Harper & Row, 1966), which includes at the end a comprehensive bibliographic essay on modernization theory.

[9] Cohen, *Civil Society* (cit. n. 5). Compare the preface, p. vii—"Our goal, rather, is twofold: to demonstrate the relevance of the concept of civil society to modern political theory and to develop at least the framework of a theory of civil society adequate to contemporary conditions"—to p. 19—"Far from viewing social movements as antithetical to either the democratic political system or to a properly organized social sphere . . . , we consider them to be a key feature of vital, modern civil society."

handing them functions within that system. This transforms into an empirical phenomenon what is arguably merely an analytical category, and clothed in the mantle of this objectivity, civil society can be assessed either as actually promoting a healthy polity or as somehow being prevented from doing so.[10]

The conceptual slippage by which civil society is routinely rendered as an empirical term undeniably constitutes the most significant problem we face when dealing with it. Moreover, our difficulties are only increased by the fact that most of us readily subscribe to at least some version of the ideology of democratic society that civil society represents. Because we believe that the institutions composing what we call "civil society" have something to do with democracy, it becomes all the more difficult to separate the social role of those institutions and an analysis of civil society *as an ideology* from the assumed contributions by civic institutions to the development of democracy. One can see this mixture of normative and descriptive inclinations at work in the recent collection edited by Nancy Bermeo and Philip Nord, *Civil Society before Democracy.* This volume brings together essays studying civic institutions in various European countries during the nineteenth century and describing the effect of those institutions on the evolution of democratic politics in those countries. The question animating these articles is this: In light of the various institutional forms assumed by civil society in different European countries and the corresponding diversity of political systems, can any general conclusions be drawn about the ability of civil society to promote democracy? Both Nord, the historian, and Bermeo, the political scientist, answer in the affirmative.[11] When such normative expectations are read into nineteenth-century history, the negative case of a stunted civil society failing to foster democracy is furnished, not surprisingly, by Germany. In an article proving (Blackbourn and Eley notwithstanding) that the *Sonderweg* thesis remains alive in contemporary German historiography, Klaus Tenfelde writes that "in the Tocquevillian scheme of things, associations are meant to function as consensus-building little republics. They did not work that way in Germany." To hammer this point home, he adds that "a civil society never developed in full, not in 1918 and not under the Weimar Republic."[12]

[10] As Frank Trentmann puts it, "[C]ivil society has always stood both for norms and for social realities." That is, the term has always represented a certain model of how society ought to work, as well as describing how in fact it does work. Leaving aside here the question of whether the concept of civil society offers an accurate description of social interaction, what might be added to Trentmann's statement is that both the normative and descriptive parts of the doctrine are informed by an ideology concerning the structure of social life and its relationship with political power. Civil society is an ideology masquerading as an empirical category, and it is precisely the slippage between empirical and normative elements described by Trentmann that makes it so powerful. See Frank Trentmann, "Introduction: Paradoxes of Civil Society," in *Paradoxes of Civil Society: New Perspectives on German and British History,* ed. Frank Trentmann (New York: Berghahn Books, 2000), pp. 3–46, on p. 3.

[11] See Philip Nord, "Introduction," and Nancy Bermeo, "Civil Society after Democracy: Some Conclusions," in *Civil Society before Democracy: Lessons from Nineteenth-Century Europe*, ed. Nancy Bermeo and Philip Nord (Lanham Md.: Rowman & Littlefield, 2000), pp. xiii–xxxiii, 237–60.

[12] Klaus Tenfelde, "Civil Society and the Middle Classes in Nineteenth-Century Germany," in *Civil Society before Democracy* (cit. n. 11), pp. 83–108. For a critique of the *Sonderweg* thesis, which holds that Germany's failed road to modernization is the explanation for its disastrous fate in the twentieth century, see David Blackbourn and Geoff Eley, *The Peculiarities of German History* (Oxford: Oxford Univ. Press, 1984). Blackbourn and Eley's book actually appeared first in German in 1980, under the provocative title *Mythen deutscher Geschichtsschreibung* (Myths of German historiography). Needless to say, it touched off a massive controversy. For an interesting commentary on this entire imbroglio, see Richard Evans, "The Myth of Germany's Missing Revolution," in *Rethinking German History* (London and Boston: Allen & Unwin, 1987). For a highly readable general

It would be carrying this criticism too far to insist that historians examining civil society leave aside completely any attention to the connections between the institutional forms of civil society and the development of democratic polities. Such connections, after all, are what give civil society much of its interest. At the same time, however, we can attempt to set aside our normative expectations and treat those connections circumstantially, and not as the inevitable result of democratic modernization. In fact, historians of science should be in an advantageous position to do just that, because they are habituated to setting aside the powerful ideological claims of their topic, science, and evaluating it independently of whatever they may believe is true in it. In each case, the point is not to renounce our belief that civil society may indeed be important to democracy, or that the natural sciences do make claims about the world that we all subscribe to, but to distance ourselves from those beliefs in order to gain a critical and analytical perspective on their place in history. Thus the essays in this volume treat civil society analytically, as an ideology that has had an important impact on history and has interacted in significant ways with science. When they speak about "civil society" more or less empirically, they do so as a shorthand designation for the social institutions that are commonly identified as belonging to civil society, but without assuming that those institutions have anything at all to do normatively with the development of democracy. Of course, this is not to deny that in many instances—such as those described by Andreas Daum (on scientific popularization in Germany), Elizabeth Hachten (on Russian scientists in the public sphere), and Zuoyue Wang (on the Science Society of China)—historical actors may have deliberately incorporated the political and ideological aims of civil society to promote reform or modernization. But the articles avoid passing judgment on these events in terms of some absolute scale of democratic fulfillment.

With this separation between the analytical and the empirical senses of civil society firmly in mind, we should now take a closer look at some of the dominant themes in the doctrine as it has evolved since the late seventeenth century. No pretension to completeness is claimed for the survey offered here.[13] Instead, the aim is to direct the reader's attention to some of civil society's distinctive characteristics as well as to highlight those parts of the doctrine that will feature in this volume. After this preliminary survey is completed, the discussion will address the relationship between civil society and the public sphere, focusing on science's role in the latter.

---

overview of the immense struggles over Germany's past, see Charles S. Maier, *The Unmasterable Past: History, Holocaust, and German National Identity* (Cambridge, Mass.: Harvard Univ. Press, 1988).

[13] On the history of the concept of civil society, see Cohen, *Civil Society* (cit. n. 5), chap. 2; John Keane, "Despotism and Democracy: The Origins and Development of the Distinction between Civil Society and the State, 1750–1850," in *Civil Society and the State: New European Perspectives*, ed. John Keane (London and New York: Verso, 1988), pp. 35–71; and John Ehrenberg, *Civil Society: The Critical History of an Idea* (New York: New York Univ. Press, 1999). Ehrenberg's treatment is probably the most accessible one for those making their first study of this topic, but it suffers from a tendency to create a normative definition of civil society and then to situate various writers on a scale of how correctly they understood the concept. Thus Locke's theory of civil society represented a step forward over Hobbes's ideas, we are told, because Locke "knew better than Hobbes that property had become a necessary condition of human life" (p. 85). For outstanding interpretations of David Hume and Adam Smith, two other seminal theorists of civil society, see John Robertson, "The Scottish Enlightenment at the Limits of the Civic Tradition," and Nicholas Phillipson, "Adam Smith as Civic Moralist," in *Wealth and Virtue: The Shaping of Political Economy in the Scottish Enlightenment*, ed. Istvan Hont and Michael Ignatieff (Cambridge: Cambridge Univ. Press, 1983), pp. 137–78, 179–202.

Finally, the essay closes with a discussion of the professions and other institutional forms of civic life, and their confrontation with the state.

THE DOCTRINE OF CIVIL SOCIETY: ORIGINS AND PRINCIPAL THEMES

The concept of civil society, as it has been understood since the late seventeenth century, depends essentially on the idea of "society" as something distinct from "politics" and from the exercise of political authority. By means of this distinction, there has emerged the well-known "public-private" boundary that features so prominently in analysis of civil society and its relations with the state. As obvious as this distinction seems to us, however, let us note that the separation between society and politics is scarcely a natural one, and in fact ancient Greek writers treated civil society as the collection of men in a town who exercised political functions. To be a member of civil society in the ancient Greek, and later Roman, understanding was to be a citizen and to engage in political life.[14] This identification of the social with the political mirrored, as Harold Cook's article tells us, a similar identification of knowledge of the good and knowledge of the true for Plato and other ancient philosophers. For this reason, Cook writes, natural and moral philosophy were distinguished only heuristically; "they were simply different ways to a knowledge of the good and the true." Virtue consisted of the application of reason informed by philosophy (referred to in Cook's article as "right reason") in restraint of the irrational passions. By extension, the same set of principles applied to the wise governance of the *polis*, the body politic, as applied to the well-being of an individual.

According to Cook, this understanding of reason's mastery over the passions broke down during the sixteenth and seventeenth centuries, in large measure because of the scientific revolution. In its wake, the understanding of "reason" was radically narrowed to what was knowable by means of experience, rather than what has been revealed to the intellect by God. This created a problem for the understanding of virtue. For if reason now could only be based on experience and consequently could no longer be seen as some transcendent legislator of virtue, then what could serve as the ethical compass for human action? The rejection of right reason as a guide to virtue, Cook claims, led to a wholesale reevaluation of the role of the passions in guiding human action. Understood in a natural-philosophical context, the passions became the foundation for arguing that a polity consisting of the mutual adjustment of interests guided by the passions was the most equitable form of politics and the true source of virtue.[15]

None of the seventeenth-century writers discussed by Cook here presented any explicit analysis of society apart from politics. But that would not be long in coming. One of the earliest—and undoubtedly the most influential—formulations of this kind came from John Locke, whose anonymously published *Two Treatises of Government* (1690) became the indispensable starting point for all later theorists. In the

[14] On the origins of this distinction, see Keane, "Despotism and Democracy" (cit. n. 13).

[15] In addition to Hont, *Wealth and Virtue* (cit. n. 13), the relationship between virtue and the passions are treated in Albert O. Hirschmann, *The Passions and the Interests: Political Arguments for Capitalism before Its Triumph* (Princeton, N.J.: Princeton Univ. Press, 1977); and J. G. A. Pocock, *The Machiavellian Moment: Florentine Political Thought and the Atlantic Republican Tradition* (Princeton, N.J.: Princeton Univ. Press, 1975). Neither Hirschmann nor Pocock deals with the writers addressed in Cook's article, however.

second of the two treatises, Locke posed a simple yet powerful question: Under what conditions do humans enter into society? The key to this development, Locke believed, lies in the notion of property. Whether one starts from the principles of natural reason or the text of Scripture, it is evident that humanity has a right to appropriate the earth's abundance to preserve life and health. We might note in passing that Locke follows earlier writers in grounding property on the satisfaction of physical needs and desires. Here, too, it seems, the passions have an important role to play. But if such natural products are given to humanity collectively for its preservation, what legitimate basis can there be for a narrower sense of "property" as pertaining to particular individuals? Locke's answer held that the investment of labor in acquiring or improving the earth's resources makes those things the property of the person who has performed the work.

> Whatsoever then he [i.e., "man"] removes out of the State that Nature hath provided, and left it in, he hath mixed his *Labour* with, and joyned to it something that is his own, and thereby makes it his *Property*. It being by him removed from the common state Nature placed it in, it hath by this *labour* something annexed to it, that excludes the common right of other Men.[16]

For Jean-Jacques Rousseau, writing more than a half century after Locke, the establishment of property was the defining moment in the origin of civil society. "The first man who, having enclosed a piece of ground, to whom it occurred to say *this is mine*, and found a people sufficiently simple to believe him, was the true founder of civil society."[17] Locke, as always, was less peremptory in his pronouncements. While property alone did not suffice to induce people to enter into society, property coupled with scarcity did promote the formation of such relations, as people became accustomed to the idea of exchanging property through barter and, ultimately, by means of money.[18]

Locke then turned to the question of what motivates people to enter into what he called "political or civil society." He opened his chapter on the formation of political societies with this oft-quoted passage:

> Men being, as has been said, by Nature, all free, equal and independent, no one can be put out of this Estate, and subjected to the Political Power of another, without his own *Consent*. The only way whereby any one devests himself of his Natural Liberty, and *puts on the bonds of Civil Society* is by agreeing with other Men to joyn and unite into a Community, for their comfortable, safe, and peaceable living amongst one another, in

[16] John Locke, *Two Treatises of Government*, ed. Peter Laslett (Cambridge: Cambridge Univ. Press, 1988), p. 288. This text is based on the third edition, which was published in 1698. The italics are in the original.

[17] Jean-Jacques Rousseau, "Discourse on the Origin and the Foundations of Inequality among Men," in *Rousseau: The Discourses and Other Early Political Writings*, ed. and trans. Victor Gourevitch (Cambridge: Cambridge Univ. Press, 1997), p. 161. Italics in original.

[18] Locke, *Two Treatises* (cit. n. 16), p. 299. Nowhere does Locke explicitly address himself to the question of what drives people to begin exchanging property. However, he seems to suggest on this page that the scarcity of resources and their uneven distribution among different communities create the impulse to exchange. Moreover, Locke's concept of civil society supposes, in quite deliberate contrast to Hobbes's model of the state of nature as a condition of unremitting violence, that humans have a natural tendency toward sociability. "God having made Man such a Creature," he says at one point, "that, in his own Judgment, it was not good for him to be alone, put him under strong Obligations of Necessity, Convenience, and inclination to drive him into *Society*, as well as fitted him with Understanding and Language to continue and enjoy it" (pp. 318–9).

> a secure enjoyment of their Properties, and a greater Security against any that are not in it. This any number of Men may do, because it injures not the Freedom of the rest; they are left as they were in the Liberty of the State of Nature. When any number of Men have so consented to make one Community or Government, they are thereby presently incorporated, and make one body Politick, wherein the majority have a Right to act and conclude the rest.[19]

Elsewhere, Locke notes that slaves are incapable of holding property and therefore "cannot in that state be considered as any part of *Civil Society*; the chief end whereof is the preservation of Property."[20]

There are at least three significant features of Locke's conception of civil society that deserve comment here. First, it is voluntaristic. People are not compelled to enter into civil society; rather, they choose to do so freely, in order to better regulate their property and enjoy it more fully. Second, the mode of formation of civil society, by means of the "social contract" (as Rousseau would later famously label it), was a thoroughly egalitarian one. Whatever forms of social differentiation and hierarchy might characterize society as it was actually constituted, the point of departure for the doctrine of civil society, as Locke formulated it, was a group of free and equal people spontaneously agreeing to submit themselves to authority. Third, the model of property and self-interested economic exchange that animated the concept of civil society placed a stout wall between civil society and the home. Relationships outside the home between "all men" could be conducted, in theory at least, as encounters between equals. Such models of sociability did not pertain to relationships between spouses or between parents and children, however, and the home became the most private of private spaces in the emerging thicket of public-private dichotomies that the doctrine of civil society put in place.[21] Needless to say, this made the role of women in civil society problematic, but not perhaps for the reason—the sequestration of women inside the domestic sphere of the home—that might first come to the modern reader's mind as an anticipation of Victorian domesticity.[22] Thus, as Shelley Costa's article on the *Ladies' Diary* illustrates, the spectrum running from inclusion to exclusion of women from the domain of "all men" in civil society permitted well-born women to be the audience for some extremely sophisticated mathematical conundrums. But the ability of that same female audience to contribute to the solution of those problems was more contested and variable.

More perhaps than any other feature of Locke's conception of civil society, the establishment of "society" as something separate from government and political authority furnished an exceptionally powerful tool for the understanding of human relations. Characteristically, according to this and later formulations, neither society

[19] Ibid., pp. 330–1.

[20] Ibid., p. 323. Locke believed that slavery was justified in situations where prisoners have been taken in the prosecution of a "just war."

[21] Indeed, Locke devotes a considerable effort in the early chapters of the "Second Treatise" to arguing that certain forms of social and political relations, such as families, slaveholding, and absolute monarchy, lie outside the boundaries of civil society.

[22] For a fascinating study of how the understanding of gender contributed to the doctrinal formulation and institutional evolution of civil society, see Isabel V. Hull, *Sexuality, State, and Civil Society in Germany, 1700–1815* (Ithaca, N.Y.: Cornell Univ. Press, 1996), especially chap. 5, where Hull engages most directly with the doctrine of civil society as it was being developed during the eighteenth century. See also Suzanne Desan, "Reconstituting the Social after the Terror: Family, Property, and the Law in Popular Politics," *Past Present* 164 (1999): 81–121, which analyzes debates over the nature of family and the social order in France after the fall of the Jacobin regime.

nor government constituted or preceded the other. Instead, both came into being simultaneously and necessarily as part of the other's formation. On the one hand, this made the analysis of the forms of political authority and their relation to society a matter of supreme importance. Montesquieu devoted one of the eighteenth century's most fascinating and widely read treatises, *The Spirit of the Laws* (1748), to just this issue, and of course, many other writers have also treated it. On the other hand, the creation of society as a topic of discussion opened up a whole new world for scientific analysis. As Lorraine Daston has pointed out, much of the work in the emergent fields of probability and social statistics was informed by the assumption of the "reasonable man" as the core object of study, an assumption fully consistent with the doctrine of civil society.[23]

The same possibilities for an empirical science of society informed the discussion of "talent" and "merit" in the eighteenth century, as described in John Carson's article. Carson directs our attention to a crucial feature of the doctrine of civil society, which was its distinction between a basic civil or political equality asserted for all human beings (or at least, for that same group of "all Men" referred to in the Declaration of Independence) and a conspicuously *un*equal distribution of talents among the members of society. As Carson puts it, talk about talents furnished "one way of speaking like a democrat" while continuing to "justify social distinctions, though ones based on decidedly different grounds than the heretofore standard differentiations according to birth and tradition." Most powerfully of all, the distribution of talent was understood by many writers as based in human nature, and although this distribution pertained to humans in the state of nature before the creation of civil society, it had nonetheless profound social consequences. As suggested by Carson's description of the debate between John Adams and Thomas Jefferson, the unequal distribution of talent pointed directly to the powerful question of who shall rule in a democracy. Moreover, it might be added, in a social ideology based on property, the discourse of talent and virtue permitted (and continues to permit) massive social and economic inequalities to thrive underneath a veneer of political equality.

The line of thinking by which the needs of property become the impulse for the formation of civil society resonated through a line of prominent thinkers, including David Hume, Rousseau, Smith, Hegel and, of course, Marx, whose devastating analysis of the role of property in civil society led him to advocate the abolition of private property as a condition for the achievement of true democracy. A rather different line of thinking was developed by Alexis de Tocqueville in *Democracy in America* (1835, 1840). Tocqueville displayed little interest in analyzing the conditions that engendered civil society or in understanding how trade and manufactures produced social progress, which had been Smith's concern as well as that of Adam Ferguson, whose *Essay on the History of Civil Society* (1767) offered one of the earliest treatments of this subject. Tocqueville instead wanted to show how Americans practiced democracy in their society and, in a question reminiscent of Montesquieu, how democratic institutions shaped American manners.

The central problem for a democratic political system, Tocqueville claimed, is how to set proper limits to individual liberty and equality, without extirpating one or the other. The tendency in young democracies is for people to emphasize their

[23] Lorraine Daston, *Classical Probability in the Enlightenment* (Princeton, N. J.: Princeton Univ. Press, 1988), especially chap. 2.

individualism, because in such conditions it is difficult to see beyond their own selfish interests to the common good. This shortsightedness, in turn, is conducive to the growth of a tyrannical despotism, because such a regime can take root only where the people are divided and unable to unite in resisting the growth of a despotic government. But Americans, Tocqueville believed, overcome the atomizing effects of democratic systems by entering into a multitude of voluntary associations. He elaborated on this point by resorting to one of his favorite rhetorical tools, the comparison between democracies and aristocracies. In aristocracies, he declared, the multitude of people are weak while a few are very powerful. These few have in their own hands the ability to "achieve great undertakings" and do not require the assistance or even the passive compliance of others to reach these goals:

> Among democratic nations, on the contrary, all the citizens are independent and feeble; they can do hardly anything by themselves, and none of them can oblige his fellow men to lend him their assistance. They all, therefore, become powerless if they do not learn voluntarily to help one another. If men living in democratic countries had no right and no inclination to associate for political purposes, their independence would be in great jeopardy, but they might long preserve their wealth and their cultivation: whereas if they never acquired the habit of forming associations in ordinary life, civilization itself would be endangered.[24]

Thus for Tocqueville it was the formation of voluntary associations, both what he labeled "civil" as well as "political" associations, that together characterize the distinctive patterns of American democracy.

Tocqueville's analysis of civil society stands as an antidote both to Rousseau's gloomy assessment of the role of property in the establishment of social inequality, quoted above and laid out in the *Discourse on the Origin of Inequality*, as well as to Hegel's claim that the state was the ultimate solvent for the radically particularized interests of civil society. Tocqueville arrived at this position by ignoring the role of property in the formation of civil society altogether. Instead, he fashioned a picture emphasizing how private life acquires a public character through institutions such as civic organizations and the press and how this "public" based in civil society becomes an effective check on the growth of tyranny.[25] Not coincidentally, by framing his treatise as a discussion of democracy as it actually existed in America, Tocqueville greatly contributed to the conflation of the empirical, analytical, and normative aspects of the doctrine.

SCIENCE, CIVIL SOCIETY, AND THE PUBLIC SPHERE

One obvious question that could be raised in response to the foregoing discussion is why did this idea of civil society take hold among so many eighteenth-century writers? Locke alone certainly did not create the discourse of civil society—no single individual could have. Instead, what gave this discourse its circulation and

[24] Alexis de Tocqueville, *Democracy in America*, ed. Phillips Bradley, 2 vols. (New York: Vintage Books, 1945), vol. 2, p. 115.

[25] Tocqueville was, however, greatly concerned about the tyranny of the majority in American politics and the suffocating effects of majority consensus on public opinion. "In America," he wrote, "the majority raises formidable barriers around the liberty of opinion; within these barriers an author may write what he pleases, but woe to him if he goes beyond them." Ibid., vol. 1, p. 274.

application was a remarkable set of cultural and social transformations that first took form during the last decades of the seventeenth century and continued through the eighteenth. For some time now, historians have been describing the rapid expansion of economic activity during this period and the corresponding emergence of what has been labeled "consumer society." The phenomenon has been most closely associated with Great Britain, but historians of continental Europe have been almost as active in describing it.[26] Alongside these new forms of consumption and clearly associated with them, new patterns of social life also developed, patterns commonly described under the rubric of "new forms of sociability." This phenomenon is represented by a host of diverse institutions—among them coffeehouses, reading societies, salons, Masonic lodges, and societies of useful knowledge and improvement—that began to appear in many parts of Europe and North America. The characteristic that united this rather heterogeneous group of meeting places was the voluntary nature of participation in them and their tendency to reduce social distinctions.[27] Even Frederick the Great (1712–1786), king of Brandenburg-Prussia between 1740 and 1786, was just another brother initiated into the Berlin lodge of the Secret and Fraternal Order of Masons in 1738, when he was still crown prince.[28]

John Locke and his many readers as well as later commentators were people who inhabited this world, the world of commercially driven town life and the open sociability of the clubs and coffeehouses, and they formulated and engaged with the doctrine of civil society as a reflection of their day-to-day social experience. Their patronage of coffeehouses, their subscription to journals and newspapers, either individually or via membership in a reading society, gave them a taste of just the kind of voluntarist and nonhierarchical society that the doctrine meant to portray. We might appropriately call civil society the "Whig view of society," in recognition of the cultural ascendancy of the urban "middling sort" and the entrepreneurial lower gentry, and their hostility toward the upper echelons of the aristocracy.

Shelley Costa's article on the *Ladies' Diary* during the first half of the eighteenth century can be read against this background of the coffeehouses and the literary culture they inspired. Echoing Habermas's point about how coffeehouses and the other new forms of sociability fostered among participants a new collective sense of themselves as "the public," Costa frames her essay around a consideration of women's place in this public. The story is a paradoxical one. She shows how the

[26] The wellspring for much of this story is Neil McKendrick, John Brewer, and J. H. Plumb, eds., *The Birth of a Consumer Society: The Commercialization of Eighteenth-Century England* (Bloomington: Indiana Univ. Press, 1982). Among the best new treatments of this subject are Ann Bermingham and John Brewer, eds., *The Consumption of Culture, 1600–1800: Image, Object, Text* (London and New York: Routledge, 1995); Margaret R. Hunt, *The Middling Sort: Commerce, Gender, and the Family in England, 1680–1780* (Berkeley and Los Angeles: Univ. of California Press, 1996); and Daniel Roche, *France in the Enlightenment*, trans. Arthur Goldhammer (Cambridge, Mass.: Harvard Univ. Press, 1998). Roche's book, which was originally published in 1993, sounds from its title like a traditional cultural/intellectual history but is in fact a much more broadly conceived interpretation of the period. For a brief summary of these cultural and social changes, see Dorinda Outram, *The Enlightenment* (Cambridge: Cambridge Univ. Press, 1995), chap. 2.

[27] On the new forms of sociability, see Outram, *Enlightenment* (cit. n. 26), chap. 2, and the secondary literature cited there.

[28] On Freemasonry in the eighteenth century, see Margaret C. Jacob, *Living the Enlightenment: Freemasonry and Politics in Eighteenth-Century Europe* (New York: Oxford Univ. Press, 1991); and Richard van Dülmen, *Die Gesellschaft der Aufklärer: Zur bürgerlichen Emanzipation und aufklärerischen Kultur in Deutschland* (Stuttgart: Fischer, 1986), pp. 55–66, which summarizes a large body of secondary literature on Freemasonry.

Ladies' Diary became a thumping publishing success, selling out its first issue of 1704 within weeks of its appearance and remaining popular throughout the century. Such an achievement was all the more remarkable in light of the fact that the *Ladies' Diary* featured in each yearly issue increasingly sophisticated mathematical problems for which readers were invited to submit solutions. Indeed, Costa claims, the *Ladies' Diary* became the "first printed forum for mathematical exchange" and an example of the participatory public sphere described by Habermas. Yet by the 1730s the role of women in this exchange over mathematics had disappeared entirely, although obviously the almanac remained oriented in other respects toward a female audience. Women dropped out of the mathematical discussion (at least to the extent of being identified in print as women) because by the middle decades of the century mastery of the increasingly technical subject matter of mathematics was seen as too far removed from the polite accomplishments in dancing and music that women were expected to cultivate.

Nor was it only with respect to mathematics that women were barred from participation in the public sphere. As Joan Landes and others have pointed out, the eighteenth-century public sphere was largely closed to women's voices.[29] In part, the reason for this exclusion can be traced to the belief that women lacked the intellectual power to engage in public debate. But there was another factor at work, too, one that had nothing to do with such forms of active exclusion. According to the prevailing assumptions of the public sphere, women were too completely categorized and marked by their social and cultural position to attain a sufficient level of impartiality and detachment to participate in public debate. Characterized as they were by the ineluctable determinations of their gender and social position, whatever their intellectual resources, women could never achieve the level of generality required to speak as members of the disinterested and universal public. "The male is male only at certain moments," wrote Rousseau in *Emile*, stating baldly the viewpoint at work in the public sphere, "the female is female her whole life."[30]

The public sphere's great trick, its deceptive sleight of hand, was to create a form of discourse that was in fact rather narrow in its origins and point of view and yet unlimited in its claims for itself. As Craig Calhoun has pointed out (and Costa mentions in her article), we must take seriously Habermas's claim, made explicitly in the title of his work on the public sphere, that it is a bourgeois (*bürgerliche* in the original German) public sphere. Its bourgeois character lies not in the fact that the public sphere and its attendant civil society were based in a certain social class, although this may also have been largely true. More to the point, Calhoun argues that the very conception of society as defined by the doctrine of civil society is itself bourgeois.[31] It is in part the viewpoint represented by Emanuel Joseph Sieyès in his

[29] Joan Landes, *Women and the Public Sphere in the Age of the French Revolution* (Ithaca, N.Y.: Cornell Univ. Press, 1988); for a penetrating critique of Landes's work and an analysis of the gendering of the public sphere, see Dena Goodman, "Public Sphere and Private Life: Toward a Synthesis of Current Historiographical Approaches to the Old Regime," *Hist. Theory* 31 (1992): 1–20. Mary Terrall has also addressed the question of whether eighteenth-century French women were participants in science or merely furnished an audience for it. Terrall, "Gendered Spaces, Gendered Audiences: Inside and Outside the Paris Academy of Sciences," *Configurations* 2 (1995): 207–32.

[30] Jean-Jacques Rousseau, *Émile, ou de l'éducation* (1762; reprinted, with introduction by François and Pierre Richard, Paris: Editions Garnier Frères, 1961), p. 450.

[31] Craig Calhoun, "Introduction: Habermas and the Public Sphere," in *Habermas and the Public Sphere*, ed. Craig Calhoun (Cambridge, Mass.: MIT Press, 1992), p. 7.

famous pamphlet *What Is the Third Estate?*, published early in 1789, in which the author heaped scorn on the French aristocracy and articulated the interests of the commercially minded bourgeoisie, whose representatives would soon be debating a new French constitution in the Estates General. But even more than this, the public sphere's overwhelming power lies in its seeming ability to transcend mere social or class interest and simply stand for "everyone."[32]

The power to speak for everyone is undeniably a potent rhetorical platform to occupy. Here, the claim to possess scientific knowledge offers a powerful aid because, from one perspective, science offers itself as the most open, nondiscriminating, and public form of knowledge. Anyone, potentially, can learn science and understand it. But who is qualified to speak as a scientist? As we will see, the scramble to occupy the position of speaking for science was an issue of no small significance in nineteenth- and twentieth-century civil society. With respect to how this issue was framed in the eighteenth century, when the public sphere was developing, we can point to an essay by Immanuel Kant, "The Conflict of the Faculties," in which Kant argued that of the universities' different faculties (theological, legal, medical, and philosophical), only the philosophical faculty was capable of representing exclusively the interests of science (*Wissenschaft*) for its own sake. The other, more career-oriented faculties were too tied to careers and social position to achieve this disinterested position. The implication of this claim—one that Kant would not have found displeasing—was that on the basis of this disinterested pursuit of truth, philosophers and other scholars (including nonmedical natural scientists) were freed from social determination, becoming thereby qualified to speak for "everyone."[33]

While it may be true that the public sphere permits various statements, scientific and otherwise, to be maintained by privileged speakers as "what everyone knows," we should guard against believing that this condition alone furnishes a sufficient explanation for the subsequent history of professionalization and the resulting tyranny of experts. This is the basic theme of Theodore Porter's article on the statistician Karl Pearson and his vision of an aristocracy of science around 1900. Porter reminds us that the history of professionalization is all too often cast in the light of a functionalism that makes the triumph of professional experts out as the inevitable by-product of an increasingly complicated society's requirements. The picture as it actually unfolds, he counters, is much more complex and interesting. Porter's article shows how the place of scientists and other experts was highly contested in British society and government at the close of the nineteenth century. Pearson believed fervently in the power of the state to direct social progress, and to direct the state he called for training an aristocracy of science. The idea of an aristocracy holding such authority may sound anathema to the doctrine of civil society, almost as if Pearson were an arch-Tory railing against the corrosion of modernity. And, as Pearson's nostalgic medievalism suggests, in part it was. But Pearson's aristocracy was anything

[32] Thomas Broman, "The Habermasian Public Sphere and 'Science in the Enlightenment,'" *Hist. Sci.* 36 (1998): 123–49; and Anthony J. La Vopa, "Conceiving a Public: Ideas and Society in Eighteenth-Century Europe," *J. Mod. Hist.* 64 (1992): 79–116. For the most recent commentary on the public sphere and a critique of attempts by historians to talk about it in terms of multiple or differentiated "publics," see Harold Mah, "Phantasies of the Public Sphere: Rethinking the Habermas of Historians,"*J. Mod. Hist.* 72 (2000): 153–82.

[33] Immanuel Kant, "Der Streit der Fakultäten," in *Kants gesammelte Schriften*, 24 vols. (Berlin: G. Reimer, 1907), vol. 7, pp. 18–9.

but the hoary old plutocracy he saw silting up England's House of Lords. This new aristocracy would be perpetuated according to the most up-to-date eugenic principles to guarantee its continued qualification for leadership, and its members would receive a thorough scientific education, including, of course, the principles of statistics. In many ways, Pearson's eugenic vision of the new aristocracy of science, peopled by men of high vision and broad training, represents the fulfillment of the social distinctions naturalized as talent and merit that are described in John Carson's article. What Jean-Jacques Rousseau and Thomas Jefferson regarded largely as a theoretical point about the division of labor in society became in Karl Pearson's hands the potential cornerstone of a new social elite.

SCIENCE BETWEEN CIVIC ASSOCIATIONS AND THE STATE

During the nineteenth century, European civil society began to experience a number of transformations that would stamp it as distinctly different from its eighteenth-century predecessor, while preserving much of the same ideological basis. No less than the partisans of Enlightenment a century earlier, nineteenth-century popularizers of science saw an essential component of their mission as combating prejudice and ignorance and bringing the public as it actually presented itself into closer alignment with the image of a rational and civically oriented population as posited by the doctrine of civil society. Yet the differences are striking. The institutions of nineteenth-century civil society were far more numerous than their counterparts a century earlier in both absolute terms and variety. The number of people engaged in this form of social life was much greater, too. It would not be correct to say that nineteenth-century civil society was truly a mass phenomenon, in which factory laborers happily rubbed elbows with older established urban elites. But the ranks of middle-class participants had swelled remarkably, as cities and economies grew with astonishing speed.[34]

The rapid urbanization of the nineteenth century and the spectacular growth of the industrial base of European and North American economies brought to cities something else as well: a self-conscious working class with its own ideology. The consequences of this for our understanding of civil society and the public sphere could not be greater. During the Enlightenment, the bourgeois ideology of the public sphere recognized only dimly outside its own glow an inchoate mass of humanity, waiting to be brought into the circle of light; the same could no longer be said by the end of the nineteenth century. The appearance of a self-conscious socialist movement, loudly proclaiming its own ideology, gave a particular urgency and orientation to campaigns for popular education, elements that had been entirely lacking previously. In the new, more complex social and cultural environment of nineteenth-century civil society, the question of who spoke for science thus took on social contours—and often a political significance—that moved it far beyond the somewhat theoretical assertions by Kant that scholars could speak as representatives of the

[34] For the most comprehensive treatment of this phenomenon, see Jürgen Kocka, ed., *Bürgertum im 19. Jahrhundert: Deutschland im europäischen Vergleich*, 3 vols. (Munich: DTV, 1988). A selection of the articles in this massive collection was published in Jürgen Kocka and Allan Mitchell, eds., *Bougeois Society in Nineteenth-Century Europe* (Oxford: Berg, 1993). Kocka sketches the main elements of the story in Jürgen Kocka, "The Middle Classes in Europe," *J. Mod. Hist.* 67 (1995): 783–806.

public. Among the older higher professions of law, theology, and medicine, physicians, of course, could claim a significant authority over the study of nature, and physicians' attachment to science provided a powerful motor for their own professional development.[35] But physicians were now joined by a host of other men (and they were nearly all men) claiming to represent science before the public as either teachers and popularizers or as experts in certain kinds of technical work.

One place to begin examining this more complex picture is with Andreas Daum's article on Emil Rossmässler and the popularization of science in mid-nineteenth-century Germany. Rossmässler was a thoroughly Tocquevillian figure: democratic agitator, scientific lecturer, and journal editor, and the guiding spirit behind the natural history–oriented Humboldt Associations, which were organized in various parts of Germany. As Daum makes clear, the establishment of science as a form of public knowledge was part of a broader transformation of German civil society in the period. We can comprehend this transformation as an increasingly dense network of civic associations and institutions in German towns and cities, all of which promoted a distinctly bourgeois form of civic identity. In this respect, the German story can be read as illustrative of a more general trend discernible elsewhere in Europe.[36]

Where Germany may have been distinct, if not unique, as Glenn Penny's article on ethnology museums suggests, was in the assertively local and civic character of these associations. The cities that provide the setting for Penny's story—Berlin, Hamburg, Leipzig, and Munich—are significant in themselves. They experienced rapid population growth during the last half of the century, and their elites competed keenly with each other to give their hometowns a prominent position on the national stage. Even after its unification in 1871, Germany remained a country where provincial and town loyalties were still in the process of merging with national ones. Penny shows how this rapid growth led to a professionalization of city management, as a new group of administrators displaced the older urban elites from their governing positions. In parallel with these changes, the museums became more thoroughly integrated into the cities' administrative structures and began receiving regular operating funds from the city governments.

It is significant for both Daum's and Penny's stories that their protagonists were not part of the scientific elite. Although university educated and thus a member of the educated middle class (*Bildungsbürgertum*), Rossmässler lacked a prestigious university position from which he could command the kind of attention that prominent scientists of the next generation, such as Rudolf Virchow and Hermann Helmholtz, enjoyed. The same holds for the ethnologists described by Penny who, like Daum's popularizers of natural history, were outsiders to the university-based scientific elite.[37] Consequently, civic museums and not university faculties became the primary point of orientation for those attempting to establish ethnology as a science.

It is no matter of coincidence that the science presented in those German civic museums was ethnology, the science of empire par excellence. The idea of comparing European society and its evolution to other societies was certainly not new in

[35] Claudia Huerkamp, *Der Aufstieg der Ärzte im 19. Jahrhundert* (Göttingen: Vandenhoeck & Ruprecht,1985).

[36] On these developments, see Trentmann, "Introduction" (cit. n. 10), and Nord, "Introduction" (cit. n. 11).

[37] See also Lynn K. Nyhart, "Civic and Economic Zoology in Nineteenth-Century Germany: The 'Living Communities' of Karl Möbius," *Isis* 89 (1998): 604–30.

the latter nineteenth century. Indeed, as Michael Adas has argued, certain roots of this line of thinking can be traced back as far as the seventeenth century.[38] But what was novel, as the discussions of ethnology by Penny and Alice Conklin make clear, was the ability to assemble, transport, and display artifacts of other cultures in metropolitan centers such as Paris and Berlin. Such displays represented in the most concrete form possible the claims by ethnologists to comprehend and categorize other peoples of the world. And although by the latter nineteenth century anthropological theory had rejected any simple models of linear cultural progress running from primitives at one end to Germans or French at the other, the mere availability of those artifacts and representations of other cultures manifested to the museums' visitors a reflection of their own social progress and technological superiority.

Conklin's article on the Musée de l'homme in Paris shows just how deeply tied the work of French ethnologists was to the sinews of France's considerable overseas empire in the 1920s and 1930s. But Conklin also demonstrates that the kind of racism present in the *musée* was highly paradoxical. Sponsored by leftist anthropologists acting in concert with the socialist Popular Front national government elected in 1936, the *musée* sought to inculcate in its visitors a specific vision of "tolerance and respect for non-Western peoples," as Conklin puts it. This stance was mounted in deliberate contrast to the racism of the French far right and, of course, to the noisy and repugnant racism being promulgated by the National Socialists in Germany, who had come to power in 1933 and promptly dismantled Germany's system of electoral politics in the bargain. The democratically minded socialists of the Popular Front clearly had something else in mind, and they wanted to make sure that everyone knew it. Yet in the end even a socialist and democratic form of ethnology fell prey to many of the same paradoxes of appropriation and domination that would befall a more avowedly racist or imperialist ideology.

It is a conspicuous feature of Conklin's and several other articles in this collection that the same sciences that took hold in civil society could serve state interests as well. Indeed, it could well be one of the outstanding features of modern science and a source of its powerful presence in public life that science can serve the "public" in both senses of the term. On the one hand, the support of scientific knowledge can directly serve the interests of the state as the locus of public authority. On the other hand, as the articles demonstrate repeatedly, science also has considerable value in the public sphere. Thus, although science can be seen to serve the interests of state power, it appears almost never to do so exclusively. Were that to happen, we might suppose, science would lose much of its legitimacy among the members of civil society as "what we all know" about the world.

Zuoyue Wang's account of the formation and history of the most important scientific society in pre-Communist China, the Science Society of China, highlights the partnership between private initiatives and the national government. The Science Society was created by a group of American-trained Chinese scientists who deliberately sought to use science to modernize China and expand civil society. In part, this was clearly an effort to establish the scientists' own professional status, featuring many of the appurtenances that historians commonly associate with such movements, such as publication of a journal and the institution of national meetings. Yet

[38] Michael Adas, *Machines as the Measure of Men: Science, Technology, and Ideologies of Western Dominance* (Ithaca, N.Y.: Cornell Univ. Press, 1989).

there is much in this story that prevents its easy assimilation under familiar categories such as "professionalization." First, the founders of the Science Society were staunch nationalists, who had grown up during a period when the Chinese state, weak and disorganized, suffered under European and Japanese domination. Thus they did not conceive of their project as the creation of a free zone of activity independent of state control. Second, the effort to professionalize scientific work necessarily included, again as a quite deliberate choice, an attempt to popularize science for a broader segment of Chinese society. Because the state was so weak, the members of the Science Society recognized the need for mobilizing public support for their organization and for scientific research. Third, the Science Society managed to promote its agenda by deploying a rhetoric of criticism toward the Republican regime, while also securing a measure of material support from it for the establishment of libraries and research institutes.

The situation in China thus led to what Wang describes, quoting the Chinese historian Philip Huang, as a "third realm" of state/society interaction. Whereas Western political tradition establishes private and public interests as separable and distinct—and even the French example of "partnership" in Conklin's article implies this distinction—according to Chinese traditions the two are far less easily separated. The end result of such reflections, according to Wang, should not be to deny that China developed elements of civil society recognizable to Westerners, but to broaden our conception of the possible forms of civil society. From this perspective, it makes little sense to judge the expansion of civil society in China during the period before the Communist takeover in 1949 as either a success or a failure. Such judgments are only conceivable in the first place when they are made along the gradient of social and historical progress that the (Western) doctrine of civil society itself supports.

The same straddling by science across the interface of public and private interests also features prominently in Elizabeth Hachten's article on Russian science during the second half of the nineteenth century. Hachten describes how Russian scientists sought to expand the role of science in Russia by appealing not to the state, which had been the principal source of support for science before this time, but to a broader alliance of urban elites and improving rural landlords. In contrast to China, where the notion of popularizing science required importation of the idea of a "public" as well and the creation of the mechanisms of public discourse, Russians had a long history of substantial cultural dialogue with western Europe, and the ideology of civil society took hold in Russia as a not completely foreign import. The Russian term for civil society (*obshchestvo*, literally, "society") may have been more distinctly colored as an aristocratically dominated "high society" than Locke and certainly Jefferson would have felt comfortable with, Hachten tells us, but its usage as a binary opposite to *gosudarstvo* (state) during the nineteenth century makes it clear that *obschchestvo* demarcated a field having conceptual overlap with Western understandings of civil society.

The quality that made science seem like such an obvious medium for the modernization of Russia—a matter of especially pressing concern in the wake of its embarrassing defeat in the Crimean War—was science's presumed ability to serve the public interest in both of the senses that we have been discussing. Traditionally, the pursuit of scientific research had been supported directly by the tsars. Peter the Great had organized the St. Petersburg Academy of Sciences in 1725 and staffed it largely with Germans, this at a time when Russia barely had anything resembling secondary schools, not to mention universities where natural sciences were taught.[39] By the nineteenth century, the perceived need to promote the growth of Russia's economy led to the chartering of new scientific societies and regionally based associations for agricultural improvement by the autocratic tsarist government. But these benefits were believed to pertain not merely to the state as the locus of public authority—one reason why the tsarist government was so ambivalent about science as a vehicle for modernization. Scientists also attempted to invoke the benefits of science for other segments of Russian society, whose interests could conveniently be represented in the semi-autonomous regional assemblies (zemstvos) that were permitted to organize after 1864. Scientists rushed to exploit the space thus created for their own professional interests by articulating a language that spoke alternately of the fruits of science for economic progress and of the edifying benefits of pursuing science for its own sake. The latter took a rhetorical stance strikingly similar to Germans' talk of academic study for self-cultivation, or *Bildung*, during the same period and appears to have marked off the interests of Russian scientists in much the same way that Kant's talk about *Wissenschaft* in the philosophical faculty had done in the 1790s.

The same three-cornered relationship between professional, state, and private or civil interests can be seen in Lynn Nyhart's paper on notions of biological community in the work of the nineteenth-century German schoolteacher Friedrich Junge. Nyhart's paper introduces the topic of education to our discussion, bringing an essential, but unfortunately all too uncommon, perspective on the history of civil society. Much like science itself, schooling serves the public interest in two ways. From the perspective of society, education acts to reproduce society and its cultural values by taking a cohort of half-savage little people and molding (or trying to mold) them into responsible and hard-working members of society. The reason why education is not entirely a private initiative, however, and why taxes usually support it, is that formal schooling serves a larger collective and public function. Educational institutions reinforce the stability of the political regime by teaching citizenship to students. Moreover, this is not simply a matter of regimentation and externalized discipline, for schools attempt to teach enlightenment by means of the students' own discovery and self-enlightenment. In doing so, educational institutions reinforce the ideology of civil society and help direct students' self-enlightenment along the proper paths.

Within these institutional priorities, Nyhart argues, German teachers, especially primary school teachers, harbored a complicated set of interests. Although technically state employees, schoolteachers numbered among the lowest members of the German civil service. Their place in the civil service scarcely counted at all in terms of status, and as Nyhart shows, much of their professional effort involved trying to raise their status to that of the occupational group on the rung above them. For this reason, local and regional teachers' associations were among the most active such groups in German civil society. For the same reason, Junge's presentation of the concept of biological community as manifested in the village pond found a receptive audience because it merged with teachers' claims for pedagogical expertise. Against the attempts by German state governments to prescribe curriculum and methods of

[39] On the founding of the St. Petersburg Academy, see Michael D. Gordin, "The Importance of Being Earnest: The Early St. Petersburg Academy of Sciences," *Isis* 91 (2000): 1–31.

teaching, the cultivation of pedagogy as a form of *Wissenschaft* represented the teachers' counterstroke. Finally, Junge's concept of biological community, in which the role played by each organism could be shown as coordinated into the well-being of the entire community, transmitted metaphorically a comfortable message of social conservatism while simultaneously bearing the more liberal associations of self-enlightenment through knowledge of modern science.

In many respects, an understanding of the distinctive location of scientists, poised as they are between civil society and the state, helps us to better comprehend the kind of public roles scientists play in modern society. Jessica Wang's article, appropriately the last in the collection, addresses this question directly. Historians' interpretations of the scientist's role in society, she points out, have often focused attention on the potential conflicts between science and democracy. In doing so, the historians have overlooked the role of scientists in civil society as private actors on a public stage whose own interests may sometimes bring them into direct conflict with government. Wang's article demonstrates the power of scientists' authority in the public sphere as well as the perils. She describes how in the aftermath of World War II and the shocking power of nuclear weapons unleashed by the United States, American nuclear physicists launched the Federation of Atomic Scientists (FAS). In a number of ways, the FAS, which consisted of many leading contributors to the Manhattan Project, resembled other interest groups that had been formed by scientists, going all the way back to the American Association for the Advancement of Science, established in 1848. But the FAS aimed at something quite different: it sought to educate the public about the perils of nuclear warfare and influence U.S. government policy. It was the FAS's supreme misfortune—although hardly an accident—that it formed during one of those recurrent flare-ups of America's long-standing paranoia over socialism and communism. In the quasi-totalitarian conditions that ensued, FAS members were spied upon, blacklisted, and otherwise hounded into silence.

The importance of Wang's story lies not merely in its exploration of how institutions of civil society might fare under conditions in America that resembled those in the Soviet Union under Stalin or Germany under Hitler more than we might like to admit. More significantly, she situates her story in terms of the concerns over the role of experts in the modern world already voiced by John Dewey in his 1927 book, *The Public and its Problems*. Dewey believed that the proliferation of such experts had eviscerated the public's ability to engage meaningfully in important matters, leaving it bewildered by the issues and impotent. In what reads like a preview of the criticisms that would be delivered by the young Habermas thirty-five years later, Dewey located one cause of the public's situation in what he labeled the "machine age," a time of immensely complex relationships "formed on an impersonal rather than a community basis." In such conditions, he concluded, the public could neither identify nor distinguish itself.[40]

From one perspective at least, the scientists who enrolled in the FAS were Dewey's worst nightmare come to life: experts possessed of a knowledge that was dauntingly abstruse and frighteningly powerful. Yet these same scientists were no mere technocrats, manipulating their slide rules and inserting fuses into their bombs. They

[40] John Dewey, *The Public and Its Problems* (1927; reprinted, Chicago: Gateway Books, 1946), p. 126.

energetically sought to use their authority to engage the public and mobilize its support for greater control over nuclear weapons. They became political actors in civil society, precisely the sort of people Tocqueville had valorized a century earlier. Therein, of course, lay the almost tragic irony of their situation, for by engaging directly in political debate the members of the FAS undermined the very claims to universality and disinterestedness that had been the bulwark of speakers in the public sphere since at least the eighteenth century.

Ultimately, then, the place of science (and of scientists) in civil society is a deeply ambiguous one. Insofar as the members of civil society are convinced that society can progress materially from the cultivation of scientific knowledge and can benefit from the spread of enlightenment (both of which beliefs have attended the doctrine of civil society almost from its first formulation), those who claim dominion over science receive a substantial portion of social status and cultural capital. Furthermore, by associating themselves with the apparent impartiality and disinterestedness of science, particular social groups can promote their own very real interests in the form of increased social status. Yet at its core, as John Dewey perhaps recognized, scientific knowledge never really belongs to the public sphere of civil society. Just as the idea of spreading enlightenment in the eighteenth century contained within itself the problem of who had the right to call themselves enlightened and teach the practice of reason to others, so, too, does the idea of science as a constitutive element of enlightened civil society merely throw a mask over the fact that certain members of civil society determine not only what scientific knowledge is but also who has the right to make such determinations.

SCIENCE AND THE EMERGENCE OF CIVIL SOCIETY

Body and Passions:
Materialism and the Early Modern State

By Harold J. Cook[*]

ABSTRACT

A group of works written in the mid-seventeenth-century Netherlands shows many defenders of commerce and republicanism embracing some of the most unsettling tenets of the new and experimental philosophy. Their political arguments were based on a view consonant with Cartesianism, in which the body and its passions for the most part dominate reason, instead of the prevailing idea that reason could and should dominate the passions and through them the body. These arguments were in turn related to some of the new claims about the body that flowed from recent anatomical investigations, in a time and place comfortable with materialism. If ever there were a group of political theorists who grounded their views on contemporary science, this is it: Johann de Witt, the brothers De la Court, and Spinoza. They believed that the new philosophy showed it was unnatural and impoverishing to have a powerful head of state, natural and materially progressive to allow the self-interested pursuit of life, liberty, and happiness.

INTRODUCTION

ESTABLISHING THE RELATIONSHIP between knowledge of nature and political systems is a classic problem for twentieth-century historians and philosophers of science. To mention only a few of the many notable arguments along these lines: science has been invoked as the keystone in the arch of truth supporting liberal democratic society against religious doctrine,[1] against the failures of capitalism,[2] Marxism,[3] and totalitarianism.[4] Herbert Butterfield's *The Origin of Modern Science* (1957) embraced the Whiggism of progressive truth, reversing his more relativistic prewar attack on it—or so A. Rupert Hall has argued.[5] During the Cold War the semireligious importance of mental insight in scientific discovery seemed a key counterargument to Stalinist materialism.[6] For the English seventeenth century, it has been argued that science grew from a politico-religious movement allied either with

[*] Wellcome Trust Centre for the History of Medicine at University College London, Euston House, 24 Eversholt St., London NW1 1AD; h.cook@ucl.ac.uk.

[1] Andrew Dickson White, *A History of the Warfare of Science with Theology in Christendom*, 2 vols. (London and New York: Appleton, 1896).

[2] Boris M. Hessen, "The Social and Economic Roots of Newton's *Principia*," in *Science at the Cross Roads* (1931; reprinted, London: Frank Cass, 1971), pp. 151–212.

[3] G. N. Clark, *Science and Social Welfare in the Age of Newton* (Oxford: Clarendon Press, 1937).

[4] Karl R. Popper, *The Open Society and Its Enemies*, 5th ed., 2 vols. (Princeton, N.J.: Princeton Univ. Press, 1966)

[5] A. Rupert Hall, "On Whiggism," *Hist. Sci.* 21 (1983): 45–59.

[6] Alexandre Koyré, "Les Origines de la science moderne," *Diogene* 16 (1956); Michael Polanyi, *Personal Knowledge: Towards a Post-Critical Philosophy* (Chicago: Univ. of Chicago Press, 1958).

 0369-7827/02/1701-0002$10.00

Puritanism or Anglicanism,[7] or emerged with the foundations of an "open and liberal" society.[8] For early modern Italy and France, it has been argued that science developed from the princely courts or the centralizing state.[9] From postmodernists have come powerful statements about how science is part of the system of modern domination or a means of shattering the public use of reason into nonpublic specialisms.[10] In short, many of the most notable arguments about science have been concerned with exploring the connections between it and politics.

If in light of such discussions one reexamines a group of works written in the mid-seventeenth-century Netherlands, it becomes clear that defenders of commerce and republicanism there embraced some of the most unsettling tenets of the new and experimental philosophy. As they did so, they offered no consolation for either classicists or clerics who believed that humankind should strive to be good. In the view of human nature advanced by the Dutch republicans, the state should not, because it could not, try to make people behave well. The best kind of civil society was the most natural rather than the most virtuous; it was one in which the material betterment of the whole progressed despite the trials and tribulations of individual members. The political arguments of the republicans were based on a view consonant with Cartesianism, in which the body and its passions for the most part dominate reason, rather than on the prevailing idea that reason could and should dominate the passions and through them the body. The views of these republicans were related to some of the new claims about the body that flowed from the recent anatomical discoveries and theories, in a time and place comfortable with materialism. I do not claim that the ideas of the new philosophers, including the anatomists, caused the political arguments advanced in the Dutch Republic, only that those ideas became crucial elements in the works defending republicanism. The republicans fundamentally divided the study of how things are from the pronouncements of moralists in ways that echo down the centuries.

REASON VERSUS THE PASSIONS

The republicans held that the new philosophy had destroyed the main premises of the old. Since at least Plato's time, two key concepts had been regularly employed for understanding the public good. One was the idea that an analysis of the public good depends on the individual good: what is good for the person is good for the public. The crucial metaphor here is that of the "body politic," suggesting that the collectivity is corporeal. A second concept was that of "right reason," which supposes that

[7] Charles Webster, *The Great Instauration: Science, Medicine, and Reform, 1626–1660* (New York: Holmes & Meyer, 1976); James R. Jacob and Margaret C. Jacob, "The Anglican Origins of Modern Science: The Metaphysical Foundations of the Whig Constitution," *Isis* 71 (1980): 251–67.

[8] Steven Shapin and Simon Schaffer, *Leviathan and the Air Pump: Hobbes, Boyle, and the Experimental Life* (Princeton, N.J.: Princeton Univ. Press, 1986).

[9] David S. Lux, *Patronage and Royal Science in Seventeenth-Century France: The Academie de Physique in Caen* (Ithaca, N.Y.: Cornell Univ. Press, 1989); Mario Biagioli, *Galileo Courtier: The Practice of Science in the Culture of Absolutism* (Chicago: Univ. of Chicago Press, 1993).

[10] Jürgen Habermas, *The Structural Transformation of the Public Sphere: An Inquiry into a Category of Bourgeois Society*, trans. Thomas Burger and Frederick Lawrence (Cambridge, Mass., MIT Press, 1989); originally published in German as *Strukturwandel der Öffentlichkeit* (Darmstadt and Neuwied: Hermann Luchterhand Verlag, 1962). Michel Foucault, *The Order of Things: An Archaeology of the Human Sciences* (New York: Random House, 1973); originally published in French as *Les mots et les choses: une archéologie des sciences humaines* (Paris: Gallimard, 1966).

truth contains more than a distinction between correct and incorrect: it also contains a distinction between good and bad. Coupled together, these notions suggested that the font of personal wisdom contained directions for both individual and collective well-being. A well-regulated state would be composed of people governed by virtue. The so-called scientific revolution, however, altered classical ideas of the body, sometimes challenging concepts of the body politic; it also radically narrowed the view of reason to judgments of true and false, undermining the notion that mortals could know the good by the faculty of reason. It therefore raised deep questions about the nature of private and public virtue, even at times threatening to do away with the metaphor of the body politic.

The classical union of the personal and the political is most clearly expressed in the works of Plato. For him, Aristotle, the Stoics, and numerous other philosophers and their heirs, true knowledge of the *logos*—the eternal meaningful order that lay beneath the appearances of things—revealed both the good and the true; the good and the true were elements of the same ultimate universal. Put another way, natural philosophy was separate from moral philosophy only heuristically: they were simply different ways to a knowledge of the good and the true. Consequently, an identity existed between the knowledgeable life and the moral life. Plato famously distinguished the philosopher from the mere sophist by declaring the former to have a true love of wisdom, or a true desire to gain a knowledge of the *logos* rather than of opinion. His teachings about the true were equally teachings about the good. At the same time, because of the identity between the true and the good, to exercise reason one had to become good. Knowledge transformed. The wise person and the good person were the same; the wiser the better, the better the wiser. For Aristotle, too, to actualize our potential we must gain knowledge and act accordingly. This would make us truly good, for "true pleasures are what seem to [the good man] to be pleasures, and the really pleasant things those which he finds pleasant." The measure of what is "real and truly human" is "the good man's pleasures." Since the good man necessarily sought theoretical wisdom, the quest for wisdom was the highest and most perfect good, and contemplative happiness was therefore the most perfect of pleasures.[11] Stoic philosophers, too, argued for the identity of the moral and the rational. As one recent commentator on Stoicism has noted, the Stoics taught that "a moral person is one who has suppressed irrational movements and who lives in perfect conformity with the divine Logos," so that "a wise person is a sign, a symbol that elucidates the deepest roots of the universe and its history."[12] Christian theologians had to wrestle with the problem of the relationship between reason and grace (for they had to allow the good Christian fool to gain salvation when the philosopher might not); but they also took the Creator to be a rational being, which meant that one's own rational soul participated, at least to a certain extent, in the nature of God. Vigorous and sometimes violent argument went into determining what that "to a

[11] Quotations from Aristotle's *Nichomachean Ethics*, trans. J. A. K. Thomson (Baltimore: Penguin Books, 1955), 10.5; the definition of perfect happiness is at 10.7–8. See also Nancy Sherman, *The Fabric of Character: Aristotle's Theory of Virtue* (Oxford: Clarendon Press, 1989); and Martha C. Nussbaum, *The Therapy of Desire: Theory and Practice in Hellenistic Ethics* (Princeton, N.J.: Princeton Univ. Press, 1994), pp. 48–101.

[12] Gerard Verbeke, "Ethics and Logic in Stoicism," in *Atoms, Pneuma, and Tranquillity: Epicurean and Stoic Themes in European Thought*, ed. Margaret J. Osler (Cambridge: Cambridge Univ. Press, 1991), p. 24.

certain extent" meant. But from at least Augustine forward, most theologians took the view that the cultivation of virtue through reason would at least help on the path toward salvation. Thus virtue came from right reason: from knowing the *logos* and acting in accord with it.

At the same time, Plato and his successors felt free to move from an analysis of the individual good to the public good by employing the concept of the body politic. As Plato's Socrates put it in the *Gorgias*, "any regularity of the body is called healthiness, and this leads to health being produced in it, and general bodily excellence." By extension, the regular and orderly states of the soul, "called lawfulness and the law," are states of "justice and temperance."[13] A person should strive both for bodily excellence and for justice and temperance in the soul, which meant restraining one's desires (ἐπιθῡμία) according to reason so that no correction from others is necessary for one's own good. "[B]ut if he have need of it [i.e., correction], either himself or anyone belonging to him, either an individual or a city, then right must be applied and they must be corrected." In other words, both the body and the *polis* needed governance: the body politic, as well as the personal body, needed to be restrained or, if necessary, corrected by reason. Following Plato, thinking about the *polis* as a person became common, with powerful consequences. Ideas of the body in turn reinforced the sense that the body politic needed a governor. As Shigehisa Kuriyama has recently put it in writing of Greek medicine: "The motions within a person had to spring from some ultimate source. There had to be a ruler."[14] Following Galen, the governor of the body came to be the head, where reason's organ, the brain, held sway; hence a just and temperate public order began with the "head of state." For Christians, the metaphor was altered to indicate that the body of Christ was the gathering of the faithful into the true church, which was opposed to the world of deceit and corruption. But with the political establishment of the Christian church as an arm of Rome, and the former bishop of Rome assuming the headship of the Latin-speaking church, many theologians felt enabled to apply notions of the collective Christian body to the entire body politic, with the Pope as its head. After the revival of Roman law in medieval Europe, the idea of the body politic even took on the color of law, as "corporations" came to obtain charters from the twelfth century onward: these were legal fictions representing in law collective groups—from guilds to whole cities—as if they were a single body *(corpus)*.[15] The corporatist nature of medieval and early modern society has been stressed particularly for France, where the notion of the body politic remained closely associated with the king's own body,[16] but it can be found everywhere in Europe.

Yet everyone understood that both knowing via right reason and acting in accor-

[13] *Gorgias*, in *Plato*, 12 vols., vol. 3: *Lysis, Symposium, Gorgias*, trans. W. R. M. Lamb, Loeb Classical Library (Cambridge Mass.: Harvard Univ. Press, 1925), 504 C-D.

[14] Shigehisa Kuriyama, *The Expressiveness of the Body and the Divergence of Greek and Chinese Medicine* (New York: Zone Books, 1999), p. 160.

[15] Antony Black, *Guilds and Civil Society in European Political Thought from the Twelfth Century to the Present* (New York: Methuen, 1984).

[16] Ernst H. Kantorowicz, *The King's Two Bodies: A Study in Mediaeval Political Theology* (Princeton, N.J.: Princeton Univ. Press, 1957); Jeffrey Merrick, "The Body Politics of French Absolutism," in *From the Royal to the Republican Body: Incorporating the Political in Seventeenth- and Eighteenth-Century France*, ed. Sarah E. Melzer and Kathryn Norberg (Berkeley: Univ. of California Press, 1998), pp. 11–31; and Susan McClary, "Unruly Passions and Courtly Dances: Technologies of the Body in Baroque Music," in *Republican Body*, pp. 85–112.

dance with it are very difficult for mortals. The major problem is that the "passions"—in most representations the middle part of the soul, intimately associated with the bodily spirits and desires[17]—keep one from knowing or acting according to the dictates of reason. For most philosophers and theologians, then, reason needed to dominate, check, or eliminate the passions.[18] In Plato's virtuous person, "the ruling principle of reason, and the two subject ones of spirit and desire are equally agreed that reason ought to rule."[19] Aristotle argued for tempering the passions so that they were not too low or too high; since they caused motions in the soul that could lead to good actions, the passions were not bad if properly checked by reason.[20] Stoics usually took a harder line, teaching that peace of mind and social harmony could come only through the extirpation of the passions by reason.[21]

In contradistinction to classical notions, however, the concept of "reason" was radically narrowed in the late Renaissance and early modern period: it came to depend on knowledge that entered the mind via the senses—experience (and later "experiment")—rather than either innate or transcendent understanding. It was a knowledge of "thingness." This had the advantage of making knowledge equally communicable among those with healthy senses. Proper reason in such people could apprehend the stuff of nature (after pinning down Proteus in order to discard mere appearance), and it could calculate what would happen provided a true account had been apprehended. Reason therefore understood truths about being, or existence ("objective" truths). The foundation for reasoning about truths concerning the good, however, had to be found in something outside nature (subjective experience), for which there could be no shared sensation, or in nature itself (which raised the danger of pantheism, paganism, and other heresies).

When this narrowed sense of reason was applied to the public, one obtained a view known as *raison d'état*. Despite the term's being best known in its French form, French thinkers owed a great deal to Italian discourses on *ragione degli stati*; the Italians in turn owed much to the infamous Machiavelli.[22] Machiavelli believed that the state acts from the same causes as a person does, including the ways in which astral emanations affect the humors of the body politic.[23] Nevertheless, as the phrase makes clear, the main goal of *raison d'état* remained that of setting out the means by which reason could dominate the body politic. As Cardinal Richelieu, that well-known exponent of *raison d'état*, put it: because man's nature[24] possesses reason,

[17] Katharine Park, "The Organic Soul," in *The Cambridge History of Renaissance Philosophy*, ed. Charles B. Schmitt, Quentin Skinner, Eckhard Kessler et al. (Cambridge: Cambridge Univ. Press, 1988), 464–84.

[18] Juda Sihvola and Troels Engberg-Pedersen, eds., *The Emotions in Hellenistic Philosophy* (Dordrecht, South Holland: Kluwer, 1998).

[19] *Republic*, 4.442, in *The Dialogues of Plato*, trans. Benjamin Jowett, 3d ed. (New York and London: Oxford Univ. Press, 1892).

[20] *Nichomachean Ethics* (cit. n. 11), bk. 2. Aristotle's work on *Rhetoric*, for instance, stressed the importance of the deliberative and epideictic forms, by which one could achieve the good by putting the passions to the service of reason: *Rhetoric*, 1.3–9.

[21] For a lively discussion of Stoic views of the passions, see Nussbaum, *Therapy of Desire* (cit. n. 11), pp. 316–401; on the early modern interpretations of the same, see Jill Kraye, "Moral Philosophy," in *Cambridge History of Renaissance Philosophy* (cit. n. 17), pp. 364–7.

[22] Peter Burke, "Tacitism, Scepticism, and Reason of State," in *The Cambridge History of Political Thought, 1450–1700*, ed. J. H. Burns and Mark Goldie (Cambridge: Cambridge Univ. Press, 1991), especially pp. 479–84.

[23] Anthony J. Parel, *The Machiavellian Cosmos* (New Haven, Conn.: Yale Univ. Press, 1992).

[24] He meant this to apply only exceptionally to women.

"he ought to make reason sovereign, which requires not only that he do nothing not in conformity with it, but also that he make all those who are under his authority reverence it and follow it religiously."[25] By "making" the governed reverence the reason known to their governors, however, we are in a world quite different from that in which people can agree on the true and the good discovered by right reason. Moreover, the politico-jurisprudential ideas of Hugo Grotius and his successors were grounded in the fundamental principle of self-preservation, which became the first "natural right" and the basis of "natural jurisprudence."[26] Many French authors in particular took the discussion of self-preservation further by considering the passions associated with it: most notably Jean-François Senault, François de La Rouchefoucauld, and Blaise Pascal developed the notion of *amour-propre*,[27] in which self-preservation was intimately tied to the love of praise, or pride.[28]

According to this new sense of reason, then, it could calculate and follow arguments of logic or interest, but when it came to understanding the good it could only demonstrate material truths, such as the fundamental necessity for self-preservation and other passions in driving human conduct. The new science therefore posed a predicament for anyone who wished to grasp the human condition: the good had to be either limited to objective "flourishing" and other physical goods, or founded on some subjective principle open to doubt. If the latter, people had to be subject to a power they could not reason about (or at least not always agree with), leading to tyranny; if the former, then materialist self-indulgence rather than virtue seemed the consequence.

The best example of the former argument is Thomas Hobbes's invocation of a powerful sovereign, echoing Richelieu's view. Hobbes took a more explicitly materialist line, however, developing his ideas about how "whatever we experience, whether in sleep or waking, or at the hands of a malicious demon, has been caused by some material object or objects impinging upon us." According to Richard Tuck's recent analysis of the origins of Hobbes's natural philosophy, Hobbes turned seriously to a study of the new philosophy during his visit to France in 1634–1637, where he became a member of the Mersenne circle and gained an acquaintance with the work of René Descartes and Pierre Gassendi, among others. His materialism owed much to Galileo's discussion of heat in *Il Saggiatore* (1623) and possibly to Gassendi's Epicureanism, in which sense-perceptions are signs of some material cause.[29] Hobbes therefore also accepted the narrowing of reason. Hobbesian reason depended on being sure of first principles and the evidence gathered from the senses from which one could calculate outcomes. This view of reason gave one access to

[25] *The Political Testament of Cardinal Richelieu*, ed. and trans. by Henry Bertram Hill (Madison: Univ. of Wisconsin Press, 1961), pp. 70–1. The treatise was probably extant by the late 1630s, although it was not published until 1688.

[26] Richard Tuck, *Philosophy and Government, 1572–1651* (Cambridge: Cambridge Univ. Press, 1993).

[27] Anthony Levi, *French Moralists: The Theory of the Passions, 1585 to 1649* (Oxford: Clarendon Press, 1964), pp. 225–33; see also the fine recent work of Susan James, *Passion and Action: The Emotions in Seventeenth-Century Philosophy* (Oxford: Oxford Univ. Press, 1997).

[28] Arthur O. Lovejoy, *Reflections on Human Nature* (Baltimore: Johns Hopkins Univ. Press, 1961), pp. 129–215; Nannerl O. Keohane, *Philosophy and the State in France: The Renaissance to the Enlightenment* (Princeton, N.J.: Princeton Univ. Press, 1980).

[29] Richard Tuck, "Hobbes and Descartes," in *Perspectives on Thomas Hobbes*, ed. G. A. J. Rogers and Alan Ryan (Oxford: Clarendon Press, 1988), pp. 11–41, on p. 40; for an attack on Tuck's views, see Perez Zagorin, "Hobbes's Early Philosophical Development," *J. Hist. Ideas* 54 (1993): 505–18.

physical truths, but did not give one access to innate moral truths, causing differences of opinion to arise. As a consequence, the state needed a powerful sovereign to make his subjects conform to his views.

In his *Leviathan* (1651), Hobbes developed the implications of these premises. In brief, he rooted his view of human nature in the idea that everything could be reduced to matter and motion.[30] "Life is but a motion of Limbs, the beginning of which is in some principall part within," he declared in the second sentence of *Leviathan*.[31] The signs of life within, in turn took their origin from the passions, which responded to stimuli. As he put it in his work on human nature, "the Passions of Man . . . are the Beginning of voluntary Motions." Later, he says, "Sense proceedeth from the Action of external Objects upon the Brain, or some internal Substance of the Head; and . . . the Passions proceed from the Alteration there made, and continued to the Heart."[32] He began his science of politics traditionally enough, with the Delphic (and Socratic) *nosce teipsum* (read thyself). This teaches "that for the similitude of the thoughts, and Passions of one man, to the thoughts, and Passions of another, whosoever looketh into himself, and considereth what he doth, when he does think, opine, reason, hope, fear, &c., and upon what grounds; he shall thereby read and know, what are the thoughts, and Passions of all other men, upon the like occasions."[33] Hobbes continued by explaining the way sense, imagination, speech, and reason worked in the personal body. Having laid out these fundamental preliminaries, Hobbes moved in chapter six of *Leviathan* to deal with "the Interiour Beginnings of Voluntary Motions; commonly called the PASSIONS. And the Speeches by which they are expressed."[34] He divided the passions into those stemming from appetite or desire and those stemming from aversion, from attraction and repulsion, love and hate (with the addition of "contempt" for "those things, which we neither Desire, nor Hate"). Hobbes therefore defined "good" as simply another word for "whatsoever is the object of any man's Appetite or Desire," and "evil" as "the object of his Hate, and Aversion"[35] (just as had the opponents of Socrates in the *Gorgias*). Hobbes acknowledged that the passions tended to make people disinclined toward civil duties, censorious, and subject to whims of fancy and rash deliberation; human life tended to "consisteth almost in nothing else but a perpetuall contention for [personal] Honor, Riches, and Authority." As he famously concluded from this, the natural condition of man was that of war.[36]

Nevertheless, Hobbes's message was that while conflicting passions pose "indeed

[30] For some of Hobbes's further views on matter and motion, see his *Dialogus physicus* (1661), translated in Shapin and Schaffer, *Leviathan and the Air Pump* (cit. n. 8), pp. 345–91; and Thomas Hobbes, *Thomas White's 'De mundo' Examined*, trans. Harold Whitmore Jones (London: Bradford Univ. Press, 1976).

[31] Thomas Hobbes, *Leviathan*, ed. Richard Tuck (Cambridge: Cambridge Univ. Press, 1991), p. 9.

[32] Thomas Hobbes, *Humane Nature: Or The Fundamental Elements of Policy. Being A Discovery of the Faculties Acts and Passions of the Soul of Man, From their Original causes; According to such Philosophical Principles As are not commonly known or asserted. The Third Edition, Augmented and much corrected by the Authors own hand* (London: Printed for Matthew Gilliflower, Henry Rogers, and Tho. Fox, 1684), pp. 30, 63.

[33] Hobbes, *Leviathan* (cit. n. 31), p. 10.

[34] He had announced in Aug. 1635 "his intention of being the first writer to speak sense in plain English on the 'faculties and passions of the soul,'" and he dealt with the passions in his first political treatise, in 1640: Johann P. Sommerville, *Thomas Hobbes: Political Ideas in Historical Context* (New York: St. Martin's Press, 1992), pp. 13, 14.

[35] Hobbes, *Leviathan* (cit. n. 31), p. 120.

[36] Ibid., pp. 86–90 (chap. 13).

great difficulties," by "Education, and Discipline, they may bee, and are sometimes reconciled."[37] Both reason and a few of the passions (fear of death, desire to live commodiously, and hope) inclined people to peace, for reason taught that the ends desired by our passions can best be achieved not through war but through cooperation. All the virtues could therefore be shown to be rooted in actions that help people to achieve peaceful, sociable, and comfortable living. But because reason had power to judge only the true rather than the good, it could not be trusted always to anticipate the future correctly: it could not know whether some action actually furthered peace. Consequently, people would differ and so remain in a state of conflict unless judgment rooted in an agreed-upon authority (an arbitrator or a sovereign) could be applied.[38] This new view of reason made a powerful monarch necessary to peace and order. Indeed, one of the most famous depictions of the body politic is an engraving, published on the title page of the first edition of Hobbes's *Leviathan*, with a body of people and the head of a king.

The most philosophical English alternative to the Hobbesian vision of a powerful sovereign revived the argument for right reason.[39] James Harrington's *Oceana* (1656) countered Hobbes's vision by backing off from Hobbes's materialism. Harrington believed that a republican senate could exercise right reason and thereby control the passions in the body politic. To support his views, he borrowed heavily from vitalist views of the body, particularly those of Jan Baptista van Helmont and William Harvey.[40] Like Hobbes, Harrington believed that the passions needed to be dominated by reason: when first defining government, Harrington took note of the classical view that when government of reason degenerated into government of passion, the three good kinds of government degenerated into the three bad kinds.[41] Distinguishing between the internal and the external principles of government (virtue and wealth), Harrington famously went on to treat the goods of wealth and fortune first. But when he shortly returned to a discussion of the internal principles of authority—virtue—he took up the common distinction between right reason and the passions. When the passions take over the mind, one falls into "vice and the bondage of sin"; when reason takes over, one finds "virtue and freedom of soul." Since "government is no other than the soul of a nation or city," the question Harrington posed is how to obtain government by the use of right reason. His answer was that the reason of humankind as a whole comes "the nearest unto right reason." Hence, he began by treating popular government, in which those who are considered the wisest—who are also the most virtuous,[42] as we know from the classical definition of right reason—will become the senators. A debating senate therefore became the embodi-

[37] Ibid., p. 483.

[38] Quentin Skinner, "Thomas Hobbes: Rhetoric and the Construction of Morality," *Proc. Brit. Acad.* 76 (1990): 1–61.

[39] For other views, see Sarah Barber, *Regicide and Republicanism: Politics and Ethics in the English Revolution, 1646–1659* (Edinburgh: Edinburgh Univ. Press, 1998).

[40] Wm. Craig Diamond, "Natural Philosophy in Harrington's Political Thought," *J. Hist. Phil.* 16 (1978): 387–98; I. Bernard Cohen, "Harrington and Harvey: A Theory of State Based on the New Physiology," *J. Hist. Ideas* 55 (1994): 187–210. Sir Matthew Hale also borrowed heavily from Van Helmont: see Alan Cromartie, *Sir Matthew Hale, 1609–1676: Law, Religion and Natural Philosophy* (Cambridge: Cambridge Univ. Press, 1995), pp. 206–8, 218–9, 223, 227.

[41] That is, monarchy, aristocracy, and democracy degenerated into tyranny, oligarchy, and anarchy.

[42] James Harrington, *The Commonwealth of Oceana and a System of Politics*, ed. J. G. A. Pocock (Cambridge: Cambridge Univ. Press, 1992), pp. 10, 19, 22, 35.

ment of right reason in a popular government.[43] Right reason, dominating the passions, provided the necessary foundation for good government: true liberty is rooted in "the empire of [right] reason."[44] Or so English opponents of Hobbes and other materialists contended.

DESCARTES AND THE GOODNESS OF THE PASSIONS

Almost all seventeenth-century commentators, then, continued to believe that reason must control the passions in the body politic, whether that be virtuous right reason or calculating *raison d'état.* But one great exception lay behind the views of the Dutch republicans: the analysis of René Descartes as expressed in his last work, *On the Passions of the Mind (*Les Passions de l'ame*)* (1649). His was a most powerful alternative, much more accepting of the passions as expressions of nature in us and therefore as good. In any case, in only a few people could reason control the passions. Therefore, understanding them and embracing them was far better than attempting to control them and suffering the anxieties that the failure to do so brought on. Descartes was led to this position through his conversations with the princess Elizabeth.

Like Hobbes and so many other contemporaries, Descartes employed the limited view of reason that ascribes to it the ability to take in information from the senses and calculate the consequences, but not the ability to grasp innate ideas about the good. Consequently, "when Descartes speaks of the judgments of the reason which the will should follow he does not identify reason with 'right reason' . . . but insists only on the attempt to judge correctly while acknowledging the fallibility of the human speculative faculty." Moreover, "when he is on the point of undertaking the method, he is clearly prepared to dissociate in practice the principles governing the conduct of life from those which govern the quest for intellectual certainty," unlike someone guided by classical right reason.[45] Hence the radical doubt about all former opinions with which he begins his *Meditations of First Philosophy* (1641). While Descartes's famous discussion of *cogito* might satisfy some that God existed, it was not clear from that proof that ethical consequences followed. Indeed, Descartes himself wrote no work on ethics or politics (perhaps because he died too soon to reply to Hobbes). To questioners he simply made it clear that he had already done the metaphysical work necessary for people to get on with the "most desirable" business of studying "physical and observable things," which would yield "abundant benefits for life."[46]

[43] In his *System of Politics* (probably composed c. 1661, published in 1700), Harrington further declared: "Formation of government is the creation of a political creature after the image of the philosophical creature, or it is an infusion of the soul or faculties of man into the body of the multitude"; so that "The more the soul or faculties of man . . . are refined or made incapable of passion, the more perfect is the form of government." Quoted from ibid., p. 273 (chap. IV, 10 and 11).

[44] Quoted in Mark Goldie, "The Civil Religion of James Harrington," in *The Languages of Political Theory in Early Modern Europe*, ed. Anthony Pagden (Cambridge: Cambridge Univ. Press, 1987), p. 211; see also J. A. I. Champion, *The Pillars of Priestcraft Shaken*: *The Church of England and Its Enemies, 1660–1730* (Cambridge: Cambridge Univ. Press, 1992), pp. 198–208.

[45] Levi, *French Moralists* (cit. n. 27), pp. 295, 246. See also Steven Shapin, "Descartes the Doctor: Rationalism and Its Therapies," *Brit. J. Hist. Sci.* 33 (2000): 131–54.

[46] *The Philosophical Writings of Descartes*, ed. and trans. John Cottingham, Robert Stoothoff, Dugald Murdoch et al., 3 vols. (Cambridge: Cambridge Univ. Press, 1985–1991), vol. 3, pp. 346–7 (hereafter cited as CSMK).

Even more, Descartes famously took the study of the human body seriously and developed the consequences of his study in many of his works. One of his first major studies, written about 1629–1633 (but published posthumously), was the *Treatise on Man*. He began by analyzing the body as "a statue or machine made of earth, which God forms with the explicit intention of making it as much as possible like us." It contained all the bones, nerves, muscles, veins, and other parts necessary. "I assume that if you do not already have sufficient first-hand knowledge of them, you can get a learned anatomist to show them to you—at any rate, those which are large enough to be seen with the naked eye."[47] Among the learned anatomists with whom he himself was acquainted, and with whom he had observed anatomical investigations, were the Dutch professors François dela Boë Sylvius of Leiden and Henricus Regius of Utrecht. It is likely to have been Sylvius (one of whose students first defended William Harvey's views on the circulation of the blood)[48] who introduced Descartes to the new anatomy. Descartes may have gotten the cause of the circulation of the blood wrong, but he was an early defender of the concept.[49] It is also likely that it was contemporary Dutch anatomists who exposed the pineal gland, an organ famously made much of by Descartes, as the central part of the brain.[50] He clearly believed that his new philosophy and physiology would lead to improvements in health: "the maintenance of health," he wrote in *Discourse on Method* (1637), "is undoubtedly the chief good and the foundation of all other goods in this life," and better health can be achieved through improvements to medicine made by his useful philosophical system.[51] In 1645 he repeated to the Marquess of Newcastle that "the preservation of health has always been the principal end of my studies," and in 1646 he wrote to Hector-Pierre Chanut (the French resident in Stockholm) that because of this, "I have spent much more time" on medical topics than on moral philosophy and physics.[52] For Descartes, "the good" therefore appears to be bodily well-being: "in his *Discours*, Descartes seems . . . to envisage the spiritual perfection of man as a function of medicine, a practical application of the exact deductive physics."[53] At the same time Descartes was occupied with medical concerns, local physicians became some of his most important allies: Regius's physiological interpretations of Descartes's views led to the first major controversy over Cartesianism; the notes of a medical student, Frans Burman, on his conversation with Descartes in 1648 are some of the most revealing explanations for his views; it was another Leiden physician, Florentius Schuyl, who discovered, translated, and brought out Descartes's *Treatise on Man* (in Latin) in 1662; and physicians remained among

[47] Ibid., vol. 1, p. 99.

[48] Roger French, "Harvey in Holland: Circulation and the Calvinists," in *The Medical Revolution of the Seventeenth Century*, ed. Roger French and Andrew Wear (Cambridge: Cambridge Univ. Press, 1989), pp. 46–86; M. J. van Lieburg, "Isaac Beeckman and His Diary-Notes on William Harvey's Theory on Bloodcirculation," *Janus* 69 (1982): 161–83.

[49] Geoffrey Gorham, "Mind-Body Dualism and the Harvey-Descartes Controversy," *J. Hist. Ideas* 55 (1994): 211–34.

[50] This suggestion was shared with me by Harm Beukers and is based on the large remaining part of Rembrandt's *Anatomy Lesson of Dr. Joan Deyman* (1656), which represents a method of dissecting the brain that would have highlighted the pineal gland.

[51] CSMK (cit. n. 46), vol. 1, pp. 142–3.

[52] Descartes to the marquess of Newcastle, Oct. 1645, and Descartes to Chanut, 15 June 1646, CSMK (cit. n. 46), vol. 3, pp. 275, 289.

[53] Levi, *French Moralists* (cit. n. 27), p. 248.

the most numerous and important Cartesians in the middle decades of the century.[54]

Descartes, then, became famous—or infamous—for incorporating fresh discoveries in anatomy and physiology into a view of the human body as a machine-like system. (The clear and distinct difference between body and soul has ever since been known as Cartesian dualism, and ideas of the mechanical body have been associated with Cartesianism, even when their authors have clearly argued against Descartes's views.) The rational soul, as an incorporeal entity, could direct some aspects of the corporeal body through the means of the pineal gland if it chose, but it was not necessary. In this way, the human body functioned just like other animal bodies, even though the latter had no souls. Descartes himself held complex views on the relationship between body and soul, however, which came out clearly in his work on the passions. As he noted at the beginning of his *Treatise on Man*, to fully explain our beings he had to "describe the body on its own; then the soul, again on its own; and finally I must show how these two natures would have to be joined and united in order to constitute men who resemble us."[55] When it came to analyzing the union of body and soul he, like his contemporaries, turned to the passions.

Although Descartes had already given some thought to the passions earlier, his close consideration of them was pressed on him by Princess Elizabeth, daughter of the ill-fated "Winter King," Prince Frederick of the Palatinate, and his wife, Elizabeth, sister of Charles I of England. Descartes had heard through mutual acquaintances in late 1642 that the princess, then residing in The Hague, was reading his *Meditations*, and he managed an introduction, which led to a life-long relationship.[56] In a letter of 6 May 1643, Elizabeth pointed out that while Descartes had discussed how matters of soul and body had to be distinguished from one another and considered according to different clear and distinct notions, he had left an important question unanswered: How do the two interact? More precisely, how could the soul—in his view a thinking substance only—get the bodily spirits to exhibit voluntary actions? Descartes replied by letter two weeks later, arguing that the soul had two aspects we could know: it thinks, and it acts on and is acted on by the body. "About the second I have said hardly anything," he confessed, since his first philosophical aim had been "to prove the distinction between the soul and the body, and to this end only the first was useful, and the second might have been harmful." Pressed further by Elizabeth, he felt compelled to take up a discussion of cognition, but he left the problem vague for the moment. Understandably confused, Elizabeth nevertheless persisted, causing him to put her off: it was when one refrained from philosophy that one understood the union of soul and body most clearly, he wrote. "It does not seem to me that the human mind is capable of forming a very distinct conception

[54] Theo Verbeek, *La querelle d'Utrecht: René Descartes et Martin Schoock* (Paris: Les impressions nouvelles, 1988); idem, *Descartes and the Dutch: Early Reactions to Cartesian Philosophy, 1637–1650* (Carbondale: Southern Illinois Univ. Press, 1992); idem, ed., *Descartes et Regius* (Amsterdam: Rodopi, 1994); G. A. Lindeboom, *Florentius Schuyl (1619–1669) en zijn betekenis voor het Cartensianisme in de geneeskunde* (The Hague: Martinus Nijhoff, 1974); C. L. Thijssen-Schoute, *Nederlands Cartesianisme* (1954; reprinted, Utrecht: Hes Uitgevers, 1989); for Burman, see CSMK (cit. n. 46), vol. 3, pp. 332–54.

[55] CSMK (cit. n. 46), vol. 1, pp. 99.

[56] Descartes to Princess Elizabeth, 6 Oct. 1642, CSMK (cit. n. 46), vol. 3, pp. 214–5; Léon Petit, *Descartes et la Princesse Elisabeth: Roman d'Amour Vécu* (Paris: Editions A.-G. Nizet, 1969).

of both the distinction between the soul and the body and their union" at the same time. The notion of a union of body and soul was something "everyone invariably experiences." He then more or less told her to forget about the problem, instructing her to "feel free" to think what she wanted, that she should understand "the principles of metaphysics" once during her lifetime, but thinking about them too long and hard would be "very harmful."[57] Rather than explain further, then, he begged off. Understandably, Descartes's answers did not satisfy Elizabeth.

But a year later he came back to a discussion of the relation between the soul and the body due to medical considerations.[58] Elizabeth was not well and had decided on a course of diet and exercise, which Descartes approved. But he clearly agreed with her that the cause of her ill health was the result of a troubled mind. "There is no doubt that the soul has great power over the body, as is shown by the great bodily changes produced by anger, fear and other passions."[59] In giving advice on how thinking could restore one to health, Descartes was making arguments about reason, the passions, and the bodily spirits that almost every learned physician of the day would second, for affections of the mind constituted one of the six non-naturals. He seemed to hold generally that the passions greatly affected the actions of the heart and other organs, thereby causing the putrefactions in the blood that gave rise to fevers.[60] In advising Elizabeth as he did, then, Descartes was returning to one of the main points of his work: its utility for medicine. But in doing so he was again forced to confront the relationship between soul and body. When he did, he minimized the power of reason to affect the passions directly:

> The soul guides the spirits into the places where they can be useful or harmful; however, it does not do this directly through its volition, but only by willing or thinking about something else. For our body is so constructed that certain movements in it follow naturally upon certain thoughts: as we see that blushes accompany shame, tears compassion, and laughter joy. I know no thought more proper for preserving health than a strong conviction and firm belief that the architecture of our bodies is so thoroughly sound that when we are well we cannot easily fall ill except through extraordinary excess or infectious air or some other external cause.[61]

Elizabeth's health nevertheless grew increasingly delicate, which Descartes attributed mainly to the continued bad news regarding her family's fortunes: it was increasingly clear in these last years of what became known as the Thirty Years' War that her family would not be restored to the princely throne of the Palatinate; more immediately, despite the brilliant efforts of her brother Rupert, the war in England against her uncle Charles I had been going badly. Elizabeth felt distresses of such a sort that "right reason does not command us to oppose them directly or to try to remove them," wrote Descartes. He took up the traditional point of view, writing that "I know only one remedy for this: so far as possible to distract our imagination and senses from them [i.e., misfortunes], and when obliged by prudence to consider

[57] CSMK (cit. n. 46), vol. 3, pp. 217 n. 1, 217–20, 218, 217–20, 226–9, especially pp. 227 and 228.

[58] Descartes to Princess Elizabeth, July 1647, *Oeuvres de Descartes*, ed. Charles Adam and Paul Tannery (Paris: Librairie Philosophique J. Vrin, 1971–1975), vol. 5, pp. 64–6 (hereafter cited as AT).

[59] CSMK (cit. n. 46), vol. 3, p. 237.

[60] Theo Verbeek, "Les passions et la fièvre: L'idée de la maladie chez Descartes et quelques cartésiens néerlandais," *Tractrix* 1 (1989): 45–61.

[61] CSMK (cit. n. 46), vol. 3, p. 237.

them, to do so with our intellect alone."[62] Descartes went on to distinguish between the intellect on the one hand and the imagination and senses on the other. The imagination and senses governed the passions and affected the spirits and body; the intellect was separate and had the power to direct the imagination (and so the passions). Descartes had even cured himself of ill health by looking at things "from the most favorable angle," he declared. In a following letter a few days later, Descartes tried to soothe his distressed patient by sympathizing with her and by declaring that "the best minds are those in which the passions are most violent and act most strongly on their bodies." But following a night's sleep, one can "begin to restore one's mind to tranquillity" by concentrating on the good news, "for no events are so disastrous . . . that they cannot be considered in some favorable light by a person of intelligence." Perhaps study would help to distract her, he suggested.[63]

Elizabeth agreed, and they began an epistolary conversation about the Stoic philosopher Seneca's *De vita beata*. For the next several months, their discussion focused on reason and the passions and continued until the end of Descartes' life.[64] Descartes began by defending a neo-Stoic position (which he thought consistent with what he had put forward in the *Discourse on Method*). One should employ reason to discover what should and should not be done in all circumstances; one should resolve to do as reason directs "without being diverted by . . . passions or appetites. Virtue, I believe, consists precisely in sticking firmly to this resolution," and one should acknowledge that all goods that one does not possess are beyond one's power and so not worth thinking about. "So we must conclude that the greatest felicity of man depends on the right use of reason" and the controlling of the passions by this. Put another way, "happiness consists solely in contentment of mind . . . but in order to achieve contentment which is solid we need to pursue virtue—that is to say, to maintain a firm and constant will to bring about everything we judge to be the best, and to use all the power of our intellect in judging well." Elizabeth objected that many, including those who are ill, do not have the free use of their reason that Descartes's views assumed. He agreed that "what I said in general about every person should be taken to apply only to those who have the free use of their reason and in addition know the way that must be followed to reach such happiness." That is, some people do not know how to think properly about happiness, and others have a bodily indisposition that prevents them from acting freely. But he came back to the neo-Stoic view that the passions are vain imaginings, or distortions of reason—that is, errors of mind—so that "the true function of reason . . . is to examine and consider without passion" one's true good and to "subject one's passions to reason."[65]

Still not satisfied, Elizabeth asked Descartes, in her letter of 13 September 1645, to give "a definition of the passions, in order to make them well known."[66] In order

[62] Descartes to Princess Elizabeth, [May or June] 1645 [I believe it a good guess to date it at about the time of the battle of Naseby, which took place on 14 June], CSMK (cit. n. 46), vol. 3, p. 249, except that they translate *la vraye raison* (AT [cit. n. 58], vol. 4, p. 218) as "true reason" rather than "right reason."

[63] Descartes to Princess Elizabeth, June 1645, CSMK (cit. n. 46), vol. 3, pp. 251, 253–4.

[64] Ibid., 12 July, 4 Aug., 18 Aug., 1 Sept., 15 Sept., 6 Oct., 3 Nov. 1645; Jan., May, Sept., Oct./Nov., Dec. 1646; March, 10 May 1647; 31 Jan. 1648; 22 Feb., June, 9 Oct. 1649, CSMK (cit. n. 46), vol. 3, pp. 255–73, 276–8, 281–3, 285–8, 292–5, 296–8, 304–5, 314–5, 317–9, 323–4, 328–30, 367–8, 378–9, 382–3.

[65] CSMK (cit. n. 46), vol. 3, pp. 257–8, 262, 264–5.

[66] Robert Stoothoof, preface to *The Passions of the Soul*, in CSMK (cit. n. 46), vol. 1, p. 325.

to comply with her request, Descartes returned to a consideration of animal physiology. At the same time, his physiological views, as expressed by Regius in Utrecht, were leading to condemnations of Descartes's philosophy. Occupied with concerns on at least these two fronts about how to express his views on the body, Descartes began to dig deeper and arrived at some new conclusions. He began by discarding a number of common associations of the word "passion" and limiting his investigations to "the thoughts that come from some special agitation of the spirits, whose effects are felt as in the soul itself,"[67] writing that he had begun to consider these in detail. In his letters, he began to make excuses for not continuing along these lines, while engaging in a long discussion with Elizabeth about free will. But by early 1646, he had drafted a work on the passions, which he sent to Elizabeth for comment.

In this draft treatise, Descartes explained that the movements of the blood accompanying each passion were grounded in physical and physiological principles, and that "our soul and our body" are very closely linked. But he also acknowledged that "the remedies against excessive passions are difficult to practise" and "insufficient to prevent bodily disorders." He still believed that such remedies might free the soul of domination by the passions so as to enable "free judgement." But now "it is only desires for evil or superfluous things that need controlling"; certainly "it is better to be guided by experience in these matters than by reason." A few months later, in writing to Hector-Pierre Chanut about how to present his philosophical views to Queen Christina (in an attempt to secure her patronage), Descartes declared that despite Chanut's expectations, "in examining the passions I have found almost all of them to be good, and to be so useful in this life that our soul would have no reason to wish to remain joined to its body even for one minute if it could not feel them."[68] In late 1647, Descartes sent copies of his 1645 letters to Elizabeth and the draft treatise on the passions to the queen;[69] he returned to working on his treatise on animals in 1648. *Les Passions de l'ame* appeared in November 1649, just three months before Descartes's death in Stockholm.

The published version of the treatise began by noting that "The defects of the sciences we have from the ancients are nowhere more apparent than in their writings on the passions." While he continued to treat topics of soul and body separately, and to see the passions as acting on the soul in the same way that objects made themselves known through sight, he also wished to show that one of the two kinds of thought proceeding from the soul is the passions (the other being volition). Moreover, "[T]he various perceptions or modes of knowledge present in us may be called [the soul's] passions." These perceptions may be caused by the soul or by the body. Because the passions were products of body as much as of mind, they could not be directly controlled by volition (this reiterated his epistolary exposition of 1644). Reason could control volition, and hence people possess free will, but volition cannot control the passions directly. "Our passions . . . cannot be directly aroused or

[67] CSMK (cit. n. 46), vol. 3, p. 271.

[68] Descartes to Princess Elizabeth, May 1646, CSMK (cit. n. 46), vol. 3, pp. 285–8; Descartes to Chanut, 1 Nov. 1646, CSMK (cit. n. 46), vol. 3, p. 290. That he was then beginning to lobby a convinced neo-Epicurean may also have had its effects in convincing him of the goodness of the passions: see Susanna Åkerman, *Queen Christina of Sweden and Her Circle: The Transformation of a Seventeenth-Century Philosophical Libertine* (Leiden: Brill, 1991).

[69] CSMK (cit. n. 46), vol. 1, p. 327.

suppressed by the action of our will, but only indirectly through the representation of things which are usually joined with the passions we wish to have and opposed to the passions we wish to reject." The body affected the passions greatly: "the soul cannot readily change or suspend its passions" because the passions "are nearly all accompanied by some disturbance which takes place in the heart and consequently also throughout the blood and the animal spirits."[70]

Descartes held out the possibility that some people might be able to master these powerful forces: "undoubtedly the strongest souls belong to those in whom the will by nature can most easily conquer the passions and stop the bodily movements which accompany them." And "Even those who have the weakest souls could acquire absolute mastery over all their passions if we employed sufficient ingenuity in training and guiding them." Yet the only hope for mastering the passions came from habits and mental exercises that anticipated events beforehand, what one's own reactions would be, and what one hoped to do in the circumstances. "I must admit that there are few people who have sufficiently prepared themselves" by the constant use of forethought and diligence "for all the contingencies of life," and "no amount of human wisdom is capable of counteracting these movements [in the body] when we are not adequately prepared to do so." Therefore, it was actually by strengthening one of the passions themselves that one achieved inner freedom: the "key to all the other virtues and a general remedy for every disorder of the passions" is that of generosity. "True generosity . . . causes a person's self-esteem to be as great as it may legitimately be" and has two parts: "The first consists in his knowing that nothing truly belongs to him but this freedom to dispose his volitions," while the second "consists in his feeling within himself a firm and constant resolution to use it well." Generosity can be generated as a thought in the soul, "but it often happens that some movement of the [bodily] spirits strengthens them, and in this case they are actions of virtue and at the same time passions of the soul." He concluded by arguing that "the chief use of wisdom lies in its teaching us to be masters of our passions and to control them with such skill that the evils which they cause are quite bearable, and even become a source of joy."[71] Trying to maintain control of the passions through reason was still a laudable goal, then, but it could not be done directly, only through anticipation and indirection; and this was impossible for almost everyone, although those with generous souls had the best chance of acting in a manner that blended the goals of reason and body.

Most importantly, Descartes suggested that one should not worry too much about controlling the passions—a complete departure for him. First, the passions show us how to remain alive and how to live well. "The function of all the passions consists solely in this, that they dispose our soul to want the things which nature deems useful to us, and to persist in this volition." Therefore the passions were good—and not just some of the passions, but all of them: "we see that they are all by nature good, and that we have nothing to avoid but their misuse or their excess." Second, they bring us pleasures. While the soul (or rational faculty) "can have pleasures of its own," the pleasures "common to it and the body depend entirely on the passions." People who can be moved deeply by the passions "are capable of enjoying

[70] Ibid., pp. 328, 333, 335, 345. For a fine discussion of Descartes's treatise on the passions, see Levi, *French Moralists* (cit. n. 27), pp. 257–98; James, *Passion and Action* (cit. n. 27), passim.

[71] CSMK (cit. n. 46), vol. 1, pp. 347, 348, 403, 388, 384, 387–8, 404.

the sweetest pleasures of this life."[72] This new view placed Descartes in a line of argument that included Lorenzo Valla, who argued in *De voluptate* (1431) that the true virtues came not from "faith in one's own reason and in a form of wisdom which derives from it," but rather from a faith "in the miracle of life as we are allowed to live it in harmony with the universe, with the plants and animals, with our fellow human beings." This faith—uniting the senses to the external world—was, Valla declared, the same as *caritas*, "Jesus Christ's love for humanity, expressed through the Redemption."[73] Perhaps Descartes had been convinced of Epicureanism by his close reading of Seneca, who although usually taken as a Stoic also introduced Epicurean concepts to a great many people in the seventeenth century: it was after all in Seneca's treatise *De vita beata*, which Elizabeth and Descartes had examined together, that the Epicurean notion of *voluptas* was introduced as sober and austere.[74] Or perhaps like others Descartes, too, was persuaded of the coherence of Epicurean ideas by his correspondent Pierre Gassendi, despite their firm disagreements.[75]

Descartes's general conclusion from his studies, then, was that "we have much less reason for anxiety about [the passions] than we had before." More powerfully, all the pleasures that are common to both soul and body, such as love, "depend entirely on the passions." This is a large step beyond Aristotle's view that some of the passions can be good; this new view would have been almost unthinkable for Descartes's neo-Stoic predecessors; and it went considerably further than all but the Epicureans in making the passions into forces for good instead of irrationality and vice. One need not fear the passions, only avoid "their misuse or their excess." Virtue lies not in the conquest of the passions by reason, but in a person's living "in such a way that his conscience cannot reproach him for ever failing to do something he judges to be the best." That way, he will have a tranquil soul (one of the chief goals of Epicureanism), which "the most violent assaults of the passions will never have sufficient power to disturb." If one pursues this course toward a virtuous life, then the rational faculty or soul will remain free of being a slave to the passions. One can therefore enjoy the pleasures the passions bring while turning the ills they cause into "a source of joy."[76] As one recent commentator has declared, it was Elizabeth's persistence in grounding the thinking subject in the body that forced Descartes to come to terms with the passions;[77] this seems to have forced him to reconsider Epicurean options; and he came to take a view that emphasized the passions, rather than reason, for maintaining life and bringing happiness.[78]

[72] Ibid., pp. 349, 403, 404.
[73] Maristella de P. Lorch, "The Epicurean in Lorenzo Valla's *On Pleasure*," in *Atoms, Pneuma, and Tranquillity* (cit. n. 12), pp. 93, 99.
[74] Louise Fothergill-Payne, "Seneca's Role in Popularizing Epicurus in the Sixteenth Century," in *Atoms, Pneuma, and Tranquillity* (cit. n. 12), pp. 115–33; see also Lynn Joy, "Epicureanism in Renaissance Moral and Natural Philosophy," *J. Hist. Ideas* 53 (1992): 573–83.
[75] On Gassendi, see especially Lisa T. Sarasohn, "Motion and Morality: Pierre Gassendi, Thomas Hobbes and the Mechanical World-View," *J. Hist. Ideas* 46 (1985): 363–79; idem, "Gassendi and the Creation of a Privatist Ethic in Early Seventeenth-Century France," in *Atoms, Pneuma, and Tranquillity* (cit. n. 12), pp. 174–95; Lynn Sumida Joy, *Gassendi the Atomist: Advocate of History in an Age of Science* (Cambridge: Cambridge Univ. Press, 1987).
[76] CSMK (cit. n. 46), vol. 1, pp. 403, 382, 404.
[77] Erica Harth, *Cartesian Women: Versions and Subversions of Rational Discourse in the Old Regime* (Ithaca, N.Y.: Cornell Univ. Press, 1992), pp. 75–8.
[78] See also J. Dankmeijer, "De biologische studies van René Descartes," in *Leidse Voordrachten*, no. 9 (Leiden: Universitaire Pers, 1951); Richard B. Carter, *Descartes' Medical Philosophy: The Organic Solution to the Mind-Body Problem* (Baltimore: Johns Hopkins Univ. Press, 1983); G. A.

FREEING THE PASSIONS: DUTCH REPUBLICANISM

Shortly after Descartes's death in 1650, a republic came into being in the Dutch world in which he had moved, and this new world's defenders argued for a view of the passions very similar to his. The republicans retained the narrowing of reason rather than trying to reassert right reason. They also highlighted the positive value of the passions. To allow for good government, then, this viewpoint argued, the passions need not be controlled by reason but could be allowed free reign in a system of laws, which would establish agreed-upon rules for letting private passions check one another. When opposite passions meet, they would cancel each other out, thus allowing peace and stability to emerge in the public realm. Instead of being the embodiment of reason, the state would be a kind of referee enforcing methods of negotiation.

As elsewhere in Europe, early modern political theory in the Netherlands had generally remained rooted in ideas in which the goal of public virtue remained uppermost; humanity obtained that goal by controlling the passions, primarily with monarchical government. Humanist education in the Netherlands had been directed "towards developing civic virtues and preparing its pupils for a life of responsible leadership"[79] just as much as elsewhere. Orthodox Calvinists hated anything that smacked of Cartesianism, while even the more liberal Arminians and Coccejians battled outright materialism, borrowing support from their English latitudinarian colleagues.[80] But in the middle of the century, at the same time Descartes was studying the passions closely, theories that spoke openly of republicanism became more widely available. Just as the horrors of civil war in England had confirmed Hobbes in his monarchism, the period known as "True Freedom" in the Netherlands, from 1650 to 1672, confirmed many in republicanism. In the republican body politic, the passions were as important as reason for directing humanity toward the good, while authority was diffused among the members and organs of the constitution, not needing the head to have the final say in all things.

The death of William II, Prince of Orange and Stadhouder (captain-general), from smallpox, on 6 November 1650, allowed the restoration of political control by the States Party, or the party of True Freedom as they sometimes styled themselves. Just before his death, Prince William had forced Amsterdam into obedience by surrounding it with his army and threatening a siege, allowing him and his conservative Calvinist allies to seize control of almost all institutions of government. For the next twenty-two years following his death, however, the oligarchy of the rich and powerful members of the city and provincial governing councils (the *regenten*, or regents) reasserted its control. The skilled author and pensionary (chief minister) of the province of Holland Jacob Cats opened the Great Assembly that ratified the return to republican rule with a discourse on the superiority of republics to monarchies. Although Orangists and anti-Remonstrant Calvinists continued to advocate a return to

---

Lindeboom, *Descartes and Medicine* (Amsterdam: Rodopi, 1979); Geneviève Rodis-Lewis, *Descartes: His Life and Thought*, trans. Jane Marie Todd (Ithaca, N.Y.: Cornell Univ. Press, 1998), pp. 186–7.

[79] E. H. Kossmann, "The Development of Dutch Political Theory," in *The Development of Dutch Political Theory in the Seventeenth Century* (London: Chato & Windus, 1960), p. 101.

[80] See especially Verbeek, *Descartes and the Dutch* (cit. n. 54); Rosalie L. Colie, *Light and Enlightenment: A Study of the Cambridge Platonists and the Dutch Arminians* (Cambridge: Cambridge Univ. Press, 1957).

a system with a princely head that would control republican and heterodox passions in the Dutch body politic, a period of true republicanism flourished at the height of the Dutch Golden Age.

For most of that period, from 1652 until his brutal murder in 1672, the chief representative of the republic was Johann de Witt. Described by one historian as a *raison d'état* politician,[81] De Witt saw himself as a skeptical realist who tried to live without illusions and without faith in the ability of humanity to save itself in this turbulent world of sin: as another historian declared, De Witt was "a Calvinist of an unmistakable neo-stoic type." De Witt noted that "nothing so much inspires men to love and affection as the feeling in the purse,"[82] and he managed the country accordingly. While a marvelously skilled politician (and mathematician) more than a theoretician of politics, De Witt did have some clear principles, which are most evident in his "Deduction" of 1654 (defending the exclusion of the House of Orange from public office). In it he supported the multiple republican constitutions of the United Provinces, which he believed best acted on the principle that "the welfare of the inhabitants of the country must be the supreme law." The basis for union in a republic was the shared interest of its people:

> But do not the present seven United Provinces have the same single interest in their own preservation? A same single fear of all Foreign Powers? Are they not so bound to each other by mutual alliances and marriages among both regents and inhabitants, by common bodies, companies and partnerships in trade and other interests, by intercourse, possession of property in each other's lands, common customs and otherwise, are they not indeed so bound and interwoven together that it is almost impossible to split them from each other without extraordinary violence, which will not occur unless there are eminent Heads [i.e., princes]?[83]

His solution to the inherent corruption of humanity, then, was to insist on the sharing of power as the best means to prevent abuse and misgovernment. In the case of the United Provinces, he defended the rights of cities and the provincial states even above those of the States General (the assembly of states), leading him to ask that his nation be referred to not as a *respublica* but as a plural: *Respublicae Foederatae*.[84] Such an outlook inherently rejected a view of a body politic ruled by a head.

De Witt clearly believed that monarchies in church and state were unnatural, the enemies of knowledge and virtue, of liberty and property. As he put it in the preface to the book of fables he published:

> But not to speak of past Ages, we may observe, that Monarchical Government has, in the Age in which we live, made such Progress both in Church and State, to the Oppression of many free Republicks, that if the Tyrants continue to tread in the same Paths but for one Age longer, all the liberal Arts and Sciences, all Virtue, and the Liberties and Properties of Men, will throughout all *Europe* dwindle away to nothing; nay the Men

[81] J. G. Boogman, "The 'Raison d'État' Politician Johan de Witt," *Low Countries History Yearbook* 11 (1978): 55–78.

[82] Quoted in ibid., p. 71.

[83] Herbert H. Rowen, *John de Witt, Grand Pensionary of Holland, 1625–1672* (Princeton, N.J.: Princeton Univ. Press, 1978), pp. 388–9.

[84] Jonathan I. Israel, *The Dutch Republic: Its Rise, Greatness, and Fall, 1477–1806* (Oxford: Clarendon Press, 1995), p. 706.

> themselves will be lessen'd in Number, as we already see it had happen'd in *Muscovy, Greece, Turky, Persia, India, &c.* unless God in his Infinite Mercy prevent it.[85]

The most telling evidence for his outlook is, however, his probable editing of (and ghostwriting of chapters 29 and 30 in) one of the most powerfully developed theories of republicanism, Pieter de la Court's *The Interest of Holland* (1662).[86] As that work put it: "people here are right to constantly pray to God, 'oh Lord, protect us from a monarch.'"[87]

There is a story that de Witt first encountered Pieter de la Court's *The Prosperity of Leiden* soon after its publication in 1659 and was so taken with it that he encouraged the author to expand it into *The Interest of Holland*.[88] It certainly uses many of the same arguments as the *Prosperity*, although it is far clearer in attacking the institution of monarchy. De la Court asserted that the province of Holland was not one single country (*Land*) but a region of multiple aspects and interests. It was not a place favored by nature, however, so the inhabitants had to rely on fishing and trading, and after those manufacturing, far more than agriculture. He estimated that perhaps 450,000 people made their living from sea fishing, 200,000 from farming and produce from the land, 450,000 from manufacture, 450,000 from trade, and 650,000 from supplying needs of residents, and that there were 200,000 professionals and civil servants—a total of 2,400,000. From this estimate it was clear that "not the eighth part of the population of Holland can find their means of living [*nooddrust*] from their own land."[89] But a brief economic history revealed why Amsterdam was a richer and greater trading city and Holland a richer country than the world had seen before: freedom. Here people had the freedom to follow whatever religion they chose; the freedom to live and work where they wanted, without the constraints of gilds or monopoly companies, giving rise to a very flexible and adaptable workforce. Freedom for people to follow their own passions and interests created a wealth of material goods. In contrast, the strict Calvinists sought to limit religious choice; the growing power of institutions narrowed the freedom of the fisheries, trade, and crafts, and the growing weight of import and export taxes plus convoy-money threatened to reduce trade. History showed that the country had flourished when it had a thoroughly free government in contrast to the times when the Prince of Orange had been powerful. "People must understand that a good government is not where the subjects fare well or badly depending on the virtue or vice of the governors, but . . . where the fate of the governors necessarily depends on whether the governed fare well or badly."[90] In *Considerations of State* (1661) Johan de la Court (Pieter's recently deceased brother) argued that popular government is the most natural, the most rational, and the fairest. Only in a popular government, moreover, did the saying *Vox populi, Vox Dei* (the voice of God is the voice of the people) hold true. In

[85] *Sinryke Fabulen* (1685), translated as *Fables Moral and Political, with Large Explications*, 2 vols. (London, 1703), unpaginated preface. The work is attributed to De Witt but may be by Pieter de la Court, on whom see below.

[86] On de Witt's involvement with the publication, see Rowen, *John de Witt* (cit. n. 83), pp. 391–4.

[87] v. d. H. [Pieter de la Court or "van der Hoeven"], *Interest Van Holland, Ofte Gronden Van Hollands-Welvaren* (Amsterdam: Joan. Cyprianus vander Gracht, 1662), p. 23: "Zulks men hier te recht altijds God behoorde te bidden, *à furore Monarcharum libera nos Domine, ô God bewaar doch Holland voor een Hoofd*."

[88] Rowen, *John de Witt* (cit. n. 83), p. 391.

[89] v.d. H. [van der Hoeven], *Interest Van Holland* (cit. n. 85), p. 21.

[90] Ibid., preface.

such a government, all the knowledge, passions, and abilities *(bequaamheid)* of the inhabitants were turned to use.[91] Finally, such a country will have no wish to make war for the benefit of its rulers, nor, if it is well ordered, can any monarchy or aristocracy ovérthrow it.

The De la Courts' are some of the century's most remarkable statements on capitalism and republicanism, and they depended on a view of the power of the human passions that could be best developed in a republic. As one historian remarked almost thirty years ago, the brothers De la Court "regarded self-interest and passion as the basis of human conduct, but at the same time they developed the concept of the harmony of self-interests, possible only in a democratic community."[92] The small democracies of the Dutch cities could be combined (along the lines de Witt argued) into a republic such as Holland, or even the United Provinces, and still maintain general harmony in the counterbalancing of personal interests. In arguing directly that the good state did not depend upon a virtuous monarch, the brothers De la Court set themselves directly against those who wanted to restore or enhance the power of the House of Orange.[93] The brothers outlined an original republican theory probably held by many advocates of True Liberty. They self-consciously sought to persuade the magistrates (*politici*) that they could establish their views not on the theories of the schools, but on experience and an analysis of the passions. The noted Dutch historian of political thought E. H. Kossmann has noticed that with its many digressions on physiology and the passions and citations to Descartes, the theory of the brothers De la Court was based on the latest psychology of *Passions de l'ame*.[94]

The De la Courts thought even less than Descartes that the power of reason could control the passions. In his discussion of their views, Kossmann describes the brothers as being more "pessimistic" and "cynical" in their description than Descartes, while also depicting the passions in an even more utilitarian manner.[95] In their view, the passions of individuals should be allowed to express themselves; provided that the political-economic system in which people operated was well ordered—rooted in the law of contracts—opposite passions would balance one another, yielding public harmony and tranquillity.[96] The De la Courts therefore held that "the public inter-

[91] V. H., [Johan de la Court or "van der Hoeven"], *Consideratien Van Staat* (Amsterdam: Jacob Volckerts, 1661), p. 443.

[92] J. W. Smit, "The Netherlands and Europe in the Seventeenth and Eighteenth Centuries," in *Britain and the Netherlands in Europe and Asia*, ed. J. S. Bromley and E. H. Kossmann (London: Macmillan; New York: St. Martin's Press, 1968), pp. 23–4.

[93] G. O. van de Klashorst, "'Metten schijn van monarchie getempert': De verdediging van het stadhouderschap in de partijliteratuur, 1650–1686," in *Pieter de la Court in zijn tijd: Aspecten van een veelzijdig publicist, 1618–1685*, ed. H. W. Blom and I. W. Wildenberg (Amsterdam: APA-Holland Univ. Press, 1986), especially pp. 119–20, 136; I. W. Wildenberg, "Appreciaties van de gebroeders De la Court ten tijde van de Republiek," *Tijdschr. Gesch.* 98 (1985): 540–56.

[94] E. H. Kossmann, "Politieke theorie in het zeventiende-eeuwse Nederland," *Verhandeling der Koninklijke Nederlandse Akademie van Wetenschappen*, Afd. Letterkunde (Nieuwe Reeks), Deel 67, No. 2 (Amsterdam: N. V. Noord-Hollandsche Uitgevers Maatschappij, 1960), pp. 38–48.

[95] Ibid., 39.

[96] Ibid., p. 37. See also E. H. Kossmann, "Volkssouvereiniteit aan het begin van het Nederlandse Ancienne Régime," *Bijdragen en mededelingen betreffende de geschiedenis der Nederlanden* 95 (1980): 1–34 (reprinted in his *Politieke theorie en geschiedenis: Verspreide opstellen en voordrachten* [Amsterdam: Bert Bakker, 1987], pp. 59–92; trans. as "Popular Sovereignty at the Beginning of the Dutch Ancien Regime," *Low Countries History Yearbook* 14 [1981]: 1–28). See also Eco O. G. Haitsma Mulier, *The Myth of Venice and Dutch Republican Thought in the Seventeenth Century*, trans. Gerard T. Moran (Assen, Drenthe: Van Gorcum, 1980), pp. 120–69, who argues for the influence of Machiavelli as well as Descartes and Hobbes, since Machiavelli also grounded politics on the pas-

est [is] the sum of individual interests" and that the true expression of the public interest was possible only in a democratically rooted and commercial republic.[97] Needless to say, those who believed that the country needed a head (the Orangists and strict Calvinists) were outraged.[98] After publishing the *Interest*, Pieter de la Court was suspended from communion by the church council of Leiden, while the magistrates "agreed to forbid the book and seize all copies already published." In 1666 he moved to Amsterdam to join the business of his second wife's family, supplying naval stores to the admiralty in the Second Anglo-Dutch War and trying to break into the trading monopoly of the Dutch East India Company.[99]

Even more striking than the De la Courts' theory was the democratic one developed a few years later by Baruch Spinoza, too radical even for de Witt.[100] Spinoza took further the argument that only in a democratic republic could people live together harmoniously and in keeping with the most authentic expression of "nature or God" (*natura sive deus*). Deeply immersed in the works of Descartes and responding to Machiavelli and Hobbes (whose *Leviathan* he studied carefully), Spinoza may also have been indebted to the De la Courts, although there is only tentative evidence that they were acquainted.[101]

Spinoza, like those discussed above, established his system on naturalistic grounds, but since he did not accept a dualistic division between an incorporeal reasoning soul and a corporeal body (or between God and nature), his naturalism (and determinism) was even more radical. To Spinoza, reasoning and acting, thought and extension, were two expressions (modes) of the same being. He therefore began his *Ethics* with a section on God, identifying him with all that exists (versus superstition in the guise of religion, which separates God from nature and anthropomorphizes him). After dealing with mind, Spinoza went to greater lengths to portray the "affects" (which can be grouped into three: desire, joy, and sadness) and their powers. The opening remarks in this third section castigate almost all previous treatments of the passions, which treat them as vices escaping the control of free will and reason. Other theorists, he wrote,

> seem to conceive man in Nature as a dominion within a dominion. For they believe that man disturbs, rather than follows, the order of Nature, that he has absolute power over his actions, and that he is determined only by himself. And they attribute the cause of human impotence and inconstancy, not to the common power of Nature, but to I know not what vice of human nature, which they therefore bewail, or laugh at, or disdain, or

sions. Also idem, "A Controversial Republican: Dutch Views on Machiavelli in the Seventeenth and Eighteenth Centuries," in *Machiavelli and Republicanism*, ed. Gisela Bock, Quentin Skinner, and Maurizio Viroli (Cambridge: Cambridge Univ. Press, 1990), pp. 247–63; and Martin van Gelderen, "The Machiavellian Moment and the Dutch Revolt: The Rise of Neostoicism and Dutch Republicanism," in *Machiavelli and Republicanism*, pp. 205–23. For a recent argument about Machiavelli's view of the working of the body and the body politic, see Anthony J. Parel, *The Machiavellian Cosmos* (New Haven, Conn.: Yale Univ. Press, 1992), especially pp. 86–112, 122–39.

[97] Smit, "The Netherlands and Europe" (cit. n. 92), p. 24.

[98] See Rowen, *John de Witt* (cit. n. 83), pp. 391–8; Israel, *Dutch Republic* (cit. n. 84), pp. 759–60.

[99] Rowen, *John de Witt* (cit. n. 83), pp. 395–6.

[100] For a general study of Spinoza and his circle, see Steven Nadler, *Spinoza: A Life* (Cambridge: Cambridge Univ. Press, 1999).

[101] Kossmann, "Politieke theorie in het zeventiende-eeuwse Nederland" (cit. n. 94), pp. 50–8; Haitsma Mulier, "Language of Seventeenth-Century Republicanism in the United Provinces: Dutch or European?" in *The Language of Political Theory in Early-Modern Europe* (cit. n. 44), pp. 191–3; idem, "Controversial Republican" (cit. n. 96), p. 256.

> (as usually happens) curse. And he who knows how to censure more eloquently and cunningly the weakness of the human mind is held to be godly.[102]

Even Descartes, who "sought to explain human affects through their first causes, and at the same time to show the way by which the mind can have absolute dominion over its affects," revealed in this "nothing but the cleverness of his understanding." But the affects are expressions of nature just as much as is the mind, and "nothing happens in Nature which can be attributed to any defect in it."[103] Hence, Spinoza believed—even more than Descartes and the De la Courts—that the passions are good and natural, and that human society ought to accord with them rather than to try to fight or dominate them.

Like Hobbes and many others, Spinoza began with self-preservation: "Each thing, as far as it can by its own power, strives to persevere in its being."And, "The striving by which each thing strives to persevere in its being is nothing but the actual essence of the thing." He also echoed Descartes on generosity in declaring: "Since reason demands nothing contrary to Nature, it demands that everyone love himself, seek his own advantage, what is really useful to him, want what will really lead man to a greater perfection, and absolutely, that everyone should strive to preserve his own being as far as he can. This, indeed, is as necessarily true as that the whole is greater than its part."[104] But while Hobbes had found human happiness in the reason of a strong sovereign restraining the passions of the body politic, Spinoza found true happiness in following our most authentic nature, which leads to a mode of life "that largely transcends merely transitory desires and which has as its natural consequences autonomous control over the passions and participation in an eternal blessedness."[105] This is because adequate ideas have as much motivational power as inadequate ideas (the passions). Ethical knowledge is both produced from nature and motivates us to act in accord with our true nature.[106] In Spinoza's state of nature, then, each person is the one most capable of achieving his or her authentic self. Consequently, Spinoza argued for democracy because "in it everyone remains equal, as they were in the state of nature, and because democracy approaches most nearly to the freedom of the state of nature."[107] He did not think, as did Hobbes, that people transferred their natural rights to another (a sovereign) without reserving the right to be consulted about their use.[108] In a democracy, one also had the freedom to think and say what nature spoke through oneself.[109] Finally, when it came to state power, although the power to do something made it right, the further consequence is "that

[102] *Ethics*, preface to pt. 3, quoted from Edwin Curley, ed. and trans., *A Spinoza Reader: The Ethics and Other Works* (Princeton, N.J.: Princeton Univ. Press, 1994), pp. 152–3.

[103] Ibid., p. 153; for a modern criticism of Spinoza's analysis, see Michael della Rocca, "Spinoza's Metaphysical Psychology," in *The Cambridge Companion to Spinoza*, ed. Don Garrett (Cambridge: Cambridge Univ. Press, 1996), pp. 192–266.

[104] *Ethics*, pt. 3, P6, P7, pt. 4, P18s, quoted from Curley, *Spinoza Reader* (cit. n. 102), pp. 159, 209.

[105] Don Garrett, "Spinoza's Ethical Theory," *Cambridge Companion to Spinoza* (cit. n. 103), pp. 267–8.

[106] See especially ibid., pp. 296–7.

[107] Edwin Curley, "Kissinger, Spinoza, and Genghis Khan," in *Cambridge Companion to Spinoza* (cit. n. 103), p. 317.

[108] Spinoza, *A Theologico-Political Treatise*, trans. R. H. M. Elwes (1883; reprinted, New York: Dover, 1951), title of chap. 17: "It is shown, that no one can or need transfer all his Rights to the Sovereign Power."

[109] Ibid., chap. 20.

rulers govern with right just to the extent that their subjects consent to their rule by obeying their commands."[110]

By 1670, when Spinoza's *Theologico-Political Treatise* saw print for the first time, De Witt and the States Party were finding it more and more difficult to retain their authority.[111] The young William III of Orange (born shortly after his father's death) reached the age of majority, and his allies became increasingly vociferous in arguing for his being made stadhouder. Now led by the anti-Cartesian leader from Utrecht Voetius, the anti-Remonstrant Calvinists, as they had done earlier in the century, gave their full support to such moves. Against this theologico-political coalition the party of True Freedom held a shaky grip on power. Then in 1672 Louis XIV invaded the country overland in alliance with the bishop of Münster and the English (who waged war against the Dutch at sea). Many Calvinists and Orangists rose up against the republican *regenten* and installed William III as head of the army and virtual head of state. In The Hague, an Orangist mob brutally butchered De Witt and his brother Cornelis, strung their corpses upside down on hooks like hogs, and handed out or sold pieces of their bodies to members of the crowd. According to a later report from his acquaintance Leibniz, Spinoza had to be locked in his room by his landlord to prevent him from posting a sign at the nearby scene of the incident (on which he had written *ultimi barbarorum* [the ultimate barbarity]), fearing that he, too, would be torn apart.[112] Never again would the Netherlands have a completely republican form of government, and there is little evidence of any Dutch political treatises explicitly adopting the positive views of Spinoza or the De la Courts about allowing the passions freedom in a republic.[113]

CONCLUSION

Explicitly republican theories may have gone underground, Descartes may have been tamed, and Hobbes (like Machiavelli and Spinoza) may have become a name synonymous with damnation. But the attempt to draw conclusions about the body politic from new notions of the body natural has remained with us, much to the consternation of those who wish to find virtue in the power of reason coupled with the transcendent or immanent.[114] David Hume famously argued in his *Treatise of Human Nature* (1739) that reason is only "the slave of the passions"; Immanuel Kant countered by saying that practical reason has motivational force equal to or greater than the passions, although he admitted that it is "beyond our capacity to explain *how* it is that reason can have this motivating force."[115] Dutch republican theories about the positive value of the passions resurfaced in England in the work of Bernard Mandeville, an emigré physician from an anti-Orangist family of

[110] Curley, "Kissinger" (cit. n. 107), p. 326.

[111] Israel, *Dutch Republic* (cit. n. 84), pp. 785–95.

[112] On the butchery, see Rowen, *John de Witt* (cit. n. 83), pp. 861–84, on Spinoza's reaction, pp. 885–6.

[113] On the continuing influence of Hobbes and Descartes in the Netherlands, see Kossmann, *Politieke theorie en geschiedenis* (cit. n. 96), pp. 59–103. But even he admitted in his "Development of Dutch Political Theory": "As far as I know, there are no traces of any direct influence exercised by De la Court and Spinoza on Dutch political theory" (cit. n. 79), p. 105.

[114] See especially Albert O. Hirschman, *The Passions and the Interests: Political Arguments for Capitalism before Its Triumph* (Princeton, N.J.: Princeton Univ. Press, 1977).

[115] Garrett, "Spinoza's Ethical Theory" (cit. n. 105), p. 295.

Rotterdam. The famous slogan to his *Fable of the Bees* (1714), "private vices, public benefits," said it all, as long as one recognizes the tongue-in-cheek substitution of "vices" for passions.[116] It may be that one can trace a path from the Dutch republican underground to the theories of Hume and Adam Smith via Mandeville.[117]

In the mid-seventeenth century, then, a group of political theorists in northern Europe took up new views of the human body and their implications for ideas about the body politic. Whether Hobbes or Spinoza, these mid-seventeenth-century political theorists referred the origins of everything, even the attributes of reason, back to nature—unless like Descartes they tried to escape the box by reserving some small, separate, and (politically) unimportant spark to a more divine sphere. Their reason was no longer the "right reason" that had allowed Plato to counter the rhetoricians. For those such as Hobbes, who still believed that reason had to control the passions, this meant that a powerful sovereign had to arbitrate competing claims to the good, threatening tyranny. For republicans, however, the passions expressed the strivings of nature, and so to be true to our nature we need to be true to our passions.[118] The head need not rule the body. Just as the passions led to material goods, so they lead to public benefits. As Mandeville later explained, if everyone acted according to the dictates of virtue as promulgated by the priests, powerful nations would rapidly decline and we would all be back to a state of nature, spending our time gathering acorns. If ever there were a group of political theorists who grounded their views on contemporary science, this is it: the famous Descartes, the first-rate mathematician de Witt, the merchant-lawyers De la Court, and the lens-grinder Spinoza.[119] Only the less mathematically astute Hobbes, opposed to the experimental science of semi-republicans such as Robert Boyle, still thought that the passions needed to be controlled by a dominating reason. The rest placed their faith in the natural law urging us toward life, liberty, and the pursuit of happiness.

[116] See M. M. Goldsmith, *Private Vices, Public Benefits: Bernard Mandeville's Social and Political Thought* (Cambridge: Cambridge Univ. Press, 1985); Harold J. Cook, "Bernard Mandeville and the Therapy of the 'Clever Politician,'" *J. Hist. Ideas* 60 (1999): 101–24.

[117] M. M. Goldsmith, "Liberty, Luxury and the Pursuit of Happiness," in *The Languages of Political Theory in Early-Modern Europe* (cit. n. 44), pp. 225–51; idem, "Regulating Anew the Moral and Political Sentiments of Mankind: Bernard Mandeville and the Scottish Enlightenment," *J. Hist. Ideas* 49 (1988): 587–606.

[118] The important work of Otto Mayr, *Authority, Liberty and Automatic Machinery in Early Modern Europe* (Baltimore: Johns Hopkins Univerity Press, 1986) draws attention to the variety of political views that sought support in mechanical theories. Views about the passions, however, seem to have limited the conclusions of rigorous authors.

[119] Alan Gabbey, "Spinoza's Natural Science and Methodology," in *Cambridge Companion to Spinoza* (cit. n. 103), pp. 142–91.

The *Ladies' Diary*:
Gender, Mathematics, and Civil Society in Early-Eighteenth-Century England

*By Shelley Costa**

ABSTRACT

The influential *Ladies' Diary, or Woman's Almanack* (1704–1840) owed its long affiliation with mathematics to the contributions of an active readership. This article describes how the *Ladies' Diary*'s first correspondents—in addition to maintaining contemporary hierarchies of class and gender—adhered to rules of polite conversation in generating the periodical's scientific content. The *Diary*'s early mathematical dialogue is analyzed in terms of a dualistic conception of civil society, one that encompasses both the activity of a print-based public sphere and the conventions of civility. This dualistic civil society, it is argued, was particularly suited to women's participation, and its degeneration in the second quarter of the eighteenth century served to masculinize the magazine's mathematical forum. The centrality of readers in forming—and in dismantling—the *Ladies' Diary*'s civil society illustrates the danger of relying on top-down models of scientific dissemination in the study of polite science.

INTRODUCTION

EARLY IN 1704—when inexpensive, calendar-containing almanacs were omnipresent in English society[1]—John Tipper (1663–1713), a Coventry schoolmaster, published an almanac for women. Prompted by the author's belief that reference to the queen would encourage "a good issue, most ladies following her example," the *Ladies' Diary, or Woman's Almanack* was ushered on to the seemingly saturated calendrical market with a portrait of Queen Anne (Figure 1).[2] While many almanacs

* Department of History, Xavier University, Cincinnati, OH 45027; sac22@cornell.edu.
I would like to thank all the participants of the workshop on Science and Civil Society associated with this volume for their feedback on the first draft of this article. Co-organizer Thomas Broman's insightful advice on restructuring the original article was particularly helpful. Additional thanks are due to the Cornell University Women's Studies Program and to the National Science Foundation (grant number SBR-9811414) for funding the research on which this article is based, and to Jodi Wyett of the Xavier University English Department for her comments on a previous draft and for many lively conversations on women, gender, and print in eighteenth-century England.

[1] Production statistics from the 1660s indicate that between 300,000 and 400,000 almanacs were sold annually in England at that time, enough "to reach between one in three and one in four families in the land." Bernard Capp, "The Status and Role of Astrology in Seventeenth-Century England: The Evidence of the Almanac," in *Scienze, Credenze Occulte, Livelli di Cultura* (Florence: L. S. Olschki, 1982), pp. 279–90, on p. 280.

[2] "Upon the title-page is the picture of the Queen in copper, which I am promised shall be (and I hope now is) very well performed." John Tipper to Humphrey Wanley, 8 Nov. 1703, *Original Letters of Eminent Literary Men of the Sixteenth, Seventeenth, and Eighteenth Centuries*, ed. Sir Henry Ellis (London: The Camden Society, 1843), pp. 307, 308. The custom of depicting the queen of England

 0369-7827/02/1701-0003$10.00

at this time were imbued with special themes or targeted at politically defined audiences, Tipper's was the first such annual intended for a female readership.[3] Here he joined an old almanac publishing tradition to a lively, burgeoning print economy, for women had just begun to be recognized as readers by a new genre of periodical literature.[4]

As a periodical dedicated to women, the *Ladies' Diary* came third, after the *Ladies Mercury* (1693–1694) and the *Lady's Journal* (1693), both offshoots of other publications.[5] In contrast to its short-lived predecessors, however, the *Diary* presented a comprehensive medley of contents to the female reader. Instead of following the *Mercury*'s moralistic question-and-answer format or the *Journal*'s literary slant, the first year's issue of the *Ladies' Diary* offered

> a preface to the fair sex, containing the happiness of England under the reign of Queen Elizabeth and the present Queen; . . . Then the calendar it self . . . and also medicinal and cookery receipts collected from the best authors. . . . After this is the second part of the almanack, which contains the praise of women in general, with directions of love and marriage, intermixt with delightful stories[.][6]

Tipper capped off the collection with his own special rhyming riddles, which he called "enigmas."

The resulting publication enjoyed immediate success with readers of both sexes.[7] The Company of Stationers[8] had initially been conservative in the number of copies

---

on the cover of the *Ladies' Diary* continued throughout its publication, which ended in the early years of Victoria's reign.

[3] Two very short-lived seventeenth-century almanacs had used the title *The Woman's Almanack*, but both had been authored by women rather than being addressed to women. Bernard Capp, *Astrology and the Popular Press: English Almanacs, 1500–1800* (London: Faber; Ithaca, N.Y.: Cornell Univ. Press, 1979), pp. 87, 109, 122–6, 160, 323, 373.

[4] See Kathryn Shevelow, *Women and Print Culture: The Construction of Femininity in the Early Periodical* (London and New York: Routledge, 1989).

[5] John Dunton's twice-weekly periodical of the 1690s, the *Athenian Mercury*, had been extremely popular with both women and men, and scores of issues had been themed for female readers before a separately titled *Ladies Mercury* appeared from 1693 to 1694. Peter Anthony Motteux followed suit first in a gallant preface to his *Gentleman's Journal*, then with a specially titled edition, the single-issue *Lady's Journal*, in 1693. On female readers of the *Athenian Mercury*, see Bertha-Monica Stearns, "The First English Periodical for Women," *Mod. Philol.* 28 (1930–1931): 45–59. For critical readings of the *Athenian Mercury* and the *Ladies Mercury*, see Shawn Lisa Maurer, *Proposing Men: Dialectics of Gender and Class in the Eighteenth-Century English Periodical* (Stanford, Calif.: Stanford Univ. Press, 1998), pp. 47–58. On Motteux's publications see Maurer, pp. 39–47, and Shevelow, *Women and Print Culture* (cit. n. 4), pp. 33–4.

[6] Tipper to Wanley, 8 Nov. 1703, *Original Letters* (cit. n. 2), pp. 306–7.

[7] The large proportion of men who read and submitted items to this early women's magazine can confuse the modern reader, who cannot help but see a contradiction between the almanac's status as a publication for women and the mixed-sex composition of its actual audience. There are two important responses to this. First, the experimental nature of print culture in this era discouraged a rigid definition of readership. A publication for women was at this time a profound curiosity; as such it attracted the curious of both sexes. Second, as Shawn Lisa Maurer has recently argued in *Proposing Men* (cit. n. 5), the periodical literature that codified women's place in society simultaneously, if implicitly, shaped a social role for men. When we consider that both gender roles were in flux during this period of substantial economic change, we are less surprised to find men reading publications targeted at the opposite sex.

[8] From its incorporation in 1557, this London-based guild of printers and booksellers controlled all of the financial details of the almanac trade, including the payment of those (such as Tipper) who composed new almanacs. The Stationers' Company's monopoly over all printed material in England had lapsed in 1695, but a previously issued patent allowed the guild to retain a legal authority over English almanacs that has remained intact to the present day.

The Ladies Diary:
OR, THE
Woman's ALMANACK,
For the Year of our LORD, 1711.
Being the Third Year after the Leap-Year.
Containing many Delightful and Entertaining Particulars, peculiarly adapted for the Use and Diversion of
The FAIR-SEX.
Being the Eighth Almanack ever Publish'd of that kind.

1. GREAT *ANNA's* Pious Name, tunes every Lyre,
And do's each *Muse* with boundless Thoughts inspire.

2. HER Sacred Breath from Royal Race *SHE* drew,
And *Britain* well *HER* Great Forefathers knew.

3. *Divinely Bright HER* Glorious Actions shine,
Such as descended from *HER* Ancient Line.

QUEEN ANNE

4. UPON *HER* Brow a Thousand Graces meet,
Where they in Thrones of spotless Goodness sit:
So ev'ry Heart with Pleasure *SHE* Commands;
No Heart, no Soul, her *Rightful-Pow'r* withstands. (*W C.*)

Printed by *J. Wilde*, for the Company of *Stationers*, 1711

Figure 1. *Cover of an early issue of the* Ladies' Diary.

of the *Diary* it had printed, but rapid sales of the debut issue encouraged the company to order a larger print run of the second edition. That edition also sold extremely well; in the third issue Tipper informed his readers that "the speedy sale of my last year's *Diary*, surpass'd my utmost expectation, when, of several thousands printed, the whole impression was sold off by New-Year's-Tide, and no doubt but a great many more would have been sold, if they had been printed" (1706, p. 2).[9] Specifically, four thousand copies of the 1705 *Ladies' Diary* had been printed and sold before January 1705.[10] This meant that as early as the second year of this calendrical entertainment, one issue had been sold for every five hundred literate individuals in England.[11] When sharing and borrowing are taken into account, the early popularity of the *Ladies' Diary* is even more striking. Its annual print runs were duly increased, and the *Diary* remained enormously popular throughout its publication, selling thirty thousand copies at midcentury and numbering among the country's five best-selling almanacs until at least the end of the eighteenth century.[12]

John Tipper had not planned a central role—indeed, any role—for mathematics in his almanac, but in 1707 one male reader's rhyming arithmetical questions inspired a succession of similar puzzles from others. Within two years, Tipper—now acting as the *Diary*'s editor as much as its author—informed readers that recipes and other contents would be sacrificed for these entertaining items:

> Having observed by a multitude of letters I have received from all parts of the kingdom, that the enigmas, and arithmetical questions, above all other particulars, give the greatest satisfaction and delight to the obliging fair, I shall in this Diary insist the longer upon them, and for that reason defer the receipts of cookery, &c. to a more favourable opportunity. (1709, p. 21)

In fact, recipes never appeared again in the *Ladies' Diary*, and verbal enigmas and mathematical questions maintained a central place. In this way, the *Diary* became the first printed forum for mathematical exchange as well as one of the very first

[9] For the sake of consistency, I have not repeated the *Diary*'s original system of page numbering, which passed over calendrical and other preliminary pages (whose number varied from year to year) and began at 1 for the first page that followed this material. Instead I count the cover, which was the first physical page of the almanac, as page 1. Since the *Diary* consisted of five sheets printed in octavo, every issue amounted to exactly forty pages. When quoting from the text of the *Ladies' Diary*, I have drawn from the extensive collection of issues owned by the Bodleian Library, Oxford. For the text of issues missing from this collection, I have consulted Charles Hutton's five-volume *Diarian Miscellany* (London, 1775), retaining parenthetical notation with the addition of square brackets: for example, "(1713 [*DM* i, p. 87])" for page 87 of the first volume of the *Diarian Miscellany*.

[10] "My almanacks sold this year beyond mine and the Company of Stationers expectation, so that of 4000 which they printed, they had not one left by New-Years-Tide." Tipper to Wanley, spring or summer 1705 (undated), *Original Letters* (cit. n. 2), p. 314. (Although this period predated the acceptance of the Gregorian system in England, calendars in all English almanacs began the year on January 1.)

[11] There were probably between two and two and a half million literate individuals in England at this time (46 percent of a total population of about five million). For an excellent summary and review of the best literacy research on this period, see J. Paul Hunter, *Before Novels: The Cultural Contexts of Eighteenth-Century English Fiction* (New York: W. W. Norton, 1990), pp. 61–81. I have drawn the figure of 46 percent from a 35 percent literacy rate for women and a 55 percent rate for men, both rough interpolations of statistics given by Hunter on pp. 72–3 and 80.

[12] Capp, *Astrology and the Popular Press* (cit. n. 3), pp. 246–7. By 1761 the annual sales of the *Ladies' Diary* had dropped to fifteen thousand, but it was still the Stationers' Company's third best-selling almanac in that year. In 1789 it was still in the top five.

periodicals for women. These contributions surpassed their arithmetical origins within a few years, and from 1713 a new editor placed even greater emphasis on the *Diary*'s mathematical challenges. By 1720 the almanac had presented two problems involving Newtonian infinitesimal calculus—or "fluxions," as it was then known in Britain—truly a cutting-edge technique at this time. Notwithstanding the continuing popularity of the *Diary*'s verbal enigmas, advanced mathematics was still considered the almanac's defining characteristic in 1840, the final year in which it was published.[13]

This influential almanac embodies the complex interdependence between science and society. The strength of the early-eighteenth-century reading public is particularly crucial here, as the most salient characteristics of the *Diary* derived from the collective will of readers who had been given voice by a newly dynamic capitalism of print. We have already glimpsed the role of turn-of-the-century print culture in inspiring its dedication to women. Similarly, the unsolicited mathematical submissions that transformed this almanac came from readers who had been trained to be active by the question-and-answer format of the *Athenian Mercury* and its associated periodicals. Numerous other social and economic structures influenced readers' correspondence, and the resulting publication in turn molded cultural views of mathematics.

In this article, I draw from two notions of civil society to interpret the reader-based mathematical discourse of the early *Ladies' Diary*. The first, inspired by Jürgen Habermas's description of the rise of a public sphere in late-seventeenth-century England, posits civil society as a conglomerate of private citizens (or subjects) who unite for the purposes of discussion and judgment.[14] Although in the original formulation such discourse was political, a mathematical version of this participatory civil society—dependent, as was the first, on the rise of a market economy—can be seen to exist in the printed dialogue of the *Ladies' Diary*. A second usage of "civil society," and one which would have been understood by the *Diary*'s readers, describes the etiquette and expectations of the English peerage and gentry.[15] As we will see, the *Diary*'s mathematical exchange quickly took on a tone of civil conversation, blurring the distinction between the two notions of civil society for this forum.[16]

[13] In 1843, for example, Sir Henry Ellis touted the *Ladies' Diary* as "a work which, however humble in its beginning, has exerted a great and beneficial influence upon the state of mathematical science in this country for near a century and a half" (Ellis, *Original Letters* [cit. n. 2], p. 304). In 1841 the *Ladies' Diary* had been merged with the younger (but by then very similar) *Gentleman's Diary*, marking the end of the existence of the *Ladies' Diary* as an independent publication.

[14] Jürgen Habermas, *The Structural Transformation of the Public Sphere: An Inquiry into a Category of Bourgeois Society*, trans. Thomas Burger with Frederick Lawrence (Cambridge: Polity Press, 1989).

[15] Steven Shapin has drawn on this organic sense of "civil society" to interpret social dimensions of knowledge in *A Social History of Truth: Civility and Science in Seventeenth-Century England* (Chicago: Univ. of Chicago Press, 1994).

[16] For the benefit of this volume, it is worth emphasizing that I do not view civil society as an idealization articulated by historical actors. While such a treatment is appropriate to other historical projects—for example, those focusing on political philosophy or on civic activism—my aim is to draw attention to the ways in which existing social and economic patterns shaped the mathematical dialogue of the *Ladies' Diary*. The power of the construct for this purpose rests in its ability to identify such patterns as part of a larger whole. The notion of civil society thus anchors aspects of economy and society that have special explanatory force for the *Diary*'s mathematics. While I would be gratified if this article threw light on civil society per se, my intent is to understand science as generated and maintained by early-eighteenth-century English civil society.

After laying this groundwork in the first half of the article, I turn to the importance of gender to this dualistic civil society. While the Habermasian version has been repeatedly criticized for obscuring women's contributions,[17] I am concerned not with the inadequacies of his original formulation, but with the gendering of the participatory civil society of the *Ladies' Diary*. Gendered patterns of address, for example, softened the edge of the intellectual female contributor, while the cultural norm of feminine modesty simply kept many women out of the *Diary*'s public mathematical forum. Here again, civil society in the second sense—manners and mores—is inseparable from the first, activity-based notion of civil society. An important point here is that the second form does not raise the same specter of gender exclusion as the first, since even contemporaries saw polite, "civilizing" behavior as feminized.

I conclude by evaluating the *Ladies' Diary* in terms of a model interface between science and civil society—the contemporary passion for things scientific. Public lectures, attended by both men and women, proliferated in England during the eighteenth century. Textbooks of all kinds were written for beginners and were often addressed specifically to female readers. Genteel enthusiasts displayed collectors' versions of instruments such as telescopes and celestial globes in their homes. These trends—understood under the rubric of "polite science"—have been productively seen within a broad historiography that emphasizes the consumerist culture of the period.[18] Much of this work, however, has also tended to portray polite science as a top-down phenomenon in which specialized lecturers, authors, and instrument makers distributed knowledge to a relatively passive audience. The example of the *Ladies' Diary* defies this image, demonstrating a need to reconceptualize the category of polite science and the role of gender within it.

THE *LADIES' DIARY* AND CIVIL SOCIETY

In *The Structural Transformation of the Public Sphere*,[19] Jürgen Habermas argued that a formative, critical-rational discourse arose in seventeenth- and eighteenth-century Europe as a significant amount of economic power shifted from the state to a rising merchant class. While the state attempted to control the newly powerful group through taxes and legal reform, its private members met in public spaces and spread ideas through print, fostering a critical public sphere—a new seat of influence and judgment based on reasoned argument.[20] In this way, writers, readers, and

[17] See, for example, the collected essays in Johanna Meehan, ed., *Feminists Read Habermas: Gendering the Subject of the Discourse* (New York and London: Routledge, 1995); Belinda Davis, "Reconsidering Habermas, Gender, and the Public Sphere: The Case of Wilhelmine Germany," in *Society, Culture, and the State in Germany, 1870–1930*, ed. Geoff Eley (Ann Arbor: Univ. of Michigan Press, 1996); Dena Goodman, "Public Sphere and Private Life: Toward a Synthesis of Current Historiographical Approaches to the Old Regime," *Hist. Theory* 31 (1992): 1–20.

[18] See Alice Walters, "Conversation Pieces: Science and Politeness in Eighteenth-Century England," *Hist. Sci.* 35 (1997): 121–54; Larry Stewart, *The Rise of Public Science: Rhetoric, Technology and Natural Philosophy in Newtonian Britain, 1660–1750* (Cambridge: Cambridge Univ. Press, 1992); Simon Schaffer, "Natural Philosophy and Public Spectacle in the Eighteenth Century," *Hist. Sci.* 21 (1983): 1–43.

[19] Habermas, *Structural Transformation* (cit. n. 14).

[20] For Habermas, the use of reason distinguished civil society from arenas in which disputes were settled by means such as rank, wealth, or birth. Although this ideal of inclusivity has often been partly realized, including within the discourse of the *Ladies' Diary*, it is clearly impossible for a real society in any period to be run solely on reason. Frank Trentmann holds that the consistent inability of the Habermasian ideal to model real civil society should inspire historians to "revise the theoretical

discussants composed a public separate from the state, a society that was "civil" in the sense of "civilian."

In a more specialized version of this process, readers of the *Ladies' Diary* formed a public that generated its own mathematical discourse. Economics was key to this development, reflecting its importance to the larger critical-rational public sphere Habermas has described. The subtitle, *An Inquiry into a Category of Bourgeois Society*, of his influential book reveals the importance of economics to the Habermasian notion of civil society.[21] Although participants in the new critical-rational discourse were often members of this increasingly powerful group, the term "bourgeois" also refers to the global dynamics of exchange that facilitated the space for this discourse. It was in coffee shops, not in the houses of Parliament or in the country seats of peers, that this public sphere arose. As Craig Calhoun has explained, "Habermas does not mean to suggest that what made the public sphere bourgeois was simply the class composition of its members. Rather, it was *society* that was bourgeois, and bourgeois society produced a certain form of public sphere."[22]

In the same vein, the print culture that gave rise to the *Ladies' Diary* was part of a famously fast-growing capitalist economy. In the recently deregulated, turn-of-the-century publishing scene, new items appeared constantly. Inventive, often sensational pieces were produced in London on a daily basis, first competing for readers in the public arenas of the capital, later circulating in the provinces. Although the Company of Stationers firmly controlled factors such as length and price, the *Diary*'s founding editor eagerly altered its contents in response to the preferences of its readership. As in the larger public sphere, the importance of a reader-driven market in shaping the *Ladies' Diary*—indeed, in permitting its existence—characterizes this publication as an object of "bourgeois society" regardless of the social positions of its readers.

Although this aspect of the *Diary* transcended social rank, contents that appealed directly to readers' preferences took social position into account. Originally, John Tipper cast a wide net, placing contents for the leisured female reader alongside those designated for working-class women. These first few issues contained

> something to suit all conditions, qualities, and humours. The ladies may here find their essences, perfumes, and unguents; [and] the waiting-women and servants, excellent directions in cookery, pastry, and confectionary. (1704 [*DM* iv, p. 2])

---

picture," arguing that depictions of "a public sphere defined by discursive communication and disinterested reasoning . . . fail to encompass much of the intellectual and cultural work of civil society" (Trentmann, "Introduction: Paradoxes of Civil Society," in *Paradoxes of Civil Society: New Perspectives on Modern German and British History*, ed. Frank Trentmann [New York: Berghahn Books, 2000], pp. 3–46, on pp. 25, 27). While I agree with Trentmann on the inability of the Habermasian model to do full justice to working civil society—hence the two conceptual frameworks used in this article—I nonetheless find Habermas's connections between economics and participatory discourse useful for understanding the *Ladies' Diary*.

[21] In "The Great Chain of Buying: Medical Advertisement, the Bourgeois Public Sphere, and the Origins of the French Revolution," (in *The French Revolution: The Essential Readings*, ed. Ronald Schechter [London: Blackwell, 2001], pp. 138–74, on p. 144), Colin Jones claims that those who do not recognize the centrality of economics to the rise of the public sphere "cut [Habermas] off at the knees."

[22] Craig Calhoun, "Introduction: Habermas and the Public Sphere," in *Habermas and the Public Sphere*, ed. Craig Calhoun (Cambridge, Mass.: MIT Press, 1992), p. 3.

While both medicinal and culinary recipes were within the purview of well-born women in this period,[23] designating "cookery" as a servant's task allowed Tipper to extend his almanac's appeal to as many groups as possible.[24] Slapstick romantic fiction also catered to this diverse audience, and the *Diary*'s low price (three pence) made it, like other printed ephemera of its day, very widely accessible. Yet the breadth of Tipper's imagined readership quickly narrowed. This, too, was a response to market forces: as we have seen, in 1709, after a few years of being flooded with submissions, Tipper restricted the focus of the almanac to verbal enigmas and mathematical questions. In doing so he created a publication that, despite having drawn much of its original contents from the realm of domestic work, addressed itself to a leisured reader.

Because economic privilege underlay not only the acquisition of mathematical and verbal skills but also the luxury of using written correspondence as entertainment,[25] the participatory civil society to which Tipper responded overlapped from the beginning with a cultivated notion of civil society. Given Steven Shapin's observations about the tension between genteel and scholarly values in early modern England, it is striking that these polite readers would have introduced something as technical as mathematics into the *Diary.* Shapin has noted that the specificity of scholarly argument was at odds with the self-conscious inclusivity of polite conversation. In its capacity to emphasize intellectual differences among social peers, therefore, mathematical discussion was implicitly insulting in a polite context: "Discourse which was too precise and which demanded too much accuracy in following it was a violation of the presumed equality of civil society."[26]

On the other hand, it is precisely this tension that explains the lighthearted format of the *Ladies' Diary*'s early mathematical discourse. John Tipper had intended the *Diary* to be pleasant above anything else, and he balked when readers sent quite difficult problems:

> I have received several arithmetical questions which are very unfit for this place; my design being not to puzzle, but to please; not to perplex the understanding, but to exercise the wit, and a moderate knowledge in numbers; and therefore those who are pleased

[23] See, for example, Lynette Hunter and Sarah Hutton, eds., *Women, Science and Medicine, 1500–1700: Mothers and Sisters of the Royal Society* (Thrupp, Stroud, Gloucestershire: Sutton Publishing, 1997); Barbara Griggs, *Green Pharmacy: A History of Herbal Medicine* (New York: Viking, 1981).

[24] The eclectic style of the original *Ladies' Diary* seems to have imitated that of late-seventeenth-century English literary anthologies. Barbara Benedict has noted that after the English Civil Wars of the 1640s, "booksellers produced anthologies that mediated between the traditional manuscript culture of the court and the new public by providing a mixture of classical and contemporary material. . . . [T]hese anthologies provided a space, if only symbolically, for the productions of all members of society." *Making the Modern Reader: Cultural Mediation in Early Modern Literary Anthologies* (Princeton, N.J.: Princeton Univ. Press, 1996), p. 5.

[25] The post was expensive, and letters to the *Ladies' Diary* were not sent simply within London (where postage was relatively economical) but, as cited in the introduction to this article, "from all parts of the kingdom" to John Tipper in Coventry. Many correspondents expected Tipper to cover the expense, so from 1709 he tried to avoid this problem by instructing readers to write directly to Stationers' Hall in London. Even then he had trouble: "all [contributors] must pay the postage of their letters; for having been abused very much in that kind, I have given orders to my correspondent, to refuse and send back all letters, directed to me, which are unpaid for" (1710, p. 3).

[26] Shapin, *A Social History of Truth* (cit. n. 15), p. 117; see also idem, "'A Scholar and a Gentleman': The Problematic Identity of the Scientific Practitioner in Early Modern England," *Hist. Sci.* 29 (1991): 279–327.

> to send me any arithmetical questions, I desire they may be very pleasant, and not too hard; and likewise that they may be proposed in verse; which will still be the more taking among the ladies. (1709, p. 29)

Tipper's wish that the mathematical problems in his almanac be "not too hard" so that they might be "more taking among the ladies" indicates less a concession to potentially inferior female intellects than respect for the etiquette of civil society.[27] Just as it would be rude to "perplex the understanding" in polite conversation, Tipper regarded it as impertinent to publish very advanced problems.

In this regard, the *Ladies' Diary*'s early mathematical discourse follows the observations of Lawrence Klein, who has noted that "where Shapin emphasizes the irreconcilability of learning and gentlemanliness, I see learning being regulated by the standards of gentlemanliness."[28] Although Klein and Shapin both use the masculine idiom of the gentleman, the rules of politeness that guided the *Diary*'s mathematics were clearly feminine as well.[29] The widespread notion of civility as feminized—partly reflected in Tipper's persistent habit of writing as if all readers were female—gave women an important role in shaping the early *Ladies' Diary.*

POLITE AND COMMERCIAL MATHEMATICS

Understanding the *Ladies' Diary*'s mathematics in terms of our two notions of civil society requires a simultaneous view of the bourgeois economic basis of participatory discourse and the subtleties of polite discourse. This scope is justified by the close links that historians have recognized between capitalism and civility in early-eighteenth-century England. Paul Langford, for example—quoting contemporary legal author William Blackstone—has characterized the enfranchised population of eighteenth-century England as "a polite and commercial people."[30] The phrase sums up a key focus of eighteenth-century British historiography: the equivalence of polite society with consumer society.[31]

This scholarship has focused on the ways in which polite culture fostered consumerism, but just as importantly, consumerism also reshaped politeness. Printed ephemera—itself receiving a new identity as a polite commodity—played a large role in this process, most famously through the essays of Joseph Addison and Richard Steele in the *Tatler* (1709–1711) and the *Spectator* (1711–1712 and 1714), which

[27] As an indication that Tipper did not think his female readers eminently incapable, this edition (1709) included not only arithmetic, but also Euclidean geometrical figures; one problem even required a mathematical understanding of the motion of pendulums. In addition, the next few issues introduced problems involving knowledge of astronomy, trigonometry and solid geometry, and Tipper consistently praised the abilities of the female contributors who solved them.

[28] Lawrence Klein, *Shaftesbury and the Culture of Politeness: Moral Discourse and Cultural Politics in Early Eighteenth-Century England* (Cambridge: Cambridge Univ. Press, 1994), p. 7 n. 8.

[29] See G. J. Barker-Benfield, *The Culture of Sensibility: Sex and Society in Eighteenth-Century Britain* (Chicago: Univ. of Chicago Press, 1992); Mary Catherine Moran, "'The Commerce of the Sexes': Gender and the Social Sphere in Scottish Enlightenment Accounts of Civil Society," in *Paradoxes of Civil Society* (cit. n. 20); Goodman, "Public Sphere and Private Life" (cit. n. 17).

[30] Paul Langford, *A Polite and Commercial People: England, 1727–1783* (Oxford: Clarendon Press, 1989), p. 1.

[31] See, e.g., John Brewer and Roy Porter, eds., *Consumption and the World of Goods* (London: Routledge, 1993); Lorna Weatherill, *Consumer Behavior and Material Culture, 1660–1790* (London: Routledge, 1988); Neil McKendrick, John Brewer, and J. H. Plumb, eds., *The Birth of a Consumer Society: The Commercialization of Eighteenth-Century England* (Bloomington: Indiana Univ. Press; London: Hutchinson, 1982).

posited manners, morality, and taste as the subject of ongoing civil discourse. Michael Ketcham has argued that the immense popularity of these periodicals derived from readers' need to anchor themselves in the fluid, image-conscious ("bourgeois" in Habermasian terms) society that was beginning to replace agrarian-based hierarchies.[32] By guiding readers toward appropriate behavior in this new milieu, the *Spectator* in particular salvaged the stratification of British society at a time when it seemed at risk.

A similar process occurred in the *Ladies' Diary*'s mathematical dialogue. Just as Addison and Steele sought decorum in the chaos of a new consumerism, the *Diary*'s contributors lent a distinctly English civility to the mathematics of commerce. In early-seventeenth-century Oxford, Savilian professor of geometry John Wallis remembered, "Mathematicks . . . were scarce looked upon as accademical studies, but rather mechanical; as the business of traders, merchants, seamen, carpenters, surveyors of the lands, or the like."[33] Unperturbed by the low status that still tinged such studies a century later, the *Diary*'s correspondents embraced "mechanical" contexts, translating them through polite discourse and formal mathematics to make them both palatable and entertaining. Instead of ignoring or rejecting the origins of early modern British mathematics, the *Diary*'s readers used these commercial associations to reinforce their own civilized status.

Figure 2 shows the social topics that recurred most frequently in the *Ladies' Diary*'s mathematical problems between 1704 and 1724. We see here that money and land occupy the first two positions, representing the rhetorical importance of material wealth and land ownership to the *Diary*'s reading public. Yet the table also shows that artisanal assistance was the third most important social theme in the *Diary* during that period. Indicative of the importance of skilled crafts to the eighteenth-century British economy, the artisanal theme also highlights the role of civility in regulating the *Diary*'s participatory dialogue.

The first problem in the *Ladies' Diary* that involved artisanal work was Question 17, submitted by a male reader:

> I happen'd one ev'ning with a tinker to sit,
> Whose tongue ran a great deal too fast for his wit.
> He talked of his art with abundance of mettle;
> So I asked him to make me a flat-bottomed kettle,
> That the top and the bottom diameters be
> In such just proportion as five is to three:
> Twelve inches the depth I would have, and no more,
> And to hold in ale-gallons, sev'n less than a score.
> He promis'd to do it, and to work he strait went;
> But when he had done it, he found it too scant.
> He alter'd it then, and too bigg he had made it,

[32] Michael Ketcham, *Transparent Designs: Reading, Performance, and Form in the* Spectator *Papers* (Athens: Univ. of Georgia Press, 1985), pp. 1–5. As an indication of its popularity, the *Spectator* alone appeared in well over six hundred issues during its two original years, and an eight-volume republication was in its eleventh edition by 1729. Its focus on female readers was also substantial, significant enough for Jonathan Swift to complain, "I'll not meddle with the Spectator—let him fair-sex it to the world's end." Swift, *Journal to Stella*, 8 Feb. 1712, cited in Shevelow, *Women and Print Culture* (cit. n. 4), pp. 1, 98.

[33] Christoph Scriba, ed., "The Autobiography of John Wallis, FRS," *Notes Rec. Roy. Soc.* 25 (1970): 17–46, on p. 27; cited in Shapin, *A Social History of Truth* (cit. n. 15), p. 316.

Note: Any problem can be in more than one social category, so percentages do not add up to 100.

General category	Total	% of probs
Money	25	19.7
Land (non-construction)	17	13.4
Artisanal assistance	14	11.0
Ale or wine	14	11.0
Private gardens	12	9.4
Measurement of structures	12	9.4
Measurement of goods	11	8.7
Construction	10	7.9
Age	9	7.1
Distance/position at sea	7	5.5
Marriage	7	5.5
Religious or historical	7	5.5
Distance over land	6	4.7
Agriculture	5	3.9

Figure 2. *Most frequently occurring social topics in the* Ladies' Diary*'s mathematical problems, 1707–1724.*

And when it held right, the diameters fail'd it:
So that making't so often, too big, or too little,
The tinker at last, had quite spoiled the kettle:
Yet he vows he will bring his said purpose to pass,
Or, he'll utterly spoil ev'ry ounce of his brass.
To prevent him from ruin, I pray help him out,
The diameters length else he'll never find out.
(1711, p. 33)

Not all readers of the *Ladies' Diary* condoned such an unsympathetic depiction of craftsmanship. A respondent to Question 17 apportioned the blame here quite differently:

Before the last Christmas there came to my house,
A jolly fine tinker, who seem'd very chouse[34];
I askt him, if store of good work he had got,
He reply'd, that he had; but that he like a sott,
Undertook t'other day for a critical ass,
To make him a flat-bottom'd kettle of brass[.]
(1712, p. 33)

[34] Gullible; easily cheated.

The perspective of the dissatisfied patron, however, was more typical of the *Diary*'s discourse, and exaggeratedly incompetent artisans made frequent appearances throughout this period.

While genteel patronage of craftsmen played an essential role in these problems, the importance of artisanal work was not restricted to this one economic relationship. Proprietors of public houses, for example, were taxed on the volume of their kegs.[35] This means that gauging—the measurement of containers, especially for ale or wine—affected the price of a drink at the local pub. It also means that gaugers were also excisemen, as Question 106 illustrates:

> Being taking a glass with some friends of the town
> The exciseman came in, whom we ask'd to sit down;
> He did so, and as we did freely converse,
> Our new friend began the following discourse.
> "That spheroidal cask (says he) on that stiller,
> Is more artfully made than any i'th' cellar;
> For whether you work by Mr. Oughtred's canon,[36]
> Or multiply the difference by seven tenths (as is common
> For us gaugers, when hasty, to do in the case)
> 'Twill bring out the same equal cylinder's base.["]
> (1724, p. 31)

Here we have an indication that gaugers have their own computational shortcuts. However, the problem ultimately emphasizes the superiority of mathematical rigor over the trade practices of artisans:

> I question'd his skill; this nettled the gauger,
> Who presently offer'd to lay me a wager,
> That he'd tell to a spoonful the liquor's content.
> 'Tis done, sir, (says I) for a crown to be spent.
> . . . And having gone thro' with the whole operation,
> I found none of his methods would bear demonstration:
> So demanding the crown, he refuses to pay,
> Till I shew him a more mathematical way.
> (1724, p. 32)

As usual, this problem ended with a request for the reader's assistance: "Now, for sake of the gains, I'd soon do't if I could, / But seeing I cannot, I beg that you would" (1724, p. 32).

Indeed, nearly all speakers in the fictional contexts of the *Ladies' Diary*'s mathematical problems explicitly requested readers' assistance. This was sometimes done in the name of romance, a putatively feminine interest. In Question 29, for example, a man asks for readers' help in finding the age at which his fiancée will agree to be married: "I pray you, ladies, help me find / What age she'll be when we are joyn'd"

[35] See Judith Grabiner, "'Some Disputes of Consequence': Maclaurin among the Molasses Barrels," *Soc. Stud. Sci.* 28 (1998): 139–68.

[36] William Oughtred (1575–1660), a parish vicar and mathematics teacher whose students included John Wallis and Christopher Wren, published the influential *Clavis Mathematicae (Key to Mathematics)* in 1631. The work subsequently went through several editions and was translated into English by the end of the seventeenth century. Readers of the *Ladies' Diary* therefore would have had access to Oughtred's text whether or not they were skilled in Latin.

(1713 [*DM* i, p. 87]). Here the speaker exerts moral pressure on the reader to assist him based on the strength of his passion. Another suitor, whose lover's agreement to marriage was contingent upon his solving a mathematical problem, was even more desperate:

> But ah! with pains I've try'd all ways about,
> And all in vain; I cannot find it out.
> Assist, fair ladies, one with grief opprest,
> And send an answer to this one request.
> (1718, p. 34)

Conversely, a "country spark" who put his foot in his mouth in Question 15 by asking a woman's age was not consumed enough by passion to elicit pity—nor did he merit aid based on respectable behavior. As a result, the decision to help him was left up to the reader: "So you're desir'd t'assist him, or perchance / The spark must still remain in ignorance" (1710, p. 36).

Artisanal problems shared this feature, but instead of soliciting readers' help on moral grounds, almost all of them followed Question 17 in presenting professional incompetence as the reason for seeking reader intervention. In these problems, sometimes explicitly and sometimes implicitly, the abstract mathematical techniques enjoyed by the *Ladies' Diary*'s readers were posited as superior to the trade methods of artisans. Thus—although the epithets "sons of art" and "artists" gave them a certain metaphorical status as artisans—the *Diary*'s readers were assumed to be members of an epistemologically privileged group.

This epistemological privilege derived from an honorary social privilege. Regardless of their actual circumstances, readers of the *Ladies' Diary* were treated in that printed forum as members of the gentility. In Question 44, for example, a landowner hired a mason to take away a stone:

> Within my garden, lying on the ground,
> A stone I've got, that weighs 300 pound:
> A little barrow likewise, six foot long,
> I've got; tho' little, it is very strong.
> Now, sirs, because the stone lay in my way,
> I sold it to a mason yesterday;
> Who begg'd my barrow for to take it home,
> And call'd his two men; presently they come;
> But one (being lately sick) was very weak,
> More than five score he could not undertake.
> How on the barrow must this stone be laid?
> Pray tell me; for this fellow is afraid,
> If he should carry much, 'twould do him more
> Diskindness, than his sickness did before.
> (1715 [*DM* i, p. 116])

The mathematical problem posed here is how the mason's two workmen will balance the weight of the stone. This issue is primarily the mason's concern, since the stone's removal is his responsibility. Why, then, did the speaker not take the voice of the mason—or, indeed, of the recently ill workman, whose personal well-being is at stake—to petition the *Diary*'s readers for help?

The reason lies in the presumed gentility of the *Ladies' Diary*'s readership. The speaker took the voice of a landowner so that his social position would allow him to mediate between the mason and the readers of the *Ladies' Diary.* If either the mason or the workman had spoken for himself, he would have been directly addressing his betters—or, more to the point, a group characterized as his betters—without a proper introduction. A laborer never pleaded his own case in the *Ladies' Diary,* and in only one mathematical problem during this period (the 1714 prize question) did an artisan—in this case a gauger—risk speaking for himself.

Readers, on the other hand, were entitled by their assumed position to address those of inferior social status directly. In answering Question 17, the first artisanal problem, one regular female contributor did exactly this:

> Well, bonny brave tinker, to save thee from ruin,
> The kind British lasses are active and doing,
> Because that thou art a brave fellow of mettle,
> Take here the diameters both of thy kettle.
> (1712, p. 33)

Note that this respondent directly addressed the tinker despite the fact that the tinker had not spoken for himself in the original question. Moreover, the respondent—who, as it happens, was entitled to use such condescension in the reality of English society as well as in the pretext of the *Ladies' Diary*[37]—underscored her social superiority by using the familiar form of "you."

As this response to Question 17 suggests, the ability of "kind British lasses" to salvage fictional artisans depended in no small part on social rank. In their capacity to act as patrons for artisanal work, women as well as men could assume a benevolent role, replacing monetary payment with mathematical advice. Gender and social rank were thus intimately linked in the social scenarios of the *Ladies' Diary.* Accordingly, several readers' comments reinforced a distance between female gentility and paid artisanal work. William Crabb drew attention to the playful, fictional context that excused female correspondents' associations with trade: "Why must ladies turn gaugers, unless't be for wagers[?]" (1715 [*DM* i, p. 106]). Elizabeth Dod emphasized the unpaid nature of female readers' assistance to gaugers: "Let 'em no more their boasted art / Extol since 'tis our aid, / That does the tubs content impart, / Tho' they for that are paid" (1723, pp. 23–4).[38]

While the impropriety of receiving payment is certainly partly a gender issue, its importance here pertains mainly to social position, as the generality of Question 33 suggests:

> Come artists inform me how it is to be done;
> There are many can tell me, but ask a round sum:

[37] This respondent was Barbara Sidway, whose family tree is replete with men described as "yeoman" and "gent." from at least the mid-sixteenth century.

[38] Elizabeth Dod, born in 1689, lived at the sixteenth-century estate of Edge in Cheshire from at least 1702, when her father (then rector of the nearby parish of Malpas) inherited it after the deaths of his two elder brothers. Elizabeth submitted this solution to the *Ladies' Diary* in 1723, when the estate was owned by her older brother, William. Another brother, Thomas, who later inherited the estate, was an even more frequent contributor to the *Ladies' Diary* than Elizabeth. Parish register of Malpas, Cheshire County Record Office, Chester, U.K.; "The Dods of Edge," *Cheshire Life* (Sept. 1936): 6–11, on p. 10.

But knowing there generous persons may be,
Who will do it for nothing, such will oblige me.
(1713 [*DM* i, p. 89])

Ultimately, then—despite its implicit importance as an economic foundation for so much of the *Diary*'s mathematics—the artisanal context emphasized the idealized reader's independence from monetary need.

The subtext of social rank within the *Diary*'s mathematical discourse corresponded with the widespread status consciousness that was evident in the new genre of essay periodicals. Just as the *Spectator* helped to redefine civility in a changing economy, the *Ladies' Diary* imbued the mathematics of commerce with the values of gentility. In principle and in practice, both periodicals were open to all; their wide distribution and minimal cost made them accessible to readers from a wide variety of backgrounds. Nonetheless, early eighteenth-century readers of the *Spectator* and of the *Ladies' Diary*—aspiring, respectively, to good taste and to good mathematics—were bound by the conventions of civility.

GENDER AND CIVIL CONVERSATION

Women were the acknowledged arbiters of civility in early-eighteenth-century English society. Polite conversation—and, relatedly, written correspondence—was considered largely a feminine enterprise. As a result, the *Ladies' Diary*'s polite forum was ideally suited to women's participation, an attribute enhanced by the founding editor's desire for a style of discourse that would be "taking among the ladies." Accordingly, a large number of women read and contributed to the *Ladies' Diary*, and the wit, style, and mathematical abilities of these women were praised in the text of the almanac throughout the first several decades of the century. At the same time, gender-based restrictions within polite society with regard to authority and publicity conflicted with women's portrayal (particularly in editorial comments) as equals to men within the *Diary*'s participatory civil society. These conflicts can be identified in the very processes of genteel conversation that invited women into the *Diary*.

Most problems in the *Ladies' Diary* were not explicitly addressed to a particular sort of reader, but from 1710 (when such problems first appeared) to 1724, 12.6 percent of all questions—an average of one each year—were addressed to men. Some of these addressed a generic male individual. "Sir" or "good sir" were especially popular, and creative variations such as "good Mr. Gauger" also appeared from time to time. Some proposers hailed a collective male group, most popularly "sons of art." (The group terms "artists" [artisans] and "gaugers" were also used, but for the purposes of this article, those terms of address are not considered gendered.)

Women were addressed in 17.3 percent of mathematical problems during this period—an average of 1.5 times each year—the majority of the time simply as "ladies." The second most common term of address among male proposers was "fair ladies," but this complimentary variation was not used by female proposers. One man addressed "ingenious ladies" (1712). There was no feminine group equivalent to "sons of art," but two male proposers did specify female correspondents' mathematical activities by posing questions to "you ladies, that in art are skilled" (1714) and "you ladies who in number sports are known" (1722).

Women never authored questions addressed to men, while men almost always authored those addressed to women.[39] Although this is partly an artifact of an overrepresentation of men among the *Diary*'s named contributors, it nevertheless reveals the extent to which the act of addressing a question to men was part of a masculine discourse (the 1710 prize question, for example, addressed a "good sir" and claimed that it was easier to find "a pritty plump girl, or a good glass of wine" than the height of a church steeple). The act of dedicating a question to a woman, on the other hand, became a gallant, gentlemanly gesture. Only two women used gendered forms of address during this period. Since they were not in a position to be gallant toward the opposite sex—and were probably more conscious of the *Ladies' Diary* as a forum for their own sex—they both directed their questions to "ladies," addressing the same group to which they belonged.

Perhaps the most striking difference in these patterns of address is that the feminine form of nonspecific individual address ("madam" or "miss") was never once used, while variations on "sir" ("sir," "good sir," etc.) composed by far the most common masculine form of address.[40] The absence of the singular form of feminine address in the *Ladies' Diary*'s mathematical questions—in stark contrast with the predominance of the corresponding form of masculine address—highlights two essential differences between male and female gender roles. First, since these questions petitioned readers for their assistance, the singular form of address imparted an authority to the addressee that group address did not. This authority, in relying implicitly on a social and intellectual autonomy denied to women as a whole, was perceived as masculine and was therefore personalized as masculine. Second, in an era when women lacked the social and economic freedom of movement that men enjoyed, a question for (as it were) "the man on the street" quite literally had to be addressed to a man to have the intended connotation of generality. That is, the device of consulting an unnamed, fictional male individual—for example, "Now, good sir, inform me, how high is the steeple?" (1710, p. 36)—asserted a generality that would not have been connoted by "Now, madam, inform me."

Addressing women as a group, on the other hand, came more naturally to the *Diary*'s correspondents. The chivalrous gesture of directing a question to the whole of "the fair sex" may have nominally excluded men,[41] but such a gesture was appropriate to the overall theme of the almanac. Moreover, group address did not confer the same sort of personal intellectual authority to the addressee that individual address did. Gallantly asking the advice of "ladies" was not nearly as subversive of gender hierarchies as consulting a highly skilled "miss" or "madam" for assistance.

Finally, despite the general acceptability of these patterns of address among both sexes, the exclusionary potential of gendered forms of address was not lost on the *Diary*'s readers. In her response to a question involving fluxions that had been addressed to "sons of art," one regular female correspondent asked:

[39] Between 1707 and 1724, sixteen proposers—fifteen male and one unnamed—used masculine terms of address. Of the twenty-two proposers of questions addressed to women during this period, eighteen were men, two were women, and two were unnamed.

[40] From time to time respondents used a form of address that matched the gender of the proposer of the question they answered, but I have not counted responses here, as I am primarily interested in the ways in which the *general* reader was gendered in newly proposed questions.

[41] In practice, forms of address had no bearing on the "real-life" gender of respondents, and questions dedicated to women were always answered by several men.

Did you think it beyond our sex's parts,
That assistance you crave from the sons of arts?
(1719 [*DM* i, p. 166])

MODESTY VERSUS AUTHORSHIP

The traditional hierarchies of gender and social rank embedded in the *Diary*'s mathematical discourse highlight an important relationship between the two faces—"civilized" and "civilian"—of its civil society. While both of these forms of civil society were intertwined in the *Diary*—and both were closely bound up in England's growing bourgeois economy—one nonetheless disciplined the other. Civility, while intrinsic to the early *Diary*'s participatory civil society, also served to regulate it.

The values of "civilized" society also governed real participation in the *Diary*'s public sphere. Specifically, the cultural ideal of feminine modesty severely limited women's desire to publicize their mathematical activities, even in a forum designed and named for them. This is shown by the discrepancies in different sources of information on contribution statistics. Indexes of contributors printed at the back of each issue imply that men made up 84.4 percent of mathematical correspondents—and 68.4 percent of all correspondents—during the first quarter of the eighteenth century (Figure 3).[42] However, women's greater tendency to submit material under pseudonyms suggests that even the boldest women—that is, those who had allowed their work to be published in the *Diary*—had reservations about letting their names appear in print, calling into question the accuracy of the proportions reflected in the indexes (Figure 4).[43] Indeed, the *Diary*'s indexes could report only officially acknowledged correspondence, and as we will see in this section, a great many of those who sent letters remained unacknowledged.

In his fourth year as editor, Henry Beighton celebrated the large number of letters he had received from mathematically skilled women:

> [In order] that the rest of the fair sex may be encourag'd to attempt mathematicks and philosophical knowledge, they here see [in the *Ladies' Diary*], that their sex have as clear judgments, a sprightly quick wit, a penetrating genius, and as discerning and sagacious faculties as ours, and to my knowledge do, and can, carry them thro' the most difficult problems. I have seen them solve, and am fully convinc'd, their works in the Ladies *Diary* are their own solutions and compositions. This we may glory in as the Amazons of our nation; and foreigners would be amaz'd when I shew them no less than 4 or 5 hundred several letters from so many several women, with solutions geometrical, arithmetical, algebraical, astronomical and philosophical.[44] (1718, pp. 17–8)

[42] I have counted repeat contributors only once when calculating these percentages. Nonetheless, this data harbors a few unavoidable sources of error. For example, a contributor's use of more than one false name (or more than one last name, in the case of a married woman) could make various submissions by one individual seem to come from different sources.

[43] Unlike the broader public forum of print, where—with the notable exception of novel writing later in the century—a woman's assumption of a masculine name increased her credibility as an author, the *Ladies' Diary* supported female authorship and intellectual achievement. A woman who wished to disguise her name in the *Ladies' Diary* would therefore have less incentive also to conceal her gender than she would have had if publishing in another forum. Similarly, the fact that the *Diary* encouraged male contributions—and that so many men actually contributed—meant that a male correspondent would have had no reason to take on a false feminine identity. For these reasons, I have taken female pseudonyms as representing real women; male pseudonyms, actual men. This simplifying assumption is suited to the special characteristics of the *Ladies' Diary*.

[44] "Philosophical" here refers to the problems we would now categorize as mathematical physics.

All statistics on contribution have been taken from the *Ladies' Diary*'s index of contributors for the years 1711, 1716-18, 1720, and 1722-24. (Years not accounted for are missing from the collection of issues of the *Ladies' Diary* at the Bodleian Library, Oxford University.)

Table 3.1: Number of acknowledged contributions, selected years

	1711	1716	1717	1718	1720	1722	1723	1724
Total contributions	102	102	103	65	128	133	183	123
by men	67	75	64	51	94	95	127	87
by women	30	22	32	12	22	28	41	30
by gender unclear	5	5	7	2	12	10	15	6
Mathematical	62	51	39	36	63	74	80	49
by men	55	47	31	33	57	60	66	46
by women	7	3	7	2	3	9	12	3
by gender unclear	0	1	1	1	3	5	2	0
Enigma-related	69	71	81	42	89	99	125	107
by men	34	47	47	28	65	70	81	76
by women	30	19	31	12	18	24	35	25
by gender unclear	5	5	3	2	6	5	9	6

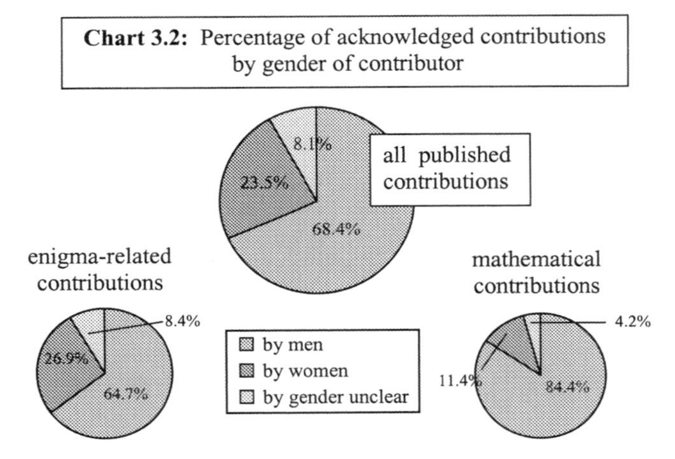

Figure 3. *Published contributions to the* Ladies' Diary *during selected years, 1711–1724.*

These "4 or 5 hundred" mathematical letters would have accumulated if an average of forty to fifty women per year had sent in these sorts of submissions after the debut of mathematical problems in 1707. Since we know that in the years preceding 1707 the *Diary* sold more than four thousand copies a year,[45] and that when Beighton made this claim in 1718 it sold "above 6 or 7 thousand yearly" (1718, p. 18), forty to fifty female contributors represent less than 1 percent of the total number of annually sold issues during that period. In all likelihood, this estimate represents an even lower percentage of the number of readers, since sharing would have meant that more people read the almanac than bought it. All told, the minuscule fraction (less than 1 percent) of the total readership represented by "4 to 5 hundred" contributors

[45] See n. 10.

Table 4.1: Number of pseudonymous contributions per year during selected years

	1711	1716	1717	1718	1720	1722	1723	1724
Total anon./pseud.	9	21	21	10	26	22	32	25
male pseud.	1	6	3	2	2	4	8	10
female pseud.	5	12	16	6	12	9	16	10
gender unclear (incl. anon. & initials)	3	3	2	2	12	9	8	5
Total contributions	102	102	103	65	128	133	183	123

Chart 4.2: Pseudonymous contributions as a proportion of all contributions during selected years, 1711–1724; repeating contributors counted only once

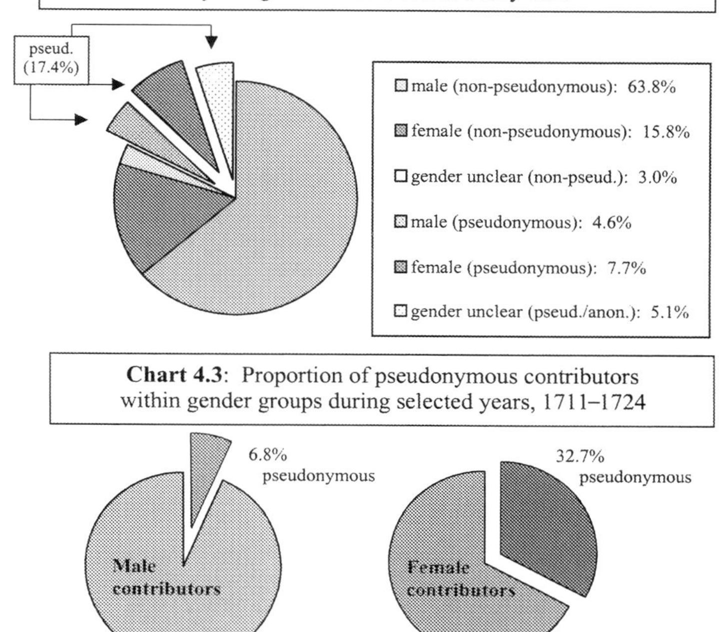

Figure 4. *Number and proportion of pseudonymous contributions.*

gives us no reason to doubt that Beighton possessed at least that many mathematical letters from women.[46]

Taking this statistic together with the data found in the indexes of contributors, we find that 88 to 90 percent of the mathematically skilled women who corresponded with Beighton were not named in the *Ladies' Diary.* That is, instead of revealing the identities (or pseudonyms) of forty to fifty women who annually submitted mathematical items, the *Diary*'s indexes report an average of only five female

[46] In addition, the value that Beighton personally placed on numerical precision in his capacities as land surveyor, engineer, and fellow of the Royal Society makes it even less likely that he exaggerated any numbers he reported. See Shelley Costa, "Marketing Mathematics: Henry Beighton, Certainty, and the Public Sphere," forthcoming in *Hist. Sci.* (2002).

mathematical contributors a year between 1707 and 1717 (Figure 3).[47] This means that no more than 12 percent of the *Diary*'s female mathematical correspondents were individually acknowledged in the almanac.

The most likely explanation for this is that the vast majority of the women who submitted these "4 or 5 hundred several letters . . . with solutions geometrical, arithmetical, algebraical, astronomical and philosophical" did not wish to be acknowledged in print. This is consistent with the acknowledgment policy adopted by the first editor of the *Diary.* John Tipper had implied that many correspondents would balk at appearing in print when he first declared that he would publish readers' contributions, "together with the names of the authors, if they think fit" (1706, p. 39). He repeated the conditional clause for two more years, but in 1709 he clarified the request: "if [correspondents] have a desire to have their names printed, upon notice given, it shall be done" (1709, p. 2). This new stipulation confirms that he assumed that most of his readers would not want their names printed; if he had assumed the opposite, special notice would have been required from those who did *not* want to be acknowledged. In the latter, hypothetical case, those who would not have minded publication but who were not bold enough to request it would still have been acknowledged in the *Diary.* As it happened, only the most confident women appeared in print.

Henry Beighton—who kept himself anonymous in the *Diary* during his thirty years of editorship—diverged somewhat from this policy. Although he initially followed Tipper by printing a stipulation that contributors' names would be inserted only "at their request" (e.g., 1716, p. 2), he decided in at least one case to acknowledge a contributor without prior permission.[48] This correspondent had explicitly requested *not* to be acknowledged:

> Sir,
> Had you studied generosity and good nature as much as the spangles of night, you would not have had the trouble of this complaint, nor I been forc'd so oft to stain my cheeks with a guilty die, as I shall this year, every time I see a Diary in any body's hands. Why would you put me in after my repeated desires to the contrary? Indeed 'twas unkindly done; and did you but know half the uneasiness it has given me (had you an inch of charity) you would at least be sorry for't. (1719 [*DM* iv, p. 164])

Beighton viewed this diffidence as a form of propriety, publishing the letter in the *Diary* expressly to honor it "as an instance of the facetious humour and modesty, as well as for a demonstration of the qualifications of the fair sex" (1719 [*DM* iv, p. 163]). The way the letter concluded encouraged him to take the complaint lightly: "Now read this over twice, laugh at it, then let the flames consume it, and set me down for an impertinent wretch, with a soul full of angry maggots" (1719 [*DM* iv, p. 164]).

Significantly, the writer of this letter had reacted against being named in the *La-*

[47] Detailed indexes are available for only three of these ten years (1711, 1716, and 1717), but the average of five female mathematical contributors per year holds over five more available indexes through 1724, as shown by the table in Figure 3.

[48] In 1719 Beighton gave her away as "the author of the 70th, 79th, and 87th enigmas" (1719 [*DM* iv, p. 163]). In the previous issues, the first of these was attributed to Indamora, the last two to Adrastea. Since the pseudonym Adrastea appeared frequently after this, while Indamora did not, I refer to this correspondent as Adrastea.

dies' Diary despite the fact that she was only acknowledged under her pseudonym, Adrastea, indicating the strength and pervasiveness of expectations for modesty in this period. As suggested by her lighthearted conclusion to the letter, she eventually overcame her aversion to public acknowledgement in the *Diary*. She had threatened in the letter to stop contributing ("I have several more [submissions] of the same nature by me, but will take care how any of them fall into your hand; being composed chiefly to divert melancholy"), but she nonetheless resumed correspondence in the following year and continued to appear—always pseudonymously—in the index of contributors for more than twenty years.

Although Adrastea preferred enigmas, she had strong mathematical skills. In addition to solving several mathematical problems and submitting new ones, she was listed in the index of contributors in 1720 as having calculated the dates and times for that year's eclipses[49] and for the phases of the moon ("lunations" or "limitations") listed in the calendar. Because of this, she must have been the person whom Beighton had in mind when he attempted to pass on the *Diary*'s editorship in the same year:

> [T]he calculations of the eclipses and limitations, are the actual performance and contributions of one of the fair-sex; who here merits my grateful acknowledgement. It was with difficulty, some years since, I discover'd her true name from that feigned one she assum'd, but that stroke of modesty I cannot quarrel at I confess, having not yet to the world discover'd my own. But having now a desire of her assistance, and being convinced of her ability and good-will to the design, perhaps the succeeding *Diaries* may come abroad under hers. (1720, p. 2)

Beighton followed this offer with a comment supporting female authorship: "whoever considers how acceptable the writings of the female sex have been, may hope this [almanac] in its sphere, will lose nothing of its original influence. 'Tis true, a Dacier in France, a Behn at home, are comets not appearing in every age, tho' we have many feminine stars . . . that shine splendidly" (1720, p. 2). Beighton could not, however, coax Adrastea or any other woman into taking over the editorship of the *Ladies' Diary* during his lifetime.[50]

Henry Beighton's encouragement of the *Diary*'s "feminine stars" in 1720 was followed the next year by a lengthy defense of women's pursuit of mathematics. In the preface to the 1721 *Diary*, Beighton argued that this endeavor would in no way interfere with the status quo. Once again, the dominance of genteel values was evident:

> [W]e often find men of so base a spirit, that either from the height of ignorance, or envious spleen, they still dare to despise [the mathematical sciences]. . . . But most generally their envy is levell'd at the female sex, for (say they) learning makes a woman proud and impertinent; 'tis not the business of their sex: the management of their families, and religion, ought only to be their study. But for my part, I cannot see that the search after truth and reason . . . should any more conduce to the bringing a woman into errors, than the search after virtue should be thought the only way to teach us vice. Nor can I conceive, that a woman who does mathematically demonstrate, that the whole is

[49] The dates, times, and locations for all solar and lunar eclipses were announced each year in the *Diary*, whether or not they would be visible in England.

[50] Beighton's wife, Elizabeth, who retained control of the *Diary* for sixteen years after her husband's death in 1743, is the only woman to have officially edited the *Ladies' Diary*.

> equal to all its parts, or 2 and 2 make 4, is less capable of managing her houshold, governing her servants, and giving her children due education. . . . [I]t might as well be said, he who understands Horace and Virgil, is incapable to manage his estate; or because he understands geometry and mathematics, he is unfit to plow, sow, or buy and sell the products of the same. Whoever should assert these things, must by all the world be thought guilty of the greatest madness or stupidity. (1721 [*DM* i, p. 189])

Despite Henry Beighton's energetic support of their cause, however, the *Diary*'s mathematical "ladies" stopped contributing—or, more plausibly, stopped allowing their contributions to be acknowledged—in the mid-1720s. The resulting silence lasted for more than two decades. Most of Beighton's thirty-year editorship was thus marked by a conspicuous absence of women's mathematical accomplishments, which he continued nonetheless to praise. Although female readers—including those such as Adrastea who had previously been mathematically active—still regularly submitted new enigmas and their solutions, women did not publicly resume participation in the *Diary*'s mathematical forum until the late 1740s, several years after Beighton's death.

A SPLINTERING OF CIVIL SOCIETY

How can we explain the abrupt and long-lasting interruption in women's public mathematical activity between 1725 and the late 1740s? It was certainly not due to disapproval or hostility on the part of the *Diary*'s first two editors, as their consistently supportive commentary shows. Nor was it due to a declining readership, since the *Diary*'s sales continued to grow throughout the entire first half of the eighteenth century, and women continued to submit enigmas. Instead, the cultural ideal of feminine modesty, combined with notable changes in the *Diary*'s style of discourse, seems to have discouraged women from continuing to participate in this mathematical forum. By midcentury the *Diary* had changed from a forum where women's "geometrical, algebraical, astronomical and philosophical" skills were displayed and praised to one nominally dedicated to women and to mathematics, but in which polite banter had been displaced by an increasingly skill-oriented, even confrontational, discourse.

During the second quarter of the eighteenth century, fewer and fewer mathematical correspondents adhered to the conventions of civil conversation. Instead of following John Tipper's suggestion that their submissions be "very pleasant, and not too hard" so as best to fit into polite society, later contributors—largely encouraged by Henry Beighton's passion for advanced mathematics—submitted more difficult problems in a less playful manner. While thorough explanations had previously been omitted in favor of rhyming solutions with minimal mathematical detail, more and more lines of equations and figures—exactly the type of discourse that, in Shapin's words, "violat[ed] the presumed equality of civil society"—were printed as midcentury approached. Editors also eventually eliminated the rhyming format that had, according to Tipper, made the *Diary*'s mathematics "the more taking among the ladies," although rhymes were retained for the *Diary*'s enigmas.

As an indication of the dilemma faced by genteel women in this new style of discourse, consider the rather unsympathetic presentation in 1751 by the *Diary*'s then editor, Robert Heath, of the mathematical efforts of someone who had recently begun to contribute regularly under the pseudonym "Sophia Western" (the female protagonist of Henry Fielding's 1749 novel, *Tom Jones*):

> This lady takes the velocity to be as [the square root of] the ordinate y, which is known to be contrary; and does the same in her last supposition, by which we apprehend her solution will not be true. Nor does it appear whether she makes the centre of the oscillation, gravity, or end of the cylinder describe the curve; yet we would not deprive her of publick judgment.[51]

The exposure of Sophia Western to the glare of uncharitable "publick judgment" would certainly have deterred more than one mathematically skilled woman who prized the appearance of grace and modesty.

With these changes, contributing mathematically to the *Ladies' Diary* lost the surface characteristics of "polite accomplishment" that it had formerly shared with acceptably feminine activities such as music, dancing, horseback riding, and poetry. According to Lawrence Klein, learning in eighteenth-century England was genteel when "it did not demand technical or specialist knowledge. Rather, it was generalist in its orientation, tending to the development of the whole person and keeping the person and his [or her] social relations in view."[52] While mathematics required "technical or specialist knowledge," the *Ladies' Diary* had originally allowed its skills to be performed for the appreciation and amusement of a genteel audience. In this regard, solving a problem in astronomy or fluxions had been no less polite—or less feminine—than demonstrating one's expertise on the harpsichord or harp.

In short, the civilized sense of "civil society" in the early *Ladies' Diary* was crucial to eighteenth-century women's comfort in its participatory civil society. Whether a gradual breakdown in civility led to women's withdrawal, or whether a lack of female participation led to a breakdown in civility, or some mutually reinforcing combination, the separation of civility from the *Diary*'s increasingly technical discourse accompanied the retreat of femininity. This had implications both for the future mathematical participation of women—who had now to be comfortable performing not only in print, but in a competitive sphere outside the bounds of civil conversation—and for mathematics, which was no longer constrained by politeness. In separating mathematical activity from civility, the *Ladies' Diary* fostered a reduced, distinctly masculine mathematical public discourse.

CONCLUSION: CIVIL SOCIETY AND POLITE SCIENCE

Because the early *Ladies' Diary* offered mathematics to a largely genteel, nonspecialist audience, it is tempting to fit it into the tradition of polite science described in the introduction. The fact that the *Diary* was explicitly addressed to the prototypical amateur—woman—reinforces this interpretation; Alice Walters has pointed out that "polite science works [in eighteenth-century England] are perhaps most distinguished by their conspicuous emphasis on developing and sustaining a female audience."[53] Public lectures also welcomed women, a trend that the *Diary*'s second

[51] Robert Heath, *The Gentleman and Lady's Palladium* (an annually published supplement to the *Ladies' Diary*) (1751): 27.

[52] Klein, *Shaftesbury and the Culture of Politeness* (cit. n. 28), pp. 5–6.

[53] Walters, "Conversation Pieces" (cit. n. 18), pp. 121–54, on p. 131.

editor, Henry Beighton, linked to the purposes of his almanac: "when I consider how many of the fair-sex have of late successfully gone thro' a course of mathematicks and experimental philosophy, I cannot but be perswaded [sic] the continuance of this *Diary* for their use may some way be thought serviceable" (1720, p. 2).

Walters views this focus on female readers and auditors as a "strategy" by which natural philosophical topics were perpetuated in the conversation of mixed company, thus diffusing more completely into the polite culture of the era. Similarly, John Mullan regards this practice as serving the ends of Newtonianism, although he emphasizes an idealized, epistemological result rather than a social one: if "even" a naive, undereducated woman could accept Newtonian knowledge, then it was eminently intelligible. "If she is ignorant and modest she can hardly ask the wrong question, which only the misinstructed could supply. . . . Having less to unlearn, a woman is a better auditor than many men."[54]

Women's active mathematical correspondence with the *Ladies' Diary* in the first quarter of the eighteenth century, however, does not sit well with the passivity of Mullan's idealized auditors. Nor does Walters's model precisely fit, for although the high production and low cost of this mathematical almanac facilitated its circulation in eighteenth-century society, the *Diary* was not a top-down disseminator of scientific material. It was not originally conceived as a science text, and its mathematics was never directly instructional. In addition, polite science texts addressed to ladies did not proliferate in England until the 1730s. This was the middle of the dormant period discussed in the section "A Splintering of Civil Society," and two decades after women had begun taking part in the *Diary*'s mathematical exchange.

The problem here is with the notion of strategic, top-down diffusion that has dominated historiography on eighteenth-century polite science. This emphasis is understandable, since many contemporaries portrayed public science in this light—for example, Beighton's description of the *Ladies' Diary* as "serviceable" to women who had attended scientific lecture series—and such descriptions increased as the century progressed. Polite culture, however, was the generator as much as the recipient of the *Ladies' Diary*'s mathematics. The reasons for this were primarily economic: the participatory requirements of masterful literacy, access to mathematical education, and ease of epistolary exchange all ceded the early *Ladies' Diary* to the English gentry. This group introduced mathematics to the *Diary*, and throughout the first few decades of the century, all correspondents—regardless of their actual social situations—performed in the *Diary* according to its social expectations.

Strong cultural connections between civility and femininity gave a significantly feminine character to much of the *Diary*'s early discourse. Yet as a social expectation tied to civility, modesty limited the extent to which women allowed their mathematical activities to be acknowledged in print. When the *Diary*'s discourse changed in the second quarter of the eighteenth century, however, women dropped out of its mathematical dialogue for more than twenty years, resurfacing only in very reduced numbers and in a much more competitive context. It thus seems that the codes of polite society—while restricting the extent to which women could comfortably per-

[54] John Mullan, "Gendered Knowledge, Gendered Minds: Women and Newtonianism, 1690–1760," in *A Question of Identity: Women, Science, and Literature*, ed. Marina Benjamin (New Brunswick, N. J.: Rutgers Univ. Press, 1993), pp. 41–56.

form in a public forum—had also allowed women to participate in a "safe" space in which mathematics and femininity were not in contradiction.

Given the example of the early *Ladies' Diary*, it may be the case that both the notion of women's scientific passivity and the top-down notion of scientific dissemination are as much products of midcentury trends in polite science as they are present-day historiographical descriptors. If so, the power of civil society to generate and perpetuate notions of science has been even stronger than we have realized. It may have misled even us.

Differentiating a Republican Citizenry:
Talents, Human Science, and Enlightenment Theories of Governance

*By John Carson**

ABSTRACT

This essay explores how the Enlightenment preoccupation with nature and reason, and the concomitant desire to restructure civil and political society according to these principles, served simultaneously to write certain speculations within mental philosophy into the heart of the republican project and to orient the emerging human sciences toward embracing those social formations most consonant with the developing notions of the republican citizen and the enlightened society. Using the development of the language of talents in the eighteenth century as its focus, the essay examines how Enlightenment political writers and mental philosophers—including Locke, Hartley, Condillac, Cabanis, Rousseau, Helvétius, Godwin, Paine, Wollstonecraft, Jefferson, and Adams—elaborated a vision, on the one hand, of a new social-political order founded on merit and, on the other, of human nature as an object of both scientific and political interest.

INTRODUCTION

> For I agree with you that there is a natural aristocracy among men. The grounds of this are virtue and talents. . . . There is also an artificial aristocracy founded on wealth and birth, without either virtue or talents; for with these it would belong to the first class. The natural aristocracy I consider as the most precious gift of nature for the instruction, the trusts, and government of society. . . . May we not even say that that form of government is the best which provides the most effectually for a pure selection of these natural *aristoi* into the offices of government? The artificial aristocracy is a mischievous ingredient in government, and provision should be made to prevent its ascendancy.[1]

So Thomas Jefferson wrote his old friend and political rival John Adams on 28 October 1813 in response to two letters from Adams discussing the kind of political leadership the new nation needed and could expect. Ever the more cynical of the two in his view of human nature, Adams had remarked in one of those letters that

* Department of History, University of Michigan, 1029 Tisch Hall, Ann Arbor, MI 48109-1003; jscarson@umich.edu.

I particularly wish to thank David Bien, Tom Broman, Michelle Craig, Eric Daniels, Dena Goodman, Ray Grew, Lynn Nyhart, Kathy Olesko, Dan Rodgers, Andrea Rusnock, the anonymous reviewers of the article for *Osiris*, the participants in the "Science and Civil Society: Historical Perspectives" workshop, the students in my "Human Natures and Social Orderings" graduate seminar at the University of Michigan, and the members of the European History Colloquium at Cornell University for their suggestions, encouragement, and careful readings of earlier drafts of this article.

[1] Thomas Jefferson to John Adams, 28 Oct. 1813, *Adams-Jefferson Letters*, ed. Lester J. Cappon (Chapel Hill: Univ. of North Carolina Press, 1988), p. 388.

 0369-7827/02/1701-0004$10.00

"Birth and Wealth together have prevailed over Virtue and Talents in all ages."[2] Later he responded to the lofty sentiments in Jefferson's October letter by providing his own rather less refined definition of an aristocrat: anyone able to command or influence votes other than his own, whether obtained "by his Birth, Fortune, Figure, Eloquence, Science, learning, Craft Cunning, or even his Character for good fellowship and a bon vivant."[3] Having brought Jefferson's vision of the *aristoi* back down to earth, Adams then made clear his own distaste for an aristocracy of any sort:

> And both artificial Aristocracy, and Monarchy, and civil, military, political and hierarchical Despotism, have all grown out of the natural Aristocracy of 'Virtues and Talents.' We, to be sure, are far remote from this. Many hundred years must roll away before We shall be corrupted. Our pure, virtuous, public spirited federative Republick will last for ever, govern the Globe and introduce the perfection of Man, his perfectability being already proved by Price Priestly, Condorcet Rousseau Diderot and Godwin.[4]

Adams's sarcasm notwithstanding, Jefferson's conception of rule by the natural aristocracy clearly struck a nerve, reminding Adams vividly of the doctrines of all those other high priests of Enlightenment political speculation he had spent a career debating. By 1813 such concerns would have already seemed more than a little fusty in a country moving headlong toward dramatic expansions of the suffrage and toward dismantling some of the intricate codes of deference that had marked eighteenth-century social life. But clearly both ex-presidents were looking as much backward as forward and continuing debates that had been pressed since the middle of the eighteenth century in France and Britain as well as America about the nature of society and governance in a republic.[5]

Three aspects of their exchange highlight important features of this transatlantic Enlightenment conversation. First, it is striking that the author of the Declaration of Independence, with its grand commitment to the equality of all human beings, could so blithely assert that some individuals were by nature superior to others and thus better suited to rule. Jefferson was not simply being contradictory, nor had he merely become more conservative with age. Rather, like many of his peers in the Atlantic community who had been puzzling over the nature of a republican society, Jefferson distinguished between certain basic political rights, which he believed applied broadly to whole categories of citizens, and allocations of civic and political power, which he argued should accrue to individuals as a result of their particular sets of virtues and talents.[6] In this Jefferson was by no means alone. The language

[2] Adams to Jefferson, 9 July 1813, *Adams-Jefferson Letters* (cit. n. 1), p. 352.

[3] Adams to Jefferson, 15 Nov. 1813, *Adams-Jefferson Letters* (cit. n. 1), p. 398.

[4] Ibid., p. 400.

[5] On the issue of the "natural aristocracy" and its place in late-eighteenth-century politics, see J. G. A. Pocock, *The Machiavellian Moment: Florentine Political Thought and the Atlantic Republican Tradition* (Princeton, N.J.: Princeton Univ. Press, 1975), pp. 513–26; and Gordon S. Wood, *The Creation of the American Republic, 1776–1787* (New York: W. W. Norton, 1969), especially chap. 12.

[6] I have spoken here of rights being ascribed to categories of people to attempt to capture some of the complexity of a political world in which only white, propertied, adult males were accorded full citizenship rights, while women, children, indentured servants, slaves, native peoples, recent immigrants, to name only a few categories, had various limitations placed on the kinds of rights they could claim. In a letter to Henri Grégoire on 25 Feb. 1809, Jefferson himself hinted at this distinction between rights and the rewards of talent, discussing his assessment of African Americans. "Be assured that no person living wishes more sincerely than I do, to see a complete refutation of the doubts I have myself entertained and expressed on the grade of understanding allotted to them [blacks] by nature, and to find that in this respect they are on a par with ourselves. . . . [B]ut whatever be their

of universal rights, historian Daniel T. Rodgers has pointed out, took on enormous power and importance in America during the second half of the eighteenth century and was equally warmly embraced throughout many parts of the Atlantic world.[7] At the same time, from plans to reform the French army to projects to educate the masses, numerous attempts were made to wed commitment to a universalistic right, such as equality of opportunity, with systems that would award positions of power and authority on the basis of individual ability. Thus lower-level nobles argued that officerships in the French army should be determined by competence, not social rank, and natural philosophers that positions such as those at the Académie des Sciences or prizes awarded by such groups should be open to all and determined solely on the basis of talent, regardless of social origins.[8]

Second, both Jefferson and Adams acknowledged, albeit from very different perspectives, that stratifications in civil and political society would inevitably persist, even within the most privileged class of white, propertied males. "Was there, or will there ever be," Adams wondered in 1787, "a nation, whose individuals were all equal, in natural and acquired qualities, in virtues, talents, and riches? The answer of all mankind must be in the negative."[9] While certainly more optimistic than Adams about the potential of education and abundant land to produce rough equality (at least for adult, white males) within a politically engaged republic of yeoman farmers, Jefferson conceded that society would remain divided; he hoped, however, to ensure that such divisions were based on the right criteria, the reward for an individual's virtues and talents rather than the legacy of family or rank.[10] Few indeed among the educated classes of the eighteenth century, whether in America or Europe, imagined the possibility of either a civic society or a political culture in which stratification and difference were not an integral part of the social order. Even Alexis de Tocqueville's much later ruminations on the nature of an egalitarian society, *Democracy in America* (1835, 1840), sketch a nation in which the logic of democracy

---

degree of talent is no measure of their rights. Because Sir Isaac Newton was superior to others in understanding, he was not therefore lord of the person or property of others." *The Portable Thomas Jefferson*, ed. Merrill D. Peterson (New York: Penguin Books, 1977), p. 517. For an earlier moment in the history of virtue as a social-political concept, see Hal Cook's article, "Body and Passions: Materialism and the Early Modern State," in this volume.

[7] Daniel T. Rodgers, *Contested Truths: Keywords in American Politics since Independence* (New York: Basic Books, 1987), pp. 45–79.

[8] On such projects see, e.g., Ken Alder, "French Engineers Become Professionals: Or, How Meritocracy Made Knowledge Objective," in *The Sciences in Enlightened Europe*, ed. William Clark, Jan Golinski, and Simon Schaffer (Chicago: Univ. of Chicago Press, 1999), pp. 94–125; David D. Bien, "The Army in the French Enlightenment: Reform, Reaction and Revolution," *Past Present* 85 (1979): 68–98; Roger Hahn, *The Anatomy of a Scientific Institution: The Paris Academy of Sciences, 1666–1803* (Berkeley: Univ. of California Press, 1971); James L. McClellan, *Science Reorganized: Scientific Societies in the Eighteenth Century* (New York: Columbia Univ. Press, 1985); Jay M. Smith, *The Culture of Merit: Nobility, Royal Service, and the Making of Absolute Monarchy in France, 1600–1789* (Ann Arbor: Univ. of Michigan Press, 1996).

[9] John Adams, *Defence of the Constitutions of Government of the United States* (1787), reprinted in *The Works of John Adams*, ed. Charles Francis Adams, 10 vols. (Boston: Little, Brown & Co., 1850–1856), vol. 4, p. 391, and for more details on Adams's views, pp. 391–8.

[10] As Jefferson explained to Adams in his letter of 28 Oct., his proposals at the time of the Constitutional debates (1788–1789) to abolish entails and primogeniture and to establish a comprehensive system of education had not had as their purpose the destruction of all social distinction, but only of the hold of the pseudo-aristocracy in favor of those born into every class who by their own merits should prosper and lead the nation. See *Adams-Jefferson Letters* (cit. n. 1), pp. 387–92. For further details on Jefferson's system of education, see Thomas Jefferson, *Notes on the State of Virginia* (1781–1785), reprinted in *Portable Thomas Jefferson* (cit. n. 6), pp. 193–8.

produced not complete social leveling, but rather continuous social jostling between the more and less elite, and a consequent deep-seated instability in the social order.[11] Eighteenth-century writers, however radical their attacks on the privileges of birth and blood, routinely assumed the persistence of a whole range of hierarchies—be it men over women, adults over children, or Europeans over other peoples—whether they saw these stratifications arising on the basis of nature or custom.

Third, Adams believed no less than Jefferson that one of the main reasons complete equality could never be achieved even in the ideal republic was nature: some people were simply better endowed with particular talents or virtues than others, and these superior abilities either entitled them to (in Jefferson's eyes) or helped them to attain (in Adams's view) positions of power in government and influence in civil society. Jefferson celebrated this fact about human nature; indeed he saw such true aristoi as the guardians of the nation's republican legacy and measured the success of the system of government by its ability to place such people in positions of leadership. Adams was warier, finding neither natural nor artificial aristocracy particularly to his liking. But he, too, had to concede that, because individuals were by nature different, distinctions would inevitably emerge in society and some would have power over others.[12] The rhetorical power of "nature," of course, can scarcely be overemphasized for Enlightenment authors. A touchstone for both natural and moral philosophers, and one of the juncture points for their often separate endeavors, "nature" served as a key means of legitimating claims to knowledge and plans for reform, whether it was the nature of Isaac Newton's *Principia* (1687) or of Adam Smith's *Wealth of Nations* (1776).[13]

The move by Jefferson and Adams and other Enlightenment writers to turn to "nature" and link particular distributions of power and influence to certain "natural" facts about human beings is the subject of this essay. Using Jean-Jacques Rousseau's widely renowned exploration of the origins of inequality in human societies to set the stage, I begin by establishing the centrality of the existence of natural human differences to Enlightenment political speculation. I then investigate how people in the eighteenth century understood such differences, focusing on their place within the discourse of mental philosophy, and conclude by examining three kinds of response—typified by the arguments of Claude-Adrien Helvétius, Thomas Paine, and Mary Wollstonecraft—that highlight the variety of ways in which the sciences of human nature and theories of republican governance informed one another in the late eighteenth century. By using the language of talents as a focus, I examine how Enlightenment political writers and human scientists, in the process of theorizing the foundations of civil society and articulating a new understanding of human nature

[11] See Alexis de Tocqueville, *Democracy in America*, 2 vols. (New York: Knopf, 1994), especially vol. 2 (1840), bks. 2 and 3.

[12] See Adams, *Defence of the Constitutions of Government of the United States* (cit. n. 9), pp. 391–8.

[13] Isaac Newton, *The Principia: Mathematical Principles of Natural Philosophy*, trans. I. Bernard Cohen and Anne Whitman (Berkeley: Univ. of California Press, 1999); and Adam Smith, *An Inquiry into the Nature and Causes of the Wealth of Nations* (1776; reprinted, New York: Modern Library, 1994), vol. 1, bk. 1, chap. 2. See also David Carrithers, "The Enlightenment Science of Society," in *Inventing Human Science: Eighteenth-Century Domains*, ed. Christopher Fox, Roy S. Porter, and Robert Wokler (Berkeley: Univ. of California Press, 1995), pp. 232–70; Peter Gay, *The Enlightenment, an Interpretation*, 2 vols., vol. 2: *The Science of Freedom* (New York: Knopf, 1969); Thomas L. Hankins, *Science and the Enlightenment* (Cambridge: Cambridge Univ. Press, 1985); and Roger Smith, "The Language of Human Nature," in *Inventing Human Science*, pp. 88–111.

established according to the dictates of nature, elaborated a vision of a social order at once egalitarian and stratified and of human talents and virtues as products of both nature and experience. This connection between understandings of human nature and notions of civil society was so well established that even radical critiques in the name of greater inclusivity—such as those promulgated by Helvétius and Wollstonecraft—had to be couched in terms of individual talents and the rights that could be derived from their possession. The conceptions of political and social order that thus emerged were ones in which openness to the reasoned opinion of all was tempered by the conviction that some, through good education or the gift of nature or both, possessed talents that privileged their voices in the cacophony of public opinion and the struggle for political power.

HUMAN NATURE AND ENLIGHTENMENT GOVERNANCE

The extent of an individual's talents or virtues had not always been an issue of great importance to either political writers or political actors. They began to cede significance to such characteristics in the latter half of the eighteenth century, when Enlightenment investigations into the foundations of state and society merged with scientific explorations into the fundamentals of human nature and the realities of an increasingly market-oriented consumer society. As a consequence, Enlightenment political writers came to interrogate the very conception of a social order legitimated on the basis of birth and privilege. Few of these authors rejected outright the highly stratified, mostly aristocratic societies they knew so intimately. Far from it. But there was a group of natural philosophers, moral philosophers, and what might anachronistically be called political theorists, ranging from Thomas Hobbes in the early seventeenth century to Thomas Paine in the late eighteenth century (and including all the figures about whom Adams was so chary), who sought to place the foundations of civil and political society on new grounds. To such philosophers and theorists, these grounds seemed anchored in the realities of nature and the dictates of reason, not to mention the experiences of everyday economic life, rather than in the conventions of custom and/or the precepts of Christian religion.[14]

Because they considered hereditary aristocracy to lie at the heart of the social-political structure of the *ancien régime*, many of these writers, mostly from what has been called the radical wing of the Enlightenment, targeted hereditary aristocracy in their attempts to imagine society reconstituted on a basis that could be deemed "rational." In general, they sought to demonstrate that rule by hereditary aristocracy violated reason because such rule was too susceptible to instability and corruption, the natural result, they suggested, of being unable to ensure that individuals with the appropriate attributes reached positions of power. In place of such an "irrational" approach to governance, some writers argued for a form of democratic republic, one in which they imagined "the people," however construed, as the foundation of the state, and social distinction and hierarchy, to the degree they would persist at all, as based on a rationalized system of merit rather than on the accidents of birth. For

[14] As Mary Wollstonecraft remarked: "that the society is formed in the wisest manner, whose constitution is founded on the nature of man, strikes, in the abstract, every thinking being so forcibly, that it looks like presumption to endeavour to bring forward proofs." Mary Wollstonecraft, *A Vindication of the Rights of Woman* (1792; reprinted, with an introduction by Miriam Brody, London: Penguin Books, 1992), p. 92.

most of these authors exploring the possibilities of republican governance and a republican social order, this meant anchoring the system of merit in human nature, specifically in those attributes of human beings that the writers deemed "naturally" valuable—an individual's virtues and talents.

With the significant exception of John Locke at the very beginning of the period, few of the writers who produced the most important examinations of the foundations for a republican polity were serious students of such entities as virtues or talents. Nonetheless, to all of these authors, a proper understanding of human nature was critical, for "nature" in all of its forms served as their means of grounding their speculations on a new and, in their eyes, much firmer foundation. Most thus paid particular attention to the explanations for the human mind and how it operated that were being developed in England by Locke and his disciples (associationism), in Scotland by the Scottish Common Sense philosophers, and in France by the abbé de Condillac and his followers (sensationalism). These systematic investigations into what would begin, by the end of the eighteenth century, to be termed "psychology," sought to provide an understanding of the operations of the human mind that paralleled those being offered by natural philosophers for the physical world. Basing their work on a form of experiment that had as its object of inquiry the universal human mind, the practitioners of associationism, Scottish Common Sense, and sensationalism turned to the facts of nature—sensations and their combinations—as the means to create an empirical and naturalistic understanding of the human intellect and the knowledge it could produce.[15] Their "science of mind," because it claimed to be able to determine the fundamental characteristics of human nature, thus seemed to many exploring the emerging "science of republican governance" to be of central relevance.[16]

This connection between the articulations of republican theory and the findings of investigators exploring the nature of the human mind, especially around the issue of talents, has been the subject of only limited historical inquiry. While there have been a number of excellent studies of the historical roots of modern psychology, and of the human sciences more generally, few scholars have examined the specific political implications of the theories and practices of their eighteenth-century subjects.[17] Michel Foucault and his followers are, of course, important exceptions, but even Foucault proved more interested in macrostructural homologies between the human sciences and the liberal polities than in the detailed connections between the

[15] See John Carson, "Minding Matter/Mattering Mind: Knowledge and the Subject in Nineteenth-Century Psychology," *Stud. Hist. Phil. Biol. Biomed. Sci.* 30 (1999): 345–76.

[16] Both terms in quotes are coined in full knowledge that the word "science" was not widely used until the nineteenth century.

[17] See Marina Frasca-Spada, "The Science and Conversation of Human Nature," in *Sciences in Enlightened Europe* (cit. n. 8), pp. 218–45; Gary Hatfield, "Remaking the Science of Mind: Psychology as Natural Science," in *Inventing Human Science* (cit. n. 13), pp. 184–230; Ronald Meek, *Social Science and the Ignoble Savage* (Cambridge: Cambridge Univ. Press, 1976); Sergio Moravia, "The Capture of the Invisible: For a (Pre)History of Psychology in Eighteenth-Century France," *J. Hist. Behav. Sci.* 19 (1983): 370–8; idem, "The Enlightenment and the Sciences of Man," *Hist. Sci.* 18 (1980): 247–68; idem, "From *Homme Machine* to *Homme Sensible*: Changing Eighteenth-Century Models of Man's Image," *J. Hist. Ideas* 39 (1978): 45–60; Richard Olson, *The Emergence of the Social Sciences, 1642–1792* (New York: Twayne Publishers, 1993); Graham Richards, *Mental Machinery: The Origins and Consequences of Psychological Ideas, 1600–1850* (Baltimore: Johns Hopkins Univ. Press, 1992); Roger Smith, "The Background of Physiological Psychology in Natural Philosophy," *Hist. Sci.* 11 (1973): 75–123; and P. B. Wood, "The Natural History of Man in the Scottish Enlightenment," *Hist. Sci.* 27 (1989): 89–123.

two.[18] Most scholars outside the Foucauldian tradition have chosen to explore the historical origins and articulations of various approaches to understanding the nature of the mind and the human being. For example, the wonderful collection of essays in the Christopher Fox, Roy Porter, and Robert Wokler volume *Inventing Human Science* (1995), though rich in detail and nuance, is oriented more toward laying out the complicated historical trajectories of the human sciences than toward linking those patterns to specific political projects.[19] More successful in this regard is Simon Schaffer's 1990 essay, "States of Mind: Enlightenment and Natural Philosophy." Wide ranging and suggestive, Schaffer's essay argues that the connection between mind and body in the English Enlightenment was a question as charged politically as it was philosophically, and thereby makes clear the rich rewards to be gained from mining the domain where social theory and explorations of human nature met.[20]

A second approach, largely found in older works, has focused on the place of various strains of eighteenth-century "psychology" in the thought of important American political leaders. Studies of Jefferson, James Madison, Benjamin Rush, and John Witherspoon, for example, have sought to explain how aspects of their political or intellectual outlooks can be traced to the influence of particular approaches to human psychology.[21] One of the most historically grounded recent pieces in this vein has been Daniel Walker Howe's *Making the American Self* (1997).[22] Like such important earlier interpreters of the American Enlightenment as Henry May and Garry Wills, Howe emphasizes the centrality of the Scottish Common Sense philosophers to the intellectual development of the leading American political thinkers.[23] Howe departs from both May and Wills, however, by stressing the influences of the Scottish form of faculty psychology. In Howe's view, Scottish Common Sense philosophy's commitment to the existence of an innate moral sense and its optimism about the possibilities of human improvement, mixed with its depiction of a continuous struggle between reason and the emotions, were relied on not only by Jefferson but by the authors of the *Federalist Papers*—Madison, Alexander

[18] See, e.g., Michel Foucault, *Madness and Civilization: A History of Insanity in the Age of Reason*, trans. Richard Howard (New York: Vintage Books, 1988); and idem, *Discipline and Punish: The Birth of the Prison*, trans. Alan Sheridan (New York: Vintage Books, 1979).

[19] See particularly the essays by Smith, "Language of Human Nature" (cit. n. 13), and Hatfield, "Remaking the Science of Mind" (cit. n. 17).

[20] Simon Schaffer, "States of Mind: Enlightenment and Natural Philosophy," in *The Language of Psyche: Mind and Body in Enlightenment Thought*, ed. George S. Rousseau (Berkeley: Univ. of California Press, 1990), pp. 233–90.

[21] See, e.g., Donald J. D'Elia, "Benjamin Rush, David Hartley, and the Revolutionary Uses of Psychology," *Proc. Amer. Phil. Soc.* 114 (1970): 109–18; Peter J. Diamond, "Witherspoon, William Smith, and the Scottish Philosophy in Revolutionary America," in *Scotland and America in the Age of Enlightenment*, ed. Richard B. Sher and Jeffrey R. Smitten (Princeton, N.J.: Princeton Univ. Press, 1990), pp. 115–32; Ronald Hamowy, "Jefferson and the Scottish Enlightenment," *William Mary Quart.* 36 (1979): 503–23; Ralph L. Ketcham, "James Madison and the Nature of Man," *J. Hist. Ideas* 19 (1958): 62–76; Patricia S. Noel and Eric T. Carlson, "The Faculty Psychology of Benjamin Rush," *J. Hist. Behav. Sci.* 9 (1973): 369–77; and Robert G. Weyant, "Helvétius and Jefferson: Studies of Human Nature and Government in the Eighteenth Century," *J. Hist. Behav. Sci.* 9 (1973): 29–41.

[22] Daniel Walker Howe, *Making the American Self: Jonathan Edwards to Abraham Lincoln* (Cambridge, Mass.: Harvard Univ. Press, 1997), especially pp. 48–103.

[23] Garry Wills, *Inventing America: Jefferson's Declaration of Independence* (New York: Vintage, 1979), especially pp. 167–255; Henry F. May, *The Enlightenment in America* (New York: Oxford Univ. Press, 1976). See also Morton G. White, *The Philosophy of the American Revolution* (New York: Oxford Univ. Press, 1978).

Hamilton, and John Jay—and proved critical constituents of their visions of the kind of politics and society possible in a democracy.

Taken together, such works on the relations between philosophy and political theory have established conclusively the importance of Enlightenment investigations into human nature for a full understanding of how some of the major figures in the American revolutionary era articulated their ideas about republican democracy. In her recent book, *A History of the Modern Fact* (1998), Mary Poovey has broadened this story in important ways by arguing that a key problem for eighteenth-century Britain was to produce the self-governing liberal subject within the emerging market system, a citizen needing to be much more regulated by the self than by the state.[24] A form of moral philosophy strongly indebted to the methods of natural philosophy, she maintains, became a central knowledge project in Britain then, and did so because it proved a valuable way to theorize, and bring into actuality, the self required for a liberal social-political order and consumer-driven economy. "To appreciate the epistemological contributions of eighteenth-century experimental moral philosophy," Poovey observes, "we need to understand that it was a practice devised in the image of natural philosophy but designed as an account of human motivation and thus, indirectly at least, as an instrument of liberal governmentality."[25]

Poovey's contention that a new kind of social order required a new kind of citizen and was interwoven with a new way of understanding the mind is suggestive. In this article, I take up her problematic by examining a related phenomenon: how notions of mind and the emerging language of talents informed one another.[26] Except in the work of Gordon Wood, little attention has been paid to the political valence of the "talents" half of one of the most common phrases in republican political rhetoric throughout the Atlantic community: "virtues and talents."[27] And even Wood, as thoughtful as his analysis is, has not examined the ways in which the language of talents was infused with notions of human nature being shaped, in part, by philosophers on both sides of the English Channel. Because the major political and philosophical treatises were widely distributed and even more broadly known among the educated elite, I do not keep tightly in this essay to national borders or chronology. Rather, by tracing out a number of lines in the development of this language of talents and how it drew together systematic explorations of mind and theories of governance, I seek to understand what work the notion of talents was performing for the advocates of republicanism during the eighteenth century, particularly in terms of limiting some of the (theoretical anyway) egalitarian possibilities inherent in the vision of a republican society founded on virtue. Because these foundational speculations were at once about the remaking of political culture and of civil society, not to mention the managing of the tension between them, this essay examines how

[24] Mary Poovey, *A History of the Modern Fact: Problems of Knowledge in the Sciences of Wealth and Society* (Chicago: Univ. of Chicago Press, 1998). See also Michel Foucault, *The Foucault Effect: Studies in Governmentality, with Two Lectures and an Interview with Michel Foucault*, ed. Graham Burchell, Colin Gordon, and Peter Miller (Chicago: Univ. of Chicago Press, 1991).

[25] Poovey, *History of the Modern Fact* (cit. n. 24), p. 175.

[26] Also very helpful in thinking through my approach to the issues I explore in this article was Carroll Smith-Rosenberg, "Dis-Covering the Subject of the 'Great Constitutional Discussion,' 1786–1789," *J. Amer. Hist.* 79 (1992): 841–73.

[27] For an extended discussion of the political ramifications of "talents," see Gordon S. Wood, *The Radicalism of the American Revolution* (New York: Knopf, 1992), especially pp. 229–43.

the science of human nature served as a resource for the understanding of both, and, indeed, as Poovey has argued, was one of the means instrumental in keeping them closely allied.

FROM ROUSSEAU'S EXCURSION TO THE REVOLUTION'S COMPACT: SITING THE NATURAL INEQUALITIES

One of the most striking and public articulations of the vision of republican civil society as derived from a combination of democracy and merit, with human nature at the center, was provided in Article VI of the Declaration of the Rights of Man and of Citizen (1789), which set out the basic compact between the people and the government on which the new French republic was to be established:

> VI. Law is the expression of the general will; all citizens have the right to concur personally, or through their representatives, in its formation; it must be the same for all, whether it protects or punishes. All citizens, being equal before it, are equally admissible to all public offices, positions, and employments, according to their capacity, and without any other distinction than that of virtues and talents.[28]

As a succinct summation of the tangle of often antithetical ideas at the center of late-eighteenth-century political discourse on both sides of the Atlantic, Article VI had few rivals. In just one short paragraph, it suggested that the people were the source of all law, that some sort of representative democracy was the preferred form of government, that law should apply the same standards to all citizens, and that merit was the only source of distinction. Article VI captured some of the tensions between the general or communal and the individual, between governmental power and personal liberty, and between sameness, equality, and difference, while at the same time occluding the many everyday distinctions of gender, class, and nationality assumed under the penumbra "citizen."

Article VI's most telling feature, however, with regard to the place of human nature in the discourse of civil society, was that distinctions between individuals remained fundamental, albeit resting on a new footing—that of virtues and talents. In this regard, Article VI was characteristic of late-eighteenth-century political writing in general. Desiring to construct a state founded on rational principles, believing that the original compact creating society was fashioned among free (white male) individuals without distinctions of rank or wealth or birth, hostile to the notion of binding the future to the traditions of the past, and convinced that some rational mechanism for distributing social roles and political power was crucial, Enlightenment political theorists seized on merit as the rational counterweight to demands for universal rights.[29] All citizens might have an equal opportunity to vie for political power and an equal voice in selecting their leaders. But, the theorists contended, the "reality" that government and the economy required particular talents and virtues not necessarily distributed evenly among all members of the populace meant that

[28] "The Declaration of the Rights of Man and Citizen" (1789), reprinted in *The Portable Age of Reason Reader*, ed. Crane Brinton (New York: Viking, 1956), p. 201.

[29] See, e.g., Thomas Paine, *Rights of Man* (1792; reprinted, with an introduction by Eric Fouer, New York: Penguin Books, 1985).

certain individuals would be better suited to positions of power than others.[30] In a world where distinctions of birth still constituted the primary means by which the social hierarchy was established and justified, the idea of granting political leadership and apportioning occupations and social rewards primarily on the basis of an individual citizen's attributes was in many ways revolutionary. Nevertheless, while Article VI and most advocates of republican politics were firmly committed to the claims of universal rights—at least for white, propertied males—in the political and the public spheres, few desired to level all social distinctions and to create a society that was radically egalitarian.[31] This tension, in fact, between the allocation of rights and the privileges of merit, and thus between the languages of equality and of difference, constituted one of the major dynamics of eighteenth-century debate about how to constitute civil society. And it was a debate that pushed issues of the nature of virtues and talents to center stage.

While Enlightenment authors were often quite vigorous in their attempts to specify exactly what they meant by "virtues," they were much more circumspect about the word "talents."[32] Almost always used in the plural, "talents" was often connected with other words—such as "abilities" or "capacities" or "faculties"—signifying some set of attributes that different individuals might possess to different degrees.[33] Whether these differences were many or few, whether they resulted from education or nature, whether they were a product of heredity or chance, however, were all issues on which there was pronounced disagreement. In the main, the specific variations seemed to matter less than the general signification of the term: an avowedly *natural* criterion by which human differences could be delineated and discussed. "Talents" was one way of speaking like a democrat and yet still being able to justify social distinctions, though ones based on decidedly different grounds than the heretofore standard differentiations according to birth and tradition. One consequence of this attempt to anchor such a language of merit in the natural attributes of human beings, however, was the creation of a link between the newly emerging sciences that sought to define and explore human nature and republican speculations about the structure of state, society, and economy. If Enlightenment authors justified distinctions and even stratifications in republican civil society and allocations of power on the basis of differences in individual virtues and talents understood as natural objects, then scientific investigations able to substantiate the reality of such

[30] Thomas Jefferson's proposal for a plan of education for the state of Virginia, for example, was predicated on educating just such an elite of the most talented. See Thomas Jefferson, *Notes on the State of Virginia* (cit. n. 10), pp. 193–8.

[31] Gordon Wood observes that "equality was thus not directly conceived of by most Americans in 1776, including even a devout republican like Samuel Adams, as a social leveling." Wood, *Creation of the American Republic* (cit. n. 5), p. 70. Jean Starobinski notes much the same of Rousseau: "he does not ask for social leveling but simply for a proportioning of civic inequality to the natural inequality of talents." Jean Starobinski, *Jean-Jacques Rousseau: Transparency and Obstruction*, trans. Arthur Goldhammer (Chicago: Univ. of Chicago Press, 1988), p. 302.

[32] There is an extensive literature on the concept of "virtue" in the early modern period. See, e.g., Ruth H. Bloch, "The Gendered Meanings of Virtue in Revolutionary America," *Signs* 13 (1987): 37–58; Carol Blum, *Rousseau and the Republic of Virtue: The Language of Politics in the French Revolution* (Ithaca, N.Y.: Cornell Univ. Press, 1986); James T. Kloppenberg, "The Virtues of Liberalism: Christianity, Republicanism, and Ethics in Early American Political Discourse," *J. Amer. Hist.* 74 (1987): 9–33; Pocock, *Machiavellian Moment* (cit. n. 5); Charles Royster, "'The Nature of Treason': Revolutionary Virtue and American Reactions to Benedict Arnold," *William Mary Quart.* 36 (1979): 163–93; and Wood, *Creation of the American Republic* (cit. n. 5).

[33] For more, see Wood, *Radicalism of the American Revolution* (cit. n. 27).

attributes and give them form could shape the horizon of possibilities for the new societies Enlightenment political thinkers were theorizing. The widespread preoccupation with nature and reason, along with the development of a growing class of people defined more by their accomplishments than by their birth, spurred many of these authors to imagine restructuring the polity according to such principles. Their works served simultaneously to write certain speculations about human nature into the very heart of the republican project and to orient the emerging human sciences toward embracing those social formations most consonant with the developing notions of the republican citizen, the enlightened society, and the self-interested economic actor.

However, as the Jefferson-Adams debate reveals, the eighteenth century had no single canonical interpretation of the relations between the structure of republican societies, the language of merit, and understandings of the natural characteristics of human beings. Thus there are any number of places to begin an investigation into the relations between notions about the fundamental characteristics of the human understanding and the proper constitution of the social order. None, however, may be more instructive than Jean-Jacques Rousseau's *Discourse on the Origin and Foundations of Inequality among Men* (1755), his first serious use of the state of nature to examine contemporary civilization.[34] In a manner similar to Hobbes, Locke, and numerous philosophers before him, Rousseau set out to discover the origins of human society by taking his readers on a journey to the state of nature to discover humanity in its pristine "original condition" and then back to civilization to describe the slow evolution of the state of nature into the state of society. His main point was a deceptively simple one: in a state of nature, outside the "corrupting" influences of civilization, the natural condition of human beings, or at least the adult males of the species, was to be free, independent, and equal. As Rousseau would remark a few years later in a famous passage from *On the Social Contract* (1762), "Man is born free, and everywhere he is in chains."[35]

The very title of his discourse suggests Rousseau's quest: to determine the source of the extreme and extensive inequalities he found manifest in eighteenth-century French society. The *Discourse on Inequality* was a detailed argument that civilization itself produced these differences and that a properly organized society could eliminate many, if not most, of them. He thus began the work of imagining a new civil order based on the free association of equals; seven years later, he would present a more elaborate version of this new order in *On the Social Contract*. Rousseau, however, acknowledged the existence of one significant exception in his account, one that made him uneasy throughout. His exploration of the state of nature convinced him that inequality had not one source, but two—civilization and nature itself:

> I conceive of two kinds of inequality among the human species, one I call natural or physical, because it is established by nature and consists of differences in age, health, physical strength, and qualities of mind *[esprit]* or soul; the other may be called moral or political inequality, because it depends upon a kind of agreement and is established or at least authorized, by the consent of men. The latter consists of the different privi-

[34] Jean-Jacques Rousseau, *Discourse on the Origin and Foundations of Inequality among Men* (1755), reprinted in *Rousseau's Political Writings*, ed. Alan Ritter and Julia Conaway Bondanella and trans. Julia Conaway Bondanella (New York: W. W. Norton, 1988).

[35] Rousseau, *On the Social Contract* (1762), reprinted in *Rousseau's Political Writings* (cit. n. 34), p. 85.

> leges that some enjoy to the detriment of others, such as being more wealthy, more honored, more powerful than they, or even to make themselves obeyed by them.[36]

This second type of inequality, being wholly reliant on the existence of civilization, could not be found in the state of nature. But the first, which included the natural, as opposed to the developed, inequalities of mind, could. Rousseau did argue that human beings in a state of nature were essentially equal, concluding that however great the differences might be between one person and another, they mattered little outside the confines of civilization. Equality prevailed in nature, he believed, because there was scant opportunity in the day-to-day struggle to survive for differences to manifest themselves, because the solitary mode of life characteristic in nature prevented whatever differences might arise from being noticed, and because without civilization there was no chance for rudimentary differences to be further developed and passed on to others.[37] With the arrival of even the most rudimentary forms of civil society, however, everything changed. Able to compare themselves to each other, individuals discovered that their talents differed, that two people engaged in the same task did not necessarily perform at the same level. Some were better than others. Not just difference but, in Rousseau's view, inequality became manifest, and with this inequality came further pressures for the development of civilization and the fashioning of social hierarchies:

> Things in this state might have remained equal, if talents had been equal. . . . [However,] the strongest did more work; the most skillful turned his to better advantage; the most ingenious found ways to curtail his work; the farmer needed more iron, or the blacksmith more wheat; and, by working equally, one earned a great deal, while the other had barely enough to live on. Thus, natural inequality spreads imperceptibly along with contrived inequality, and the differences among men, developed by differences in circumstances, make themselves more obvious, more permanent in their effects, and begin, in the same proportion, to influence the fate of individuals.[38]

There are two important points to be noted about Rousseau's juxtaposition of the state of nature with the state of society. First, he posited a certain irreducible naturalness about variations in human mental and physical characteristics. Unlike the inequalities of wealth, family, and power, differences of physical constitution and quality of mind persisted in Rousseau's account even outside the nurture of civilization. They were, in a sense, bedrock attributes of human nature. And second, differences in these "natural" human talents could have profound social consequences. They formed the basis for the development of many of the inequalities and hierarchies that Rousseau found most characteristic of, and deplorable about, civilization.

Rousseau's analysis of the origins of human differences was in no way atypical of Enlightenment discussions of human nature. From such midcentury philosophers as David Hartley and the abbé de Condillac to late-century political writers William Godwin, Mary Wollstonecraft, and Thomas Paine, not to mention Jefferson, there was a common tendency at least among the more liberal to radical Enlightenment authors to depict human mental and physical characteristics as somehow more fundamental, more real, and ultimately more natural than the "accidents" of birth,

[36] Rousseau, *Discourse on Inequality* (cit. n. 34), pp. 8–9.
[37] Ibid., pp. 31–2.
[38] Ibid., pp. 41–2.

wealth, or class.[39] While there might have been intense, and important, disagreement over exactly what these essential human attributes were, whence they were derived, and whether some were more important than others, there was little dispute that such characteristics were anchored in "nature" and that this connection had significant political implications. This is the first point to draw from Rousseau's investigation into the sources of human inequality: that he linked the articulation of civil society to conceptions of the nature of human beings, and thus gave speculations about mind and behavior a particular political and social valence.[40]

At the same time, Rousseau's assertion that life in the state of nature revealed the existence of inequality as fundamental to human nature suggested that those who dreamed of a society shorn of all marks of distinction and difference were misguided. While the social contract might well be fashioned out of independent agents coming together for mutual good, the result would not be the simple equality of the republican agora. Though various political rights might be distributed equally among those deemed full citizens, political power and civil society would continue to be defined by stratifications and differences. Some people would naturally be better than others, be it at farming or blacksmithing or governing or perhaps reasoning itself, and these superior abilities could readily be translated into political and social advantage. Indeed, one of the attractions of such a conception of human nature to Rousseau and many other philosophes—almost all members of the middling classes or lower—may have been precisely that it confirmed their own senses of deserving to be rewarded for having minds far superior to the common run of humankind.[41] Rousseau himself, having articulated the problem, provided little in the way of a solution in *Discourse on Inequality* and indeed was still wrestling with its implications in *On the Social Contract* and *Emile* (1762). Others were equally engaged, confronted by the twin puzzles of first understanding just what the talents were and how natural they might be, and then trying to imagine a system of society and governance that would accommodate these fundamental characteristics of human nature while still providing space for the demands of equality.

"FASHION HAS INTRODUCED AN INDETERMINATE USE OF THE WORD 'TALENTS'"

In his first major work, the *Enquiry Concerning Political Justice and Its Influence on Morals and Happiness* (1793), William Godwin—high on Adams's list of most suspect Enlightenment political authors—boiled his principle for the proper organization of society and government down to its essence: "The thing really to be desired, is the removing as much as possible arbitrary distinctions, and leaving to tal-

[39] As James Madison argued in Federalist No. 57: "Who are to be the electors of the federal representatives? Not the rich, more than the poor; nor the learned, more than the ignorant; not the haughty heirs of distinguished names, more than the humble sons of obscure and unpropitious fortune. . . . No qualification of wealth, of birth, of religious faith, or of civil profession is permitted to fetter the judgment or disappoint the inclinations of the people." Alexander Hamilton, James Madison, and John Jay, *The Federalist Papers* (New York: Mentor Books, 1961), pp. 350–1.

[40] Michel Foucault has elaborated on the consequences of this point at length in *Discipline and Punish* (cit. n. 18).

[41] I am indebted to my colleague Dena Goodman for this point. For more on the culture of the philosophes, see Dena Goodman, *The Republic of Letters: A Cultural History of the French Enlightenment* (Ithaca, N.Y.: Cornell Univ. Press, 1994).

ents and virtue the field of exertion unimpaired."[42] Like Jefferson and many other republican theorists at the end of the century, Godwin was clear on this much: merit should be preeminent and talents and virtues were integral to any conception of merit. Such writers were much less clear, however, about two related issues: First, just what constituted talents and virtues and who decided? Second, how exactly did these attributes originate and vary, and could they be altered? While late-eighteenth-century authors would invoke both terms frequently, they rarely attempted to provide precise definitions. Instead, "talents" especially tended to be used mainly as a placeholder, able to suggest a multitude of positive characteristics without being restricted to any particular denotation. In *Rights of Man*, for example, Paine littered his text with references to talents and how government required a variety of them for its successful operation. At the same time, however, he provided few concrete examples and never a specific definition. He simply assumed that his readers would know what he meant by the term and would themselves fill in the relevant attributes.[43]

Nevertheless, given the centrality of talents to the project of determining the boundaries that human nature set on the making of republican government and society, the term must be defined a bit more precisely. Paine did provide one important insight in this regard: he often suggested in *Rights of Man* that talents were intimately connected with powers or faculties of the mind, and that they were the critical variables in choosing good leaders.[44] Contemporary dictionaries corroborate this sense of the word. Samuel Johnson, for example, defined "talent" in his *A Dictionary of the English Language* (1755) as "faculty; power; [or] gift of nature," a meaning little different from that provided by *Le Grand vocabulaire François* (1773) and the *Dictionnaire de l'académie Françoise, nouvelle édition* (1786), both of which stated that "talent" figuratively referred to a "gift of nature, natural disposition or aptitude for certain things, capacity, [or] ability."[45] As both the English and French dictionaries suggested, talent was taken to refer to some sort of potential for superior achievement, with the strong implication in both languages that one was born with it rather than developed it. In addition, the French dictionaries suggested, although only slightly, that *talent* referred to characteristics that had specific external manifestations, particularly those involving the ability to accomplish a particular task. *Le Grand vocabulaire François*, for example, distinguished between qualities and talents—*qualités* made one good or bad, *talents* useful or amusing—and between genius and talents—*génie*, it noted, was more interior and partook of invention, *talent* more exterior and provided for brilliant execution.[46]

[42] William Godwin, *Enquiry Concerning Political Justice and Its Influence on Morals and Happiness*, 3 vols., ed. F. E. L. Priestley (1793; reprinted, with an introduction by F. E. L. Priestley, Toronto: Univ. of Toronto Press, 1946), vol. 1, p. 147.

[43] Paine's closest attempt at a definition of talents may have been the following: "if we examine, with attention, into the composition and constitution of man, the diversity of his wants, and the diversity of talents in different men for reciprocally accommodating the wants of each other, his propensity to society, and consequently to preserve the advantages resulting from it, we shall easily discover, that a great part of what is called government is mere imposition." Paine, *Rights of Man* (cit. n. 29), pp. 163–4.

[44] Ibid., pp. 175–6.

[45] Samuel Johnson, *A Dictionary of the English Language* (London: W. Strahan, 1755), s.v.; *Le Grand vocabulaire François* (Paris: n. p., 1773), vol. 27, pp. 324–5; and *Dictionnaire de l'Académie Françoise, nouvelle édition* (Nismes: Pierre Beaume, 1786), vol. 2, p. 554.

[46] *Le Grand vocabulaire François* (cit. n. 45), vol. 27, pp. 324–5. On genius in relation to natural philosophy in the late eighteenth and early nineteenth centuries, see Simon Schaffer, "Genius in

In the world of eighteenth-century dictionaries, therefore, "talents" denoted particular attributes of mind or body, probably present from birth, with at least a hint in French that the term referred more to externalized accomplishments than to internal potentials. Left at this level of generality, "talents" could refer to just about any operation that an individual could successfully perform. Indeed Adams exploited just this openness in the word when responding to Jefferson's long missive on the natural aristocracy:

> We are now explicitly agreed, in one important point, vizt. That 'there is a natural Aristocracy among men; the grounds of which are Virtue and Talents.' . . . But tho' We have agreed in one point, in Words, it is not yet certain that We are perfectly agreed in Sense. Fashion has introduced an indeterminate Use of the Word 'Talents.' Education, Wealth, Strength, Beauty, Stature, Birth, Marriage, graceful Attitudes and Motions, Gait, Air, Complexion, Physiognomy, are Talents, as well as Genius and Science and learning.[47]

Adams, of course, was having a bit of fun at Jefferson's expense by twitting his faith in the judgment of the common citizen and idealized vision of democratic politics. Nonetheless, in at least two regards Adams's remarks are quite revelatory. First, his very acknowledgment of the political implications of the meaning of talents suggests the extent to which aspects of human nature had become central elements in republican discourse by the end of the eighteenth century. And second, Adams's cynicism highlighted a critical feature of Jefferson's conception of talents: he restricted it to those attributes he deemed most essential to the continued flourishing of a democratic republic—genius, science, and learning. In this regard, indeed, Jefferson was not alone. If "science" were dropped or broadened to denote systematic thinking of any sort, and if "reason" were specifically included, then Jefferson's version of talents would come very close to what was routinely assumed about politics and society by the whole cadre of Enlightenment theorizers that Adams held in such contempt. All considered themselves to be the final judges of what was, and was not, a true talent, and most sought, in one form or another, to legitimate their own right to speak authoritatively by establishing that talents such as theirs were the ones of critical importance.

Mary Wollstonecraft, another leading light in Godwin's circle of radical Enlightenment authors, clearly manifested this usage of talents in her work. Desirous of establishing a role for women in democratic republics, Wollstonecraft was concerned throughout *A Vindication of the Rights of Woman* (1792) with delineating which mental attributes made a good republican citizen and which did not. Beauty, cunning, coquettishness, sensualism, and dissimulation—indeed most of the skills commonly associated, as she pointed out, with both women and aristocrats—were not abilities that had merit in republics.[48] Rather, she valorized talents that gave evidence of independent minds well stocked with experience and knowledge, of

---

Romantic Natural Philosophy," in *Romanticism and the Sciences*, ed. Andrew Cunningham and Nicholas Jardine (Cambridge: Cambridge Univ. Press, 1990), pp. 82–98; and Richard Yeo, "Genius, Method, and Morality: Images of Newton in Britain, 1760–1860," *Sci. Context* 2 (1988): 257–84.

[47] Adams to Jefferson, 15 Nov. 1813, *Adams-Jefferson Letters* (cit. n. 1), pp. 397–8.

[48] Wollstonecraft, *Vindication* (cit. n. 14), especially chap. 4.

reasoning faculties finely sharpened, of passions and interests dominated by reason, and of politeness, modesty, and concern for the good of the whole.[49] These attributes accorded well with Jefferson's ideal of the independent yeoman farmer and Robespierre's of the virtuous citizen—"male" characteristics, in the main, and specifically those attributed to hardworking males of the middling sort, especially those who had by means of such talents achieved success in some aspect of public or economic life. "Abilities and virtues," Wollstonecraft noted, "are absolutely necessary to raise men from the middle rank of life into notice, and the natural consequence is notorious—the middle rank contains most virtue and abilities."[50]

In a democratic republic, therefore, as even Adams would have agreed, not all talents deserved to be accorded merit, only those that contributed in some way to an individual's wisdom or virtue. And those were talents that, at least for the philosophers theorizing the republican project, involved the ability to acquire information and to generalize from it, to rise above local prejudices, and to channel the passions and desires according to the dictates of reason. James Madison explained simply the interrelations of democratic theory and (appropriate) human capacities in Federalist No. 58 (1788) when discussing the proper size for the House of Representatives. Those "of limited information and of weak capacities," he maintained, were most likely to be susceptible to demagogues and thus to allow a democratic republic to be transformed into an oligarchy. Republics, Madison concluded, required representatives who were knowledgeable and intellectually talented in order to survive.[51]

Having fashioned a central role in republican discourse for individuals with talents, and having loosely indicated what general sorts of characteristics these were and their own right to settle the question, Enlightenment political theorists confronted one further problem as they developed their language of merit: the explanation of where talents came from, why some people had more of them to a higher degree than others, and why that mattered. The question of origins and differential abilities, crucial to the explanation of just how egalitarian a republic could theoretically be, generated little consensus among those who explored it. Some theorists argued for the primacy of native endowment, others education; some believed that heredity played an essential role, others an inconsequential one; some distinguished between individuals and groups, others did not. The differences were enormous and extremely varied, although there were certain commonalities: most discussions of the origins of talent among liberal to radical Enlightenment authors focused on the polarities, and many were couched in terms of particular theories of the operations of the human mind.

Probably the most important distinction eighteenth-century writers fastened on in analyzing the origins of talent was that between talents as gifts of nature and talents as products of education. As we have seen, the definitions of the term provided by contemporary English and French dictionaries laid heavy emphasis on the "gift of nature" sense of "talent," suggesting that talents were faculties or capabilities present from birth. Writers as diverse as Denis Diderot, Pierre-Jean-Georges Cabanis, Wollstonecraft, James Mill, Paine, and Jefferson all shared this understanding of the term,

[49] Ibid., especially chap. 6.
[50] Ibid., pp. 147–8.
[51] Hamilton, Madison, and Jay, *Federalist Papers* (cit. n. 39), no. 58, pp. 360–1.

assuming that education could sharpen or dull an individual's talents but could not create them or develop them beyond a certain point. This belief lay at the heart of both Jefferson's ideal of rule by a natural aristocracy and Paine's celebration of the potentials of republican democracy.

Counterposed to this explanation of talents as primarily the product of nature stood those explanations placing equal or greater weight on the power of education. Virtually all writers, no matter where they fell on the nature-education spectrum, conceded that external influences could have some significant effect on human talents. The essence of Wollstonecraft's argument in *Vindication*, for example, was that women mostly lacked the abilities required of republican citizens because they had been denied the necessary training. Proper education, she asserted, would show that women as a group possessed the same kinds of faculties as men, though whether to the same degree she left an open question, at least rhetorically.[52] What she did not deny, however, was that individuals, male or female, naturally differed in the types of talents with which they were endowed and the degree to which development was possible. "But, treating of education, or manners," Wollstonecraft remarked, "minds of a superior class are not to be considered, they may be left to chance; it is the multitude, with moderate abilities, who call for instruction, and catch the colour of the atmosphere they breathe."[53]

Other Enlightenment writers—including Helvétius, Antoine Louis Claude Destutt de Tracy, Smith, Joseph Priestley, Godwin, and Benjamin Rush—were willing to push the power of education much further, conceding to it not just some influence but, in fact, a definitive role. For these authors the mind was, in Locke's words, a tabula rasa, empty of all ideas and potentials before being filled by the set of experiences that defined an individual's life. The arguments that Godwin presented in *Enquiry Concerning Political Justice* are an excellent illustration of what might be called the "environmentalist" position. He specifically attacked the physiological explanations of human abilities on at least three grounds. First, he maintained that while individuals might indeed differ to some degree in their original endowments, those differences were not particularly significant. "But, though the original differences of man and man be arithmetically speaking something, speaking in the way of a general and comprehensive estimate they may be said to be almost nothing."[54]

Second, he completely reversed the notion that physiological characteristics, such as skull size, might in some sense be the causes of differences in powers or abilities by suggesting that human physiognomy was plastic and was actually largely shaped by individual action.[55] And third, Godwin rejected the notion that faculties or capabilities were directly heritable. In place of native endowment, Godwin emphasized the enormous powers of education: "the question whether the pupil shall be a man

[52] "If women are by nature inferior to men," Wollstonecraft argued, "their virtues must be the same in quality, if not in degree." Wollstoncraft, *Vindication* (cit. n. 14), p. 108.

[53] Ibid., p. 162.

[54] Godwin, *Enquiry* (cit. n. 42), vol. 1, pp. 42–3.

[55] "It is not improbable, if it should be found that the capacity of the scull of a wise man is greater than that of a fool, that this enlargement should be produced by the incessantly repeated action of the intellectual faculties, especially if we recollect of how flexible materials the sculls of infants are composed, and at how early an age persons of eminent intellectual merit acquire some portion of their future characteristics. In the mean time it would be ridiculous to question the real differences that exist between children at the period of their birth." Ibid., p. 38.

of perseverance and enterprise or a stupid and inanimate dolt, depends upon the powers of those under whose direction he is placed. . . . [T]here are no obstacles to our improvement, which do not yield to the powers of industry."[56] Godwin summed up nicely the fundamental precept underlying the arguments of those writers most committed to explaining human talents on the basis of education. Negating the importance of an individual's original endowments, if indeed such entities were presumed to exist, a number of authors maintained that while individuals and groups did differ, experience alone created the differences.[57]

Not surprisingly, writers such as Godwin who emphasized the power of education in shaping individual talents had little interest in, and provided little role for, the actions of heredity in determining what abilities an individual manifested and to what degrees. But even some of those writers who adopted a more physiological approach to the origin of an individual's talents also gave short shrift to strictly hereditarian explanations of the appearance of natural capacities. Paine, for example, relied on a quite widespread conception of heredity when he remarked in *Rights of Man*:

> Experience, in all ages, and in all countries, has demonstrated, that it is impossible to control Nature in her distribution of mental powers. She gives them as she pleases. Whatever is the rule by which she, apparently to us, scatters them among mankind, that rule remains a secret to man. It would be as ridiculous to attempt to fix the hereditaryship of human beauty, as of wisdom. Whatever wisdom constituently is, it is like a seedless plant; it may be reared when it appears, but it cannot be voluntarily produced. There is always a sufficiency somewhere in the general mass of society for all purposes; but with respect to the parts of society, it is continually changing its place.[58]

Not that Paine's was the only approach to the issue of heredity. Jefferson and Adams, in their correspondence about natural aristocracy, believed just as certainly that talents could be bred, agreeing that desirable characteristics could be produced in humans if they married with the same attention to lineage as was employed in the mating of sheep.[59] Physician Benjamin Rush, though one of the most vociferous American advocates of the transformative power of education, also came to adopt something of a hereditarian stance, speculating that heredity might in the end prove to be the most significant factor in the apportionment of talent. "It is possible, the time may come," Rush speculated, "when we shall be able to predict, with certainty, the intellectual character of children by knowing the specific nature of the different intellectual faculties of their parents."[60]

What is most striking about Enlightenment discussions of heredity is that there was no orthodox position. As much intellectual respectability lay in denying the power of heredity as endorsing it, and a vast array of intermediate positions, involving

[56] Ibid., p. 44.

[57] On this general issue, see Henry Vyverberg, *Human Nature, Cultural Diversity, and the French Enlightenment* (New York: Oxford Univ. Press, 1989).

[58] Paine, *Rights of Man* (cit. n. 29), p. 175.

[59] See Jefferson to Adams, 28 Oct. 1813, *Adams-Jefferson Letters* (cit. n. 1), pp. 387–92.

[60] Benjamin Rush, "On the Influence of Physical Causes in Promoting an Increase in the Strength and Activity of the Intellectual Faculties of Man" (1799), reprinted in *Two Essays on the Mind* (New York: Brunner/Mazel, 1972), p. 119.

belief in the inheritance of acquired characteristics and/or in the powerful effects of environmental forces acting on the child while in the womb, were both perfectly acceptable and widely endorsed.[61] With no expert view conceded authority on the subject, eighteenth-century writers were largely free to construct the hereditarian dimensions of human talents according to their own experiences and understandings of heredity's operations. As a result, the writers advanced a multiplicity of views, and as a group they no more accorded talents a specific implication with regard to heritability than they did with regard to physiology.

The consequences of such an open-ended understanding of human heredity for notions of republican governance were twofold. On the one hand, individual republican theorists still had to wrestle with their own ideas about the heritability of talents, the evidence that would support their positions, and the implications of their beliefs for their conceptions of a functioning republic. Paine's vision of a republic hierarchically organized according to talent but founded on equal opportunity for all, for example, was deeply dependent on his particular understanding of the nature of talents and how they were acquired. Different "facts" about heredity might have resulted in a much different theory of the social-political order. On the other hand, the variety of credible approaches to the question of the heritability of talents meant that a number of different visions of republican governance could be brought into accord with the truths of nature. Science might dictate the terms, but much depended on which rendering of the science of human nature an author accorded credence.

POLITICAL THEORY MEETS MENTAL PHILOSOPHY

Given this confusion of possibilities, many eighteenth-century theorists, when forced to attend to the issue of the origins and nature of the human talents, turned to the precepts of mental philosophy to buttress their speculations. For some, what beckoned was the ancient language of the "faculties," Aristotelian in its origins, though updated especially as a part of the flowering of mental philosophy during the Scottish Enlightenment. Within the understanding of the Scottish Common Sense school, talents were faculties believed to be either present from birth or developed through education and exercise.[62] For a growing minority of authors, however, this approach seemed problematic. Helvétius, Godwin, Rush, Diderot, and Cabanis, among others, sought to ground their political beliefs, at least in part, in the new experiential psychologies fashioned out of Locke's observations in *An Essay Concerning Human Understanding* (1690 and especially the fourth edition in 1700).[63] In so doing, they helped to forge links between notions of republican social-political order and mental philosophy, and revealed the new psychology's ultimate inability to prescribe any one reading of the nature of talents.

The two major Enlightenment experiential approaches to the mind were associationism, codified in England by David Hartley, and sensationalism, the French ver-

[61] On ideas about heredity in America, see Charles E. Rosenberg, "The Bitter Fruit: Heredity, Disease, and Social Thought," in *No Other Gods: On Science and American Social Thought* (Baltimore: Johns Hopkins Univ. Press, 1978), pp. 25–53. For France, see Carlos López-Beltrán, "'*Les Maladies héréditaires*': 18th Century Disputes in France," *Rev. Hist. Sci.* 48 (1995): 307–50.

[62] Rousseau and Paine are two examples of eighteenth-century writers who relied on a faculty understanding of the human mind.

[63] John Locke, *An Essay Concerning Human Understanding*, 2 vols. (New York: Dover, 1959).

sion of Lockeanism articulated by Etienne-Bonnot de Condillac. Both were characterized by a commitment above all to the constitutive power of experience.[64] In brief, associationism and sensationalism each sought to ground human knowledge and abilities on sensations and the associations of ideas produced by sensations. Subsuming and exceeding Locke's rejection of innate ideas in favor of the tabula rasa, Hartley and Condillac stressed the passivity of the mind in all of its operations. Experience alone, both contended, furnished the mind with all of its raw materials and with most of the powers by which those materials could be manipulated. Condillac's famous example of the statue, presented in his *Traité des sensations* (1752), illustrated the nature of these Enlightenment theories of the mind well. Condillac asked his readers to imagine a statue internally constructed like a human being and with a mind totally devoid of ideas. He then sequentially endowed it with each of the five senses and showed how, on the basis solely of the experiences acquired via sense perceptions and their associations, all human faculties, ideas, and abilities would arise:

> The principle determining the development of its [the statue's] faculties is simple. It is comprehended in the sensations themselves, for these being naturally pleasant or unpleasant, the interest of the statue will be to enjoy the one kind and reject the other. Now we are able to show that this interest is sufficient to explain the operations of the understanding and the will. Judgment, reflexion, desires, passions, &c. are only sensations differently transformed.[65]

While there were numerous internal inconsistencies in both the sensationalist and associationist versions of experiential psychology, the basic message was clear. As Condillac mused in the final sentences of the *Traité*: "The statue is therefore nothing but the sum of all it has acquired. May not this be the same with man?"[66]

For proponents of an education-based explanation of the origins of an individual's talents, associationism and sensationalism were easily rendered into forms that would provide strong support. Godwin, for example, referred throughout his *Enquiry Concerning Political Justice* to the general principles of associationist psychology and, at length, to Hartley's theories. He used them to buttress his claim "that the actions and dispositions of mankind are the offspring of circumstances and events, and not of any original determination that they bring into the world."[67] Physiologically oriented authors, while they had a more difficult time of it, also proved adept at grounding their accounts of talents as gifts of nature on the precepts of experiential psychology. In general, their approach was to concede that all ideas must be based on experience but to contend that different mental physiologies would result in the same set of experiences being processed in distinct ways and thus would result in the expression of different individual talents or the same talent to varying

[64] David Hartley, *Observations on Man. His Frame, His Duty, and His Expectations* (London: Leake, Frederick, Hitch, and Austin, 1749); Etienne-Bonnot de Condillac, "Essai sur l'origine des connoissances humaines" (1746), reprinted in *Oeuvres complètes de Condillac* (Paris: Houel, 1798); and idem, *Condillac's Treatise on Sensations* (1754), trans. Geraldine Carr (London: Favil Press, 1930).

[65] Condillac, *Treatise on Sensations* (cit. n. 64), pp. xxxi-xxxii.

[66] Ibid., p. 239.

[67] Godwin, *Enquiry* (cit. n. 42), vol. 1, p. 26.

degrees. Diderot for one used arguments of this kind in developing his critiques of radically environmentalist interpretations of human mental characteristics.[68] And Pierre-Jean-Georges Cabanis explored them even more fully in a series of lectures he gave before the National Institute beginning in 1796, published in book form in 1802 as *Rapports du physique et du moral de l'homme*.[69]

Like so many other eighteenth-century writers investigating the operations of the mind, Cabanis began with Locke. "It was Locke who for the first time clearly exposed and fortified with his most direct proofs, this fundamental axiom: THAT ALL IDEAS COME THROUGH THE SENSES, or ARE THE PRODUCT OF THE SENSATIONS."[70] Cabanis departed from Locke, and from his successors Condillac and Hartley as well, however, by following Diderot's route and tying sensationalist psychology to human physiology. Human physiology, Cabanis contended, taught that human beings naturally vary, that "the different organs or systems of organs do not have the same degree of force or effect in different subjects. Each person has his strong and his weak organs."[71] Much of *Rapports*, in fact, was given over to illustrating how variations—whether caused by heredity, disease, age, climate, diet, or occupation—in the various organs, especially the genitalia, produced the great differences that existed among individuals and between at least certain kinds of human groups (such as men and women, or young and old).

According to Cabanis and most other advocates of a physiological understanding of talents, the environmentalists' theory that differences arose because each individual had his or her own unique set of experiences was insufficient. Even were two individuals to have exactly the same sensations over their entire lives, Cabanis contended, they would still exhibit distinct mental characteristics because their internal structures, especially their brains, not being exactly the same, would process the sensations differently. "Men are certainly not alike in the way in which they feel," Cabanis noted, "age, sex, temperament, illness, set up notable differences between them, and in the same man the various impressions have, according to their nature and many other incidental circumstances, a very unequal degree of force or of acuteness."[72]

These two divergent ways of reading Enlightenment experiential psychology—as supporting either an environmentalist or physiological interpretation of the origin of human mental characteristics—illustrate both the newfound proximity of the language of mental philosophy to the language of social-political theory and the ambiguities that surrounded Enlightenment deployments of the word "talents." The ambiguities were heightened because either reading of the new psychology could be used to support a variety of theories about republican governance. For figures such as Godwin, committed to the radical restructuring of human society along egalitar-

[68] Denis Diderot, "Refutation of the Work of Helvétius Entitled *On Man*" (1773–1776), reprinted in *Diderot's Selected Writings*, ed. Lester G. Crocker and trans. Derek Coltman (New York: Macmillan, 1966), pp. 283–98; and idem, "Réflexions sur le livre *De l'esprit* par M. Helvétius" (1758), reprinted in *Oeuvres complètes*, vol. 9: *L'Interpretation de la nature (1753–1765), Idées III*, ed. Jean Varloot (Paris: Hermann, 1981), pp. 261–312.

[69] Pierre-Jean-Georges Cabanis, *On the Relations between the Physical and Moral Aspects of Man*, 2 vols. (1802), ed. George Mora and trans. Margaret Duggan Saidi (Baltimore: Johns Hopkins Univ. Press, 1981).

[70] Ibid., vol. 1, p. 10.

[71] Ibid., vol. 1, p. 60.

[72] Ibid., vol. 1, pp. 51–2.

ian lines, the potential benefits of notions of mind as enormously malleable were obvious: every putative difference between individuals or groups that could be used to justify a particular social hierarchy was vitiated if all differences were deemed the result of education.[73] Almost as readily, however, the same theory about the nature of the human mind could be used to support a position such as that of Adam Smith, who also contended that differences in talent were largely the product of education, but nonetheless concluded that a highly differentiated, and stratified, workforce was essential to a nation's prosperity and a society's happiness.[74]

Similarly, while arguments in favor of hierarchy and the status quo could be sustained relatively easily by reference to human mental characteristics as the products of native endowment, neither Paine nor Diderot, for example, had any difficulty in reconciling such a position on the origins of the faculties with their own interest in the radical transformation of society. Convinced that talents, while gifts of nature, were spread equally throughout all social classes and that republics required a diversity of talents to prosper, the two men concluded that a republican society must make opportunity equally available to all.[75] What was less possible for those who adopted a physiological orientation, however, was to conceive of a society in which hierarchy of every sort was deemed artificial and thus eliminable. For them, the reality of talents typically made distinctions of some kind seem both inevitable and justified. Moreover, because the meaning of "talent" typically carried with it some sense of comparative superiority, and the kinds of talents that they deemed relevant to articulations of the nature of republican societies were highly circumscribed, the slippage between superiority in one domain and superiority overall easily occurred, as Jefferson's conception of the natural aristocracy vividly demonstrates.[76]

The range of possible ways in which conceptions of merit, understandings of talents, and theories of society and governance could be rendered and combined by republican theorists in the late eighteenth century was thus quite large. Even among the group of liberal and radical writers who attracted Adams's particular ire, substantial disagreements existed over just how perfectible (to use Adams's language) human beings were, as well as over issues of the heritability of talents, their nature and number, and their relative importance in placing limits on the kind of republican state and society that seemed feasible.[77] In order to explore these connections in more depth, we will conclude with an examination of three authors—Helvétius, Paine, and Wollstonecraft—who were routinely grouped by critics as part of the same radical Enlightenment attack on the foundations of traditional society, and who can be seen, nonetheless, to have adopted very different positions on the source of human talents and their role in defining the nature of a republican polity.

[73] As Godwin argued, "A principle deeply interwoven with both monarchy and aristocracy in their most flourishing state, but most deeply the latter, is that of hereditary pre-eminence. No principle can present a deeper insult upon reason and justice. Examine the new-born son of a peer, and of a mechanic. Has nature designated in different lineaments their future fortune?" Godwin, *Enquiry* (cit. n. 42), vol. 2, p. 86.

[74] Smith, *Wealth of Nations* (cit. n. 13), vol. 1, bk. 1, chap. 2, especially pp. 16–7.

[75] See Paine, *Rights of Man* (cit. n. 29), p. 175.

[76] Thus even Paine did not advocate the elimination of all social distinctions, but simply their placement on a more rational footing.

[77] On eighteenth-century ideas about perfectibility, see Victor Hilts, "Enlightenment Views on the Genetic Perfectibility of Man," in *Transformation and Tradition in the Sciences: Essays in Honor of I. Bernard Cohen*, ed. Everett Mendelsohn (Cambridge: Cambridge Univ. Press, 1984), pp. 255–71.

HELVÉTIUS, THE EGALITARIAN CHALLENGE, AND THE ENLIGHTENMENT LEGACY

Ironically, the Enlightenment writer who may have caused late-eighteenth-century radicals their most significant problems was one of the century's great champions of equality, Claude-Adrien Helvétius. A tax farmer turned philosopher, seen as something of a protégé of Voltaire, Helvétius became one of the most notorious radicals of the French Enlightenment, a figure almost on a par with Julien Offray de La Mettrie and Rousseau. Helvétius's two major philosophical works, *De l'esprit* (1758) and *De l'homme* (1772–1773), constituted the most thoroughgoing eighteenth-century arguments in favor of the natural equality of human beings and came to be seen as representative of some of the most extreme possibilities of Enlightenment thought, influencing—if only negatively—almost all those writing about such topics after him.[78] Extremely well known, at least by reputation, throughout the republic of letters and beyond, both works generated an enormous amount of controversy, including criticism from almost every point on the political and intellectual spectrum.

There was little, on the surface, about Helvétius's theory of mind that would have seemed particularly novel to a student of Condillac or Locke. Bypassing all of the equivocations and hesitancies apparent in Condillac's formulation, Helvétius accepted without reservation Condillac's insistence on the primacy of experience, made experience the cardinal principle of his philosophy, and boldly, and perhaps bluntly, deduced the consequences. Experience alone, Helvétius maintained in both *De l'esprit* and *De l'homme*, determined every facet of the unimpaired human mind.[79] Innate ideas, innate faculties—both were banished from Helvétius's theory in favor of a mind composed solely of sensations and the associations of sensations. For Helvétius, a person was the sum of his or her experiences, no more and no less. Where Helvétius deviated from Condillac, besides in the lack of nuance, was in what he deduced from this emphasis on experience and in how he attempted to convince his readers of the truth of his deductions. Simply put, Helvétius concluded that if human minds were the result of nothing but sensations and their associations, then intrinsically all minds must be the same and all differences, even those between geniuses and ordinary people, must arise solely from variations in experience. "Genius is not the gift of nature," Helvétius remarked, ". . . a man of genius spends his time in study and application."[80]

Thus, according to Helvétius, any person could become a genius. But if that were so, Helvétius wondered, why in fact was genius so rare? The final sentence of *De l'esprit* suggested his answer, an answer that Helvétius would develop more fully in *De l'homme*—politics:

> [I]t is certain that the great men that are now produced by a fortuitous concourse of circumstances, will become the work of the legislature, and that, by leaving it less in

[78] Claude Adrien Helvétius, *De l'esprit; or Essays on the Mind and Its Several Faculties* (1758) (New York: Burt Franklin, 1970); and idem, *A Treatise on Man: His Intellectual Faculties and His Education* (1772–1773), trans. William Hooper, 2 vols. (London: Albion Press, 1810).

[79] For an excellent discussion of Helvétius's philosophy and how it changed between *De l'esprit* and *De l'homme*, see C. Kiernan, "Helvétius and a Science of Ethics," *Stud. Voltaire 18th Cent.* 60 (1968): 229–43.

[80] Helvétius, *Essays on the Mind* (cit. n. 78), p. 359.

> the power of chance, an excellent education may infinitely multiply the abilities and virtues of the citizens in great empires.[81]

In *De l'homme*, Helvétius would show himself to be much less sanguine about empires and much more enamored of republics as the way to ensure the general education of the public.[82] Nonetheless, what is critical here is not which political system Helvétius found most appropriate, but that the conclusion to an essay on the mind would be a call for civic reform. Helvétius was certainly not the only Enlightenment writer to merge philosophy and politics, but his linking of the potential equality of human minds with their actual inequality made the issue of social change particularly acute. For if all human beings were endowed with the same potential for experiencing and associating sensations, and experience produced all differences, including the whole range of current inequalities, then changing experience along the appropriate lines, especially through the state's establishment of a comprehensive system of education, should be sufficient to increase the mental abilities of the populace. All that was necessary, Helvétius contended, was both the desire to expand the mental abilities of the citizenry and the willingness to have the state be an active agent of social change.

Helvétius actually devoted little attention in either of his treatises to the practicalities of achieving such an improvement in the capabilities of the citizenry. He was also rather reticent on the issue of how some individuals were able to manifest genius at all, despite the impediments and ridicule generated by contemporary society. Nevertheless, rather than being dismissed as idle philosophical speculation, Helvétius's books, especially *De l'esprit*, generated an extraordinary amount of public controversy. As David W. Smith has noted in his excellent study of the reaction to the publication of *De l'esprit*:

> No book during the whole of the eighteenth century, except perhaps Rousseau's *Emile*, evoked such an outcry from the religious and civil authorities or such universal public interest. Denigrated as the epitome of all the dangerous philosophic trends of the age, condemned as atheistic, materialistic, sacrilegious, immoral, and subversive, it [*De l'esprit*] enjoyed an immense *succès de scandale*.[83]

While Helvétius's pronouncements about human equality added some fuel to the fire, in the main the *scandale* erupted because of his theory of human morality and his willingness, as Smith argues, to seem to resurrect the dreaded materialism of La Mettrie's *L'Homme machine* (1748). Like La Mettrie, Helvétius was read as doing away with the human soul and free will by advocating a strictly mechanical determinism driven by sensation and a sensibility potentially present in all matter.[84] Nonetheless, the dramatic events that surrounded the publication of *De l'esprit*—its appearance in late July and the annulment on 10 August of the *privilège* that allowed it to be published in France; its condemnation, along with the *Encyclopédie* and other exemplars of Enlightenment *philosophe* thought, by most Parisian religious and civil bodies; and its author's three retractions (two before the Jesuits and one

[81] Ibid., p. 498.
[82] See particularly Kiernan, "Helvétius and a Science of Ethics" (cit. n. 79).
[83] David W. Smith, *Helvétius: A Study in Persecution* (Oxford: Clarendon Press, 1965), p. 1.
[84] Ibid., p. 13.

before the Parlement as he tried to forestall further censure)—conspired to make its arguments widely known and much discussed.[85]

Broad dissemination, however, did not produce many converts. For all of the egalitarianism of much of the rhetoric of the late eighteenth century, few people were willing to go as far as Helvétius in their commitment to the full natural equality of human beings. Adams, writing to Jefferson in 1813, summed up in his typical no-nonsense manner the reaction to Helvétius common in much of the transatlantic world:

> I have never read Reasoning more absurd, Sophistry more gross, in proof of the Athanasian Creed, or Transubstantiation, than the subtle labours of Helvetius and Rousseau to demonstrate the natural Equality of Mankind. Jus cuique [Justice for everyone]; the golden rule; do as you would be done by; is all the Equality that can be supported or defended by reason, or reconciled to common Sense.[86]

Adams did not get Rousseau quite right, for Rousseau was only slightly more sympathetic to Helvétius than was Adams.[87] But Adams did capture the incredulity with which many contemporary and subsequent writers and political thinkers reacted to Helvétius's theories.[88] The visceral quality of the response to Helvétius is most illuminating. Helvétius struck a nerve, and that nerve may have had much to do with the nature of the republics that these political thinkers were so assiduously attempting to invent.

The problem was simple: What space existed for merit in Helvétius's theory? If Helvétius were right, and each human being was completely a product of education, then, as he pointed out, it ought to be possible for the state, by educating all equally, to produce the same level of talents and virtues in every citizen. Fine in theory, but in practice, what would it have meant for a social and political order to be structured around merit if there were no legitimate criteria of differentiation? If human beings were nothing but unmolded clay and society had the power to shape all to become the best conceivable, then how could anyone justify the existence of social distinctions? Either an absolutely egalitarian society would have to be established—a prospect with few proponents on either side of the Atlantic—or recourse would have to be made to arbitrary distinctions, a solution difficult to square with condemnations of hereditary aristocracy on the very same grounds. Helvétius threatened to bring the whole edifice of merit tumbling down and fashioned his arguments on the basis of a theory of the human mind conceded to have many strong arguments in its favor. While some theorists did unite experiential psychology with notions of physiological difference, these claims were at best highly disputed in the late eighteenth century. The safest approach to Helvétius, and the one most often taken, was to dismiss his ideas as chimerical, and then to move on to less troubling matters. Adams followed this solution in his letter to Jefferson, and Voltaire, Helvétius's one-time patron, did so even more pointedly in a letter to Prince Golitsin in 1773: "No one will convince me that all minds are equally suitable to science, and that they differ only

[85] For the details of the story of the response to *De l'esprit*, I have relied on the excellent account provided by Smith in *Helvétius* (cit. n. 83), especially pp. 1–3 and *passim*.

[86] Adams to Jefferson, 13 July 1813, *Adams-Jefferson Letters* (cit. n. 1), p. 355.

[87] See Jean-Jacques Rousseau, *Emile, or Education* (1762), trans. Eleanor Worthington (Boston: Ginn, Heath, & Co., 1883).

[88] On reactions to *De l'esprit* among philosophes, see Smith, *Helvétius* (cit. n. 83), pp. 157–71.

in regard to education. Nothing is more false: nothing is demonstrated more false by experience."[89] After the tumultuous events of the American and French revolutions, few authors of the time would have disagreed with Voltaire. Even radical republicans such as Paine and Wollstonecraft formulated their approaches to remaking the social order on decidedly different grounds.

TALENTS SOWN TO THE WIND: PAINE'S *RIGHTS OF MAN* AND THE ASSAULT ON ARISTOCRACY

Partisan of the American and French revolutions, one of the most noted political pamphleteers of his day, and a supporter of radical causes of all stripes, Thomas Paine carried out a tireless campaign near the end of the century against the power and, indeed, the very existence of hereditary aristocracy. Nonetheless, Paine did not follow Helvétius's lead and reject the concept of an elite entirely. Rather, committed to the idea that power and sovereignty resided in the people (however this entity was to be understood), Paine represented monarchic governments and social systems founded on hierarchies of birth as violating the ultimate principle of Enlightenment thought, rationality, and of doing so through a failure to take adequate account of one of the natural facts about human beings—the distribution of talents throughout the population. As Paine argued in his jeremiad against hereditary aristocracy, *The Rights of Man* (1791–1792):

> We have heard the Rights of Man called a levelling system; but the only system to which the word levelling is truly applicable is the hereditary monarchical system. It is a system of mental levelling. It indiscriminately admits every species of character to the same authority. Vice and virtue, ignorance and wisdom, in short, every quality, good or bad, is put on the same level. Kings succeed each other, not as rationals, but as animals. It signifies not what their mental or moral characters are.[90]

There was perhaps no greater calumny that Paine could have cast on hereditary monarchy than to suggest that the system was more associated with the practices of unreasoning brutes than with the methods of civilized and rational human beings. By so doing, he accentuated the strategic role that he accorded mental and moral characteristics in his argument because those attributes traditionally marked the dividing line between the human and animal worlds. Human mental attributes were valuable tools for Paine in the attack on aristocracy because, grounded in nature, they carried the highest legitimacy possible in the eighteenth century, and, independent of birth, they "proved" that hereditary systems of social organization were inadequate to the needs of state and society. In the process, Paine—and many other opponents of hereditary aristocracy as well—constructed a vision of government in which an individual's mental characteristics replaced birth and blood as the rhetorical criteria by which distributions of power and the restructuring of society could rationally be justified.[91]

[89] François Marie Arouet de Voltaire to Prince Dmitry Alekseevich Golitsin, 19 June 1773, *The Complete Works of Voltaire*, vol. 124: *Correspondence* (Oxfordshire: Voltaire Foundation, 1975), vol. 40, p. 29.

[90] Paine, *Rights of Man* (cit. n. 29), p. 172.

[91] As Alexander Hamilton observed in Federalist No. 68: "It will not be too strong to say that there will be a constant probability of seeing the station [the presidency] filled by characters pre-eminent for ability and virtue." Hamilton, Madison, and Jay, *Federalist Papers* (cit. n. 39), p. 414. See also Paine, *Rights of Man* (cit. n. 29), pp. 163–4.

To accomplish his goals of delegitimizing hereditary aristocracy on the basis of its violation of the principles of reason and of developing a new model for the structure of society, Paine imported into his analysis certain "truths" about two aspects of the natural history of human talents. First, he argued vociferously and repeatedly that mental and moral abilities did not follow any simple law of heredity and thus that there was no warrant for the belief that a good king in one generation would be likely to produce a talented monarch in the next:

> Could it be made a decree in nature, or an edict registered in heaven, and man could know it, that virtue and wisdom should invariably appertain to hereditary succession, the objections to it would be removed; but when we see that nature acts as if she disowned and sported with the hereditary system; that the mental characters of successors, in all countries, are below the average of human understanding; that one is a tyrant, another an idiot, a third insane, and some all three together, it is impossible to attach confidence to it, when reason in man has power to act.[92]

Second, Paine insisted that not only were the talents necessary to good government and a republican society many and varied, but from generation to generation they were scattered throughout all social classes, so that only a society that remained open to developing abilities wherever they lay could be deemed to be operating in concert with the dictates of reason and nature. "It appears to general observation, that revolutions create genius and talents; but those events do no more than bring them forward," Paine asserted. ". . . [T]he construction of government ought to be such as to bring forward, by a quiet and regular operation, all the extent of capacity which never fails to appear in revolutions."[93]

In essence, Paine combined his argument against aristocracy with an argument in favor of a form of social-political order in which he cast merit as the basis for justifying distributions of power and represented merit itself as tightly bound to particular understandings of the nature and distribution of individual talents. Paine painted the open, liberal, and nonaristocratic society that he envisioned as the logical response to a set of what he took to be natural facts about human beings: that human minds were replete with faculties, that the power of these faculties was largely a consequence of physiology, that powers of mind were rarely inherited, and that talents and abilities were distributed widely and unpredictably throughout the whole of a population. Human nature, properly understood, provided the vehicle by which he could dispense with hereditary monarchy and in its place erect a republican order open to all on the basis of merit, without losing entirely the sense that hierarchy and difference would persist.

"THE RANK THAT REASON ASSIGNED HER": WOLLSTONECRAFT AND THE VINDICATION OF WOMEN

Although closely allied with Paine and fully supportive of his attack on hereditary monarchy, Mary Wollstonecraft had a somewhat different view of the origin of talents and their distribution among the various social groups.[94] In developing her case

[92] Paine, *Rights of Man* (cit. n. 29), p. 173
[93] Ibid., p. 176.
[94] On Wollstonecraft's background, see Miriam Brody, introduction to *A Vindication of the Rights of Woman* (cit. n. 14), pp. 4–20; and Virginia Sapiro, *A Vindication of Political Virtue: The Political Theory of Mary Wollstonecraft* (Chicago: Univ. of Chicago Press, 1992), chap. 1.

for the extension to women of the claims of republican citizenship and the full rights of participation within civil society—articulated most forcefully in *A Vindication of the Rights of Woman*—Wollstonecraft argued from a kind of middle position, employing faculty psychology and an understanding of talents that conceived of them as shaped by both nature and education.

Wollstonecraft opened *Vindication* by conceding that most women, in their current degraded state, were generally frivolous, coquettish, and cunning, exhibiting few of the intellectual abilities or virtues required of active citizens in a republic.[95] She maintained, however, that this state of affairs had come to pass largely because women had been trained to act in such a manner; with educations similar to those of men, she argued, women would exhibit virtues and talents at least similar in kind:

> I will allow that bodily strength seems to give man a natural superiority over woman; and this is the only solid basis on which the superiority of the sex can be built. But I still insist that not only the virtue but the *knowledge* of the two sexes should be the same in nature, if not in degree, and that women, considered not only as moral but rational creatures, ought to endeavour to acquire human virtues (or perfections) by the *same* means as men, instead of being educated like a fanciful kind of *half* being—one of Rousseau's wild chimeras.[96]

Women had this potential, Wollstonecraft contended, because all members of the human race were, by definition, endowed with the same types of mental faculties. The question of difference was solely one of degree and could be determined only by providing women and men with an identical education. Once women were able to exercise their mental faculties to the fullest, she argued, then and only then would it be possible to determine the types of political rights and social roles appropriate for them. Wollstonecraft even accepted the possibility, at least rhetorically, that women might prove the inferior of men in those characteristics of greatest importance to a republican society. She maintained only that women deserved the right to find out: "Further, should experience prove that they [women] cannot attain the same degree of strength of mind, perseverance, and fortitude, . . . woman would then only have the rank that reason assigned her, and arts could not be practised to bring the balance even, much less to turn it."[97]

Both here and throughout *Vindication*, Wollstonecraft accepted without reservation the position that at least certain rights and opportunities should be consequent on the degree of an individual's or group's virtues and talents.[98] In addition to illustrating the extent to which civic and political rights had become linked with assessments of mental capacity, however, her exploration of the potentials of the language of merit for justifying a role for women in civil society reveals much about the

[95] One of Wollstonecraft's first rhetorical moves was to represent military officers as "idle superficial young men" because of the nature of their educations, and then to assert that women and officers were in completely similar states. See Wollstonecraft, *Vindication*, pp. 97, 104–8, also chap. 4. Brody remarks on this feature of Wollstonecraft's argument in introduction to *Vindication* (both cit. n. 14), pp. 44–5.

[96] Wollstonecraft, *Vindication* (cit. n. 14), p. 124.

[97] Ibid., p. 121.

[98] As Wollstonecraft asserted early in her discourse: "That the society is formed in the wisest manner, whose constitution is founded on the nature of man, strikes, in the abstract, every thinking being so forcibly, that it looks like presumption to endeavour to bring forward proof." Ibid., p. 92.

ways in which the discourse of merit could be employed to construct differences and distinctions. Whether in absolute terms (exemplified by their use in the *exclusion* of women from political power) or in relative terms—illustrated by Wollstonecraft's willingness to concede the justice of a social hierarchy established on rational principles—"natural" differences among groups and individuals became freighted with political meaning, and the distinction between nature and nurture began to take on powerful significance. For the core of Wollstonecraft's argument was that women, and any other group or indeed any individual, deserved just that rank in society consonant with the mental capacities with which they had been endowed.

Vindication, in fact, leaves little doubt that Wollstonecraft believed women to be equal to men in terms of native intellectual abilities. Indeed, she reminded her readers of the existence of prominent women whose skills were so manifest as to require no further justification of their possession of faculties the equal of any male.[99] But by couching her argument in terms of native endowments and faculties able to differ in their power by degree, she left open the possibility that new truths about the actual potentials of various types of human intellects could have significant political and social repercussions. If women's mental faculties were indeed proven to be physiologically inferior, for example, then there would be strong grounds, according to the logic of Wollstonecraft's argument, to limit women's role in political and civil life. The basis for this possibility of exclusion lay in her willingness, at least rhetorically, to make the facts about human nature, whatever they might be determined to be, stand as the final arbiter. Given her own middle position on the relative roles of nature and education in producing talents, she left open to question how far education could remedy inequities in the distribution of talents and thus how egalitarian a republican society could be.

CONCLUSION

The responses of Paine and Wollstonecraft to Helvétius's "provocation" illustrate the enormous number of subterranean fault lines that radiated around the concept of "talents" and its place in the Enlightenment vision of a republican society. The ambiguities surrounding "talents" gave the term enormous power and resonance. It could be used simultaneously to legitimate the persistence of social and occupational distinctions and to validate the establishment of broad-based educational systems designed to discover or foster talent across the very lines of those distinctions. It provided both a language in which to argue for greater political and social power and a means to exclude whole groups of people from all but the most basic rights. It helped hold together a system that seemed to offer opportunity to all and yet justified restricting those opportunities to only some. In other ways, however, the ambiguities surrounding "talents" presented real problems. By leaving the concept ill defined, the line determining where universal rights ended and the privileges of talent began was open to continual renegotiation. And by according mental characteristics such a prominent place in the imagined republic, the possibility was left open that changes in theories about the mind could have significant, and largely uncontrollable, effects on how the social-political order was understood and structured. As long as the system of governance remained justified in terms of claims about human nature and the

[99] See, e.g., ibid., pp. 119, 172.

illuminations of reason, it was almost necessarily linked to the knowledge systems in which such entities were given definition and meaning.

In this, Jefferson and Adams can be seen to have been looking forward as well as back. Their vision of an aristocracy of intellect, while certainly risible in America of the early twenty-first century, nonetheless captured an element in republican culture that would continue to attract the energies of political thinkers and the interest of human scientists, not to mention the passions of reformers of all stripes. Jefferson and Adams identified, if nothing else, a language through which outsiders could claim inclusion and experts positions of power and authority within a democratic system of governance. What is more, their sense that human differences were of legitimate social relevance solely to the degree they were derived from nature (because nature alone could be seen to stand outside of the partisan conflict of personal interests) meant that those theorists able to claim persuasively the ability to interpret nature would conceivably be in positions to have significant social impact. By the early twentieth century, as is well known, one such group was the army of mental testers who fanned out across the United States, offering to reorganize the social order on the basis of a rationalized distribution of a number of social goods—ranging from educational opportunities to employment possibilities—legitimated through their presumed ability to objectively assess what they argued was the most fundamental human talent, intelligence. That their claims to be able to measure merit in this way were, and continue to be, taken seriously is one small indication that the world imagined by Jefferson and Adams at the end of their lives did not completely disappear and that the desire to counterbalance a commitment to equality with a conception of merit rooted as much in science as in politics persisted. Once woven into the core of republican political theory, the science of human nature would prove difficult to remove.

EXPANSIONS AND REFORMS: THE NINETEENTH CENTURY

Science, Politics, and Religion:
Humboldtian Thinking and the Transformations of Civil Society in Germany, 1830–1870

By Andreas W. Daum*

ABSTRACT

Between 1830 and 1870 the infrastructure of civil society in Germany expanded significantly. In spite of the failure of the 1848 revolution, the culture of the *Bürgertum*, the middle classes, began to flourish through associations, educational efforts, and diverse media and institutions of public life. The study of natural history and the new natural sciences became an integral part of this seminal process. This essay examines the development of civil society through the interplay of science, religion, and politics, which is paradigmatically embodied in the biography of Emil Adolf Rossmässler. As a science teacher, free-religious activist, and democratic politician, Rossmässler became one of Germany's preeminent advocates of science as a civil practice, though one who acted outside university science and the mainstream of academia. Rossmässler and many of his peers understood the study of nature as a democratic exercise and sought to integrate the natural sciences into the German concept of education (*Bildung*). Instead of associating themselves with the philosophical materialism of the time, however, they drew heavily on the thinking of Alexander von Humboldt and prolonged a reconciliatory understanding of nature into the public sphere. A look at their efforts reveals the richness and diversity of a civil culture of nature studies outside the realm of state-supported science and university research.

INTRODUCTION

IN 1830 POLITICAL UNREST swept over Europe. Barricade fights broke out in the streets of Paris at the end of July, and public protest forced the Bourbon king to abdicate. These events questioned the setting of monarchical rule that had been restored at the Congress of Vienna in 1814–1815, and they reverberated throughout Europe, stimulating similar movements from Spain to the Russian border. Though many of the revolutionary efforts failed in the short run, their underlying ideas of democracy continued to survive among the bourgeoisie, artisans, and workers.[1] The

* Minda de Gunzburg Center for European Studies, Harvard University, 27 Kirkland St., Cambridge, MA 02138; adaum@fas.harvard.edu.

My thanks go to Tom Broman, Paul Lerner, Lynn Nyhart, Kathy Olesko, and the anonymous *Osiris* reviewer for their useful suggestions and constructive feedback while writing this article. I also would like to thank the participants of the History of Science Colloquia of Timothy Lenoir at Stanford University and Roger Hahn at the University of California at Berkeley, where I was kindly given the opportunity to present and discuss some of the arguments in this article.

[1] For the place of the Paris revolution of July 1830 in the history of Europe and its repercussions in various European states, particularly Germany, see Eric Hobsbawm, *The Age of Revolution, 1789–1848* (New York: Random House, 1996), pp. 109–31; James J. Sheehan, *German History, 1770–1866*

foundations of the monarchist system swayed under the pressure of democratic demands, which were seen by many as preconditions for the realization of a civil society: the quest for individual freedom and a rule of law, for constitutional government and a participatory model of politics, for a market-oriented economy and an abolition of social privileges based on tradition and birth.

In the German-speaking territories, the realization of civil society still seemed a utopian project around 1830.[2] Too segmented were the social strata that would potentially encompass the ideals of such a society, too diverse were their political interests, and too fragmented was the social formation of the *Bürgertum*, the middle classes including the bourgeoisie as its core group, whose social position was based on economic position and educational merits. In spite of these factors, the set of ideas that had been laid out by social thinkers in the late eighteenth century, refined by early reform movements in the wake of the French Revolution, and expressed in the revolutionary movements around 1830 would continue to reverberate in Germany in the coming decades. Political energies aiming at materializing the ideals of a civil society—and thereby to create a society based on liberal norms, practices, and institutions—could, as historian James Sheehan has written, "be frustrated, inhibited, or displaced, but they could not be destroyed."[3] One stage on which the ongoing quest for a civil society unfolded between 1830 and 1870 emerged from a peculiar interplay of science, politics, and religion. A look at one exemplary biography, that of the naturalist and democratic politician Emil Adolf Rossmässler (1806–1867), and an analysis of his activities as an agent of civil society both illuminate and explain the character and consequences of this interplay.

A LIFE IN THE AGE OF CIVIL SOCIETY: EMIL ADOLF ROSSMÄSSLER

In 1830, the same year Charles X was forced off the French throne, a young man in the southeastern German kingdom of Saxony became interested in politics. Emil Adolf Rossmässler, born in 1806 as the son of a copperplate engraver, was just at

(Oxford and New York: Oxford Univ. Press, 1994), pp. 604–53; Thomas Nipperdey, *Germany from Napoleon to Bismarck, 1800–1866* (Princeton, N.J.: Princeton Univ. Press, 1996), pp. 323–33.

[2] The term "civil society," or *bürgerliche Gesellschaft*, has always consisted of empirical characteristics as well as normative assignments; as such it has never ceased to bear some utopian meaning. That is why I am using here and in the following the term "project" in connection with civil society. For the history of this and related terms in the German context, see Manfred Riedel, "Bürger, Staatsbürger, Bürgertum," in *Geschichtliche Grundbegriffe: Historisches Lexikon zur politisch-sozialen Sprache in Deutschland*, 8 vols., ed. Otto Brunner, Werner Conze, and Reinhart Koselleck (Stuttgart: Klett-Cotta, 1974), vol. 1, pp. 672–725; Utz Haltern, *Bürgerliche Gesellschaft: Sozialtheoretische und sozialhistorische Aspekte* (Darmstadt: Wissenschaftliche Buchgesellschaft, 1985); Lutz Niethammer, "Einführung: Bürgerliche Gesellschaft als Projekt," in *Bürgerliche Gesellschaft in Deutschland: Historische Einblicke, Fragen, Perspektiven* (Frankfurt am Main: Fischer, 1990), pp. 17–38; *Bürgerschaft: Rezeption und Innovation der Begrifflichkeit vom Hohen Mittelalter bis ins 19. Jahrhundert*, ed. Reinhard Koselleck and Klaus Schreiner (Stuttgart: Klett-Cotta, 1994).

[3] Sheehan, *German History* (cit. n. 1), p. 621. For some general accounts of the history of civil society and the bourgeoisie in Germany, see Hans-Ulrich Wehler, *Deutsche Gesellschaftsgeschichte*, 3 vols., vol. 2: *Von der Reformära bis zur industriellen und politischen 'Deutschen Doppelrevolution,' 1815–1845/49*, 2d ed. (Munich: Beck 1989), pp. 174–241; David Blackbourn and Richard J. Evans, eds., *The German Bourgeoisie: Essays on the Social History of the German Middle Class from the Late Eighteenth to the Early Twentieth Century* (London: Routledge, 1991); Wolfgang Hardtwig, *Nationalismus und Bürgerkultur in Deutschland, 1500–1914* (Göttingen: Vandenhoeck & Ruprecht, 1994), pp. 34–190; Jürgen Kocka, "Das europäische Muster und der deutsche Fall," in *Bürgertum im 19. Jahrhundert*, 3 vols., vol. 1: *Einheit und Vielfalt Europas*, ed. Jürgen Kocka (Göttingen: Vandenhoeck & Ruprecht, 1995), pp. 9–84.

the stage of his life in which he was turning his lay interests in natural history into a profession. In 1827 Rossmässler had abandoned the study of theology at the University of Leipzig, Saxony's largest city, after only two years to become a teacher at a small school in nearby Thuringia. This position had allowed him enough time to devote himself to his favorite subject, botanical studies. In 1829, Rossmässler had successfully applied for a vacant teaching position at the Saxon Forest Academy in Tharandt[4] and had been assigned to instruct the students in zoology, later to be complemented by botany. Four months after Rossmässler's official installation, in September 1830, the revolutionary wave surging over Europe seized the kingdom of Saxony. In Leipzig, young craftsmen revolted against the police, and in the city of Dresden activists stormed the city hall with the *Marseillaise* on their lips. Rossmässler, "deeply shaken to my core" ("im Innersten aufgerüttelt"), immediately left for Dresden to witness the events.[5]

Though focused on academic teaching and writing over the course of the next eighteen years, Rossmässler remained affected by the ideals of the 1830 revolutionary movement and never lost touch with contemporary politics. He slowly began to develop his own ties to democratic forces in Saxony, becoming part of a growing communication network that articulated political concerns through associations, festivals, newspapers, and journals. In the coming years, this network, which emerged from local settings, would extend outward to the regional level and eventually become a national political discourse in 1848.[6] Despite his involvement, Rossmässler was not a significant political talent, nor would his political interests ever outweigh his scientific ambitions. From the 1830s on, however, political and scientific thinking merged in his worldview, a conflation both facilitated and enforced, in a way, by a specific religious affiliation. This merger marked Rossmässler's thinking and activities until his death in 1867.

The German revolution of 1848 turned the interplay of science, politics, and religion into a public affair. Many of Rossmässler's contemporaries shared his fundamental belief that any sound societal order had to be grounded in a scientific view of the world. According to this conviction, religious devotion had to be channeled away from the institutionalized churches and—through the mediation of science teachers—directed toward the admiration of the natural world. Consequently, freethinking and politically engaged scientists such as Rossmässler tried to create democratic political structures by cultivating nature studies. Their hero was a man whose affinities for democratic and liberal ideas were already well known, but whose close affiliation with leading circles of established scientists in Europe and with the royal court of the Prussian king made him, on first view, a rather surprising model to

[4] William R. Lazenby, "The Forests and Forestry of Germany," *Popular Science Monthly* 83 (1913): 590–8.

[5] According to Rossmässler's autobiography, *Mein Leben und Streben im Verkehr mit der Natur und dem Volke*, ed. Karl Ruß (Hannover: C. Rümpler, 1874), p. 60. For the repercussions of the 1830 revolutionary movement in Saxony and the state of the Saxon kingdom in this period, see James Retallack, ed., *Saxony in German History: Culture, Society, and Politics, 1830–1933* (Ann Arbor: Univ. of Michigan Press, 2000); and Michael Hammer, *Volksbewegung und Obrigkeiten: Revolutionen in Sachsen, 1830/31* (Weimar: Böhlau, 1997). Rossmässler himself wrote his name with two German *ßs* (Roßmäßler). However, even among contemporaries and in bibliographies and library catalogs the spelling varies. To facilitate the spelling for tracing Rossmässler in both German and non-German catalogs, I therefore stay with the spelling "Rossmässler."

[6] See Wolfram Siemann, *Vom Staatenbund zum Nationalstaat: Deutschland, 1806–1871* (Munich: Beck, 1995), pp. 204–49.

emulate: Alexander von Humboldt (1769–1859), cosmopolite, traveler, and Germany's towering figure in the field of natural sciences.[7]

Imbued with the spirit of Humboldtian thinking, Rossmässler and those sympathetic to his ideas left a remarkable imprint on postrevolutionary cultural and intellectual life in Germany, despite numerous obstacles and even open resistance by state and church authorities. They helped launch a broad movement to make science popular among the German population, although the legitimacy and potential intellectual authority of this field remained highly contested.[8] The growing infrastructure of civil society shaped the efforts of Rossmässler and others; their activities, in return, served to reinforce the very structures and ideological trajectories of this society: the emphasis on the value of education, the goal of enhancing an understanding of self-determination of individuals as well as social groups, the spread of free associations and print media, and the expansion of the public sphere as a crucial platform for civil discourse.[9]

The following article takes the life of Emil Adolf Rossmässler as a biographical prism for examining the interplay of science, politics, and religion as part of the project of civil society. Although far behind many university scientists in public reputation, Rossmässler was well known among contemporaries, yet later historians have barely paid attention to him.[10] Casting a fresh look at such a figure gives us the

[7] The most reliable account of Humboldt's life is still Hanno Beck, *Alexander von Humboldt*, 2 vols. (Wiesbaden: F. Steiner, 1959–1961). Humboldt's writings have not yet been made accessible in a comprehensive edition; some of his important works can be reread through the edition Hanno Beck, ed., *Alexander von Humboldt: Studienausgabe*, 7 vols. (Darmstadt: Wissenschaftliche Buchgesellschaft, 1987–1997). For modern assessments of the meaning of Humboldt for the nineteenth century and our time, see Michael Dettelbach, "Humboldtian Science," in *Cultures of Natural History*, ed. N. Jardine, J. A. Secord, and E. C. Spary (Cambridge: Cambridge Univ. Press, 1996), pp. 287–304; Nicolaas A. Rupke, "Humboldtian Medicine," *Med. Hist.* 40 (1996): 293–310; and Andreas W. Daum, "Alexander von Humboldt, die Natur als 'Kosmos' und die Suche nach Einheit: Zur Geschichte von Wissen und seiner Wirkung als Raumgeschichte," *Ber. Wissenschaftsgesch.* 23 (2000): 243–68.

[8] It is only in recent years that the development of popular science has become a topic for historians of Germany. See Alfred Kelly, *The Descent of Darwin: The Popularization of Darwinism in Germany, 1860–1914* (Chapel Hill: Univ. of North Carolina Press, 1981); Kurt Bayertz, "Spreading the Spirit of Science: Social Determinants of the Popularization of Science in Nineteenth-Century Germany," in *Expository Science: Forms and Functions of Popularisation*, ed. Terry Shinn and Richard Whitley (Dordrecht and Boston: D. Reidel, 1985), pp. 209–27; Andreas W. Daum, *Wissenschaftspopularisierung im 19. Jahrhundert: Bürgerliche Kultur, naturwissenschaftliche Bildung und die deutsche Öffentlichkeit, 1848–1914* (Munich: Oldenbourg, 1998); Angela Schwarz, *Der Schlüssel zur modernen Welt: Wissenschaftspopularisierung in Großbritannien und Deutschland im Übergang zur Moderne (ca. 1870–1914)* (Stuttgart: Franz Steiner, 1999).

[9] See Jürgen Habermas, *The Structural Transformation of the Public Sphere: An Inquiry into a Category of Bourgeois Society* (Cambridge, Mass.: MIT Press, 1989). For a critical discussion of Habermas, see the essays in Craig Calhoun, ed., *Habermas and the Public Sphere* (Cambridge, Mass., and London: MIT Press, 1992); and Belinda Davis, "Reconsidering Habermas, Gender, and the Public Spheres: The Case of Wilhelmine Germany," in *Society, Culture, and the State in Germany, 1870–1930*, ed. Geoff Eley (Ann Arbor: Univ. of Michigan Press, 1996), pp. 397–426.

[10] Rossmässler has been treated in several accounts of the history of education and the early workers' movements in Germany, but there is neither a satisfying biography nor a modern monograph dealing with his main fields of activity. For a balanced overview, see Burghard Burgemeister, "Emil Adolf Rossmässler: Ein demokratischer Pädagoge, 1806–1867," Ph.D. thesis, Humboldt University, Berlin, 1958; and Karl-Heinz Günther, *Bürgerlich-demokratische Pädagogen in Deutschland während der zweiten Hälfte des 19. Jahrhunderts* (Berlin: Volk & Wissen, 1963); for the early decades of Rossmässler's life, see Andreas W. Daum, "Emil Adolf Rossmässler als Professor in Tharandt von 1830 bis 1848," *Wissenschaftliche Zeitschrift der Technischen Universität Dresden* 42 (1993): 59–66. Still the most comprehensive source is Rossmässler's autobiography, *Mein Leben und Streben* (cit. n. 5), which was first published as a sequence of articles in his journal *Aus der Heimath*.

opportunity to follow up on recent attempts to understand the culture of a historical period not only by looking at its eminent intellectuals, but also by examining the nonelite thinkers and more popular writings.[11] Rossmässler can serve as a paradigmatic example that opens a window into the state of science and civil society between 1830 and 1870. His biography reflects in a uniquely telling way how science, politics, and religion interacted in mid-nineteenth-century Germany.[12]

BETWEEN *BÜRGERTUM* AND *BILDUNG*: ARGUMENTS ABOUT GERMAN HISTORY FROM 1830 TO 1870

The years between 1830 and 1870, which coincided with Rossmässler's professional life, have long been accepted as a crucial period of social transformation and intellectual change in Germany as well as in Europe as a whole.[13] Most of central Europe and England witnessed the breakthrough of industrialization, the rise of rationalistic, positivistic, and scientific thinking, and the emergence of the middle classes as agents of social change. These processes were accompanied by a rise in the working-class population and the persistent presence of the aristocracy in some areas of society. The precise connection between these processes and their overall relationship to political democratization and what has often been called modernization have long been disputed, nowhere more so than in the case of Germany. The potentials of civil society in Germany lie at the core of these debates.

In the debate about Germany's road to modernity, the classic argument has been that Germany failed to establish the key values of civil ideology because it lacked a strong social formation, a *Bürgertum* that would have been able to achieve the breakthrough of civil society (*bürgerliche Gesellschaft*). This was particularly the view after 1933 (and, indeed, after 1945) when historians and political scientists tried to explain the rise of National Socialism by looking for the long-term tendencies in modern German history that eventually led to Hitler. In Germany, they claimed, unlike in other Western countries, a gap widened over the course of the nineteenth century between rapid economic modernization and political democratization, which allowed state authorities, the military, and old elites to dominate the political arena and deny claims for participation raised by the middle classes and workers. This vision of a German *Sonderweg*, a "special path," into modernity began to be powerfully challenged by a new generation of historians. They rejected the notion of a model trajectory toward civil society based on the English example and warned against seeing the German revolution of 1848 as a failure and a mere indicator of bourgeois (*bürgerlich*) weakness. From different standpoints, David Blackbourn, Geoff Eley, and Thomas Nipperdey argued that throughout the nineteenth

[11] As examples for such an approach, see Fritz Stern, *The Politics of Cultural Despair* (Berkeley: Univ. of California Press, 1961); as well as Klaus Vondung, ed., *Das wilhelminische Bildungsbürgertum: Zur Sozialgeschichte seiner Ideen* (Göttingen: Vandenhoeck & Ruprecht, 1976); Rudolf Schenda, *Volk ohne Buch: Studien zur Sozialgeschichte der populären Lesestoffe, 1770–1910*, 3d ed. (Frankfurt am Main: Klostermann, 1988); Adrian Desmond, *The Politics of Evolution: Morphology, Medicine, and Reform in Radical London* (Chicago: Univ. of Chicago Press, 1989).

[12] This focus on Rossmässler as a paradigmatic figure in order to delineate the character of civil society in Germany differs, therefore, from the systematic treatment of popular science in German history that I have suggested elsewhere; see Daum, *Wissenschaftspopularisierung* (cit. n. 8).

[13] The best overview of German history across the century is provided by David Blackbourn, *The Long Nineteenth Century: A History of Germany, 1780–1918* (Oxford and New York: Oxford Univ. Press, 1998).

century Germany, rather than going through a backward process of feudalization, experienced a rise in civil values, attitudes, and societal practices—that is, a large-scale process of embourgeoisement, or *Verbürgerlichung*.[14]

Since then, various research efforts have scrutinized the complex fabric of German civil society and set it in comparative perspectives.[15] The classic argument about a "special path" has given way to a far more nuanced view. Recent accounts have shed a much more positive light on the strengths of German *Bürgertum* and the power of the process of *Verbürgerlichung*. Given the heterogeneous social composition of the German middle class, however, historiography now tends to turn away from static social definitions of civil society and to view it more as a set of intellectual attitudes, material lifestyles, and cultural dispositions—that is, to see civil society as a social and cultural practice, an attitude of *Bürgerlichkeit*. Among its characteristics are the valorization of work and a rational and disciplined way of life; an appreciation of the family as the basis of that civil life; respect for high culture, the arts, and education (*Bildung*); the pursuit of the rule of law and urban prosperity; membership in free associations as vehicles for social self-organization and intellectual activity; and the free articulation of political beliefs in the media of the public sphere.[16] This in turn facilitates an examination of the place of science, politics, and religion in the context of civil society in Germany.

Between 1830 and 1870 contemporaries primarily conceived of "science" as the natural sciences (*Naturwissenschaften*), a field only beginning to find its institutionalization in different branches at German universities.[17] Science encompassed

[14] See David Blackbourn and Geoff Eley, *The Peculiarities of German History: Bourgeois Society and Politics in Nineteenth-Century Germany* (Oxford and New York: Oxford Univ. Press, 1984); and Nipperdey, *Germany from Napoleon to Bismarck* (cit. n. 1); idem, "War die Wilhelminische Gesellschaft eine Untertanen-Gesellschaft?" and "1933 und die Kontinuität der deutschen Geschichte," in *Nachdenken über die deutsche Geschichte: Essays* (Munich: Beck, 1986), pp. 172–205. The most recent revisionist study, convincingly arguing that pre-1914 German society had a great democratic potential, is Margaret Lavinia Anderson, *Practicing Democracy: Elections and Political Culture in Imperial Germany* (Princeton, N.J.: Princeton Univ. Press, 2000). For an overview of the complex debate triggered by Blackbourn and Eley, see Helga Grebing, *Der "deutsche Sonderweg" in Europa, 1806–1945: Eine Kritik* (Stuttgart: Kohlhammer, 1986); Wolfgang Hardtwig, "Der deutsche Weg in die Moderne: Die Gleichzeitigkeit des Ungleichzeitigen als Grundproblem der deutschen Geschichte, 1789–1871," in *Deutschlands Weg in die Moderne: Politik, Gesellschaft und Kultur im 19. Jahrhundert*, ed. Wolfgang Hardtwig and Harm-Hinrich Brandt (Munich: Beck, 1993), pp. 9–31; Hans-Ulrich Wehler, *Deutsche Gesellschaftsgeschichte*, 3 vols., vol 3: *Von der "Deutschen Doppelrevolution" bis zum Beginn des Ersten Weltkrieges, 1849–1914* (Munich: Beck, 1995), pp. 449–86.

[15] Research on the German *Bürgertum* has been enormously intensified since the 1980s and has produced a wealth of publications; for an overview see Utz Haltern, "Die Gesellschaft der Bürger," *Geschichte und Gesellschaft* 19 (1993): 100–34; Lothar Gall, *Von der ständischen zur bürgerlichen Gesellschaft* (Munich: Oldenbourg, 1993); and Jonathan Sperber, "*Bürger, Bürgertum, Bürgerlichkeit, Bürgerliche Gesellschaft*: Studies of the German (Upper) Middle Class and Its Sociocultural World," *J. Mod. Hist.* 69 (1997): 271–97. Research on the peculiar educational ideology of German *Bürgertum* and the meaning of *Bildung* has culminated in the four-volume series *Bildungsbürgertum im 19. Jahrhundert* (Stuttgart: Klett-Cotta, 1985–1992).

[16] See the articles by M. R. Lepsius, "Zur Soziologie des Bürgertums und der Bürgerlichkeit," and Thomas Nipperdey, "Kommentar: 'Bürgerlich' als Kultur," in *Bürger und Bürgerlichkeit im 19. Jahrhundert*, ed. Jürgen Kocka (Göttingen: Vandenhoeck & Ruprecht, 1987), pp. 79–100, and 143–8; Kocka, "Das europäische Muster" (cit. n. 3); idem, "Obrigkeitstaat und Bürgerlichkeit. Zur Geschichte des deutschen Bürgertums im 19. Jahrhundert," in *Deutschlands Weg in die Moderne* (cit. n. 14), pp. 110–1.

[17] Everett Mendelsohn, "The Emergence of Science as a Profession in Nineteenth-Century Europe," in *The Management of Scientists*, ed. Karl Hill (Boston: Beacon, 1964), pp. 3–48; R. Steven Turner, "German Science, German Universities: Historiographical Perspectives from the 1980s," in *'Einsamkeit und Freiheit' neu besichtigt: Universitätsreformen und Disziplinbildung in Preussen als*

emerging disciplines such as chemistry and physics, but still left much room for traditional natural history (*Naturgeschichte*), whose branches (particularly zoology, botany, and geology) were only slowly developing into distinct scientific disciplines. One of the main challenges for promoters of both natural history and the natural sciences was how to cast these fields within the German conception of education. Harking back to the eighteenth century and the post-1800 educational reforms that led to the founding of the University of Berlin and the establishment of the *Gymnasium* as the main vehicle for higher education, the concept of *Bildung* relied particularly on the historical, philosophical, and philological disciplines.[18] *Bildung* primarily meant intellectual and moral self-cultivation and was separated from practical education and mere utilitarian purposes. Contemporary terminology reflected this juxtaposition in the dichotomy between humanism and realism; the latter placed a heavy burden on the promoters of education in natural sciences, forcing them to struggle against all kinds of negative stereotypes. Moreover, from the perspectives of the state and conservative politicians, utilitarianism was believed to have politically dangerous and morally perilous effects, an argument that made the situation even more difficult for scientists.[19] These were the obstacles Rossmässler faced as he sought to make science an integral part of civil life.

The political stigmatizing of the natural sciences indicates that the realm of political discourses was, in fact, fundamentally broadened during the mid-century. This trend, which originated in the Enlightenment epoch, gained strength after 1848, when a distinct proletarian society and culture began to take shape.[20] They were heavily affected by a growing interest in education as a means of achieving individual freedom and enhancing the social progress of the proletarian population. But the early workers' movement was not as attached to the ideals of humanism as the defenders of the ideology of *Bildung* were. On the contrary, the movement put special emphasis on the study of natural sciences,[21] thus creating alliances between proletarians and those members of the middle classes who saw the natural sciences as a

---

Modell für Wissenschaftspolitik im Europa des 19. Jahrhunderts, ed. Gert Schubring (Stuttgart: F. Steiner, 1991), pp. 24–36; Kathryn M. Olesko, ed., *Science in Germany: The Intersection of Institutional and Intellectual Issues, Osiris*, 2d. ser., 5 (1989); idem, *Physics as a Calling: Discipline and Practice in the Königsberg Seminar for Physics* (Ithaca, N.Y., and London: Cornell Univ. Press, 1991); Arleen Marcia Tuchman, "Institutions and Disciplines: Recent Work in the History of German Science," *J. Mod. Hist.* 69 (1997): 298–319.

[18] For the meaning of the concept of *Bildung* in German history and its link to the middle classes, the *Bildungsbürgertum*, see Rudolf Vierhaus, "Bildung," in *Geschichtliche Grundbegriffe* (cit. n. 2), vol. 1, pp. 508–51; Ulrich Engelhardt, *"Bildungsbürgertum": Begriffs- und Dogmengeschichte eines Etiketts* (Stuttgart: Klett-Cotta, 1986); Rüdiger vom Bruch, "Gesellschaftliche Funktionen und politische Rollen des Bildungsbürgertums im Wilhelminischen Reich: Zum Wandel von Milieu und politischer Kultur," in *Politischer Einfluß und gesellschaftliche Formation*, ed. Jürgen Kocka (Stuttgart: Klett-Cotta, 1989), pp. 146–79; Aleida Assmann, *Arbeit am nationalen Gedächtnis: Eine kurze Geschichte der deutschen Bildungsidee* (Frankfurt am Main: Campus, 1993); Georg Bollenbeck, *Bildung und Kultur: Glanz und Elend eines deutschen Deutungsmusters* (Frankfurt am Main and Leipzig: Insel, 1994).

[19] See Otto Brüggemann, *Naturwissenschaft und Bildung: Die Anerkennung des Bildungswertes der Naturwissenschaften in Vergangenheit und Gegenwart* (Heidelberg: Quelle & Meyer, 1967); and Manfred Eckert, *Die schulpolitische Instrumentalisierung des Bildungsbegriffs: Zum Abgrenzungsstreit zwischen Realschule und Gymnasium im 19. Jahrhundert* (Frankfurt am Main: Fischer, 1984).

[20] Thomas Welskopp, *Das Banner der Brüderlichkeit: Die deutsche Sozialdemokratie vom Vormärz bis zum Sozialistengesetz* (Bonn: Dietz Nachfolger, 2000).

[21] Karl Birker, *Die deutschen Arbeiterbildungsvereine, 1840–1870* (Berlin: Colloquium, 1973); Gerhard A. Ritter, "Arbeiterkultur im deutschen Kaiserreich: Probleme und Forschungsansätze," in *Arbeiter, Arbeiterbewegung und soziale Ideen in Deutschland: Beiträge zur Geschichte des 19. und*

proper field for enhancing understanding between different social groups. One of the main goals of bourgeois social reformers and liberals was to incorporate proletarians into civil society by means of education, and Rossmässler himself became one of the primary advocates of this strategy. These kinds of alliances helped keep the lines between proletarian and bourgeois cultures blurred until the end of the 1860s.[22]

As to the question of religion in this equation, historiography, particularly in the case of Germany, is still approaching the topic of science in culture mainly from the perspective of the *Kulturkampf*. Most accounts structure their observations along a dichotomous model that is based on the ardent conflicts between progressive liberals on the one side and conservative followers of the—primarily Catholic—church on the other side, embodying the contemporary paradigms of modern education and ecclesiastical stubbornness. Recent studies, however, have corrected this dualistic view in regard to British and North American history and have underlined those historical instances in which science and religion reacted to one another in a more differentiated manner, leaving room for mutual respect, reconciliation, or even convergence.[23] Similarities can be found in nineteenth-century Germany. Pluralization in religious life, which would reach its peak around 1900, had already led to a diversity of religious attitudes during midcentury.[24] The diverging religious positions either remained within the confines of the established denominations, mainly those of the Protestant and Catholic Churches, or found expression in dissenting, nondenominational, and nonconformist movements that rejected organized religion but nevertheless displayed religious attitudes or even created new religiously-imbued ideologies.[25] Amid these changes, scientific thinking could and did, as this article argues, find an integral place in the diverging strands of religious thinking and even infiltrated the dominant denominations.

Against this background, I would like to develop two arguments meant to open

---

20. Jahrhunderts (Munich: Beck, 1996), pp. 113–30; Kurt Bayertz, "Naturwissenschaft und Sozialismus: Tendenzen der Naturwissenschafts-Rezeption in der deutschen Arbeiterbewegung des 19. Jahrhunderts," *Soc. Stud. Sci.* 13 (1983): 355–94.

[22] See Rüdiger vom Bruch, ed., *Weder Kommunismus noch Kapitalismus: Bürgerliche Sozialreform in Deutschland vom Vormärz bis zur Ära Adenauer* (Munich: Beck, 1985); and Dieter Langewiesche, *Liberalismus in Deutschland* (Frankfurt am Main: Suhrkamp, 1988).

[23] See, e.g., Walter J. Wilkins, *Science and Religious Thought: A Darwinism Case Study* (Ann Arbor: UMI Research Press, 1987); Ronald Numbers, "Science and Religion," *Osiris*, 2d. ser., 1 (1985): 59–80; David C. Lindberg and Ronald L. Numbers, eds., *God and Nature: Historical Essays on the Encounter between Christiantity and Science* (Berkeley: Univ. of California Press, 1986); John Hedley Brooke, *Science and Religion: Some Historical Perspectives* (Cambridge and New York: Cambridge Univ. Press, 1991); Frank M. Turner, *Contesting Cultural Authority: Essays in Victorian Intellectual Life* (Cambridge and New York: Cambridge Univ. Press, 1993), pp. 1–37, 171–200.

[24] Thomas Nipperdey, *Religion im Umbruch: Deutschland 1870–1918* (Munich: Beck, 1988); Kurt Nowak, *Geschichte des Christentums in Deutschland: Religion, Politik und Gesellschaft vom Ende der Aufklärung bis zur Mitte des 20. Jahrhunderts* (Munich: Beck, 1995), pp. 181–5; Rüdiger vom Bruch, Friedrich Wilhelm Graf, and Gangolf Hübinger, eds., *Kultur und Kulturwissenschaften um 1900*, 2 vols. (Stuttgart: F. Steiner, 1989–1997); Olaf Blaschke and Frank-Michael Kuhlemann, eds., *Religion im Kaiserreich: Milieus—Mentalitäten—Krisen* (Gütersloh: Kaiser, 1996).

[25] See Hermann Lübbe, *Politische Philosophie in Deutschland: Studien zu ihrer Geschichte*, paperback ed. (Munich: dtv, 1974); Horst Groschopp, *Dissidenten: Freidenkerei und Kultur in Deutschland* (Berlin: Dietz, 1997); Frank Simon-Ritz, *Die Organisation einer Weltanschauung: Die freigeistige Bewegung im Wilhelminischen Deutschland* (Gütersloh: Kaiser, 1997); Helmut Obst, *Apostel und Propheten der Neuzeit: Gründer christlicher Religionsgemeinschaften des 19. und 20. Jahrhunderts*, 4th ed. (Göttingen: Vandenhoeck & Ruprecht, 2000).

new space for studying the intricate relationship between science, politics, and religion, as encapsulated in the life of Rossmässler. The first argument concerns the way in which science has been situated in the overall context of civil society. A peculiar state-centrism and a concentration on university science have dominated our picture of science in German history from the nineteenth century on. The success story of the German university system still eclipses alternative narratives, which have scarcely been explored. In particular, the history of the natural sciences in Germany has been studied almost exclusively in terms of institution building at universities, state funding, cooperation between state and industry, and the professionalization of scientists—that is, phenomena that undoubtedly contributed in an important way to the enhancement of civil life.[26] The roles that individual scientists and scientific disciplines have played in the formation of the German nation-state and the German public have therefore attracted considerable interest.[27] I would like to suggest a complementary approach. This article seeks to demonstrate that the commitment to science outside state-influenced educational institutions also played a crucial role within the culture of science in particular and civil culture in general; indeed, the field of nonuniversity science deserves further examination, including the vast array of lay activities in the field of natural history and what has been called amateur science in the British and American cases.[28]

My second argument deals with the ideological consequences of the interplay of science, politics, and religion in Germany's civil society between 1830 and 1870. Undoubtedly, our sense of the cultural impact that the natural sciences had on nineteenth-century society has been shaped considerably by our understanding of their role in the process of secularization and the "disenchantment of the world"

[26] Charles McClelland, *State, Society and University in Germany, 1700–1914* (Cambridge and New York: Cambridge Univ. Press, 1980); Lothar Burchardt, *Wissenschaftspolitik im Wilhelminischen Deutschland: Vorgeschichte, Gründung und Ausbau der Kaiser-Wilhelm-Gesellschaft zur Förderung der Wissenschaften* (Göttingen: Vandenhoeck & Ruprecht, 1975); David Cahan, *An Institute for an Empire: The Physikalisch-Technische Reichsanstalt, 1871–1918* (Cambridge and New York: Cambridge Univ. Press, 1989); Rüdiger vom Bruch and Rainer A. Müller, eds., *Formen außerstaatlicher Wissenschaftsförderung im 19. und 20. Jahrhundert: Deutschland im internationalen Vergleich* (Stuttgart: F. Steiner, 1990); Bernhard vom Brocke and Rudolf Vierhaus, eds., *Forschung im Spannungsfeld von Politik und Gesellschaft: Geschichte und Struktur der Kaiser-Wilhelm-/Max-Planck-Gesellschaft* (Stuttgart: Deutsche Verlags-Anstalt, 1990); Gerhard A. Ritter, *Großforschung und Staat in Deutschland: Ein historischer Überblick* (Munich: Beck, 1992); Arleen Marcia Tuchman, *Science, Medicine, and the State in Germany: The Case of Baden, 1815–1871* (Oxford and New York: Oxford Univ. Press, 1993); Margit Szöllösi-Janze, *Fritz Haber, 1868–1934: Eine Biographie* (Munich: Beck, 1998).

[27] Timothy Lenoir, *Instituting Science: The Cultural Production of Scientific Disciplines* (Stanford, Calif: Stanford Univ. Press, 1997); Rüdiger vom Bruch, "Wissenschaftspolitik, Wissenschaftssystem und Nationalstaat im Deutschen Kaiserreich," in *Wirtschaft, Wissenschaft und Bildung in Preussen: Zur Sozialgeschichte Preussens vom 18. bis zum 20. Jahrhundert*, ed. Karl Heinich Kaufhold and Bernd Sösemann (Stuttgart: F. Steiner, 1998), pp. 73–89; idem, *Wissenschaft, Politik und öffentliche Meinung: Gelehrtenpolitik im Wilhelminischen Deutschland, 1890–1914* (Husum: Matthiesen, 1980).

[28] See David Allen, *The Naturalist in Britain: A Social History*, 2d ed. (Princeton, N.J.: Princeton Univ. Press, 1994); Lynn Barber, *The Heyday of Natural History, 1820–1870* (Garden City, N.Y.: Doubleday, 1980); Sally Gregory Kohlstedt, "The Nineteenth-Century Amateur Tradition: The Case of the Boston Society of Natural History," in *Science and Its Public: The Changing Relationship*, ed. Gerald Holton and William A. Blanpied (Dordrecht and Boston: D. Reidel, 1976), pp. 173–90; Elizabeth B. Keeney, *The Botanizers: Amateur Scientists in Nineteenth-Century America* (Chapel Hill: Univ. of North Carolina Press, 1992); Lynn K. Nyhart, "Natural History and the 'New' Biology," in *Cultures of Natural History* (cit. n. 7), pp. 426–43; and idem, "Civic and Economic Zoology in Nineteenth-Century Germany: The 'Living Communities' of Karl Möbius," *Isis* 89 (1989): 605–30.

(Max Weber).[29] Accounts of this era usually depict efforts to enhance and promote a scientific education as essential elements in the universal process of secularization. The natural sciences have most often been described in terms of their mechanistic and materialistic approaches to natural phenomena independent of nonphysical and nonmeasurable influences. It has often been taken for granted that those views served as catalysts in the spread of antireligious tendencies. This understanding assumes that those who spread scientific knowledge in the public—intentionally or not—helped undermine the cultural power of religion.

Recent studies have complicated the one-dimensional picture of a linear process of secularization in which rationalization and materialistic thinking supposedly triumphed and swept away the power of religion.[30] Revisionist accounts focus primarily on the fin-de-siècle in order to point out strategies of "enchantment" in the world of science and to show ambiguities in the contemporary reception of modernity.[31] Both phenomena, however, are rooted in ideological traditions that go further back in history and, at least in part, emerged from the relationship between nonuniversity science, liberal thinking, and dissenting religion between 1830 and 1870. Taking Rossmässler's biography as a paradigmatic example, this article highlights those ideological currents based on a Humboldtian understanding of nature and aimed at a reconciliation between science, metaphysics, and even religion. This approach helps us to realize that many of the ambivalences and seemingly contradictory elements of modernity in the late nineteenth century, particularly at the fin-de-siècle, were already present in the decades between 1830 and 1870—that is, during the breakthrough of industrialization and the rise of science as an institutional power.

ON THE ROAD TO A "PEOPLE'S EDUCATION": ROSSMÄSSLER IN THE DEMOCRATIC AND FREE-RELIGIOUS MILIEUS

A drawing, published at the time of his death in 1867, shows Emil Adolf Rossmässler as a small, sober-looking man. (See Figure 1.) He looks nothing like a political agitator or a revolutionary hero. Obviously, Rossmässler preferred to be represented with the instruments and objects—such as a microscope and forceps, a shell and botanical findings—that marked his primary field of interest, natural history. From this drawing, it is hard to imagine that Rossmässler ever left his study and exchanged his microscope for a political pamphlet. Yet he did, and scientific, political, and religious interests interacted in an intriguing way in this biography.

[29] Max Weber, "Wissenschaft als Beruf," (1919), reprinted in *Über das Studium der Geschichte*, ed. Wolfgang Hardtwig (Munich: dtv, 1990), pp. 197–227, on p. 209.

[30] Nipperdey, *Religion im Umbruch* (cit. n. 24); Wolfgang Schieder, ed., *Religion und Gesellschaft im 19. Jahrhundert* (Stuttgart: Klett-Cotta, 1993); Hartmut Lehmannn, ed., *Säkularisierung, Dechristianisierung, Rechristianisierung im neuzeitlichen Europa: Bilanz und Perspektiven der Forschung* (Göttingen: Vandenhoeck & Ruprecht, 1997); Steve Bruce, ed., *Religion and Modernization: Sociologists and Historians Debate the Secularization Thesis* (Oxford: Clarendon, 1999).

[31] Ann Harrington, *Reenchanted Science: Holism in German Culture from Wilhelm II to Hitler* (Princeton, N.J.: Princeton Univ. Press, 1996); Andreas W. Daum, "Das versöhnende Element in der neuen Weltanschauung: Entwicklungsoptimismus, Naturästhetik und Harmoniedenken im populärwissenschaftlichen Diskurs der Naturkunde um 1900," in *Vom Weltbildwandel zur Weltanschauungsanalyse: Krisenwahrnehmung und Krisenbewältigung um 1900*, ed. Volker Drehsen and Walter Sparn (Berlin: Akademie, 1996), pp. 203–15. For a subtle interpretation of the amorphous antimodernism at the fin-de-siècle see T. J. Jackson Lears, *No Place of Grace: Antimodernism and the Transformation of American Culture, 1880–1920* (New York: Pantheon Books, 1981; reprinted, Chicago and London: Univ. of Chicago Press, 1994).

Figure 1. *Emil Adolf Rossmässler (1806–1867). (From* Die Gartenlaube *18 [1867]: 629.)*

During the 1830s and early 1840s, Rossmässler remained focused on his academic tasks at the Forest Academy in Tharandt and published scientific treatises in all classical fields of natural history, in addition to writing textbooks for students. He also began to publish, in installments, an iconography of European mollusks, which eventually became the most comprehensive and reliable compendium of its kind.[32] During this time, Rossmässler made several scientific trips; one led him to Berlin, where he gained the acquaintance of Alexander von Humboldt.

[32] Emil Adolph Rossmässler, *Iconographie der Land- und Süsswasser-Mollusken*, 3 vols. (18 nos.) (Dresden and Leipzig: Arnold [later others], 1837–1858). On the continuation of this project in later years, see Ruud A. Bank, "Die Veröffentlichungen der Rossmässler'schen 'Iconographie der

Starting in 1840 Rossmässler intensified his contacts with democratic groups in Saxony. Like those in many other German states, Saxon democrats were increasingly organizing themselves through the formation of political parties, free associations, public festivities, and discussions in periodicals and newspapers. Rossmässler joined a local *Bürgerverein*, an association of young democrats from the lower bourgeoisie. He also maintained contacts with members of the second chamber of deputies in Saxony and with the progressive *Sänger* (Singers) movement, a semipolitical association that grew out of the cultivation of patriotic singing.[33] In 1843 Rossmässler, still teaching at the Forest Academy, joined two young lawyers in forming the core of a local progressive party (*Fortschrittspartei*); all three began contributing political articles to the press of Saxony's opposition forces.[34] By this time, Rossmässler had already gone through the steps of political socialization that many of his contemporaries from the democratic and liberal strands of German society followed: membership in political and semipolitical associations, articulation of democratic beliefs in the periodical press, and concern with issues of law and self-organization on the grassroots level of German society.

In 1845 Rossmässler took another step into the democratic milieu, one that allowed him to unite his heterogeneous interests in a nonconformist religious ideology. He left the Protestant Church and became a member of the so-called German-Catholics. The German-Catholics had split from the Catholic Church in 1844 in protest against pilgrimage practices and other rituals that centered around the authority of the papal church. They were heavily connected to a parallel movement on the Protestant side, the *Lichtfreunde* (Friends of Light); the two groups would officially merge in the 1850s.[35] German-Catholics as well as the Friends of Light encompassed a heterogeneous set of beliefs ranging from ardent atheism and a religiously imbued socialism to a monistic pantheism and a so-called religion of humanity. Both free-religious groups also tolerated beliefs closer to Protestant and Catholic assumptions, but couched them in strong anti-dogmatic terms. In general, the free-religious movement turned against clerical orthodoxy, the authoritative claims of established theology, and institutionalized hierarchies, thus becoming a part of the democratic movement in Germany. German-Catholics and Friends of Light were joined by many members of the progressive camp, and they developed strong affinities for the natural sciences as the basis of a modern worldview and popular education. Both groups valued the rationalism of scientific analysis, pre-

---

Land- und Süßwasser-Mollusken' Europas (1835–1920)," *Mitteilungen der deutschen malakozoologischen Gesellschaft* 44/45 (1989): 49–53. For an (incomplete) bibliography of Rossmässler's publications see Burgemeister, "Emil Adolf Rossmässler" (cit. n. 10), pp. 164–74.

[33] Dieter Düding, *Organisierter gesellschaftlicher Nationalismus in Deutschland (1808–1847): Bedeutung und Funktion der Turner- und Sängervereine für die deutsche Nationalbewegung* (Munich and Vienna: Oldenbourg, 1984); idem, "Die deutsche Nationalbewegung des 19. Jahrhunderts als Vereinsbewegung: Anmerkungen zu ihrer Struktur und Phänomenologie zwischen Befreiungskriegszeitalter und Reichsgründung," *Geschichte in Wissenschaft und Unterricht* 42 (1991): 601–24.

[34] Rossmässler, *Mein Leben und Streben* (cit. n. 5), p. 111.

[35] Jörn Brederlow, *"Lichtfreunde" und "Freie Gemeinden": Religiöser Protest und Freiheitsbewegung im Vormärz und in der Revolution von 1848/49* (Munich and Vienna: Oldenbourg, 1976); Friedrich Wilhelm Graf, *Die Politisierung des religiösen Bewußtseins: Die bürgerlichen Religionsparteien im deutschen Vormärz. Das Beispiel des Deutschkatholizismus* (Stuttgart and Bad Cannstatt: Frommann-Holzboog, 1978); Peter Bahn, *Deutschkatholiken und Freireligiöse: Geschichte und Kultur einer religiös-weltanschaulichen Dissidentengruppe dargestellt am Beispiel der Pfalz* (Mainz: Gesellschaft für Volkskunde in Rheinland-Pfalz, 1991).

ferred a descriptive and analytical language over the philosophical categories of idealism, and favored—in multiple variations—pantheistic and deistic ideas over the Christian duality between an extramundane power and the earthly world.

Alexander von Humboldt and his *Cosmos* volumes, which were just beginning to appear in the mid-1840s, played a crucial role in this context and immediately attracted immense public attention. (See Figure 2.)[36] Earlier, Humboldt had published specialized works on such topics as volcanism, the geography of plants, and galvanism. He had written massive travel accounts and scientific documentations on his trips to South and Central America (1799–1804) and Russia (1829), encompassing sociological, economic, and historical observations, and he had also published a collection of scientific essays (*Ansichten der Natur*, or *Views of Nature*) with an underlying tone of aesthetic appreciation of nature greatly welcomed by many readers.[37] *Cosmos* surpassed all of his earlier works and revealed all of Humboldt's strengths (and his weaknesses, such as his elaboration on details in the third to fifth volumes that appeared to many already outdated at the time of publication.) The first two volumes, published in 1845 and 1847, in particular provoked extraordinary interest among German readers. Here, Humboldt outlined his endeavor to "comprehend the phenomena of physical objects in their general connection, and to represent nature as one great whole, moved and animated by internal forces."[38] Humboldt's ambition in his first two *Cosmos* volumes was at least threefold: to provide readers with exact scientific data on a comprehensive physical description of the earth and the sky, based on precise measurements taken during his trips and extended through decades of correspondence with fellow scientists; to reflect on the impact that human knowledge, the historicity of this knowledge, and aesthetic views of nature have on constructing such a comprehensive image; and to give himself an appealing "delineation," or, better, "painting of nature."[39]

Cosmos enjoyed a particularly enthusiastic reception among members of the free-religious communities. This compendium of the natural world, imbued in the first two volumes with historical, philosophical, and aesthetic reflections, realized a much-wanted unification of scientific analysis and aesthetic appreciation of nature harkening back to Romantic thinking. Neither an encyclopedia nor a mere holistic essay, Humboldt's magnum opus sought to derive "laws that regulate the forces of the universe" from empirical observations and to delineate the "order and harmony pervading the whole universe," thus depicting a harmonious interplay of all natural phenomena.[40] Free-religious authors quickly capitalized on the Humboldtian understanding of harmony in nature to ascribe a religious meaning to the study of nature. Nature itself became the center of religious devotion and was depicted as a

[36] Alexander von Humboldt, *Kosmos: Entwurf einer physischen Weltbeschreibung* (Stuttgart and Tübingen: Cotta, 1845–1862). In the following, I take quotations from the English translation by Elise C. Otté, which was published in five volumes (1848–1865), with the first two volumes reprinted with substantial new introductions in 1997; see Alexander von Humboldt, *Cosmos: A Sketch of a Physical Description of the Universe*, trans. E. C. Otté (Baltimore and London: Johns Hopkins Univ. Press, 1997), vols. 1–2.

[37] The complete, wide-ranging ouevre is captured in an excellent bibliography that includes foreign-language translations: Horst Fiedler and Ulrike Leitner, eds., *Alexander von Humboldt Schriften: Bibliographie der selbständig erschienenen Werke* (Berlin: Akademie, 2000).

[38] Humboldt, *Cosmos* (cit. n. 36), vol. 1, p. 7.

[39] Ibid., p. 79; translator Otté used the term "delineation" for the German *Gemälde*, the correct meaning of which—as applied by other English translators—is "painting."

[40] Humboldt, *Cosmos* (cit. n. 36), vol. 1, p. 25.

Figure 2. *Alexander von Humboldt—savant, traveler, and cosmic thinker. (From H.[ermann] Klencke,* Alexander von Humboldt. Ein biographisches Denkmal. *2d ed. [Leipzig: O. Spamer, 1852], opposite title page.)*

"cathedral," a "high altar," and a "gospel."[41] Ironically, the line between the much-acclaimed rationality beyond idealistic *Naturphilosophie* and a new descriptive, but even more emphatic and moralistic, idealism was blurred again. The enthusiasm for

[41] As in a popular treatise by the German-Catholic priest Heribert Rau, *Das Evangelium der Natur: Ein Buch für jedes Haus*, 2d ed. (Frankfurt am Main: Literarische Anstalt, 1856), pp. 12, 78, 111, 224, 297.

Humboldt's *Cosmos* on the part of free-religious groups was reinforced by the author's obvious refusal to mention the existence of a personal God and his subtle references to nature as the realm of a "free" spirit, which was immediately taken as a political allusion.[42]

Rossmässler, head of the German-Catholic community in Leipzig during most of the 1850s, was among the leaders of this trend. He contributed significantly to the propagation of a "free natural *Weltanschauung*,"[43] a philosophy based on a combination of free-religious beliefs, scientific insights, and Humboldtian cosmology. In 1851 Rossmässler expressed his convictions in a "nature sermon to the people": propagating a "religion of fraternity," he used highly emotional and descriptive language to depict nature as the all-embracing "house of God" and as the "beautiful motherly home" in which humankind should seek reconciliation with itself and see the model of a sound societal order.[44]

The question of science as an object of public discussion and even as the basis for societal order gained more prominence when the revolution of 1848 brought several democratically minded scientists to the forefront of public life. A number of scientists as well as German-Catholics and Friends of Light were elected as members of the Prussian Parliament in Berlin and the first German National Assembly, which convened in the Hessian city of Frankfurt in May 1848. Voters in Saxony elected Rossmässler to the German National Assembly, where he joined a leftist parliamentary group and contributed to the deliberations of the committee for school affairs and the subcommittee for affairs of the elementary schools. On Germany's most prominent political podium at the time, Rossmässler made himself known as a fighter against any church influence on public schools. He vehemently argued for state control of Germany's entire school system, hoping thus to reject any pressure from ecclesiastical parties.[45] His personal attitude toward religion followed this trajectory, setting a provocative example of anticlericalism in education. During his time as parliamentary deputy in Frankfurt, Rossmässler sent two of his four children to a Jewish school and joined other members of the National Assembly in a reform-minded, irenic association where Christians, religious dissenters such as the German-Catholics, and Jews convened.[46]

Although Rossmässler remained in the second rank of delegates, he was the most prominent scientist in the parliament with the exception of the geologist Karl Vogt, who later became even more famous as the advocate of a polemical scientific materialism.[47] When several German governments ordered the delegates from their regions

[42] Humboldt, *Cosmos* (cit. n. 36), vol. 1, p. 25; again, the English translation is not precise and misses some important nuances because in the German original, Humboldt talked about the "*Gefühl der freien Natur*" and "*d a s F r e i h e* [sic] *(wie wir tief bedeutsam in unserer Sprache sagen)*," which could literally be translated as "the sentiment of free nature" and "freedom, as we say with a deeper meaning in our language." (p. 7).

[43] Rossmässler, *Mein Leben und Streben* (cit. n. 5), p. 215.

[44] Emil Adolf Rossmässler, "Eine Naturpredigt an das Volk," *Neue Reform: Zur Förderung der Religion der Menschlichkeit* (1851): 43–50. *Neue Reform* was one of the publication outlets of the Friends of Light.

[45] Burgemeister, "Emil Adolf Rossmässler" (cit. n. 10), pp. 12–25.

[46] Rossmässler, *Mein Leben und Streben* (cit. n. 5), pp. 112, 121–3.

[47] Surprisingly, we still lack a systematic study of the interrelationship between scientific views and political activities during the German revolution of 1848, in spite of the wave of new publications commemorating the 150th anniversary of the revolution. For attempts to assess the relationship between science and revolutionary politics, see Andreas W. Daum, "Naturwissenschaften und

to pull back from the parliament in the spring of 1849, Rossmässler and some of the other parliamentarians chose, despite immense pressure, to continue to convene at an alternative site, the Swabian town of Stuttgart.[48] In mid-June troops ultimately forced the remnants of the National Assembly to dissolve, and its members were dispersed over the country or had to leave the German states. The dissolution of the freely elected national parliament in 1849 marked the end of the German revolution and the beginning of what contemporaries and historians have called the time of "reaction." During the succeeding decade, governments and state administrations introduced new restrictions on public life and systematically persecuted democrats.[49]

Shortly after the dissolution of the National Assembly in June 1849, Rossmässler decided that his quest for a democratic Germany should materialize in the fields he knew more intimately than politics: natural history and the emerging natural sciences. He turned "from an academic professor of natural science to a natural history teacher for the people."[50] The subtle terminological shift from natural science to natural history indicated Rossmässler's ambivalence about finding a proper label for scientific education at a time when the natural sciences themselves were in the process of becoming institutionalized and still facing opposition as a field of instruction. Significantly, this shift also refers to the conception of natural history as a descriptive science and reflects the concern of many science popularizers that natural history represented the actual testing ground for a culture of amateur science and the participation of laypeople in nature studies. The dichotomy between a (more popular) natural history and the (more academic) natural sciences helped widen the gap between the popularizers of science in Germany and university scientists who did not necessarily dismiss a broader concept of education, but pursued the latter on the basis of their own empirical and specialized scientific research and from a more secure position within civil society.

Rossmässler himself began to reconceptualize *Bildung* as *Volksbildung*, a "people's education" addressing all social strata of the population, which he envisaged as the key area in securing a future for a "morally" and "politically free" German people. Beyond the traditional understanding of *Bildung*, based on the ideology of neohumanism, Rossmässler's concept was particularly rooted in the natural sciences and asked for massively increasing technical, financial, and pedagogical support for science teachers. This concept relied on the assumption that education in

---

Öffentlichkeit in der deutschen Gesellschaft: Zu den Anfängen einer Populärwissenschaft nach der Revolution von 1848," *Hist. Z.* 267 (1998): 57–90; and Thomas Junker, "Darwinismus, Materialismus und die Revolution von 1848 in Deutschland: Zur Interaktion von Politik und Wissenschaft," *Hist. Phil. Life Sci.* 17 (1995): 271–302. See also Heinz-Elmar Tenorth, "Bildungsreform als Gesellschaftsreform: Die Revolution von 1848 als Zäsur der Bildungsgeschichte," in *Europa 1848: Revolution und Reform*, ed. Dieter Dowe, Heinz-Gerhard Haupt, and Dieter Langewiesche (Bonn: Dietz, 1998), pp. 961–84.

[48] See the new edition of Rossmässler's diary, first published immediately after the end of the revolution: E. A. Rossmässler, *Das Stuttgarter Rumpfparlament 1849: Das Tagebuch von Emil Adolph Rossmässler und das Selbstverständnis der Abgeordneten*, ed. Barbara Weiss (Stuttgart: Klett-Cotta, 1999).

[49] See Wolfram Siemann, *'Deutschlands Ruhe, Sicherheit und Ordnung': Die Anfänge der politischen Polizei, 1806–1866* (Tübingen: Niemeyer, 1985); idem, *Der 'Polizeiverein' deutscher Staaten: Eine Dokumentation zur Überwachung der Öffentlichkeit nach der Revolution von 1848/49* (Tübingen: Niemeyer, 1983).

[50] Rossmässler, *Mein Leben und Streben* (cit. n. 5), pp. 110–1.

general and science education in particular needed to be enhanced on a broad scale and would at best go hand in hand with a "true religiosity."[51] The latter, as Rossmässler wrote in a democratic journal in 1850, would need to distance itself from the official church as the "confederate" of the "police state."[52] This set of ideas became the nucleus of his personal ambitions from this point on.

THE POLITICS OF "NATURAL *WELTANSCHAUUNG*"

Rossmässler's transformation from a civil servant to an advocate of science education without institutional security and a freelancer in the field of popular science occurred as much out of necessity as out of choice. In Saxony the Ministry of Education suspended him from teaching because of his political past, and the government indicted him for high treason. The court, however, found him not guilty of the charge. Instead of resuming his academic teaching, Rossmässler started touring through a number of German cities in the winter of 1849–1850 to give widely attended public lectures on natural history. The former professor was eager to spread his new "natural *Weltanschauung*,"[53] to ingrain the spirit of Humboldtian thinking in the minds of the German population. In addition, he depended on honoraria to secure a living for himself and his family. Rossmässler shared this dual motivation with many of his colleagues who made the public sphere their field of professional enterprise.[54] Following up on his promise to enhance the people's education, Rossmässler strengthened the didactic element in his presentations and put particular weight on the visual imagery. Thus he used large, spectacular charts (which he produced himself) to show drawings of fossils and other natural objects, and he demonstrated the use of the microscope by putting enlarged slide preparations on display.[55]

Rossmässler's reputation as a democratic politician caused the police to closely monitor his lectures, but such scrutiny did not dampen Rossmässler's enthusiasm. Already part of the democratic network in Saxony before 1848, Rossmässler continued to work for his ambitious goal of spreading people's education in Germany through the vehicles and media that were essential elements of the ongoing project of expanding the realm of civil society: the print culture of journals and books, public discussions such as those generated at his lectures, and the increasing number of free associations. Beyond his public appearances as an "itinerant scientific preacher," as Rossmässler half-ironically labeled himself,[56] the former professor developed a broad range of activities. In 1852 he helped to found the first major journal in Germany explicitly devoted to popularizing science, *Die Natur* (*Nature*). Rossmässler even drafted the front cover of this journal: he put a volcanic eruption on display, a highly political metaphor for the power of unbound, dynamic forces. (See

[51] All quotes from a personal memorandum of July 1, 1849, in which Rossmässler outlined his personal and political plans for the years following the end of the National Parliament; ibid., pp. 131–5.

[52] E. A. Rossmässler, "Die Demokraten und die religiöse Bewegung," *Der Leuchtthurm* 5 (1850): 547.

[53] Rossmässler, *Mein Leben und Streben* (cit. n. 5), pp. 127, 154, 203.

[54] See also Christian Jansen, *Einheit, Macht und Freiheit: Die Paulskirchenlinke und die deutsche Politik in der nachrevolutionären Epoche, 1849–1867* (Düsseldorf: Droste, 2000).

[55] Rossmässler, *Mein Leben und Streben* (cit. n. 5), pp. 140–1.

[56] Ibid., p. 140.

Die Natur

Zeitung zur Verbreitung naturwissenschaftlicher Kenntniß
und Naturanschauung für Leser aller Stände.

Herausgegeben von
Dr. Otto Ule, in Verbindung mit Dr. Karl Müller, E. A. Roßmäßler und andern Freunden.

№ 2. Halle, G. Schwetschke'scher Verlag. 10. Januar 1852.

Figure 3. *Title page of the journal* Die Natur, *edited by Otto Ule, Karl Müller, and Emil Adolf Rossmässler. (From* Die Natur *1(2) [1852].)*

Figure 3.) *Die Natur* was meant as a general magazine for all branches of sciences, and it stressed the importance of natural sciences in post-1848 Germany and included specialized treatments of natural phenomena. The journal also offered information on the book market and on natural history associations. Rossmässler himself contributed numerous articles and illustrations; through its illustrations and visual features *Die Natur* contributed significantly to an increasing visual mass culture in journalism. (See Figure 4.) Between the lines, some of Rossmässler's articles could be read as political statements. For example, in his examination of natural and artificial systems—a problem that biologists had long been struggling with in their attempt to find a proper classification for the natural world—Rossmässler left no doubt that "artificial systems" were "arbitrary" and "enforced" and that any true "order," following Humboldt's thoughts, derived from natural needs and the inner structure of the natural world itself.[57]

Rossmässler also published his lectures and wrote books on natural history for

[57] Emil Rossmässler, "Künstliche und natürliche Systeme," *Die Natur* 1 (1852):110–1.

Figure 4. *A Borneo Butterfly. (From H. Bettziech-Beta, "Der neue Schmetterling von Borneo," in* Die Natur, *vol. 4 (1855): 312–3, illustration p. 312.)*

the lay public.[58] Those books addressed well-known phenomena of the natural world—such as the forest and its animals, the four seasons, and water—in order to guide his readers into the realm of scientific knowledge. Rossmässler's *Populaire Vorlesungen aus dem Gebiete der Natur (Popular Lectures from the Realm of Nature)* of 1852, based on public talks given in previous years and devoted to Humboldt, explicitly asserted that the people had, as the author put it, a "property right" regarding scientific knowledge.[59] In this and other books Rossmässler repeatedly asserted his firm opinion that the study of nature through an improved science education would be the only way toward a democratic renewal of Germany. He thus positioned himself between the idealism of the early nineteenth century and the radical materialism of contemporaries such as Karl Vogt and Ludwig Büchner. Rossmässler went so far as to call nature the "great reconciler" of his time.[60]

Rossmässler knew that educating the people on the basis of science education as an innovative expansion of the concept of *Bildung* meant not only lecturing to a growing population, but also applying a more dialectical approach in order to induce this population, above all the middle classes and the workers, to actively cultivate their own interests in nature studies and create concrete opportunities to practice these interests. Teaching an audience and encouraging this audience to teach itself were two equally important sides of "people's education." It was against this background that Rossmässler became the first person to promote the use of home aquariums in Germany. Following up on a wave of popular interest in aquariums in England, which began in the early 1840s,[61] Rossmässler published a widely read article on the "sea in the glass" in Germany's most popular family magazine in 1856; a year later he followed with another manual, which eventually became his most successful book, going through five editions by 1892.[62] Rossmässler conceived of aquariums as a means of promoting lay interest in zoology and establishing nature studies in the daily life of the bourgeoisie. Rossmässler realized that both the domestic sphere of private households and the public sphere of popular media and educational insti-

[58] E. A. Rossmässler, *Flora im Winterkleide* (Leipzig: Costenoble, 1854); idem, *Das Gebirgsdörfchen. Eine Perspektive in die Naturgeschichte des Volkes. Mit einer Einleitung: Über die Bedeutung der naturwissenschaftlichen Heimatkunde in Rossmässlers Sinne für die Volksbildung*, new ed. (Leipzig: Quelle & Meyer, 1909); idem, *Die Geschichte der Erde: Eine Darstellung für gebildete Leser und Leserinnen* (Frankfurt am Main: Meidinger, 1856); idem, *Der Mensch im Spiegel der Natur: Ein Volksbuch*, 5 vols. (Leipzig: Wigand and later Keil, 1850–1853); idem, *Populaire Vorlesungen aus dem Gebiete der Natur*, 2 vols. (Leipzig: Costenoble, 1852–1853); idem, *Die vier Jahreszeiten* (Gotha: Scheube, 1855); idem, *Der Wald: Den Freunden und Pflegern des Waldes geschildert* (Leipzig and Heidelberg: Winter, 1863); idem, *Das Wasser: Eine Darstellung für gebildete Leser und Leserinnen* (Leipzig: F. Brandstetter, 1858). See also Alfred E. Brehm and E. A. Rossmässler, *Die Thiere des Waldes*, 2 vols. (Leipzig and Heidelberg: Winter, 1864–1867); and B. Auerswald and E. A. Rossmässler, *Botanische Unterhaltungen zum Verständnis der heimathlichen Flora* (Leipzig: Mendelssohn, 1858).

[59] Rossmässler, *Populaire Vorlesungen* (cit. n. 58), vol. 1, p. 5.

[60] Rossmässler, *Der Wald* (cit. n. 58), p. 10. Also see idem, *Anleitung zum Studium der Tier- und Pflanzenwelt: Ein Leitfaden für Lehrer an höheren Lehranstalten und zur belehrenden Unterhaltung für gebildete Leser*, 3d ed. (Dresden and Leipzig: Arnold, 1856), pp. 4–5. Rossmässler summarized his ideas about science education as a "people's education" and the significance of teaching natural history in two volumes: see *Der naturgeschichtliche Untennicht: Gedanken und Vorschläge zu einer Umgestaltung desselben* (Leipzig: F. Brandstetter, 1860); and *Volksbildung* (Leipzig: Kollmann, 1865).

[61] David Allen, "Tastes and Crazes," *Cultures of Natural History* (cit. n. 7), pp. 404–7; idem, *Naturalist in Britain* (cit. n. 28), pp. 117–22.

[62] E. A. Rossmässler, "Der See im Glase," *Die Gartenlaube* 4 (1856): 252–6; idem, *Das Süßwasser-Aquarium: Eine Anleitung zur Herstellung und Pflege desselben* (Leipzig: Mendelssohn, 1857).

tutions offered ample opportunities for such endeavors. Consequently, he also agitated for the creation of a museum for "patriotic natural history" in Saxony.[63]

It hardly came as a surprise that in 1859 Rossmässler finally founded his own natural history magazine, *Aus der Heimath* (*From the Home Region*). He believed *Die Natur*, whose editorial board he had already left in 1854, was becoming too elitist, and he wanted to address ordinary people, particularly artisans, in a more direct way.[64] *Aus der Heimath*, whose subscription price lay significantly below that of *Die Natur*, presented itself to readers as a "people's magazine for the natural sciences" and covered a wide spectrum of topics, ranging from technology to botany, often illustrated by excellent drawings. Because there were many teachers from elementary schools among the subscribers, *Aus der Heimath* in fact served as a journal for the needs of scientific school education. But *Aus der Heimath* also vigorously advocated increased science teaching far beyond school education. The journal even called for reserving a place for the natural sciences in kindergartens and at popular festivals, and recommended scientific books and instruments for parents seeking suitable Christmas gifts.[65] Rossmässler himself left no doubt that his journal was devoted to the specific mission of explaining the close interplay of science, politics, and religion in the postrevolutionary era. He neither hesitated to use *Aus der Heimath* to write about the liberal-national movement, which gained new strength after 1859, nor shied away from addressing the highly disputed relationship between modern science and religion in its pages. Rossmässler explicitly warned against embarking on the "ugly war between the church and the natural sciences" and often appealed to the reconciliatory metaphor of "mother nature." Again and again, Rossmässler invoked the image of nature harmoniously structured according to "natural laws" and based on the ideas of Alexander von Humboldt, who became the indisputable intellectual point of reference for *Aus der Heimath*.[66]

While books and journals offered important outlets for articulating the concerns of educational reform and science education, the free associations were no less important as avenues for organizing civil society and articulating its concerns.[67] Natural history associations offered basic means of communication and incorporated even remote local and regional areas into a dense network in which information, journals, and specimens were exchanged and visits of itinerant public speakers organized. Open to nonscientists and scientists alike, these associations significantly contributed to spreading knowledge, giving laypeople an opportunity to participate in this

[63] Rossmässler, *Volksbildung* (cit. n. 60), p. 92.

[64] Emil Adolf Rossmässler, ed., *Aus der Heimath: Ein naturwissenschaftliches Volksblatt Amtliches Organ des Deutschen Humboldt-Vereins* (Glogau, 1859–1866); not published with the subtitle until 1861.

[65] "Die Naturwissenschaft im Kindergarten," *Aus der Heimath* 3 (1861): 529–32; "Die Naturwissenschaft auf Volksfesten," *Aus der Heimath* 4 (1862): 787–90; "Naturwissenschatliche Weihnachtsgeschenke," *Aus der Heimath* 4 (1862): 771–4. See Burgemeister, "Emil Adolf Rossmässler" (cit. n. 10), p. 95.

[66] *Aus der Heimath* 1 (1859): 1, 4, 171–2; see also Rossmässler, *Populaire Vorlesungen* (cit. n. 58), vol. 1, p. 111.

[67] On the crucial importance of free associations for the growth of civil society, see Thomas Nipperdey, "Verein als soziale Struktur in Deutschland im späten 18. und frühen 19. Jahrhundert: Eine Fallstudie zur Modernisierung I," in *Gesellschaft, Kultur, Theorie: Gesammelte Aufsätze zur neueren Geschichte* (Göttingen: Vandenhoeck & Ruprecht, 1976), pp. 174–205, 439–47; Otto Dann, ed., *Vereinswesen und bürgerliche Gesellschaft in Deutschland* (Munich: Oldenbourg, 1984); Wolfgang Hardtwig, *Genossenschaft, Sekte, Verein in Deutschland: Vom Spätmittelalter bis zur Französischen Revolution* (Munich: Beck, 1997).

knowledge, and practicing a sort of "science training"[68] for those not professionally linked to universities. The natural history associations thus significantly helped both to preserve and to expand the role of amateurs in German society.

Rossmässler had realized that associations and their work on the local level would provide him with an ideal means of enhancing the understanding of science in German society already by the early 1850s. On September 25, 1852, during the last session of the annual—and most prestigious—meeting of German scientists and physicians, the *Gesellschaft deutscher Naturforscher und Ärzte*, a national association founded in 1822, Rossmässler had pleaded for a "generalization of the efforts of natural history associations." He had asked the professors and scientists present at the meeting to declare their solidarity with the concerns of the people, to share their knowledge with the local population, and—in pursuit of these goals—to form local natural history associations with scientific collections.[69] This particular plea had not resonated with those assembled, but this negligence did not matter in the long run: the network of local natural history associations had already grown significantly during the previous years and continued to expand even more rapidly during the decades between 1850 and 1880.[70]

Rossmässler eventually added to this expansion himself. In 1859 he initiated local Humboldt Associations as "scientific people's associations"[71] in memory of the universal scholar who had passed away on May 6, 1859. Humboldt Associations soon spread through the border regions of Saxony, Silesia, and Bohemia; several successors were established in other parts of Germany. The associations under the name of Humboldt embodied an ambitious attempt to organize lay scientific activities. They were designed as a new type of free association for naturalists. They renounced all distinctions of class and gender, and in contrast to the existing societies for natural history, the Humboldt Associations were explicitly aimed at the lower social strata, particularly artisans and small business owners, and focused on lay activities in local settings such as the collecting of specimens, the common reading of relevant literature, and excursions into the countryside.

In 1861 Rossmässler and some of his friends proposed a German Humboldt Association in order to establish closer bonds between the small and dispersed societies and to create a popular version of the *Gesellschaft deutscher Naturforscher und Ärzte*. As the German Humboldt Association took shape in 1861, the arrangement of festivities and celebrations—already accompanying the Humboldt meetings before—moved into the foreground at the conventions. Humboldt Festivals, as Rossmässler and his supporters now called these conventions, began to orchestrate local elites, representatives of different professions such as teachers, and visitors from other regions into a dense program consisting of toasts, speeches, songs, excursions, visits to regional attractions, festive dinners, and such. This was in microcosm a culture of civil festivity that the *Gesellschaft deutscher Naturforscher und Ärzte* had

[68] See Michael D. Stephens and Gordon W. Roderick, "Science Training for the Nineteenth Century English Amateur: The Penzance Natural History and Antiquarian Society," *Ann. Sci.* 22 (1971): 135–41.

[69] Rossmässler, *Mein Leben und Streben* (cit. n. 5), pp. 164–5; R. Fresenius and Braun, eds., *Amtlicher Bericht über die neunundzwanzigste Versammlung der Gesellchaft deutscher Naturforscher und Aerzte zu Wiesbaden im September 1852* (Wiesbaden: Schellenberg, 1853), p. 74.

[70] I have documented the activities of natural history associations elsewhere in further detail; see Daum, *Wissenschaftspopularisierung* (cit. n. 8), pp. 85–118.

[71] "Humboldt-Vereine," *Aus der Heimath* 1 (1859): 419.

been practicing for many years. It connected the social world of natural history teachers and scientists with what was already an essential element of civil society: the public display of sociability that provided *Bürger* with opportunities not only to articulate their immediate concerns and ideals, but also to define their identities as participants in a widening civil culture and symbolically state their political claims.[72]

The Humboldt Associations acted in close connection with other associations within the democratic and liberal political spectrum, such as societies for workers and artisans. This politicization of public science went hand in hand with Rossmässler's willingness to return to political activities and coincided with the political and national reawakening of the so-called New Era, which had begun in 1858. In the New Era, German state governments loosened political restrictions, new liberal activities unfolded, and the idea of a nation-state gained strength.[73] Rossmässler's Humboldt Associations and the accompanying, annual Humboldt Festivals were part of this movement and had a clear national orientation, grounded in the democratic and liberal convictions of former supporters of the 1848 revolution. While his personal interests remained almost exclusively devoted to botanical, zoological, and ecological issues, Rossmässler engaged in the political arena with the same objectives that the Humboldt Associations pursued in reaching out to the lower social strata. In Leipzig Rossmässler personified the policy of integration that leading members of Germany's liberal bourgeois movement pursued in an effort to harmonize the interests of the middle class, artisans, and proletarians. He became active in several associations for the spread of *Bildung*, and he insisted that knowledge about the natural sciences would be a key element in the reform movement that sought to integrate proletarians into bourgeois society.[74]

In the winter of 1862–1863, Rossmässler became the major opponent of Ferdinand Lassalle, the socialist agitator and later founder of the first socialist party in Germany, in the debate over the best way to organize the growing proletariat. While Lassalle sought to create a distinct workers' party on a national level, Rossmässler advocated continuing the educational efforts for workers and artisans by liberal associations and imbuing these efforts with the idea of nature studies and science education based on rational analysis.[75] Both men agreed on the necessity of spreading scientific knowledge. Lassalle, however, called for scientific rationality in the articulation of a proletarian political program that clearly disassociated itself from liberal bourgeois policy and pursued a confrontational strategy; Rossmässler extrapolated

[72] On the importance of civil festivities, see also Dieter Düding, Peter Friedemann, and Paul Münch, eds., *Öffentliche Festkultur: Politische Feste in Deutschland von der Aufklärung bis zum Ersten Weltkrieg* (Reinbek bei Hamburg: Rowohlt, 1988); and Manfred Hettling and Paul Nolte, eds., *Bürgerliche Feste: Symbolische Formen politischen Handelns im 19. Jahrhundert* (Göttingen: Vandenhoeck & Ruprecht, 1993). For a comprehensive treatment of the Humboldt Associations and Humboldt Festivals, see Daum, *Wissenschaftspopularisierung* (cit. n. 8), pp. 138–67.

[73] Sheehan, *German History* (cit. n. 1), pp. 869–88; Nipperdey, *Germany From Napoleon to Bismarck* (cit. n. 1), pp. 620–27; Wolfram Siemann, *Gesellschaft im Aufbruch: Deutschland, 1849–1871* (Frankfurt am Main: Suhrkamp, 1990).

[74] E. A. Rossmässler, "Eine wichtige Zeiterscheinung," *Aus der Heimath* 4 (1862): 769–70. He developed his political ideas in programmatic publications: see E. A. Rossmässler, *Die Fortschrittspartei und die Volksbildung* (Berlin: Janke, 1862); idem, *Ein Wort an die deutschen Arbeiter (Im Einvernehmen mit dem Leipziger Central-Comité verfaßt)* (Berlin: Jonas, 1863); idem, *Unsere Lage. Ein ernstes Wort an das deutsche Volk und dessen Fürsten* (Leipzig: Priber, 1866).

[75] Shlomo Na'aman, *Die Konstituierung der deutschen Arbeiterbewegung 1862/63: Darstellung und Dokumentation* (Assen: van Gorcum, 1975); and Toni Offermann, *Arbeiterbewegung und liberales Bürgertum in Deutschland, 1850–1863* (Bonn: Neue Gesellschaft, 1979).

his natural science interests into a harmonistic vision of a just society in which the pursuit of nature studies per se would balance divergent social interests. In Rossmässler's view, every proletarian would and should become a "disciple of Humboldt."[76]

In the spring of 1863, the leading workers' committee in Leipzig opted for Lassalle's program and prepared a new institutional platform, the General German Workers Association, which was officially founded on May 23 of that year. Although Rossmässler's rather diffuse program did not succeed, one month later, opponents of Lassalle assembled around Rossmässler and other representatives of Germany's bourgeoisie to call a Convention of German Workers Associations (*Vereinstag deutscher Arbeiter-Vereine*) into being. But Rossmässler had already begun to withdraw from politics. After 1863 he refocused on nature studies, the Humboldt Associations, and his journal *Aus der Heimath*. Increasingly stricken by physical problems, however, Rossmässler could no longer pursue his projects; planned Humboldt Festivities for 1864 and 1865 had to be canceled. In 1866 he rallied enough to join the socialists August Bebel und Wilhelm Liebknecht in founding the People's Party of Saxony, but that same year he saw his *Aus der Heimath* cease publication, no longer financially competitive in the expanding market for nature journals. On April 4, 1867, Rossmässler died in Leipzig. His fame slowly decreased in a society that—paradoxically—had absorbed his ideas and made them more fruitful than ever in the years after his death. Then, institutionalized science education and professionalized popularization went far beyond Rossmässler's initial enthusiasm. Rossmässler remained well remembered by those groups who maintained a vivid interest in people's education, such as school teachers of natural history, and many of his books underwent numerous reprints and revised editions later in the century.[77]

Although Rossmässler remained marginal in terms of Germany's state-supported university system and many of his projects enjoyed only limited success, his efforts and those of his associates had a tremendous impact on Germany's cultural life. The history of nature journals is a telling example. To be sure, journals such as *Die Natur* and *Aus der Heimath* rarely had more than three thousand subscribers;[78] but their true impact lay in the fact that they pioneered science journalism and created a print market that rapidly expanded, in Germany no less than in other countries.[79] There was an extensive circulation of articles and topics, even on an international scale,[80] and numerous authors wrote simultaneously for several periodicals. By the end of

[76] E. A. Rossmässler, "Eine wichtige Zeiterscheinung," *Aus der Heimath* 4 (1862): 770.

[77] Rossmässler's *Der Wald*, for example, saw a third edition in 1881, as did *Das Wasser* in 1875; his *Flora im Winterkleide* was printed in a fourth edition in 1908, the *Vier Jahreszeiten* in a sixth edition in 1888, and the *Süßwasser-Aquarium* in a fifth edition in 1892, plus a translation into Dutch and a reprint in 1995.

[78] *Aus der Heimath* 5 (1863): 1–2; ibid. 8 (1866): 403.

[79] See Susan Sheets-Pyenson, "Popular Science Periodicals in Paris and London: The Emergence of a Low Scientific Culture, 1820–1875," *Ann. Sci.* 42 (1985): 549–72; Bernadette Bensaude-Vincent and Anne Rasmussen, eds., *La science populaire dans la presse et l'édition: XIXe et XXe siècles* (Paris: CNRS Editions, 1997); Ruth Barton, "Just before *Nature*: The Purposes of Science and the Purposes of Popularization in Some English Popular Science Journals of the 1860s," *Ann. Sci.* 55 (1998): 1–33; Daum, *Wissenschaftspopularisierung* (cit. n. 8), pp. 337–76. On science in mass-circulating magazines, see also Peter Broks, *Media Science and the Great War* (London: Macmillan, 1996).

[80] Andreas W. Daum, "'The Next Great Task of Civilization': International Exchange in Popular Science, The German-American Case, 1850–1900," in *The Mechanics of Internationalism: Culture, Society, and Politics, 1850–1914*, ed. Martin H. Geyer and Johannes Paulmann (Oxford: Oxford Univ. Press, 2001), pp. 280–314.

the century, not only could the German reading public choose from a dense array of general natural history magazines, specialized journals for amateur scientists, and popular technological journals, it could also find nature-related and scientific topics in all major cultural journals of the *Bürgertum*. In addition, many natural history associations started opening libraries, subscribing to popular science journals, and offering these journals to their members.

BEYOND THE "RESEARCH IMPERATIVE": SCIENCE AND THE INFRASTRUCTURE OF CIVIL SOCIETY

Rossmässler was certainly the most enthusiastic pioneer of a "natural *Weltanschauung*" based on Humboldtian views in mid-nineteenth-century Germany. But his ideas did not reflect only an individual disposition; their essential ingredients can be found among many of his contemporaries. Rossmässler embodied a new intellectual and social type in Germany, that of a public-oriented naturalist and advocate of science outside the institutions of science. This type deliberately replaced the "research imperative," which was essential to pursuing a successful career in university science,[81] with an imperative in the public sphere. For Rossmässler and others, popularizing science became both a vocation and a profession. These popularizers could mostly rely on their own training as science students at universities, but they were no longer embedded in the social and communicative network of universities. Professional popularizers could not rely on the security net and the social prestige that German professors enjoyed by virtue of their status as civil servants.

Members of this group came from very different backgrounds and thus evade generalization. They included sons of Protestant preachers who had a strong amateur interest in science; religious dissenters searching for a new rationality in the religious worship of nature; teachers at primary and secondary schools trying to extend their interests into the public sphere; political journalists who discovered that the reading public had a growing interest in pieces about science; medical doctors who did not succeed at universities; and democratic-minded scientists whose careers at state institutions had failed after 1848. Strikingly in the German case, no women figured prominently in this group.[82]

In general, this heterogeneous group can be seen as a generational cohort that was educated and politically socialized after the 1830 revolution, and thus exposed to democratic ideas, which they often embraced. A surprisingly high number of them had some commitment to the free-religious movement or became active during the political mobilization of 1848. Consequently, some suffered under the ensuing "reaction." Due to political or social reasons or from personal preference, hardly any of them maintained formal links to a university. They occupied, however, a crucial place in civil society. Two examples may illustrate this argument further. Like Rossmässler, his younger contemporary Otto Ule, born in 1820, had started to study

[81] R. Steven Turner, "The Growth of Professorial Research in Prussia, 1818 to 1848—Causes and Context," *Hist. Stud. Phys. Sci.* 3 (1971): 137–82.

[82] In contrast, see for the English-speaking world Ann B. Shteir and Barbara T. Gates, eds., *Natural Eloquence: Women Reinscribe Science* (Madison: Univ. of Wisconsin Press, 1997); Ann B. Shteir, *Cultivating Women, Cultivating Science: Flora's Daughters and Botany in England, 1760 to 1860* (Baltimore: John Hopkins Univ. Press, 1996); and Schwarz, *Der Schlüssel zur modernen Welt* (cit. n. 8), pp. 124–9, 251–2.

theology but switched to the natural sciences.[83] During the early 1840s he became a supporter of the Friends of Light. In 1847 Ule began holding public lectures on topics in natural history and soon devoted himself to explaining Humboldt's *Cosmos* to the lay public in the town of Frankfurt (Oder). Ule supported the democratic left in 1848 and was sent to prison after the revolution because of his political opinions. After some weeks of imprisonment, he eventually moved to Halle (Saale). The city of Halle had a well-established reputation as a center of philosophical rationalism and nature studies, dating back to the Enlightenment and built on a firm tradition of Protestant educational institutions. In Halle, Ule became coeditor of *Die Natur* and involved himself in numerous urban associations, ranging from the local society for geography to the municipal fire brigade and the progressive party (*Fortschrittspartei*), where liberal-minded *Bürger* came together. Ule also strongly supported Rossmässler's Humboldt Associations and joined the former zoology professor in propagating the Humboldtian *Weltanschauung* through numerous articles and popular books devoted to spreading the "cosmic view" of nature.[84]

While Otto Ule remained a jack-of-all-trades in Halle's *Bürgertum* until his sudden death in 1876, his contemporary Alfred Brehm, born in Thuringia in 1829, took up similar activities in a number of places. Like Ule, Brehm came from a strong Protestant family in which nature studies had always been cultivated. In fact, his father, the Protestant pastor Christian Ludwig Brehm, was one of Germany's most prominent amateur scientists and an expert in ornithology.[85] After an apprenticeship as a craftsman and the study of architecture in Dresden, Alfred Brehm spent nearly five years on an expedition in Africa (1847–1852). Soon after his return, he established links to Rossmässler and the editor of the family magazine *Gartenlaube* and eventually became Germany's best-known zoological journalist. Brehm had already published accounts of his African trip and now continued to report on his subsequent travels, bringing the "exotic" to the private homes of his readers in Germany. Not surprisingly, Brehm tried his luck as director of the zoological garden in Hamburg (1863–1866) and afterward founded and directed the first public aquarium in Berlin. Brehm's fame, though, arose from the publication of his multivolume *Tierleben* (*Animal Life*), which became a bestseller and a classic Christmas gift for children of bourgeois families.[86] In his *Tierleben*, Brehm developed a highly anthropomorphic style of animal narrations, which became extremely popular with readers from all social classes.

[83] See [Rossmässler,] "Otto Ule," *Die Gartenlaube* 6 (1858): 664–6; and Karl Müller, "Otto Ule," *Die Natur* 25 (1876): 405–6, 416–7, 431–2, 442–3.

[84] Otto Ule, *Aus der Natur: Essays* (Leipzig: Frohberg, 1871–1873), a collection of essays taken from *Die Natur*; idem, *Ausgewählte kleine naturwissenschaftliche Schriften*, 5 vols (Halle: G. Schwetschke, 1865–1868); idem, *Die Natur: Ihre Kräfte, Gesetze und Erscheinungen im Geiste kosmischer Anschauung* (Halle: H. W. Schmidt, 1851); idem, *Physikalische Bilder im Geiste kosmischer Anschauung: Allen Freunden der Natur gewidmet*, 2 vols. (Halle: H. W. Schmidt, 1854–1857); idem, *Das Weltall: Beschreibung und Geschichte des Kosmos im Entwicklungskampfe der Natur*, 2 vols. (Halle: H. W. Schmidt, 1850); idem, *Populäre Naturlehre (Physik) oder die Lehre von den Bewegungen in der Natur und von den Naturkräften im Dienste des Menschen* (Leipzig: Keil, 1867).

[85] See Hans-Dietrich Haemmerlein, *Der Sohn des Vogelpastors: Szenen, Bilder, Dokumente aus dem Leben von Alfred Edmund Brehm* (Berlin: Evangelische Verlagsanstalt, 1987); and Siegfried Schmitz, *Tiervater Brehm: Seine Reisen, sein Leben, sein Werk* (Munich: Harnack, 1984).

[86] A. E. Brehm, *Illustriertes Thierleben: Eine allgemeine Kunde des Thierreiches*, 6 vols. (Hildburghausen: Bibliographisches Institut, 1864–1869).

Rossmässler, Ule, and Brehm can be seen as protagonists among the professional mediators of science education in Germany. Beyond this group, however, were others who acted as public mediators and helped to place nature studies and scientific issues on the agenda of public discourses and make them relevant to *bürgerlich* life. Among these were academic mediators, based at universities, who contributed to the lecture programs of free associations, supported educational institutions, and wrote articles in the nonacademic press. And there was a cohort of eminent professional scientists who became opinion leaders in academia and acted as "public scientists."[87] Above all this holds true for Justus Liebig (1803–1873), who reorganized chemistry as a laboratory-based professional science; Rudolf Virchow (1821–1902), cofounder of modern medical pathology and cell theory; Emil Du Bois-Reymond (1818–1896), expert in electric physiology; and Hermann Helmholtz (1821–1894), exponent of modern physiology and physics, who was ennobled in 1883 and from then on could add "von" to his last name.[88] Clearly, all of them greatly influenced the public perception of science and became prominent examples of a *bürgerliche* elite, one that had made its way up on the social ladder toward success and prestige by pursuing key ideals of civil society: work discipline, educational merits, academic training, and respect for a rational view of the world. But their motivations for the promotion of science were not the same as Rossmässler's, Brehm's, or Ule's. Their public appearances, as recent historiography has emphasized, were aimed at securing state funding for their research, legitimizing their new disciplines and institutes, and pursuing their social interests as a new elite.[89] It would be misleading, however, to equate the public promotion of science in general during the decades between 1830 and 1870 primarily with the social interest of this rising elite. Rather, Liebig, Virchow, Du Bois-Reymond, and Helmholtz formed one group—though undoubtedly the best known—among the many scientists who reached out to the German public, and their social interest can be generalized only to a certain extent.

It is telling that Rossmässler asked Rudolf Virchow for support in the plan to establish the Humboldt Associations but failed to persuade the Berlin professor, who in principle supported the idea, to attend any of the Humboldt Festivals. Professional popularizers and the eminent "public scientists" moved in quite different social circles and seldom met in professional contexts. The meetings of the *Gesellschaft deutscher Naturforscher und Ärzte* offered them one of the rare opportunities to do so, but often Rossmässler and many of his colleagues could not afford to travel to the

[87] I borrow this term from Frank M. Turner, "Public Science in Britain, 1880–1919," *Isis* 71 (1980): 589–608; Turner has used it to describe the role of academic opinion leaders in Britain in the late nineteenth and early twentieth centuries.

[88] See R. Steven Turner, "Justus Liebig versus Prussian Chemistry: Reflections on Early Institute-Building in Germany," *Hist. Stud. Phys. Sci.* 13 (1982): 129–62; William Hodson Brock, *Justus von Liebig: The Chemical Gatekeeper* (Cambridge and New York: Cambridge Univ. Press, 1997); Tuchman, *Science, Medicine, and the State* (cit. n. 26); Byron Boyd, *Rudolf Virchow: The Scientist as Citizen* (New York and London: Garland, 1991); Heinrich Schipperges, *Rudolf Virchow* (Reinbek bei Hamburg: Rowohlt, 1994); Gunter Mann, ed., *Naturwissen und Erkenntnis im 19. Jahrhundert: Emil Du Bois-Reymond* (Hildesheim: Gerstenberg, 1981); David Cahan, ed., *Hermann von Helmholtz and the Foundations of Nineteenth-Century Science* (Berkeley: Univ. of California Press, 1973); Hermann von Helmholtz, *Science and Culture: Popular and Philosophical Essays*, ed. David Cahan (Chicago: Univ. of Chicago Press, 1995).

[89] See Bayertz, "Spreading the Spirit of Science" (cit. n. 8); Everett Mendelsohn, "Revolution and Reduction: The Sociology of Methodological and Philosophical Concerns in Nineteenth Century Biology," in *The Interaction between Science and Philosphy*, ed. Y. Elkana (Atlantic Highlands, N.J.: Humantities Press, 1974), pp. 407–26; Lenoir, *Instituting Science* (cit. n. 27), pp. 75–95.

conventions, which met in different places each year. Rossmässler did find himself in contact with someone who undoubtedly surpassed, in his mind, any living scientist. Remarkably, Alexander von Humboldt himself had corresponded with Rossmässler and praised him for writing "truly popular writings" with the "noble purpose of distributing useful knowledge and refuting the dogmatic axioms of an arrogant half-knowledge, accumulated for centuries."[90] In 1855, in an enthusiastic congratulatory letter thanking Rossmässler for his book on the *Four Seasons*, Humboldt captured the essence of Rossmässler's democratic concept of science emphasizing the local context of nature studies. Humboldt explicitly lauded the nonacademic purpose of the book and subtly supported the political implications of Rossmässler's thinking: "You have intended to stimulate, to extend knowledge, to be useful and restore the grace of local nature. Given the current condition of the German people this is gratifying in double respect: 'The Free,' . . . the enjoyment of the 'free' nature, remains to the Germans, as they put it beautifully and significantly in their language."[91]

After 1860 university scientists became increasingly involved in the business of popularization. Many of them became contributors to nonscientific journals, supported new and professionalized educational institutes for the general public, or started to give lectures at educational associations and natural history associations as in the cases of Rudolf Virchow, Hermann Helmholtz, Emil du Bois-Reymond, and the astronomer Wilhelm Foerster, in order to promote the "civilizing power of science."[92] Clearly, the general appreciation of science as an integral element of *Bildung* in Germany's civil society did not emerge from either the efforts by Rossmässler or the public statements by Virchow alone; rather it developed in a dialectical process that was neither a bottom-to-top ascendance nor a simple trickle down of knowledge.

RECONCILIATION AND NATURAL LAWS: THE HUMBOLDTIAN VISION

Most of the science popularizers of the pioneering generation between 1830 and 1870 left no doubt about the subtext of their nature writings. They depicted a world of harmony ruled not by human action, but by natural laws. In both subtle and explicit ways, their writings voiced a plea for liberty and equality to "reverberate as a reminder in the field of science." They claimed that only devotion to nature would "heal the wounds" of political battles and that only science would be the "real republican power" of intellectual life.[93] For these men, anchoring the natural sciences in the infrastructure of civil society was a means of securing a democratic future. Some caveats, however, have to be kept in mind. Though democrats and liberals dominated

[90] Alexander von Humboldt to E. A. Rossmässler, letter, 6 Nov. 1852, Manuscript Division, Houghton Library, Harvard University, Cambridge, Mass.

[91] Alexander von Humboldt to E. A. Rossmässler, letter, 16 Sept. 1855, Manuscript Division, Houghton Library, Harvard University, Cambridge, Mass. See Rossmässler, *Die vier Jahreszeiten* (cit. n. 58).

[92] David Cahan, "Helmholtz and the Civilizing Power of Science," in *Hermann von Helmholtz*, pp. 559–601; also see Helmholtz, *Science and Culture* (both cit. n. 88).

[93] Ule, *Das Weltall* (cit. n. 84), vol. 1, p. 1; idem, *Die Natur* (cit. n. 84), p. 190; Friedrich Steinmann, ed., *Volks-Kosmos. Bibliothek für Naturwissenschaft, Erd- und Himmels-, Länder- und Völkerkunde*, 2 vols. (Berlin: Röhring, [ca. 1859]), vol. 2, p. 346.

the field of science education between 1830 and 1870, it was not exclusively their realm; more moderate, apolitical, and even conservative voices could be heard among the public mediators of science. Presenting scientific knowledge to the lay public became the self-appointed task of all kinds of ideological and political groups. This forms an important background against which the role of religion and the polyvalences in the later reception of Darwinism in Germany have to be considered.[94]

Investigating the concepts and images of nature that underlay the public representation of science also reveals a diversity of contemporary views. Until now, historiography has concentrated almost exclusively on the most radical of these views, as articulated by the followers of scientific materialism, the so-called materialists. This meant essentially three authors: the medical doctor Ludwig Büchner (1824–1899), brother of the literary author Georg Büchner; the Dutch physiologist Jacob Moleschott (1822–1893); and the zoologist and geologist Karl Vogt (1817–1895), who escaped to Switzerland after his term as a member of the German National Assembly of 1848–1849. These three men shared a radical antimetaphysical approach, one that reduced natural phenomena to mechanistic, physiological processes.[95] Büchner and Vogt, in particular, expressed their convictions polemically and categorically dismissed all religious beliefs. The publication of their most important books in the early 1850s led to heated controversies and forced all three out of the German university system.

The three materialists certainly rocked the boat, and their writings were widely distributed. But they remained clearly in a minority position. Most popular writers, including Rossmässler and Ule, did not adopt the radical antimetaphysical view. They clung to the philosophical idea of science that Alexander von Humboldt had put forward in the first half of the century. They preserved Humboldt's cosmic concept of nature, which bore strong elements of Romantic nature philosophy. They revitalized the idea that empirical research had to go hand in hand with an aesthetic perception of nature and that it depended on a philosophical idea of unity. Otto Ule's immensely successful biography of Alexander von Humboldt helped to cast the eminent savant as a hero and a model for a "scientific people's literature"; Humboldt's *Cosmos*, according to Ule, had hit the "dull atmosphere" of postrevolutionary Germany like "purifying lightning." Its publication embodied the "liberation of the

[94] For the latter, see William M. Montgomery, "Germany," in *The Comparative Reception of Darwinism*, ed. Thomas F. Glick, 2d ed. (Chicago and London: Univ. of Chicago Press, 1988), pp. 81–116; Paul J. Weindling, "Darwinism in Germany," in *The Darwinian Heritage*, ed. David Kohn (Princeton, N.J.: Princeton Univ. Press, 1985), pp. 685–98; idem, "Ernst Haeckel, Darwinismus, and the Secularization of Nature," in *History, Humanity, and Evolution: Essays for John C. Greene*, ed. James R. Moore (Cambridge: Cambridge Univ. Press, 1989), pp. 311–27; Rolf Peter Sieferle, *Die Krise der menschlichen Natur: Zur Geschichte eines Konzepts* (Frankfurt am Main: Suhrkamp, 1989).

[95] For analyses of the materialistic positions and sketches of their main representatives, see Frederick Gregory, *Scientific Materialism in Nineteenth Century Germany* (Dordrecht and Boston: D. Reidel, 1977); Dieter Wittich, ed., *Vogt, Moleschott, Büchner: Schriften zum kleinbürgerlichen Materialismus in Deutschland*, 2 vols. (Berlin: Akademie, 1971); Kurt Bayertz, "'Siege der Freiheit, welche die Menschen durch die Erforschung des Grundes der Dinge errangen': Wandlungen im politischen Selbstverständnis deutscher Naturwissenschaftler des 19. Jahrhunderts," *Ber. Wissenschaftsgesch.* 10 (1987): 169–83; Carl Vogt, Jacob Moleschott, Ludwig Büchner et al., *Briefwechsel*, ed. Christoph Kockerbeck (Marburg: Basiliken-Presse, 1999).

treasures of knowledge from the gloomy walls of the study chamber and the cabinets. Science descended from its arrogant, laughable throne to arrange its place in the midst of the people."[96]

Alexander von Humboldt, not the radical materialists, inspired most of the nature-writing in Germany after the revolution. Much of this writing devoted itself to a "*Real-Idealismus*"[97] that in the following years would even survive critics of religion such as David Friedrich Strauß and Charles Darwin. (Darwin's *Origin of the Species* appeared in 1859, the same year that Humboldt died.) This "real-idealism" endured well into the fin-de-siècle. Neo-idealistic concepts of nature enjoyed immense popularity at that time. Around the turn of the century, the two most prominent German science writers, Ernst Haeckel and Wilhelm Bölsche, developed pronounced antimaterialistic positions, and popular writers propagated what I would call a cosmic evolutionary theory that was inspired far more by Humboldt's static and aesthetic imagery of nature than by Darwin's more dynamic model.[98]

These observations on the schism between materialistic and cosmic thinking again raise the question of religion. The radical, anti-Christian materialists remained a minority in the discussion of science education from 1830 to 1870; among both liberals and conservatives, there were numerous attempts to go beyond a dichotomy of science versus religion, as already indicated in the case of Rossmässler. It is a persistent myth that "devout defenders of Christianity simply surrendered to the enemy the techniques of popular science."[99] In fact, Catholics as well as Protestants began to use these techniques in postrevolutionary Germany. A striking example is provided by the Catholic journal *Natur und Offenbarung* (*Nature and Revelation*). Founded in 1855 as an "organ for the reconciliation between scientific research and faith," *Natur und Offenbarung* certainly differed from such journals as *Die Natur* or *Aus der Heimath* in fundamental ideological ways. The Catholic publishers renounced ardent materialism and Brehm's anthropomorphic animal narrations. But over time, their reports became increasingly similar to those published elsewhere in the popular science press and remained attentive to the newest research developments.[100] *Natur und Offenbarung* eventually survived all of its competitors on the print market. The fate of this journal is indicative of the highly ambiguous stand Catholicism took toward the dynamic forces of civil society.

No less important were the attempts by democrats and liberals to reconcile science and religious beliefs. Surprisingly, the cultural climate of the 1840s, 1850s, and 1860s did not simply follow the *Kulturkampf* model, although there were, of course, bitter struggles between devout supporters of the Catholic Church and freethinking liberals. Yet significant intellectual space for reconciliation also existed. The case of Rossmässler shows that this is true especially for those groups commonly placed under the vast rubric of anti-Christian freethinkers: the German-Catholics and the

[96] Otto Ule, *Alexander von Humboldt: Biographie für alle Völker der Erde*, 3d ed. (Berlin: Lesser, 1869), pp. 114, 137.
[97] Müller, "Otto Ule" (cit. n. 83), p. 406.
[98] Niles R. Holt, "Ernst Haeckel's Monistic Religion," *J. Hist. Ideas* 32 (1971): 265–80; Daum, *Wissenschaftspopularisierung* (cit. n. 8), pp. 309–23, 464–8.
[99] Kelly, *Descent of Darwin* (cit. n. 8), p. 94.
[100] *Natur und Offenbarung: Organ zur Vermittlung zwischen Naturforschung und Glauben für Gebildete aller Stände* (Münster), 1 (1855) to 56 (1910).

Friends of Light. Rossmässler and those of his colleagues known as free-religious Humboldtians clearly urged the separation of church and state. At the same time, however, they refused to incite the obvious conflict between Christianity and scientific materialism.

Through their interpretation of Humboldt's *Cosmos*, men such as Rossmässler and Ule tried to reach a conciliatory third way between religious orthodoxy and atheist materialism. They emphasized the idea of harmony, and—following the irenic attitude Rossmässler had already demonstrated in his private life during his time in Frankfurt in 1848—allowed their audiences and readers to maintain a diversity of religious views and theological positions, including Christian beliefs. Therefore, the free-religious support for the natural sciences, as much as it aroused hostility from the established churches, did not result in a clear antagonism between science and religion; rather, it offered an opportunity to blur the lines between these two fields.

In light of these observations, a central element of the common model of modernization in nineteenth-century history becomes questionable: the assumption that the spread of scientific knowledge inevitably accelerated, perhaps even caused, the process of secularization. This argument might be true if one ascribed a purely rationalistic, antimetaphysical character to science in nineteenth-century culture. However, if we note the representations of scientific knowledge in the nonacademic public sphere, we are confronted with a much more complicated picture. Secularization does not exclude the creation of a new religiosity outside established denominations and even includes the integration of science into confessional thinking, such as that of Catholics. Science between 1830 and 1870 thus can be seen as a cultural good that represented empirical research *and* the Humboldtian search for unity, a cultural good that could include the Christian search for documenting the power of God in designing nature. Science certainly marked the cultural field in which the world became disenchanted by looking at cell structures and physiological processes instead of relying on philosophical narratives. Science, however, could also represent a cultural field that produced new strategies of reenchantment aiming at a harmonious picture of nature, accessible to and acceptable by diverse strands of society, including conservatives and churchgoers. Modernity, in other words, becomes more ambiguous.

SCIENCE AS CIVIL PRACTICE AND CULTURE

Natural history and the emerging natural sciences gained an important place in the infrastructure of civil society between 1830 and 1870; they reinforced the modes of civil sociability and became part of the ideals and practices of *Bürgerlichkeit*. Science became a civil practice between 1830 and 1870 as it was institutionalized at universities, transformed into a public good, and slowly accepted, while still disputed, as an element of civil culture and *Bildung*. Inspired by democratic hopes, which could not be realized during the revolution of 1848, new groups of public mediators began to incorporate the natural sciences into civil society and its institutions, thereby promoting the very project of *Bürgerlichkeit*. In particular, the interplay of science, politics, and religion worked as a catalyst in this development and contributed to the expansion of the public sphere. In the biography of Rossmässler, these processes converged paradigmatically.

Democrats and liberals, conservatives and churchgoers, free-religious individuals and atheists contributed to the new appreciation of science in civil culture. All of them helped create a new cultural market of scientific information and nature-oriented ideologies. This market was solely dominated by neither the small group of academic leaders and "public scientists" nor the fervent followers of radical materialism. Instead, popularizers such as Rossmässler who pursued conciliatory strategies in the wake of Alexander von Humboldt's *Cosmos* played a crucial role and formed new enterprises in the service of science education outside state institutions.

These observations facilitate the integration of the history of science into the discussion on the development of civil society and its public sphere during the nineteenth century. In particular, they help to make the image of science in Germany less state centered, less bound to professionalization and academic institution building, and less framed by the establishment of scientific disciplines. Integrating the history of science into the history of civil society helps us to appreciate the immense array of nonuniversity modes that existed for the production, transformation, and consumption of knowledge in German society. The spread of this knowledge into local and regional spaces outside universities—even into leisure activities and local pubs[101]—thereby becomes more visible. This approach allows us to take into account the achievements of those mediators who were not only not supported by, but even excluded from universities, and who are therefore easily dismissed as outsiders and losers.

A shift of interest from the state-funded system of science to nonuniversity attempts to integrate science into civil society helps us to contextualize the history of science in broader terms and to expose and further explore the close link between science in German history and political, social, and religious developments in general. As Roger Cooter and Stephen Pumfrey have rightly reminded historians, it is scarcely possible to draw a rigid opposition between popularization from above and public reception from below. Ultimately, then, we need a concept of science that acknowledges historically existent hierarchies of knowledge, but at the same time is less exclusive. There might then be more space to discuss what some have called "expository science," what Steven Shapin indicated when he mentioned the need to know more about the "ethnoscience" of our own, Western societies, and what Cooter and Pumfrey reflected on when they addressed the concept of "ethno-natural knowledge" or, even more encompassing, "natural knowledge." The latter concept would include everything from "genuine," "official" science to popular science, pseudosciences, and craft knowledge.[102]

This reorientation would also mean that we have to reconsider the umbrella term "science" when entering the sphere of popular culture, as it has been conceptualized by Peter Burke, Roger Chartier, and others.[103] When Rossmässler showed graphics

[101] As a case study, see Anne Secord, "Science in the Pub: Artisan Botanists in Early Nineteenth-Century Lancashire," *Hist. Sci.* 32 (1994): 269–315.

[102] Shinn, *Expository Science* (cit. n. 8); Steven Shapin, "Science and the Public," in *Companion to the History of Modern Science*, ed. R. C. Olby et al. (London and New York: Routledge, 1990), p. 994; Roger Cooter and Stephen Pumfrey, "Separate Spheres and Public Places: Reflections on the History of Science Popularization and Science in Popular Culture," *Hist. Sci.* 32 (1994): 237–67, on p. 254; Adi Ophir and Steven Shapin, "The Place of Knowledge," *Sci. Context* 4 (1991): 3–12.

[103] See Peter Burke, *Popular Culture in Early Modern Europe* (Aldershot, Hants: Wildwood House; Brookfield, Vt.: Gower, 1988); idem, "Popular Culture Reconsidered," *Stor. Storiogr.* 17 (1990): 40–9; Steven L. Kaplan, ed., *Understanding Popular Culture: Europe from the Middle Ages*

of geological findings to local audiences in Saxony in the early 1850s, or when his Humboldt Associations organized botanical excursions into the countryside, did they popularize knowledge that had been produced elsewhere? One might well assume rather that they relied on different layers of knowledge, some that came into existence as a result of learning from current academic research, others drawing on traditional learning experiences such as everyday-life observation.

The exemplary biography of Rossmässler raises questions that are worth being discussed on a larger, transnational scale. How does knowledge acquire the character of public knowledge, and when, how, and under which methodological preconditions is science defined as a public endeavor? Recent studies, for example, have dealt with these problems in the treatment of how the new natural philosophy acquired authority and legitimacy in England from the seventeenth to the nineteenth centuries, establishing the experimental method and the notion of science as a "public science."[104] If "claims to knowledge become accepted insofar as they are embodied in effective acts of communication," as Jan Golinski has aptly written,[105] the requirements for efficiency and the power of communicative strategies—their enabling factors as well as restrictions—deserve close scrutiny. Rossmässler's successes as well as the limits of his resonance are cases in point and underscore the necessity of seeing the validity of public knowledge in the social context of those who promoted this knowledge.[106]

Finally, connecting the history of science to the history of civil society and the public sphere contributes to a reconsideration of some of the master narratives that have been constructed in both fields. A closer look at the interplay of science, politics, and religion reveals the dynamism of a period long regarded as stable and reactionary. This view highlights the ambiguities of modernity, and it underlines the paradoxes of secularization. It helps us to see elements of reenchantment and religious worship of nature in a world that was disenchanted by rationalism and

---

to the Nineteenth Century (Berlin and New York and Amsterdam: Mouton, 1984); Tony Bennett, Colin Mercer, and Janet Woollacott, eds., *Popular Culture and Social Relations* (Milton Keynes, Buckinghamshire and Philadelphia: Open Univ. Press, 1986); Lynn Hunt, ed., *The New Cultural History* (Berkeley and Los Angeles and London: Univ. of California Press, 1989); Christoph Conrad and Martina Kessel, eds., *Kultur und Geschichte: Neue Einblicke in eine alte Beziehung* (Stuttgart: Reclam, 1998). See also several contributions to *Populäres Wissen*, special issue of the journal *WerkstattGeschichte* 8(23) (1999).

[104] Larry R. Stewart, *The Rise of Public Science: Rhetoric, Technology, and Natural Philosophy in Newtonian Britain, 1660–1750* (Cambridge and New York: Cambridge Univ. Press, 1992); Jan Golinski, *Science as Public Culture: Chemistry and Enlightenment in Britain, 1760–1820* (Cambridge and New York: Cambridge Univ. Press, 1992); idem, *Making Natural Knowledge: Constructivism and the History of Science* (Cambridge: Cambridge Univ. Press, 1998); and Jan C. C. Rupp, "The New Science and the Public Sphere in the Premodern Era," *Sci. Context* 8 (1995): 487–507.

[105] Golinski, *Science as Public Culture* (cit. n. 104), p. 3.

[106] For an overiew of these questions, further bibliographical references, and a number of case studies, see Bernard Lightman, ed., *Victorian Science in Context* (Chicago and London: Univ. of Chicago Press, 1997); and Jardine, *Cultures of Natural History* (cit. n. 7). On the development of popular science in the nineteenth century, see, e.g., Roger Cooter, *The Cultural Meaning of Popular Science: Phrenology and the Organization of Consent in Nineteenth-Century Britain* (Cambridge and New York: Cambridge Univ. Press, 1984); John C. Burnham, *How Superstition Won and Science Lost: Popularizing Science and Health in the United States* (New Brunswick: Rutgers Univ. Press, 1987); Bruno Béguet, ed., *La science pour tous: Sur la vulgarisation scientifique en France de 1850 à 1914* (Paris: Bibliotheque du Conservatoire National des Arts et Métiers, 1990); and Jean-Marc Drouin and Bernadette Bensaude-Vincent, "Nature for the People," *Cultures of Natural History* (cit. n. 7), pp. 408–25.

increasingly ruled by institutionalized bureaucracies. And such an approach eventually means to accept that small, sober-looking men such as Rossmässler, excluded from the mainstream of the academic world, indeed contributed to reshaping the contemporary understanding of nature, played an important role in promoting the project of civil society in Germany, and substantially helped make science an integral part of nineteenth-century culture.

Teaching Community Via Biology in Late-Nineteenth-Century Germany

*By Lynn K. Nyhart**

ABSTRACT

In the 1880s, the German schoolteacher Friedrich Junge proposed reforming the school biology curriculum away from taxonomy and replacing it with a hands-on curriculum organized around the concept of the *Lebensgemeinschaft*, or "biotic community." Junge's ideas spread rapidly across Germany, eventually becoming the basis for an entirely reformed curriculum. This paper argues that Junge's reform scheme appealed primarily for two reasons: first, its presentation of nature as a harmonious community resonated with the goals of schoolteachers who wanted to use nature as a model for inculcating proper behavior among their pupils, especially in light of the threat of socialism; and second, it afforded teachers—especially low-status elementary school teachers—a chance to claim much-wanted autonomy for their classroom practices. The paper thus shows how educational reform may be seen as a key historical zone of interaction at the meeting points of science, civil society, and the state.

INTRODUCTION

THE HISTORY OF EDUCATION has long offered historians a fruitful means of examining the intersecting forces of society and state in steering and accommodating social change. As one of the basic institutions of modern society, schools must balance the tensions between maintaining social order and promoting change, and between reproducing elites and giving the rest of the population what they are believed to need to function in society. Where the balance points lie at any given time and place depends on the constellation of power among those invested in schooling, which may include state, local, and religious authorities, school directors, teachers, and parents. The conviction among these various participants that schools are essential to shaping society has often made contests over school reform a revealing mirror of both the beliefs held by these different groups and the relations among them.

In Germany in the 1880s and 1890s, education was a particularly intense field of struggle, as Germans sought to deal with their rapidly changing world. The country was undergoing massive physical, technological, economic, and social changes as it industrialized late and fast. Areas of heavy industry, once mere pockets of production, had begun to eat up whole portions of the countryside, especially along the Rhine, Ruhr, and Saar Rivers. Railroads now crisscrossed the country, moving food,

* Department of the History of Science, University of Wisconsin–Madison, 1180 Observatory Dr., Madison, WI 53706-1393; lknyhart@facstaff.wisc.edu.

goods, and people to the ever expanding cities, where the population of workers and would-be workers was growing ever faster. Political changes accompanied these transformations: whereas liberals dominated much of political life in the 1860s and the early 1870s, by the late 1870s their power was being increasingly challenged on both the left and the right. The National Liberal Party itself began adopting more conservative positions beginning in 1879. In the 1880s and 1890s the political landscape would become even more fractured and polarized between left and right.[1]

Within this context, both liberals and conservatives believed schooling to be a key to Germany's future. Perhaps the most urgent issue was how schools would deal with the problem euphemistically called "the social question": how to contain the bitter class differences that had emerged in the recent dramatic upswing of industrialization and urbanization and their attendant social, economic, and political dislocations. What information, behavior, and values should be taught in the schools to help hold German society together? In what direction should schools shape the future of Germany as a society and as a nation?

In 1885 Friedrich Junge, the head of a girls' school in the Schleswig-Holstein port city of Kiel, entered the reform arena with a book titled *Die Naturgeschichte in der Volksschule: Der Dorfteich als Lebensgemeinschaft* (Natural history in the elementary school: the village pond as a biotic community). The *Dorfteich*, as it became known, catalyzed a chain of pedagogical reforms that by the early twentieth century would transform the teaching of biology in German schools. This paper analyzes how and why, within the context of struggle over German schooling, Junge's reform ideas so rapidly came to stand at the center of debate among schoolteachers over the reform of biology education.

This story shows how schoolteachers could use science—in this case, the theoretical principles associated with Junge's biotic community concept and the pedagogical principles associated with hands-on investigation of phenomena—to educate their pupils toward a vision of nature and society that, while accommodating change, maintained respect for existing social, political, and economic structures. At the same time, this story illuminates the ambiguous position in late-nineteenth-century Germany of reform-minded schoolteachers, who moved back and forth between their roles as low-level servants of the state and independent advocates of their own ideas for reform. This paper thus examines two kinds of articulation between science and civil society: the uses of a particular model of nature and a particular style of science learning to inculcate civic virtues, and the means by which teachers of natural history used the tools of civil society—free associations and the independent periodical press—to assert their own (limited) autonomy in matters of curricular reform.

In tracing both these articulations it is important to bear in mind the conceptualization of civil society that predominated in nineteenth-century Germany. Germans did not see civil society (or its German rough equivalent, *bürgerliche Gesellschaft*) and the state as autonomous realms defined in opposition to one another, but as two

[1] My characterization of this period has been drawn especially from Wolfgang J. Mommsen, *Imperial Germany 1867–1918: Politics, Culture, and Society in an Authoritarian State*, trans. Richard Deveson (London: Arnold 1995); originally published as *Autoritäre Nationalstaat* (Frankfurt: Fischer Taschenbuch, 1990); and David Blackbourn, *The Long Nineteenth Century: A History of Germany, 1780–1918* (New York: Oxford Univ. Press, 1997).

tightly interconnected aspects of the social organization of life beyond the family.[2] Although this idea is often treated as peculiarly German, it is not quite as far from other concepts of civil society as it may initially seem. Indeed, the idea of a citizenry that is not itself part of the formal state bureaucratic apparatus but that makes the state possible through participation in public economic, political, and social life has been central to many conceptions of the state–civil society relationship. In this conceptualization, the behavior of the good citizen, and a shared concept of the society to which he or she contributes, is critical to the success of the larger life of the polity.[3] And in late-nineteenth-century Germany, as elsewhere, teaching the behavioral norms appropriate to good citizenship was seen as a vital function of the public schools.

Attention to these two aspects of civil society—its normative dimensions and its close relationship to the state in German conceptions—is especially appropriate for the case under discussion in this paper, for biology teaching reformers in the 1880s and 1890s devoted little effort to distinguishing between state and civil society. On the contrary, at a time when socialists and communists appeared permanently committed to persuading the rapidly growing industrial lower classes that the only solution to their plight lay in the overthrow of the existing system, a prominent *bürgerlich* response was to reassert the commonalities of civil society and the state. Biology teachers shared in this response by repeatedly making use of analogies that emphasized the similarities among different social and political realms. In fact, as I shall argue, themes of connection, mediation, and reconciliation are central to understanding the success of Junge's program, recurring as they do in the focus on the *Lebensgemeinschaft* as an integrated network of interdependent individuals and in the use of this curriculum to promulgate behavioral norms (including especially respect for others) that applied equally to all and—crucially—could be understood

[2] Georg Wilhelm Friedrich Hegel influentially articulated one version of this view in his *Philosophy of Right* in 1821, in which he presented civil society as the somewhat chaotic, unselfconscious arena for expression of individual, private, self-interest that needed to be tempered, regularized, made self-conscious, and thereby perfected in the form of the state. (For a recent gloss on Hegel's concept of civil society, see John Ehrenberg, *Civil Society: The Critical History of an Idea* [New York: New York Univ. Press, 1999], pp. 121–32.) In 1851 and again in 1864, Munich's influential conservative publisher-editor-journalist and sometime student of German folk culture and history Wilhelm Riehl offered a different assessment of *bürgerliche Gesellschaft*, arguing that civil society and the state were not two different things, but rather two different viewpoints for conceptualizing the structure of society. (Wilhelm H. Riehl, *Die bürgerliche Gesellschaft*, ed. Peter Steinbach [1887; reprinted, Frankfurt am Main and Berlin and Vienna: Ullstein, 1976]; idem, "Über den Begriff der bürgerlichen Gesellschaft" [1864], reprinted in ibid.) Nor were such views of the interconnectedness of civil society and state confined to conservatives such as Riehl. As Daniel A. McMillan has recently put it, "Even the most determined liberals abhorred conflict between state and civil society, and sought not so much the autonomy of civil society from the state, but rather harmonious collaboration between the two." Daniel A. McMillan, "Energy, Willpower, and Harmony: On the Problematic Relationship between State and Civil Society in Nineteenth-Century Germany," in *Paradoxes of Civil Society: New Perspectives on Modern German and British History*, ed. Frank Trentmann (New York: Berghahn Books, 2000), pp. 176–95, on p. 177.

[3] Such behavioral norms are implicit in political theories as different as Jean-Jacques Rousseau's social contract and Hegel's theory of civil society (*bürgerliche Gesellschaft*) and the state, in which social custom (*Sittlichkeit*) plays a prominent role. For a brief overview, see the introduction to this volume; for a more extended discussion, see William A. Dunning, *A History of Political Theories from Rousseau to Spencer* (New York: Macmillan, 1920). For a conservative nineteenth-century elucidation and critique of behavioral norms characterizing *bürgerliche Gesellschaft* see Riehl, *Die bürgerliche Gesellschaft*; and idem, "Über den Begriff der bürgerlichen Gesellschaft" (both cit. n. 2).

as deriving from nature. Overall, the concepts of *Gemeinschaft* (community) and *Lebensgemeinschaft* (biotic community) were used in a way that blurred the distinctions between different social, institutional, and political categories to present an organicist vision of society variously articulated in connection with state, economy, or civil society.

Junge's program included more than a theorized model of nature and society. Central to it was also a particular approach to learning, one which we might today call "hands-on" or investigative learning about nature. Here, too, a particular view emerges of the appropriate characteristics of the ideal citizen, this time understood not in relation to the content of the curriculum, but in relation to its pedagogy: the citizen modeled here is one who sees and thinks for himself or herself but understands that the rule of law must be obeyed. The program associated with the *Lebensgemeinschaft* thus balanced innovative subject matter and a progressive pedagogical approach with a more conservative ideological message aimed at maintaining social harmony through the inevitable processes of change. Although the biological content of Junge's program remained controversial, its pedagogy, the opportunities it offered schoolteachers to shape their own curriculum, and the ideological messages it supported, I argue, made it a powerful force for change in biology education.

The paper is divided into five sections. After laying out some basics concerning the larger educational reform context in the first section, I discuss in the second Junge's background and the basic elements of his program. The third section gives a brief overview of the spread of Junge's ideas, focusing on the various venues in which the *Dorfteich* gained attention. In the fourth section, I consider the reception of Junge's pedagogical innovations and biological concepts among both elementary and high school teachers, showing which aspects were embraced, which were more likely to undergo modification, and which tended to be rejected altogether. The fifth and final section examines more closely the ideological messages of the *Dorfteich* within the broader context of character education, arguing ultimately that Junge's program succeeded, despite its controversial scientific content, because its ideological uses were so powerful.

I. THE EDUCATIONAL SETTING AND SCHOOL REFORM

Natural history was generally taught to children between the ages of nine and thirteen. In the structure of German schools, this meant that Junge's program could be of interest to teachers in elementary, middle, and higher schools, since each school type included children of these ages. To understand the program's aims as well as its reception, then, it is useful to know something about these different school types and the larger battles over school reform, which revolved around the schools' profoundly class-based nature and their perceived potential for being a fundamental instrument for shaping society.

Although we usually think of public education as a function of "the state," in fact, when we look at the structure of German school administration in the late nineteenth century, the nation-state rapidly dissolves into lower-level entities. In the first place, no uniform national educational standards existed: each of Germany's states made its own rules. Nor were these regulations necessarily the ones that counted most, for education, especially at the lower levels, had long been largely a matter of local, regional, and provincial interest. Within each state, the structure of education varied

Table 1:
German school types, grades, and ages, ca. 1890. Shaded area represents most frequent grades for natural history instruction. The higher schools traditionally counted their grades from the oldest age downward, using roman numerals; each of the highest three grades was divided into two years, designated Oberstufe *(O) and* Unterstufe *(U). Middle-level schools numbered their grades from oldest (1) to youngest (6), while elementary schools typically numbered theirs in the opposite direction.*

Age	Higher Schools (*Gymnasium, Realgymnasium*)	Middle Schools (e.g., *Bürgerschulen*)	Elementary Schools (*Volksschulen*)	Junge's Curriculum (*Mädchenschule*)
18	OI			
17	UI			
16	OII			
15	UII	1		
14	OIII	2	8	5
13	UIII	3	7	4
12	IV	4	6	3
11	V	5	5	2
10	VI	6	4	1
9			3	
8			2	
7			1	

enormously from province to province and city to city, as well as between urban and rural districts, variations deeply intertwined with local and regional regulations and customs. Cities traditionally supported their own schools, with little or no financial or administrative involvement from the state. Both urban and rural elementary schools were also deeply shaped by local confessional practices and politics, since nearly all were run as either Catholic or Protestant schools. (Efforts to develop a secular school type in the 1870s, though a significant feature in the reform landscape of the period, did not succeed in upending the confessional schools.)[4] Even the basic structures of schooling varied wildly: although most elementary schools *(Volksschulen)* had only two or three grades, they might contain as many as seven or eight. In Prussia, reforms in 1882 reorganized that state's educational structure but did not reduce its complexity, instead producing a finely graded hierarchy beginning with the elementary schools and extending through at least six different school types with varying educational requirements and grade levels to the pinnacle, the *Gymnasium*. Many of these school types overlapped in their grade levels and curricular expectations, but each had a separate niche in the attempt to create a curricular order. (See Table 1.) Other states had equally arcane and complicated systems. Generally, the most important division was that between "higher" schools, which prepared students for university and polytechnical education; and "lower" schools, which encompassed elementary and middle-level schools—all those schools that did not prepare

[4] Frank-Michael Kuhlemann, "4.I.i. Niedere Schulen," in *Handbuch der deutschen Bildungsgeschichte*, vol. 4: *Von der Reichsgründung bis zum Ende des Ersten Weltkrieges*, ed. Christa Berg (Munich: C. H. Beck, 1991), pp. 179–227, especially pp. 180–3.

students for higher education. Throughout Germany in the imperial period, school forms were in flux and constituted a field of continued struggle over who would have access to what sorts of schooling, advanced education, jobs, and legal privileges.[5] Reform efforts aimed variously at the structure of schools, the educational requirements of the teachers, and curricular content.

Structure was a major battleground: it had long been the case that possibilities for upward mobility were tied to, and limited by, access to the right type of school. Unlike most small towns and villages, larger cities had more than one school, and the decision about what sort of school to send one's child to was, more often than not, a matter of class. For example, in the city of Kiel, where Junge taught, there were four kinds of schools. The free elementary schools, or *Volksschulen*, intended for the poorest segment of Kiel's society, taught only the most basic subjects. The regular *Bürgerschulen* (one for boys and one for girls, both founded in 1861) charged a modest tuition and had a broader curriculum. The "higher" girls' and boys' *Bürgerschulen*, also founded in 1861, were intended to serve Kiel's better citizens (though not university-bound boys); still their fees were kept unusually low to open up access to the lower orders (in line with Kiel's liberal educational policy). They sought primarily to compete with the private schools for the sons and daughters of the *Besitzbürgertum*, or propertied middle class, and offered instruction in a wide assortment of "modern" subjects, including two modern languages. At the top of the system stood the *Gelehrtenschule*, or humanistic *Gymnasium*, which prepared boys for the university and the futures opened up by it.[6] This system was replicated elsewhere, with school fees largely determining the options open to different families; because of the different curricular structures in the various schools, once a child was placed in one kind of school it was very hard to jump to another.

The efforts by reformers (often themselves from modest backgrounds) to open up structural access to the higher schools for pupils of the lower and middling classes created a major arena of struggle, in which each school type pushing to open access upward generally had to battle a certain resistance from teachers and administrators of the next higher level of school. The latter, though willing in theory to admit a trickle of the best and brightest from the lower orders, sought to maintain their own level of social exclusivity (while themselves often pushing to open up access to yet a higher school level).[7]

[5] For an overview of German school types, see Detlef K. Müller and Bernd Zymek, *Sozialgeschichte und Statistik des Schulsystems in den Staaten des Deutschen Reiches, 1800–1945*, pt. 1 of *Höhere und mittlere Schulen*, vol. 2 of *Datenhandbuch zur deutschen Bildungsgeschichte* (Göttingen: Vandenhoeck & Ruprecht, 1987). On school reform, especially at the lower and middle levels, see Hellmut Becker and Gerhard Kluchert, *Die Bildung der Nation: Schule, Gesellschaft und Politik vom Kaiserreich zur Weimarer Republik* (Stuttgart: Klett-Cotta, 1993); and Folkert Meyer, *Schule der Untertanen: Lehrer und Politik in Preußen, 1848–1900* (Hamburg: Hoffmann & Campe, 1973). For analysis of the reform movement at the higher school levels, see James C. Albisetti, *Secondary School Reform in Imperial Germany* (Princeton, N.J.: Princeton Univ. Press, 1983); also idem, *Schooling German Girls and Women: Secondary and Higher Education in the Nineteenth Century* (Princeton, N.J.: Princeton Univ. Press, 1988).

[6] Peter Wulf, "Kiel wird Grosstadt (1867 bis 1918)," in *Geschichte der Stadt Kiel*, ed. Jürgen Jensen and Peter Wulf (Neumünster, Schleswig-Holstein: Karl Wachholtz, 1991), pp. 257–9; Jens Godber Hansen, *Schule–Spiegel ihrer Zeit: Die Geschichte der Ricarda-Huch-Schule in Kiel, 1861–1986* (Mitteilungen der Gesellschaft für Kieler Stadtgeschichte, Bd. 72) (Kiel: Gesellschaft für Kieler Stadtgeschichte, 1986), especially pp. 24–31.

[7] On structural reform as a vehicle for social change, see Becker and Kluchert, *Die Bildung der Nation* (cit. n. 5), especially pp. 1–28.

The question of structural mobility in the schools applied not only to schoolchildren. Teachers, too, were segregated by educational background and thus to a great extent by class as well. To teach in the higher schools normally required a university degree as well as a state license obtained through an examination. Elementary school teachers participated in an entirely separate educational system, organized around teaching seminaries that they generally attended in their late teenage years. Most states typically had a three-year course, though some required only two years of training and a few as many as five (including a preparatory year or two).[8] Prospective elementary teachers could, with appropriate preparatory courses, attend the seminaries after having attended *Volksschulen*, without having attended a higher school; indeed, until the turn of the century the seminaries had lower status than the *Gymnasien*, since the seminary certificates did not confer admission to the universities. Graduates of the seminary courses could teach at elementary schools and perhaps eventually rise to the level of running their own teaching seminaries. They could also teach in the middle-level schools, though only in the case of girls' middle schools were they eligible to rise to the level of director. But seminary graduates could not teach in the higher schools, which required a university degree. Moreover, they drew such low salaries that most had to take on some kind of second job to survive, especially if they had families to support. Nor were they eligible for the voluntary one-year military service that would allow them the prestige-increasing entry into the reserve officer corps and access to positions in the imperial bureaucracy. In sum, their educations, salaries, and lack of legal privileges put them at or below the bottom end of the *Bürgertum*. It is not surprising that the lower schools served as a perennial wellspring of school reformers.[9]

Structural considerations were only part of the picture, though. Of what subjects should the curriculum consist, and to what ends? Traditionally, the lower schools emphasized reading, writing, arithmetic, and religion, while the higher schools offered a classical education that promoted an idealistic cultivation of the self associated with the term *Bildung*. But already in the 1860s and 1870s both levels of schooling faced new pressures to include more "modern" subjects. In Prussia, new school regulations in 1872 dramatically increased the hours elementary teachers were expected to devote to the so-called *Realfächer*, the subjects that concerned factual knowledge of the world (the natural sciences, geography, and history), seen by reformers as essential for the modern citizen. At higher educational levels, too, the development of *Realschulen* and *Realgymnasien* as alternatives to the classics-oriented humanistic *Gymnasium* marked a pronounced challenge to classical education, one that led to major and protracted battles in the field of secondary education over reducing the amount of Latin and Greek instruction and replacing those hours with other subjects, especially modern languages, German history, and natural sciences.[10]

The struggles over the content of education were not, of course, merely about how many hours of what subjects should be covered in which grades. At a deeper level,

[8] J. Tews, "Lehrer an Volksschulen," in *Enzyklopädisches Handbuch der Pädagogik*, ed. Wilhelm Rein, 7 vols. (Langensalza: H. Beyer, 1895–1899), vol. 4, pp. 370–9.

[9] Becker and Kluchert, *Die Bildung der Nation*, pp. 81–2; Meyer, *Schule der Untertanen* (both cit. n. 5), pp. 92–109.

[10] On the battle over modernizing the curriculum, see especially Albisetti, *Secondary School Reform* (cit. n. 5).

they were about what sort of future citizens the schools should be developing, how best to form such citizens, and who would get to choose the paths of citizen formation. From the 1870s on, reformers vied with one another to demonstrate the potential of different subject matter, organized in different ways, to develop not only the sorts of knowledge but also the moral and ethical characteristics required of the good German citizen. Thus in 1884 a teacher at the annual provincial teachers' meeting in Silesia argued that a main task of the elementary school was to instill in children the sense of joy in work and desire for work (*Lust an der Arbeit*), while in 1885, the same year Junge's *Dorfteich* appeared, a teacher at the annual meeting of the national teachers' association could ask in the title of a speech, "What must the school do to fulfill the moral task of natural history instruction?"[11] Physics teachers argued that the proper teaching of such scientific concepts as "work" in physics could reinforce respect for work in the sense of labor as well; a scientific understanding of "law" could teach students the limits of opinion. Proponents of particular subjects ranging from art to economics, from mathematics to geography and, of course, natural history, all made pitches for the moral, ethical, and/or practical value of their special fields.[12]

The intensity of these claims would reach a peak soon after 1889, when the new emperor, Kaiser Wilhelm II, who had ascended the throne the previous year, weighed in with his own agenda: schools should educate students toward "fear of God and love of the fatherland," and away from communistic and socialistic ideas.[13] There followed in 1890 a major national school conference and in 1892 new regulations, in which modern history, geography, and physical education—all aimed at building the spirit and body of the nation—took on new prominence, while attention to the exigencies of modern practical life came in the form of new emphasis on drawing, modern languages, and natural sciences.[14] Although the new formal guidelines for curricular reform were put into place in the early 1890s, it is important to recognize that they represented a culmination, not a starting point, of reformers' activities. Junge's reform proposal was the most prominent of these in the field of natural history.

Junge's *Dorfteich* and the curriculum organized around the *Lebensgemeinschaft*

[11] Report on Allgemeine schlesische Lehrerversammlung zu Breslau, *Deutsche Blätter für erziehenden Unterricht* 11 (1884): 155 (hereafter cited as *DB*); G. Bergemann, "Was hat die Schule zu thun, damit die sittliche Aufgabe des naturgeschichtlichen Unterrichts erfüllt werde?" *Zeitschrift für mathematischen und naturwissenschaftlichen Unterricht* 16 (1885): 536–44

[12] Kathryn M. Olesko, "The Politics of Fin-de-siècle Physics Pedagogy in Europe," *History of Physics Newsletter: A Forum of the American Physical Society* 7(3) (1998): 9–10; published in full at www.aps.org/FHP/olesko.html; on the art education movement, see Sterling Fishman, "Alfred Lichtwark and the Founding of the German Art Education Movement," *Hist. Educ. Quart.* 6(3) (1966): 3–17; on economics, see the discussion of Oskar Pache in section 5 of this article; on mathematics, see Gert Schubring, "Die Mathematik: Ein Hauptfach in der Auseinandersetzung zwischen Gymnasien und Realschulen in den deutschen Staaten des 19. Jahrhunderts," in *Bildung, Staat, Gesellschaft im 19. Jahrhundert: Mobilisierung und Disziplinierung*, ed. Karl-Ernst Jeismann (Stuttgart: Steiner Verlag Wiesbaden, 1989), pp. 276–89. Examples of essays touting the value of individual subjects and how to go about teaching them abound in the pages of journals such as the *DB*.

[13] "Erlaß Wilhelms II. zur Bekämpfung sozialistischer und kommunistischer Ideen durch die Schule und Ausführungsbestimmungen (1889)," in *Politik und Schule von der Französischen Revolution bis zur Gegenwart: Eine Quellensammlung zum Verhältnis von Gesellschaft, Schule und Staat im 19. und 20. Jahrhundert*, ed. Berthold Michael and Heinz-Hermann Schepp, 2 vols. (Frankfurt am Main: Athenäum Fischer Taschenbuch, 1975), vol. 1, p. 409.

[14] Becker and Kluchert, *Die Bildung der Nation* (cit. n. 5), pp. 74–9.

idea, then, entered a turbulent educational field. Like other reform ideas, Junge's proposals spoke to the need to modernize the curriculum as well as to the moral and ethical lessons that could be drawn from nature. To understand how he sought to do this, it is now necessary to turn to Junge himself and the exposition of his program.

II. FRIEDRICH JUNGE AND THE *DORFTEICH*

Friedrich Junge (1832–1905) was in many ways a typical German self-taught teacher-naturalist.[15] Raised in extremely modest circumstances in the Holstein town of Oldesloe by his mother (widow of the shoemaker father he never met), he was taken on at age sixteen by his elementary schoolteacher to prepare for the seminary and subsequently attended the three-year teaching seminary in the nearby town of Segeberg. Between 1854 and 1873 he worked at four different lower schools around Holstein; in 1873 he moved to the first girls' *Bürgerschule* in Kiel, where he taught until his retirement in 1899. When he published the *Dorfteich* in 1885, Junge had been rector of the girls' school for seven years.

Like most teachers of the lower school grades, Junge had learned his science, including natural history, largely on his own, filtered through the existing pedagogical and popular science literature. As a seminarian, he, along with his classmates, had been inspired by Friedrich Adolf Diesterweg, the prominent liberal *Volksschule* reformer of the 1830s–1850s, to read the newly founded popular magazine *Die Natur*. This magazine had introduced him to popular books in natural history, including those by the magazine's coeditor, Karl August Müller, whose Humboldt-inspired *Buch der Pflanzenwelt. Versuch einer kosmischen Botanik* (Book of the plant world. Attempt at a cosmic botany) Junge found particularly appealing.[16] Curiosity and an experimental nature had led him in the late 1850s to create a photographic camera "out of a cigar-box and a magnifying glass"; later borrowing another camera, he had earned money on the side taking photographic portraits. Out of these funds he had purchased a microscope to study plant development. He also had used it to moonlight as a "trichina-examiner" (presumably inspecting pork), which in turn had earned him enough money to buy his own camera. He thus bootstrapped himself into considerable scientific and technical knowledge, experience that undoubtedly strengthened his conviction that every teacher could, indeed, be a scientific researcher, as Diesterweg had argued. Only in the late 1870s did Junge take a formal scientific course, attending the public zoology lectures for teachers held by the Kiel university professor Karl Möbius (himself once a schoolteacher) and working with Möbius's assistant Friedrich Heincke in the Kiel university zoology laboratory.

It was from Möbius that Junge learned the two terms that would figure prominently in his work: *Erhaltungsmäßigkeit* and *Lebensgemeinschaft*. *Erhaltungsgemäß*

[15] Unless otherwise noted, all information on Junge's background comes from "Friedrich Junge," *Deutsches Biographisches Archiv* (DBA) Fiche 615, pp. 37–50, which includes an autobiographical sketch.

[16] The term "cosmic botany" may in fact refer to Humboldt's *Kosmos* and be more properly translated as "in the manner of the *Kosmos*." I thank Tom Broman for pointing out this possibility to me. On the influence of Humboldt's *Kosmos* on popular and amateur naturalists in the mid-nineteenth century, see Andreas Daum, *Wissenschaftspopularisierung im 19. Jahrhundert: Bürgerliche Kultur, naturwissenschaftliche Bildung und die deutsche Öffentlichkeit, 1848–1914* (Munich: Oldenbourg, 1998), pp. 138–67, 273–86; and idem, article in this volume titled "Science, Politics, and Religion: Humboldtian Thinking and the Transformations of Civil Society in Germany, 1830–1870."

and *Erhaltungsmäßigkeit* referred to the ability of organisms to maintain themselves in relation to their surroundings; these terms represented Möbius's effort to find nonteleological language to describe this phenomenon. *Lebensgemeinschaft* was a translation of Möbius's neologism *Biocönose*, a word Möbius coined in 1877 to refer to an ecological community; Möbius himself first translated the term as *Lebensgemeinde* but also referred to the same concept, right from the start, as a "*Gemeinschaft* of living beings." Both *Erhaltungsmäßigkeit* and *Lebensgemeinschaft* were central to Junge's pedagogical program.

The *Dorfteich* set out from a simple but radical idea. Instead of organizing the teaching of natural history around taxonomic categories, Junge proposed organizing it around *Lebensgemeinschaften*, or biological communities, groups of organisms that lived in a particular chemico-physical setting and were dependent on that setting and on one another for their survival.[17] As he elaborated in the book's programmatic essay, "Ziel und Verfahren des naturgeschichtlichen Unterrichts" (The goal and method of natural history instruction), this radical shift from taxonomic categories to those of the community and its members had numerous pedagogical benefits. Following reformist precepts that had been popular since at least the time of Johann Heinreich Pestalozzi in the previous century, Junge's approach started off from the familiar, material world surrounding the children. The example he worked out most thoroughly in this book was, of course, the village pond, but he emphasized that whatever was local to the children was the place to start. (This meant that teachers could not blindly follow Junge's model but would in most cases have to create their own curricula—something for which Junge was often criticized, but which appealed to other populists.) A second, closely related precept was that the teaching should take place at least in part in a natural setting. Field trips outside the walls of the school, to learn about nature *in* nature, were an essential aspect of the curriculum; students then brought back to the classroom elements of that place to study, using school aquariums, terrariums, and gardens to keep the organisms alive. But the experience of nature *in situ* was crucial, for several reasons. For one, it kept things concrete and familiar, on the assumption that children were most interested in that with which they had daily contact. For another, the learning was to be "hands-on," mediated through the various senses in accordance with pedagogically prominent ideas in psychology, such that children would learn through *Anschauung*, visual and tactile experience of the objects.

These various pedagogical methods and principles were part of a long tradition in primary education in the German-speaking world, hardly radical in themselves. Indeed, the famous early-nineteenth-century natural history pedagogue August Lüben (whose books were still being used in the 1880s) had advocated many of them, and liberal reformers such as E. A. Rossmässler in the 1850s and 1860s had been touting the same ideas (though apparently to little effect, since the need for reform

[17] As Junge put it, "A *Lebensgemeinschaft* is a totality of beings that live together in accordance with the internal law of maintenance [*Erhaltungsmässigkeit*] because they exist under the same chemical-physical influences and, moreover, frequently depend on one another, in any case on the whole, and also have an effect on one another and on the whole." Friedrich Junge, *Naturgeschichte. I. Der Dorfteich als Lebensgemeinschaft nebst einer Abhandlung über Ziel und Verfahren des naturgeschichtlichen Unterrichts*, 2d ed., (Kiel and Leipzig: Lipsius & Tischer, 1891), p. 34 (hereafter cited as *Dorfteich*).

remained).[18] What gave Junge's plan its novelty was his proposal to jettison the traditional principles of taxonomy as the essential underlying structure in favor of the concept of the community.

Junge offered numerous reasons for this extreme move. "The elementary school is an institution for general education, not for scholarly [*wissenschaftliche*] education. By contrast, . . . *the [taxonomic] system is a scholarly instrument*, which cannot be an end in itself for the school." Scientists need it, as do teachers, to help them keep the enormous volume of facts of nature in order, but regular people do not. Junge continued: if, as many naturalists and teachers have long agreed, the goal for the general school population is to gain an appreciation of the unity of nature, an emphasis on systematics will fail to achieve this goal. Children (and even many teachers) cannot gain an overview of the entire system because it is too large and complex. Moreover, the emphasis on the "logical" unity of the system, according to Junge, requires ripping apart things that belong together in nature: living beings and their settings, form and function, the linked nature of the whole. "Nature itself makes no distinction between organs and their activities, between the existence [*das Sein*] and the life of a thing." The separation of systematics from physiology and the study of anatomy and physiology based on dead, anatomized bodies make it impossible to study the phenomena of life in any natural way and will yield only broken, separate shards of knowledge, not an understanding of life as a whole. Finally, children studying systematics quickly lose sight of the whole and thus get bored with the subject, becoming uninterested in pursuing their studies any further on their own. In sum, the traditional structure of teaching natural history through systematics is "unnatural," with regard to both nature and the child.[19]

In the context of the educational reforms of the time, Junge's opening gambit criticizing existing natural history instruction as formal, unnatural, and overly dominated by taxonomic memorization shows him aligning himself with populist and modernizing reformers. These men opposed classical education as overemphasizing rational, logical aspects of learning to elitist ends, and argued instead for an approach that would truly benefit all schoolchildren. Junge's own program sought to achieve such a broadly appealing education by setting the concept of the unity of nature at its center. Repeatedly referring to Alexander von Humboldt, whose statement "The richness of science no longer lies in the abundance of facts but in their linkage" he used as an epigraph both for a chapter and for the book as a whole,[20] Junge brought specificity to the general concept of a "unified nature" in two ways: through the *Lebensgemeinschaft* principle itself and through a set of "laws" or principles of living nature. On the one hand, the manifold connectedness of nature can be found in unities at many levels. One is not required to study all of nature at once to understand its unity and connectedness; rather, these features may be found in individuals and communities. Every organism is a unit made up of coordinated parts that contribute to the whole, and similarly, organisms in a community both depend on and contribute to the maintenance of the whole. Thus by studying individual

[18] Ibid., p. 3. On Rossmässler, see Daum, "Science, Politics, and Religion;" and idem, *Wissenschaftspopularisierung* (both cit. n. 16), pp. 61–3, 138–67, and passim.
[19] Junge, *Dorfteich* (cit. n. 17), pp. 3, 7–8.
[20] "Die Reichtum der Naturwissenschaft besteht nicht mehr in der Fülle, sondern in der Verkettung der Thatsachen," quoted in ibid., on first title page and p. 8.

organisms and their composition into a community it is possible to illustrate the principle of a whole connected system in a relatively simple, direct, and concrete way.

On the other hand, the various modes of connectedness can also be described in the form of regularities or "laws of nature" governing the maintenance of life, which give order and specificity to what might otherwise be a chaos of connections. Drawing on a list published in a textbook by the Viennese zoologist Ludwig Schmarda,[21] Junge proposed eight laws that he thought should guide the teacher and—when appropriate—the students. The most important of these were the first four and the last:

> 1. The law of maintenance [*Erhaltungsmäßigkeit*]: [an organism's] living location, mode of life, and anatomical-physiological apparatus for living all correspond to one another. One can also say: what the animal (the plant) has, it needs to live in this place, and vice-versa: what it needs for its life in this setting are certain organs. . . .
> 2. The law of organic harmony: "every being is a link in the whole." . . .
> 3. The law of accommodation or adaptation: mode of life and anatomical-physiological structures adapt themselves (up to a certain degree) to changed conditions, and vice-versa. . . . This law is a special form of 1). . . .
> 4. The law of division of labor and the differentiation of organs. . . . The more the work of the whole is divided among individual organs, the more perfectly will the work be carried out. . . .
> 8. The law of economy [*Sparsamkeit*] in space and in number. Folding of leaves in the bud, of wings in the insect's pupa; the more intensive the care of the offspring [*Brutpflege*], the smaller the number of eggs.[22]

These biological precepts had been promulgated by professional biologists since midcentury and would come as no surprise to any advanced student of zoology. In the context of school teaching, however, the list took on normative dimensions that would have been far less visible in a university-level scientific textbook, for Junge explicitly sought to use his biological laws as models of lawfulness for schoolchildren. By teaching that the world is a unified whole and that all its members, including humans, are parts of the whole, Junge stated, the child will receive an answer to the most burning question of all: "Who am *I* in this diversity?" The answer is, "You are a link [*Glied*] in the whole, you receive and give, you are dependent and have an effect." In case this was not clear enough, Junge elaborated in a footnote: "If a person recognizes himself as a member [*Glied*] of a community, there follows directly from this realization his rights but also duties toward the other members—thus here toward one's fellow humans."[23]

Elsewhere in laying out his program, Junge found other occasions to make links between the natural and the human, with the goal of emphasizing harmonious mutual support in the service of a larger whole. A husband and a wife, for example,

[21] Ludwig K. Schmarda, *Zoologie*, 2d ed. (Vienna: Wilhelm Braunmüller, 1877), pp. 207–15. It should be noted that Schmarda had quite a few more laws that did not make it into Junge's list, such as the law of chemical composition, the law of conservation of energy, the law of nutrition (every animal requires protein), the law of connection of parts, and various "laws" of form. Although Junge later professed to have derived his laws from Schmarda (DBA 615 [cit. n. 15], pp. 37–50, especially p. 40), similar principles may be found in the works of Rudolf Leuckart from the early 1850s on. See, e.g., Carl Bergmann and Rudolf Leuckart, *Anatomisch-physiologische Uebersicht des Thierreichs. Vergleichende Anatomie und Physiologie* (Stuttgart: J. B. Müller, 1852).

[22] Junge, *Dorfteich* (cit. n. 17), pp. 10–13.

[23] Ibid., p. viii.

form a kind of community, because each helps out the other to maintain the household as a whole. A city can be viewed as a *Lebensgemeinschaft*, too. Although the "struggle for existence" [*Kampf ums Dasein*] and competition may exist "between two members who have absolutely equal living conditions," normally each member, through working to maintain himself or herself, will contribute to the health of the city as a whole. Conversely, when a citizen works directly for the city, this contributes back to his or her own well-being. A state is a still more complex community in which the forms of dependence are more intricate and less visible and yet the same: "give and take, the rendering of service and the dependence of the individual" are true here, too.[24] Thus although Junge did not press the point further here, it seems pretty clear that in learning that the laws of community were natural, inescapable, and benevolent, children were also supposed to apply those laws to themselves, to practice harmonious adaptation to their circumstances (even when those changed), to understand their occupations as part of a natural system of division of labor, and to exercise thrift in their daily lives.

In light of the later intensified subordination of the individual to the state in Nazi ideology, it is tempting to view these prescriptions as profoundly undemocratic and even antimodern. In the context of *bürgerlich* fears of radicals advocating the revolutionary overthrow of the state in the 1880s, however, these messages were more ambiguous. Although we lack direct evidence on Junge's politics, his entire background suggests that he stood in the liberal part of the political spectrum. His third law, "accommodation," showed acceptance of change to be part of nature. His emphasis on duty to others was balanced by an emphasis on individual development; his theory considered an individual's striving for his or her own self-maintenance to contribute naturally to the good of the whole; and his idea that all individuals contribute to the welfare of the whole could be viewed as valorizing all kinds of work, including industrial labor. In addition, as we will see, his emphasis on observation and hands-on experience aimed at teaching children that individual observation, tempered by logical judgment, was the key to secure knowledge; thus it was understood that individuals make their own knowledge, rather than simply receive it from a higher authority.[25] That his program lent itself to different socio-political readings at the time is evident from its reception, to be discussed in later sections of the paper.

In his program for teaching, then, Junge blended a number of important ingredients: an emphasis on the biology of the living organism in relation to its organic and inorganic surroundings (as opposed to focusing on formal characteristics and the memorization of taxonomic names and relationships) coupled with a pedagogy that took children away from book learning to living nature; a sustained vision of organic harmony mediated through the subordination of parts to a larger whole; and a commitment to education as a moral enterprise whose goal was to raise children into upright members of society. Each of these programmatic elements struck a chord with others in the broader teaching community.

[24] Ibid., p. 33.

[25] See, especially, Junge, *Naturgeschichte. II. Die Kulturwesen der deutschen Heimat nebst ihren Freunden und Feinden, eine Lebensgemeinschaft um den Menschen. I. Die Pflanzenwelt* (Kiel and Leipzig: Lipsius & Tischer, 1891), p. v.

III. THE SPREAD OF THE *DORFTEICH* GOSPEL

Junge's program circulated with incredible speed through the ranks of schoolteachers in the 1880s. As a curriculum that crossed over between the higher and lower schools, it made its way into the hands of reformers at both levels (though as we shall see, his program appealed more to teachers in the lower schools). Some teachers evidently picked up on Junge's initial presentations in four 1883 articles, while others read the book itself.[26] Still others learned about it either through the well-developed systems of teachers' meetings or through reading reports on the meetings in the pedagogical press, which also published numerous reviews of the book and other books that engaged with it. These forums for communication show how the tools of the public sphere could be used in the service of educational reform; they also reveal just how segregated the communities of higher and lower school teachers were from one another.

Lower school teachers and their own instructors in the seminaries received Junge's program in the context of associations devoted to pedagogy or to improving teachers' social and economic standing, such as the Allgemeiner deutscher Lehrerverein (General German Teachers' Association). In addition to discussing pedagogical issues, this association worked tirelessly to remove lower school teachers from local confessional oversight and place them under a secular school administration system, to improve wages and death benefits, and to improve the educational opportunities for teachers. In 1886 Junge's book was presented at the Leipzig Teachers' Association meeting as part of a lecture reviewing the recent *Volksschule* literature and at the Silesian Pestalozzi and Provincial Teachers' Association, where the assembled audience voted in support of the reviewer's theses, which demanded reform in natural history teaching along the lines Junge advocated. In 1887 his ideas were discussed at the annual meeting of the Verein für wissenschaftliche Pädagogik (Association for Scientific Pedagogy), which met that year in Leipzig, and at the annual national meeting of seminary teachers in Eisleben, near Halle, where the presentation of Junge's ideas elicited a lively discussion. Those who could not attend the meetings could learn about them in such pedagogical journals as the *Deutsche Blätter für erziehenden Unterricht*, which not only reported on these meetings but also reviewed the literature. Eduard Scheller gave the *Dorfteich* great press in this teaching reform journal, concluding his review of it by saying that it was an unprecedented book that would become "a milestone in the development of natural history instruction."[27]

[26] The articles were Junge, "Sollen Gesetze des organischen Lebens im Volksschulunterricht vorkommen? Welche etwa?" *DB* 10 (1883): 43–5, 52–3; idem, "Das Gesetz der Erhaltungsmässigkeit im naturgeschichtlichen Unterricht der Volksschule," ibid., pp. 107–9, 115–7; idem, "Entwurf eines Pensenplans für den Unterricht in Naturgeschichte für die erste Mädchen-Bürgerschule in Kiel," ibid., pp. 251–3, 259–62, 267–71; and idem, "Was soll naturgeschichtlicher Unterricht in der Volksschule? Aphorismen." (*Erziehungsschule* 1883, no. 7). I have not been able to obtain the latter article, but it is reviewed together with Junge's very first lecture on the topic, to the Schleswig-Holstein Teachers' meeting in Altona in 1882, "Über die Methode des naturgeschichtlichen Unterrichts," in *DB* 10 (1883): 207–9.

[27] "Die neuere naturhistorische Volksschulliteratur im Leipziger Lehrerverein," *Leipziger Tageblatt* Nr. 178 (1886), Beilage V, reprinted in *Zeitschrift für mathematischen und naturwissenschaftlichen Unterricht* 17 (1886): 626–7; H. Grabs, [report on] "General-Versammlungen des Schlesischen Pestalozzi- und Provinzial-Lehrervereins in Sagan." (Lecture by Seminaroberlehrer Waeber [Liegnitz], "Ist der Unterricht in der Naturbeschreibung in der Volksschule einer Reform bedürftig?") *DB*

Word spread quickly by other means as well. In 1887 Otto Frick, leader of the movement for unified schooling, opened his series "Schriften des Deutschen Einheitsschulvereins" with a monograph that advocated Junge's *Lebensgemeinschaft* as the best approach to teaching nature study. Already that year a newly published natural history textbook could tout itself in its subtitle as following "the new methodical principles of treatment and organization (*Lebensgemeinschaften*)," and by 1889 a tired reviewer could refer to the *Lebensgemeinschaft* principle as having "instantly become a kind of dogma."[28] In 1892, at the annual meeting of the General Saxon Teachers' Association, a group of Dresden teachers took the calls for reform one step further, displaying an exhibit that illustrated how to use visual aids to teach about *Lebensgemeinschaften*, using the River Elbe near Dresden as its key biotic community.[29] By 1894, nine years after the publication of the *Dorfteich* and eleven years after Junge's ideas first saw print, an advocate of his program could afford to skip over the meaning of *Lebensgemeinschaft* "since the majority of readers certainly know long since" about it.[30]

Although the original title of Junge's book, *Naturgeschichte in der Volksschule*, suggested an intended audience of elementary teachers, the work almost immediately began to reach teachers in the higher schools, more often, it appears, through their journals than through teachers' meetings. In 1886, in the first volume of the new review journal *Jahresbericht über das höhere Schulwesen* (Annual report on the higher school system), Junge's book was flagged with the comment that his idea of "*Lebensgemeinschaften*—that is, a group of animals that stand in biological connection to one another— . . . deserves careful scrutiny." (And indeed, in the second edition of Junge's book, the title was changed to read simply *Naturgeschichte*, reflecting a broadening of its audience.)[31] In September 1887 news of Junge's program reached perhaps the most prestigious site for higher school teachers, the annual Meeting of German Scientists and Physicians (where schoolteachers rubbed shoulders with university professors); at the section for science education there, a schoolteacher from Strasbourg gave a lecture advocating Junge's community-oriented pedagogical structure, which was then picked up by the pedagogical review literature.[32]

That same year, J. C. V. Hoffmann, editor of the *Zeitschrift für mathematischen*

---

13 (1886): 170–1; J. C. V. Hoffmann, "Ein Besuch der Jahresversammlung des Vereins für wissenschaftliche Pädagogik," *Zeitschrift für mathematischen und naturwissenschaftlichen Unterricht* 18 (1887): 460–2; Herr Rektor Bösel-Artern, "Der naturgeschichtliche Unterricht in der Volksschule," in [Report on] Seminarkonferenz zu Eisleben, *DB* 14 (1887): 305; E[duard] Scheller, [review of *Dorfteich*] *DB* 12 (1885): 322–3, on p. 323.

[28] Frick's book is reviewed in *Jahresberichte über das höhere Schulwesen* 2 (1887) [1888]: B231–3 (hereafter cited as *JbhS*); Odo Twiehausen, *Naturgeschichte I. Der naturgeschichtliche Unterricht in ausgeführten Lektionen. Nach den neuen methodischen Grundsätzen für Behandlung und Anordnung (Lebensgemeinschaften) bearbeitet. 1. Abt.: Unterstufe* (Leipzig: Ernst Wunderlich, 1887); Ernst Loew's comments on Junge's ideas having become a "dogma" appear in *JbhS* 4 (1889) [1890]: XI 36–7.

[29] Robert Mißbach, "Eine naturkundliche Ausstellung," *Aus der Heimat* 5 (1892): 9–13.

[30] Karl G. Lutz, "Für die Schule. 1. 'Nach Lebensgemeinschaften,'" *Aus der Heimat* 7 (1894): 72–7, on p. 73 n. 2.

[31] E. Loew, "XI. Beschreibende Naturwissenschaften und Chemie," *JbhS* 1 (1886) [1887]: 277. See Junge, foreword to the second edition, *Naturgeschichte. I. Der Dorfteich als Lebensgemeinschaft nebst einer Abhandlung über Ziel und Verfahren des naturgeschichtlichen Unterrichts*, 2d ed. (Kiel and Leipzig: Lipsius & Tischer, 1891), p. xii.

[32] "Bericht über die Verhandlungen der 60. Versammlung deutscher Naturforscher und Ärzte in Wiesbaden. September 1887," *Zeitschr. f. math. u. nat. Unterricht* 19 (1888): 64–7, especially p. 66; see also *JbhS* 2 (1887) [1888]: B230–1, 241–4.

und naturwissenschaftlichen Unterricht, which tracked the activities of higher school teachers in mathematics and the natural sciences (especially mathematics), crossed over to the territory of the lower school teachers to attend the annual meeting of the *Verein für wissenschaftliche Pädagogik*, lured in large part to hear the discussion of Junge's *Lebensgemeinschaft* there. Although disappointed in the lack of time allotted to discussion (as he reported to higher school science teachers in the pages of his journal), the book's warm reception later moved him to obtain a copy and read it closely. In 1889 he offered his own review of it under the title "A New Gospel of Natural History Instruction," opening his essay with the remark that "Among all the works that have been written recently on natural history, none have provoked so much attention, so many opinions, lectures and journal articles, as that of the Kiel head teacher (school director) Junge." Why, Hoffmann went on to ask, should such a work be reviewed in a journal devoted primarily to mathematical articles? Because Junge's work reaches so deeply into the issues of natural history education that "they should not remain unattended to by the higher schools. For much, indeed most, of what the author says is applicable to the lower and middle classes of the higher schools. . . ."[33] And to girls' schools and their female teachers as well, it seems: in 1893 at the second annual meeting of the Deutsche Lehrerinnenverein (German Women Teachers' Association), Helene Sumper, its Munich-based cofounder, delivered a talk on the value of the *Lebensgemeinschaft* reform idea.[34]

Within a decade of its appearance, then, the *Dorfteich* had succeeded in capturing the attention of teachers at different kinds of schools across Germany, from Kiel in the north to Munich in the south, from Strasbourg on the western front to the Silesian town of Sagan in the east. What was it that made this book such a lightning rod for discussion of reform? Despite the novelty claimed for it, the degree of interest surely must have derived from strong existing currents within the educational community. One aspect of this interest, largely absent from the content of the discussions themselves but present in their very existence, was the opportunity it gave teachers to weigh in publicly on matters of reform. In a system in which the numbers of class hours, curricular guidelines, and even sanctioned behavior were mandated from above, reform-minded teachers seized at opportunities provided by meetings and journals to assert their independence and their right to shape what went on in the classroom. In Prussia, in 1879, the new, conservative education minister had publicly discouraged teachers from attending the independent teachers' associations, promoting instead official "school conferences" under the oversight of the state; this made the act of speaking out at the teachers' associations even more clearly an assertion of political independence, indeed, defiance.[35] There and in some other states, moreover,

[33] J. C. V. Hoffmann, "Ein Besuch der Jahresversammlung des Vereins für wissenschaftliche Pädagogik," *Zeitschr. f. math. u. nat. Unterricht* 18 (1887): 460–2; idem, "Ein neues Evangelium vom naturgeschichtlichen Unterricht," ibid. 20 (1889): 246–56, on pp. 246–7.

[34] Elisabeth Meyn-von Westenholz, *Der Allgemeine Deutsche Lehrerinnenverein in der Geschichte der deutschen Mädchenbildung* (Berlin: Herbig, 1936), pp. 147–8.

[35] Meyer, *Schule der Untertanen* (cit. n. 5), pp. 156–7. See also Manfred Heinemann, "Der Lehrerverein als Sozialisationsagentur. Überlegungen zur beruflichen Sozialisation der Volksschullehrer in Preußen," in *Der Lehrer und seine Organisation*, ed. Manfred Heinemann (Stuttgart: E. Klett, 1977), pp. 39–58, especially the section titled "Der Lehrerverein als Bollwerk gegen den totalen Staatseinfluß auf das Volksschulwesen," pp. 55–7; Wolfgang Kopitzsch, "Leherorganisation in der Provinz: Weiterbildung von Volksschullehrern zur Zeit des Kaiserreichs in Schleswig-Holstein,"

teachers in the lower schools still chafed under the administrative supervision of the parish clergy; discussion of curriculum reform among schoolteachers was an assertion of autonomy not only from the state but also from the church. Junge's reform program, requiring schoolteachers to draw on their own experience and understanding of local natural communities, was especially suited to claims for the necessity of schoolteachers' autonomy from higher powers.

It would be going overboard, however, to argue that these open associational discussions constituted explicit efforts at free and democratic participation in the formation of the curriculum. These meetings and journals were almost exclusively professional venues, ones closed to other voices, such as those of parents. Not all meetings were even open to all teachers: women were excluded from many teachers' associations (resulting in the development of separate female teachers' associations). Yet despite these limitations, it still can be argued that the meetings and the publications opened up discussion of a highly political subject to a broader segment of society. Teachers lacked the respect or power accorded Germany's traditional professional classes; they had neither the social and legal standing nor the access to state influence that university professors and other professionals had. The associations thus offered one viable way of seeking to effect change.[36]

As we will see in the next two sections, the *Dorfteich* also spoke directly to two significant currents among teaching reformers: the need to bring the curriculum in natural history up to date with modern scientific and pedagogical developments, a concern expressed by reformers at all educational levels; and the interest in education for moral character, which was understood to relate to both Christian tenets and political-economic ones—concerns especially significant to *Volksschule* teachers. It is to these topics we now turn.

IV. MODERNIZING THE BIOLOGY CURRICULUM

Proponents of reforming natural history education faced an uphill battle. For reasons of both structure and ideology, natural history was something of a curricular stepchild in most German schools. Because of the great variation of grade levels in the *Volksschulen* and the lack of uniform state oversight, the curricula for these schools varied widely and patterns are hard to detect; anecdotal evidence suggests, however, that natural history generally had a limited place in a school system focusing on the basic elements of reading, writing, arithmetic, and religion. Similarly, it appears that the teaching seminaries that instructed *Volksschule* teachers paid little attention to the sciences, concentrating instead on pedagogy, basic skills, religion, and instruction in auxiliary skills, such as organ playing, that were likely to be required of a parish schoolteacher.[37] At the higher school levels, patterns are easier to detect; here the evidence is stronger that natural history was not viewed as a serious requirement.

---

ibid., pp. 93–104; and Werner Sacher, "Lehrerfortbildung im Spannungsfeld zwischen Staat und Lehrerorganisation," ibid., pp. 105–20.

[36] On the professionalizing aspect of school reform among Volksschule teachers, see Becker and Kluchert, *Die Bildung der Nation* (cit. n. 5), pp. 87–90.

[37] In his autobiography the natural history reformer Otto Schmeil described his request to teach in a city school rather than a village school so that he would not have to take up the position of organist, "which was tied to the school post in most villages." Schmeil, *Leben und Werk eines Biologen*, 2d ed. (Heidelberg: Quelle & Meyer, 1986), p. 174. Other church duties, such as bell ringer or cantor, were also typical. Tews, "Lehrer an Volksschulen," (cit. n. 8).

Among those struggling to reform the curriculum, modern languages and German history took priority over the sciences, and in the sciences, mathematics and the physical sciences generally took priority over biology. Although the situation for natural history was slightly better in the *Realschulen* and *Realgymnasien*, even there physics, chemistry, and mathematics, bolstered by the strong support of the German engineering profession, gained most of the time conceded by the classical languages.[38]

Those who wished for more time for botany and zoology in the lesson plan, and who desired coverage of these subjects to continue through the highest grade levels, found their efforts complicated by the longstanding association of biology with materialism, antireligious sentiment, and political radicalism, stretching back to the radicals of 1848 and reinforced by Darwinism since the early 1860s. The claims of botany and zoology to curricular time were not helped by Germany's most famous Darwinian, Ernst Haeckel, who in 1877 notoriously argued for replacing religious instruction in the schools with instruction in evolutionary monism. Nor did natural history teaching reformers benefit from the scandal extending from 1876 to 1879 over Hermann Müller's teaching Darwinian theory to his high school class. This widely publicized event led Prussia's cultural minister to announce that teachers were not to teach evolution in their classrooms, and when new curricular guidelines came out in 1882, biology had been removed entirely from the upper grade levels of the Prussian *Gymnasium*, strictures that came to be known collectively as the "ban on biology" (*Biologieverbot*).[39] Henceforth, natural history instruction was confined to the lower and middle school grades. Only in 1891 did the tide begin to turn: in the new Prussian regulations, although natural history instruction actually lost further ground in terms of hours taught in the official higher curriculum, for the first time it was mandated that schools take into account instruction in animal geography and "biology" (understood then as the study of the dynamic relations of organisms to their natural surroundings), and teachers were encouraged to introduce outdoor instruction, especially field trips, as part of their curriculum. These were the first formal effects of the biology teaching reform movement in the higher schools; although superficially minor, they signaled a significant change in perspective from the old approach and may be understood as the first fruits of the reform associated with Junge's name.[40] Further changes would come gradually through the first half of the twentieth century.

[38] On the history of natural history education, see Irmtraut Scheele, *Von Lüben bis Schmeil: Die Entwicklung von der Schulnaturgeschichte zum Biologieunterricht zwischen 1830 und 1933* (Berlin: Dietrich Reimer, 1981); on the role of the German engineering profession in the high school curriculum debates, see Becker and Kluchert, *Bildung der Nation*, p. 77; Albisetti, *Secondary School Reform* (both cit. n. 5), p. 92ff.

[39] Ernst Haeckel, "Über die heutige Entwicklungslehre im Verhältnisse zur Gesamtwissenschaft," *Amtlicher Bericht der 50. Versammlung deutscher Naturforscher und Ärzte* (Munich: F. Straub, 1877), pp. 14–20. On the Müller affair, see Philipp Depdolla, "Hermann Müller-Lippstadt (1829–1883) und die Entwicklung des biologischen Unterrichts," *Sudhoffs Archiv* 34 (1941): 261–344; Scheele, *Von Lüben bis Schmeil* (cit. n. 38), pp. 99–101; Daum, *Wissenschaftspopularisierung* (cit. n. 16), pp. 72–4.

[40] On the Prussian regulation changes, see Scheele, *Von Lüben bis Schmeil* (cit. n. 38), pp. 107–10. In other states, natural history teaching fared unevenly in the 1880s. See "Die realistischen Fächer nach den wöchentlichen Lehrstundensummen der höhern Lehranstalten des deutschen Reichs," *Zeitschr. f. math. u. nat. Unterricht* 15 (1884): 321–2. For an overview of the situation of the natural sciences in the higher schools, see Daum, *Wissenschaftspopularisierung* (cit. n. 16), chap. 2.

For those seeking to improve the status of natural history in the schools, a prominent appeal of Junge's program lay in his effort to enliven the subject by moving it away from memorizing names and taxonomic categories to presenting it as a science of living nature. This "modernizing" side of his program had multiple aspects that appealed differently to teachers at different levels and were subject to considerable modification by other reformers. Junge's advocacy of learning from living nature was hardly new in the 1880s, but his call to increase attention to visual and hands-on learning in natural history was in line with more recent, similar calls from educational psychologists and advocates of a more modern curriculum. The expansion of the *Realfächer* after 1872 was not just a matter of adding "modern" subjects to the school curriculum; it was also seen as developing a different set of the child's sense perceptions and mental faculties—those visual, tactile, and mental operations suited to understanding the material world. The increasing commerce in classroom visual aids, demonstration equipment, and ever more elaborately illustrated textbooks testifies to the success of this pedagogical trend.

Underlying this claim was the deeper epistemological and ethical claim that direct observation of nature could supply children with a secure moral foundation in a world filled with superstition and misinformation. Thus one reformer wrote in 1885, the same year Junge's *Dorfteich* was published, that natural history education was often mistakenly blamed for spreading materialism in the schools. *Au contraire*! The bedrock of natural history is observation, which keeps "thought in strict discipline," forms the foundation of "correct judgments," "suppresses superstition, and promotes the sense of truth of the youth." Contact with nature, moreover, has "a moderating effect, it dampens the passions and supports calm and clear judgments about ourselves and our near and dear."[41] Junge himself echoed these thoughts in the second volume of his *Naturgeschichte*, published in 1891. Natural history, he wrote, "is no place for opinion—the superstitious opine that they have seen ghosts, which then disappear upon serious investigation. How many 'scientific' ghosts in recent times have gone up in smoke upon closer research, substantial observation!"[42] Junge and other advocates of natural history instruction, then, assumed that children were not simply memorizers of information handed down to them, but individuals who exercised their own judgment in coming to their views of the true and the good. The value of an observation-based approach was that it disciplined students' thinking away from the dangers of speculation and toward the practice of sound judgment—an essential skill in an age when the temptations of Darwinism, atheism, and socialism lurked around every corner.

In addition to its greater emphasis on visual and hands-on learning, Junge's pedagogy also dovetailed with those seeking to develop patriotic and preservationist sentiments through the study of the local homeland, or *Heimatkunde*. As the *Heimatkunde* movement began to take off in the late 1880s, teachers found it a useful way of economizing in school hours, since it could be used to combine the study of local history and geography with natural history. Junge's program, with its emphasis on studying local nature, could be adapted by those seeking to develop children's knowledge and love of their natural and historical surroundings, especially through the use of field trips. Moreover, in the face of a landscape that was becoming

[41] Bergemann, "Was hat die Schule zu thun" (cit. n. 11), pp. 536–44, on pp. 536–40.
[42] Junge, *Naturgeschichte. II* (cit. n. 25), p. v.

increasingly degraded by industrial growth and pollution, the study of the natural *Heimat* in terms of *Lebensgemeinschaften* could be used to promote a sense of a natural balance of nature that should be preserved through conservation and beautification measures. In urban areas, many children's ignorance of the most basic facts of nature offered an added incentive for this method of study. An 1882 book reported that of 1,000 Berlin schoolchildren surveyed, only 364 had seen a forest or woods and only 167 recognized the song of a lark; these figures were repeated in other publications and reinforced by later surveys of urban children. On the basis of similar experience, many teachers advocated using field trips to familiarize students with nature and to teach them to value German natural places and beings as part of the German heritage.[43]

Although Junge's pedagogy thus resonated with various reform efforts among elementary school teachers, his scientific principles met with more resistance. A major reason offered was that they were too difficult. Many critics agreed that Junge's laws were simply beyond the reach of nine- to thirteen-year-olds, and the *Lebensgemeinschaft* concept itself was sometimes judged too complex. For instance, one writer, though announcing in the subtitle of his textbook that he was following the new methodological program of *Lebensgemeinschaften*, actually found Junge's notion of natural communities too difficult for students (and perhaps for teachers) and simplified it into a broader notion of grouping organisms by their natural settings or *Lebensbilder* (life pictures), such as the woods or the meadow; an 1888 pamphlet on natural history reform followed suit, the author preferring the simpler approach of presenting animals and plants in a common setting. In such ways, Junge's program was adapted and watered down. By the 1890s many elementary teachers were using *Lebensgemeinschaften* or related words such as *Lebensbilder* to refer to a nature study that de-emphasized systematics and was generally oriented toward presenting organisms in their environment, but was not necessarily focused on the biological-functional relationships at the heart of Junge's curriculum.[44]

[43] On the *Heimat* movement, see Celia Applegate, *A Nation of Provincials: The German Idea of Heimat* (Berkeley and Los Angeles: Univ. of California Press, 1990); and Alon Confino, *The Nation as a Local Metaphor: Württemberg, Imperial Germany, and National Memory, 1871–1918* (Chapel Hill: Univ. of North Carolina Press, 1997). For an example of how different topics were combined in a *Heimatkunde* curriculum, see R. Gentsch, "Heimatskunde. Lehr- und Stoffplan für eine vierstufige Volksschule," *DB* 16 (1889): 329–30, 336–7, 344–5, 353–4, 360–2, 367–9. The statistics on ignorance of nature come from Ernst Pilz, *Über Naturbeobachtung des Schülers* (Weimar: Böhlau, 1882), quoted in Bergemann, "Was hat die Schule zu thun" (cit. n. 11), p. 540. Another widely publicized survey in 1896 revealed that some 40% of 150 twelve- to fourteen-year-olds at an urban elementary school had never been in the woods and 56% had never heard a nightingale sing: Otto Schmeil, *Über die Reformbestrebungen auf dem Gebiete des naturgeschichtlichen Unterrichts*, 4th ed. (Stuttgart: Nägele, 1900), pp. 74–5; originally published in 1897. It should be noted that the *Lebensgemeinschaft* concept used by Germans easily accommodated humans—although it could be associated with *Heimatkunde* and nature preservation, it was not used primarily in the service of preserving a past "primitive" nature untouched by human hands. On the inclusion of humans in *Lebensgemeinschaften*, consider the title (and contents) of Junge, *Naturgeschichte. II. Die Kulturwesen der deutschen Heimat nebst ihren Freunden und Feinden, eine Lebensgemeinschaft um den Menschen* [Natural history II. The cultivated beings of the German home region along with their friends and enemies, a biotic community surrounding humans] (cit. n. 25).

[44] Twiehausen, *Naturgeschichte I* (cit. n. 28), reviewed by E. Schelling in *DB* 14 (1887): 377–9; F[ranz] Kiessling and E[gon] Pfalz, *Wie muss der Naturgeschichts-Unterricht sich gestalten, wenn er der Ausbildung des sittlichen Charakters dienen soll? Eine Methodik des Naturgeschichts-Unterrichts nach reformatorischen Grundsätzen* (Braunschweig: Bruhn, 1888), reviewed by E. Loew in *JbhS* 3 (1888) [1889]: B304.

A second objection focused more directly on Junge's eight laws themselves. Despite Junge's own clear interpretation that these laws would lead children to understand that laws in general were natural and therefore needed to be followed, critics worried that the presentation of laws to young students would lead immature minds down the primrose path of deductivism toward the dangerous terrain of speculation.[45] Following Haeckel's bombastic publicity stunts and the Müller affair, many schoolteachers became skittish about departing from facts; anything smacking of theoretical speculation, of knowledge not firmly grounded in facts, threatened to lead students into skepticism and wild philosophizing—ways of thinking clearly dangerous in the eyes of a religiously minded society and a politically conservative state. Indeed, Junge reported that his own school inspector believed the *Dorfteich* could be used to promote atheism and pantheism, and had therefore opposed it. Junge protested that his laws were made mainly for the teacher and should be conveyed to students only if and when the teacher thought they were ready, and then only in very simple language. In the second volume of the *Naturgeschichte*, he argued explicitly that proper natural history teaching, based on observation and induction, taught students to put the brakes on speculation and theorizing.[46] Whichever way one interpreted Junge's laws, it is clear that a central concern for elementary teachers was avoiding the potent association of natural history with radicalism.

For university-educated teachers in the higher schools, both the appeal and the limitations of Junge's program were probably somewhat different. In contrast to teachers in the elementary schools, where pedagogical goals took a front seat, teachers in the higher schools were more likely to model their instruction on the universities', paying less attention to pedagogy. Those who had taken university courses in zoology and botany would have been likely to learn a functional approach to these subjects that examined the relations between an organism's form, the functions its various structures performed, and its conditions of existence—an approach that has been called "functional morphology" and at the time was often called "scientific" botany or zoology. These instructors would undoubtedly have learned something about evolutionary theory, and they would have been oriented toward a systematics infused with evolutionary meaning. Depending on where they had studied, they might have paid some attention to biogeography as well.[47] Thus the biological-functional aspects of Junge's proposal would probably have been easier for them to view as appropriate for teaching their pupils, and his emphasis on studying the living organism *in situ*, too, would have been in line with at least one prominent strand of university-level biology.

Higher school teachers also tended to adapt Junge's *Lebensgemeinschaft* concept to their own needs. For instance, in an 1893 essay Carl Matzdorff, who taught at the Lessing-Gymnasium in Berlin, advocated field trips, school gardens, and classroom experiments to study the "biological" qualities of organisms—their structures and functions, and their relationships to their surroundings; he used the *Lebensgemeinschaft* concept to draw together material at the end of a series of lessons. In 1895 Ernst Loew, an Oberlehrer at the royal *Realschule* in Berlin who was an expert

[45] E. Loew, "Naturwissenschaft," in *JbhS* 1889 [1890], 4: XI 34.

[46] "Friedrich Junge," *DBA* 615 (cit. n. 15), pp. 47–8; Junge, *Naturgeschichte II* (cit. n. 25), p. v.

[47] Lynn Nyhart, *Biology Takes Form: Animal Morphology and the German Universities, 1800–1900* (Chicago: Univ. of Chicago Press, 1995).

in the relations between insects and blossoming plants, praised Junge while reminding readers of the less rigorous and so more practicable principle he had developed, which he called the "biocentric viewpoint," in which organisms occupying the same setting would be studied together (for example, forest plants and animals) *in situ*. Similarly, Otto Schmeil, a schoolteacher who had worked his way up the social and educational scale to receive a university degree, and who would become the most successful German biological textbook writer of the early twentieth century, liked the thrust of Junge's reforms but thought they went too far. Schmeil was unwilling to abandon the traditional taxonomic structure for school natural history teaching; rather than organizing his textbooks around biotic communities, he integrated information about an animal's way of life into the chapter sections on each animal, within a larger taxonomic chapter structure that descended from the mammals down to the single-celled animals.[48]

Reviewing the situation in 1893, Egon Ihne wrote that "in Volksschule teaching Junge's method seems to have found many followers . . . , but it does not appear that it will have such a successful arrival in the higher schools."[49] While this claim has some merit, it overstates the case. If a curriculum organized around the *Lebensgemeinschaft* idea per se met with less success in the higher schools, many of Junge's pedagogical and scientific principles were nevertheless carried over, and the attention given to the *Dorfteich* makes it seem probable that his reform program gave impetus to teaching reforms there. In particular, following upon Junge's suggestions, pedagogical discussions on both subject matter and teaching methods—especially regarding field trips, school gardens, and laboratory exercises—are significantly more visible in the 1890s and early 1900s than they were earlier. It seems plausible that the *Dorfteich* helped to strengthen the resolve of higher school teachers to reform their pedagogy. There can be no doubt, in any case, that by the turn of the century, both higher and lower schools had shifted from teaching "natural history" with a focus on taxonomy, terminology, and the separate study of plants and animals to teaching "biology" as an integrated science of living things.

V. BIOLOGY INSTRUCTION AS MORAL EDUCATION

If Junge's program appealed to some reformist schoolteachers because of its modernizing pedagogical tendencies and biological content, these aspects (especially the content) remained controversial; therefore alone they were probably not enough to ensure his program's success. Although earlier analyses of Junge's work have not called particular attention to this aspect, I will argue here that a crucial feature of his program lay in its use of nature as a model for human social behavior and thus

[48] Carl Matzdorff, "Über lebende Anschauungsmittel im naturwissenschaftlichen Unterricht," in *Wissenschaftliche Beilage zum 11. Jahresbericht des Lessing-Gymnasiums zu Berlin* (Berlin: R. Gaertner, 1893); Ernst Loew, "Naturwissenschaft," *JbhS* 13 (1898): XIII, 42. Loew had already made reference to this "biocentric approach" nearly a decade earlier; see Loew, "Naturwissenschaft," *JbhS* 2 (1887): B231; Schmeil, *Über die Reformbestrebungen* (cit. n. 43), pp. 51–2. See also A. Seybold, "Otto Schmeils Lebenswerk," in *Leben und Werk eines Biologen* (Heidelberg: Quelle & Meyer, 1986), pp. 236–7; Schmeil, *Lehrbuch für Zoologie für höhere Lehranstalten und die Hand des Lehrers. Unter besonderer Berücksichtigung biologischer Verhältnisse*, 10th ed. (Stuttgart and Leipzig: E. Nägele, 1904).

[49] E. Ihne, "Naturwissenschaft," *JbhS* 8 (1893): XIII, 28.

an especially powerful vehicle for teaching the lessons of good citizenship.[50] While most curriculum reformers would probably claim that they, too, aimed at teaching the information and values required for good citizenship, one pedagogical reform group held a particularly prominent place in its claims for "character education," and Junge's program must be understood in its somewhat complicated relationship to this reform movement. It is thus useful to understand something of the aims and methods of this program.[51]

The movement for *erziehenden Unterricht*, or "character-building instruction," was led by the theorist Tuiskon Ziller, following the precepts of the earlier educationist Johann Friedrich Herbart. In contrast to the liberal Pestalozzian tradition carried forward by Diesterweg, which sought to educate all children to reach their own potential in whatever direction they went, the Herbart-Ziller school took the conservative view that the object of primary education was to direct the developing will of children toward moral virtue, so that the child was "enabled independently to take his part in the work of the people."[52] According to Herbart, developing the will to know, or deep-seated "interestedness" in the world, rather than conferring disciplinary knowledge itself, was an essential element of this education toward virtue. In the 1860s and 1870s Ziller reoriented this approach more explicitly toward religious and moral education, combining it with another popular midcentury curricular novelty known as "concentration." This approach sought to center instruction on one goal, such as a topic or a mode of thought, and then structure different formal and disciplinary aspects of instruction, such as writing, reading, drawing, and natural history, all around this goal. Applying the concentration idea to the goal of developing moral character, Ziller took the view that all instruction must be moral-religious instruction, and many of his followers argued that the teaching of reading, writing, history, natural history, singing, drawing, and so forth, should be "concentrated" on Bible stories or other lessons in Christianity.[53] Although many educators found this position extreme and developed less one-sided classroom approaches, the Herbart-Ziller goal of character education nevertheless had a profound impact on pedagogy in the elementary schools.

This movement of education for character building, which was especially influential in Germany's Protestant schools, and which grew in prominence in response to the increasing threat of socialism, provided the predominant pedagogical milieu for Junge's program. Early in his work on the *Dorfteich*, Junge was encouraged in his reformist views by Eduard Scheller, one of the main supporters of the *Deutsche*

[50] Gerhard Trommer discusses the significance of Junge's links to the Herbart-Ziller movement for character education but concentrates on the implications for nature protection rather than its social and political implications. Trommer, "Die Dorfteich-Naturgeschichte," in *Der Dorfteich als Lebensgemeinschaft*, by Friedrich Junge (1907, 3d ed.; reprinted, with a new foreword by Willfried Janssen and introductions by the editors Wolfgang Riedel and Gerhard Trommer, St. Peter-Ording: Lühr & Dircks, 1985), pp.15–53, especially pp. 28–31. Trommer has also discussed the ecology-related educational ideas of Junge and others in his *Natur im Kopf: Die Geschichte ökologisch bedeutsamer Naturvorstellungen in deutschen Bildungskonzepten* (Weinheim: Deutscher Studien Verlag, 1990).

[51] For a discussion of the political complexities of this movement and its opposition to the more liberal Pestalozzians, see Meyer, *Schule der Untertanen* (cit. n. 5), pp. 136–7.

[52] Wilhelm Rein, "Erziehender Unterricht," in *Enzyklopädisches Handbuch der Pädagogik* (cit. n. 8), vol. 2, pp. 1–9, on p. 4.

[53] See R. Schubert, "Konzentration" in *Enzyklopädisches Handbuch der Pädagogik* (cit. n. 8), vol. 4, pp. 205–9; and Rein, "Erziehender Unterricht," in ibid., pp. 1–9.

Blätter für erziehenden Unterricht (hereafter cited as *DB*), a leading Herbartian journal, and Junge would publish numerous articles in this journal during the 1880s.[54] In 1883 he concluded his "Sketch for a syllabus for instruction in natural history for the first girls' Bürgerschule in Kiel," published in the *DB*, with the rousing statement that his curriculum organized around *Lebensgemeinschaften* showed how "natural history can be viewed as a main branch of moral-religious instruction."[55]

Although natural history could support moral instruction in many ways, the aspect perhaps most resonant with the program of character education was Junge's emphasis on teaching the child to accept limits on personal freedom in order to become a productive member of society. Even more starkly than in the *Dorfteich*, in the article sketching out his reformed natural history curriculum, Junge emphasized that the *Lebensgemeinschaft* concept helped children learn that "a properly conducted instruction in natural history shows the human, too, but not as a freely trading [*freihändelnden*] individual, but as a link [*Glied*] in a series of innumerable creations. To be sure, he stands at the top of the chain of being, but he also finds a physical world-order and feels that he is subject to the same laws as every other individual; as a being in nature, he is dependent on nature as a whole and on her often minute parts. But [the child's] self-consciousness cannot recognize a dependence upon raw forces and lower beings that are often unknown to him; [this consciousness] demands a higher, goal-oriented will, to which he voluntarily submits himself."[56] Here the shaping of the individual will toward the good of the whole, the education of the child toward a view of the self as independent and yet also dependent, may best be understood in the context of Herbartian education and its broader conservative political agenda.

Junge taught this lesson not only through the content of natural history teaching but also through instructional practices. In another early article in the *DB* on the best way to run field trips, he discussed how they could be used to underline the necessity of subordinating the individual to the group. "Give children rules of behavior, but only the most necessary ones, and avoid any appearance of imperative arbitrariness; rather show them the necessity of these measures. 'Ask yourself if each of you 40, 50, 60 pupils could do what you are thinking of doing without fear of creating disorder'—this is the main rule that each pupil must carry within himself. . . ." Children must "learn the necessity of recognizing the factual limitation of personal freedom" and learn it in their very bodies, not just in their minds. Junge also advocated "cultivating a communal spirit [*Gemeingeist*]" such that when someone makes mischief students would learn to think, not "X did that," but "one of J's students did that. You and you and I, we are all responsible."[57]

Junge's work thus provided reinforcement in the realm of natural history for teaching one of the chief social virtues of German character-building education: *Gemeinsinn*, or the "sense of community," a virtue valued by Germans across the political

[54] "Friedrich Junge," *DBA* 615 (cit. n. 15), pp. 45–6.

[55] Junge, "Entwurf eines Pensenplans" (cit. n. 26), p. 271.

[56] Ibid., p. 252.

[57] Friedrich Junge, "Was ist zur Ausführung von Exkursionen zu beachten?" *DB* 11 (1884): 37–9, 45–7, 53–5, 61–4, on pp. 38–9. At this time in Germany, elementary school classes averaged well over 60 children per teacher; in Minden, the school district with the worst student-teacher ratio, nearly 71% of the schools had 120–200 students per teacher. See Frank-Michael Kuhlemann, "Schulsystem. I. Niedere Schulen," in *Handbuch der deutschen Bildungsgeschichte* (cit. n. 4), especially pp. 195, 197.

spectrum. An article on *Gemeinsinn* in the 1896 *Enzyklopädisches Handbuch der Pädagogik*, for instance, started off by proclaiming, "Man can never be an absolute whole. What he is, even at the highest level of earthly perfection, he is thanks to—after God and his own efforts—society [*Gesellschaft*] in its widest sense, whose member [*Glied*] he is . . . . The greater the sense of community, the greater the common good. [*Je mehr Gemeinsinn, desto mehr Gemeinwohl.*]" Like other articles on various social virtues in this encyclopedia, the article went on to elaborate at length on how to "educate for the sense of community."[58]

The devotion to community was a virtue especially resonant among the broader community of schoolteachers concerned with raising a new generation of working-class children who would eventually be able to vote; the thrust of this education was to inculcate in them a sense of appropriate behavior and citizenship that would direct them away from antistate activities. At the seventh German teacher's conference in 1888, for example, in a speech on "The Introduction of Legal Studies and Economics [*Volkswirtschaftslehre*] in the Schools," the vocational-school advocate Oskar Pache argued that one of the first things a child learns in his family is "to subordinate himself to a communal [*gemeinschaftlichen*] principle and to work for the realization of the principles that animate the family. The principle of the family frequently forces its individual members, on their own volition, to deny their own rights for the good of the whole, and in this way the community achieves its own right, its own life; it becomes an independent personality that encompasses the individuals that belong to it." This principle is repeated at the level of the parish [*Gemeinde*], the state, and indeed, in any corporate entity to which people choose to belong. "No human can escape this tendency of association [*Zusammenhang*] among individuals, and community life is as indispensable for him as light, air, and the earth."[59]

Pache focused his speech on two principal *Gemeinschaften*: the state and society. Following a long tradition in German political theory, Pache argued that the object of the state is "the elevation of every individual to the greatest possible freedom"—something which was already on its way with the introduction of a constitution, an independent administration, and the more recent expansion of voting rights. But at the same time, the state can only function properly (and so guarantee the freedom of the individual) when its citizens subordinate their wills to it.[60] The point for the education of the masses was that they should learn how the state functions, how its parts work, and what role they as voters play in it. It is part of the job of the teaching profession to convey to citizens the seriousness of their role, "that the participation in public affairs is in no way an exciting and stimulating sport, but a serious job laden with meaning."[61]

The other great "organism" to which the common people belonged was the economic one, one that was no longer simply national, but global. In this regard, the job of the teacher was to help the people understand the interconnectedness of every part of the economy, to understand that "for the sure and successful operation of

[58] A. Hug, "Gemeinsinn," *Enzyklopädisches Handbuch der Pädagogik* (cit. n. 8), vol. 2, pp. 554–8, on pp. 554–5. A subsection is titled "Erziehung zum Gemeinsinn."

[59] O[skar] Pache, "Die Einführung der Gesetzeskunde und Volkswirtschaftslehre in der Schule," as reported in Wilhelm Meyer-Markau, "Der 7. (siebenten) deutsche Lehrertag," *DB* 15 (1888): 203–4, 218–20, on p. 203.

[60] See references given in note 1 above.

[61] Pache, "Die Einführung der Gesetzeskunde" (cit. n. 59), p. 203.

the great global economy the conscientious activity of every single member is required. . . ." According to Pache, the goal of the study of economics, then, is to demonstrate "how the entire working life can thrive only when the necessary unity of all the participating elements reigns, when every single member of the gigantic organism fulfills his daily work punctually and faithfully, that for the happy development of the whole every area of work is necessary and important, that there is no such thing as a superfluous and insignificant job. . . ." Finally, economics instruction should also develop "the sense for lawfulness [*Gesetzlichkeit*] and the law [*Recht*]" in two ways: it should impart knowledge of "the most important regulations that one absolutely must know to live in the civic community [*bürgerliche Gemeinschaft*]" and it should teach the "laws" of production and consumption in connection with drawing up a family budget.[62]

As in Junge's work, published three years earlier, Pache's concept of *Gemeinschaft* was a nested one that operated at several levels, from the family up to the state and the world economy. Like Junge, Pache saw the educator's job to be to help pupils answer the question "Where do I fit into the larger picture?" The answer to that question emphasized duty to others and a valorization of "simple work" through an organismal model in which all parts contributed to the welfare of the whole. Furthermore, both men also emphasized the need to impart an understanding of the necessity to subordinate one's will to "laws," where this term conjoined a philosophical sense of "law," as a principle that was unbreachable because it operated in nature, with the legal sense and the sense of moral obligation to obey the laws of the state. As the case of Pache suggests, the values articulated by Junge resonated well with those being advocated elsewhere in the community of teachers supporting character education.

Junge was hardly alone among teachers of natural history in projecting the social values of devotion to work, service to and benefit from the community, and economic interdependence onto the natural realm. Junge himself was quite restrained in how he did this, simply reminding children of the analogies between nature and human community values. In his exposition of the natural community and its members, his laws and their consequences stayed largely at the level of biological and biophysical relationships, with only pointers to "compare with human situations";[63] it was up to the teacher to make explicit the connections in the classroom—or not.

Many *Lebensgemeinschaft* enthusiasts were not so restrained and wrote in much more anthropomorphic tones. The natural history textbook author Odo Twiehausen, for instance, wrote that the starling acted as the "policeman" of the garden and field, the bullfinch was the "nobleman" among the finches, the sparrow was the beggarman. Taking such analogies even further, Junge's own colleague Groth in Kiel described the woodpecker as a "carpenter," the owl as a "night watchman," the heron as a "fisherman," and the domestic chicken as a "careful housewife." Each lived in its own particular setting; each had its own job, for which it was equipped with appropriate tools (its body structures); and each received appropriate rewards for its work: "To the heron the fish, to the hen the seed corn, to the owl the mouse, to the woodpecker the larva is a piece of gold." Moreover, none could do the work of another: "How could the heron manage the work of a woodpecker and the hen do

[62] Ibid., pp. 203, 219.
[63] See e.g., Junge, *Dorfteich* (cit. n. 17), p. 95, on the salamander.

the service of the owl! Each must come itself. Among them it is said: he who doesn't work, doesn't eat. But he who works will have enough food for himself and his family."[64]

Junge's program, then, could easily be used by those seeking to hammer home moral lessons about humans by turning nature into a field of conservative parables, and it is tempting to see this as a main feature of his appeal. But we must be cautious about viewing Junge himself as too completely a follower of the Herbart-Ziller school. Although he shared with it the goal of educating children toward a moral upbringing, Junge rejected a more extreme tenet associated with the school in the 1880s: complete concentration around religious subject matter. In the first place, he wanted natural history instruction to maintain an independent place in the curriculum. His adaptation of "concentration" lay in combining botany with zoology and (to a lesser degree) the study of the inorganic aspects of the environment, and he opposed the subordination of natural history to religious instruction.[65] In addition, he appears not to have wanted to alienate teachers who held more liberal worldviews. In a provocative footnote in the *Dorfteich*, Junge wrote that he was not surprised that some people had read into his earlier publications a bent toward materialism or pantheism. However, he continued, such people would be mistaken, for he was writing only about natural history teaching, "which as such has nothing to do with religious instruction. And here the *results could* certainly be used to justify a materialistic, pantheistic, darwinistic—and who knows, what else—worldview, but also a deistic one. That depends entirely upon the teacher. But whoever maintains that my instruction *must* lead to religious error isn't paying any attention [*steht gar nicht in der Sache*]; such a person can say anything about anything."[66] Thus Junge's curriculum by no means ruled out its use by people who wanted to adapt it to a variety of ideological ends; indeed, Junge's statement suggests that in his view, *as science* the curriculum was neutral.

This footnote offers an important clue to understanding Junge's impact and suggests how his program served a mediating function. Andreas Daum has written of the "element of reconciliation" evident in popular natural history around 1900, arguing persuasively that a conciliatory interpretation of evolution, which emphasized the perfection resulting from evolution and de-emphasized struggle and competition, was an important feature of the decades around the turn of the century.[67] Junge's program may be viewed as another, earlier kind of "biology of reconciliation"—one that made the reconciliation of worldviews possible through its silence on the subject of evolution. Despite the importance of the context of moral education for Junge's work, his silence regarding the causes of adaptation allowed it to move beyond the (already wide) circle of religious proponents of character education. Both those

[64] Groth, "Aus meinem naturgeschichtlichen Tagebuche," *DB* 14 (1887): 85–8, 95–6, 103–6, 111–3, on p. 112; on Twiehausen, see Loew, "Naturwissenschaft," *JbhS* 4 (1889): XI 43.

[65] Junge, "Entwurf eines Pensenplans" (cit. n. 26), pp. 271, 252–3. On the independence of natural history instruction from religious instruction, see especially p. 252 n. 2.

[66] Junge, *Dorfteich* (cit. n. 17), p. 20.

[67] Andreas Daum, "Das versöhnende Element in der neuen Weltanschauung: Entwicklungsoptimismus, Naturästhetik und Harmoniedenken im populärwissenschaftlichen Diskurs zur Naturkunde um 1900," in *Vom Weltbildwandel zur Weltanschauungsanalyse. Krisenwahrnehmung und Krisenbewältigung um 1900*, ed. Volker Drehsen and Walter Sparn (Berlin: Akademie Verlag, 1996). See also Daum, *Wissenschaftspopularisierung*, pp. 314–8; and, for an earlier type of reconciliation, see Daum, "Science, Politics, and Religion" (both. cit. n. 16).

leaning toward natural theology and those leaning toward either a deist or a materialist understanding of evolution could embrace Junge's program and teach it as they chose. Junge's program, though existing comfortably within the context of Protestant character-building education, provided a basis from which people could drive home quite different lessons concerning the ultimate causes of the regularities to be found in nature.

Character education provided a significant touchstone for Junge's program, and, I would argue, the context of moral education, as a powerful force for stability against the socialist threat, goes a long way toward explaining his appeal in the 1880s and the 1890s. By placing natural history within this context, Junge seems to have hoped to rescue natural history from the taint of socialism and antistate political activity, placing it instead in the service of creating dutiful citizens and workers. Thus one "reconciling" quality of his program was to reconcile children to their place in life by showing their value to the system as a whole. At the same time, by asserting the need for an independent place for natural history in the curriculum, and by refusing to abjure evolutionary ideas, he kept his program open to those who might want to use it for more liberal ends. Here again, I would argue, we can see an act of mediation, one that provided common ground for many kinds of natural history teachers and contributed in turn to the program's success.

CONCLUSION

The story of Junge's *Lebensgemeinschaft* program and its reception in Germany invites three kinds of concluding reflections. First, it is worth underscoring that in the discourse of biology teaching reformers in late-nineteenth-century Germany, although the terms "civil society" (*bürgerliche Gesellschaft*) (or more often, simply "society") and "state" had a place, reformers seem to have been less interested in maintaining or using the distinctions between them than in searching out metaphors of mediation and analogies that emphasized their interpenetration. The concepts of *Gemeinschaft* and *Lebensgemeinschaft* served this function well, for they were presented as embodying natural relationships that could be understood to operate at many levels, from the unit of husband and wife to the state or even the world economy. This nested quality of analogies between microcosm and macrocosm emphasized what was similar among all of these relationships over what was different about them and served to reinforce the idea that the stability of society and that of the state went hand in hand.[68] This emphasis on commonality and mediation is especially noteworthy since it was occurring just as Ferdinand Tönnies was developing his contrast between *Gemeinschaft* as an "organic" form of social organization that was older, more primitive, and closer to nature and *Gesellschaft* as a "mechanical" form that was, however regrettably, indelibly modern.[69] If Tönnies's *Gemeinschaft* resembles Junge's *Lebensgemeinschaft* in its organic web of relationships, there appears to be no contrasting equivalent to Tönnies's *Gesellschaft* in the sphere of teaching reform. Rather, children were to be taught to act and live in a society that retained

[68] It is worth noting that exactly this same ambiguity holds for another popular movement of the time, Heimatkunde. See references in note 43.

[69] Ferdinand Tönnies, *Community and Society* (*Gemeinschaft und Gesellschaft*), trans. and ed. Charles P. Loomis (East Lansing: Michigan State Univ. Press, 1957).

the moral features of a *Gemeinschaft*, the most prominent of which was the *Gemeinsinn*, or "sense of community," and the virtue of service to others.

Second, the ways in which the *Gemeinschaft* concept glossed over differences between state and society invite us to reflect on the ambiguous position of teachers themselves vis-à-vis these two categories. Because they were officially civil servants, it would be easy to view teachers as simply working in the interests of the state and doing its bidding; certainly the sort of citizen to be raised by means of the Herbart-Ziller style of character-building education would seem to be just the kind of member of society the state would like to have. Yet teachers were also among the most active participants in the associational life of Germany's public sphere. The *Lehrervereine* (teachers' associations) were emphatically not mere extensions of the state; they were where teachers organized themselves as members of a profession with autonomous interests, which could range from pedagogical novelties to active resistance to normalization and regulation by the state. Teachers thus moved back and forth between their roles as part of the state apparatus and as free advocates of their group professional aspirations and their own individual ideas about teaching. We may speculate that for teachers part of the appeal of the concepts of *Gemeinschaft* and *Lebensgemeinschaft* may have lain in their reflection of this very ambiguity: teachers, especially those active in reforms, embodied the tension between the interests of the state and those of civil society, and the *Gemeinschaft* concept made those less contradictory. To take another angle on the same idea, the success of Junge's reform program may have derived in part from its fortuitous balance of liberal pedagogy, with every teacher and student learning to be his or her own independent observer of nature, and the conservative, state-serving ideology that placed the student firmly within a larger whole (or set of wholes). This fit very well with the broader blend of conservative and liberal views shared by many educational reformers, who often advocated a structural liberalizing of the schools (such as a common elementary school for all children, greater mobility among different school types, and higher educational and social standing for teachers) in combination with a conservative social view that laid great stress on subordinating the individual to the good of the society as a whole.

Finally, and most broadly, this story suggests how science could be used to maintain space for the variety of viewpoints deemed desirable in an open civil society. We have seen that Junge's curriculum appealed to an organicist view of society because of the easy analogy between natural *Lebensgemeinschaften* and human communities. But Junge also insisted that his subject—the study of nature—was neutral to politics, standing in principle above and apart from any theoretical and political position. The naturalizing of political positions by claiming such neutrality for science is well known among historians of science, and one could argue that this is yet another case of the projection of a particular value system onto nature. And yet, I would argue, Junge's position was somewhat subtler than that; he was claiming something more for science, something different. According to Junge, it will be recalled, *Lebensgemeinschaften could* be used "to justify a materialistic, pantheistic, [or] darwinistic . . . world view, but also a deistic one."[70] The reason it could be taken in any of those directions, I believe Junge would have argued, was that nature itself was a neutral field: only as it was glossed by humans did it gain meaning.

[70] Junge, *Dorfteich* (cit. n. 17), p. 20.

Herein lay the saving grace of natural history. As individuals pursued the scientific truths of natural history, nature itself would provide the neutral common ground upon which proponents of different political, philosophical, and religious views could come together and hash out their differences. From this perspective, the reasonable investigation of nature provided a promising field of action for civil society itself and a site for the possibility of civil and social progress.

In Service to Science and Society: Scientists and the Public in Late-Nineteenth-Century Russia

*Elizabeth A. Hachten**

ABSTRACT

This paper examines the historically unprecedented reciprocal relationship between science and civil society in late Imperial Russia, focusing on the period from 1855 to 1905. It argues that the pursuit of theoretical and applied natural sciences came to be seen as contributing both ideologically and practically to the reform and renewal of Russian public life, and to the construction of the institutions and social networks of emerging civil society. Within those same structures, Russian scientists increasingly pursued their own professional agendas and social duties. These developments in turn had significant impact on the evolution of scientific disciplines, career patterns, and the formation of scientists' professional and social identities. Since these processes can be most easily traced within a specific locale and discipline, the paper concludes with a case study of the interactions between bacteriologists and elements of Russian civil society in the 1880s. Scientists and the interested public, particularly in southern Russia (now Ukraine), interacted in the civic realm to exploit this new science for their own purposes in the name of building the material bases for a modern society. In this instance, the extension of civil society in the late nineteenth century had produced the civic institutions and sphere of critical discourse that allowed their varying interests to be mediated independently of the state.

INTRODUCTION

OVER THE PAST three hundred years, the most striking feature of Russian science has undoubtedly been the predominant role of the state in directing, funding, and regulating scientists, their institutions, and their activities. It is therefore not surprising that the Russian and Soviet states have loomed so large in the historiography of Russian science. But there was at least one period when this pattern shifted and society became a more important factor in shaping scientific culture

* Department of History, University of Wisconsin–Whitewater, Whitewater, WI 53190; hachtene@uww.edu.

I would like to thank Daniel Alexandrov, Ol'ga Elina, David Joravsky, Mark Adams, Doug Weiner, Jim Andrews, Nathan Brooks, Joseph Bradley, David McDonald, and the participants in the April 2000 conference on Science and Civil Society for stimulating discussions and comments on earlier versions of this essay. Special thanks are due to Lynn Nyhart, Tom Broman, Kathryn Olesko, and the anonymous referees for their many helpful criticisms and suggestions. The research on which this essay is based was carried out with the generous financial assistance of the International Research and Exchanges Board, the National Research Council, and the University of Wisconsin–Whitewater; the writing was carried out thanks to a fellowship from the Institute for Research in the Humanities at the University of Wisconsin–Madison. I am grateful to all of these institutions for their support.

 0369-7827/02/1701-0007$10.00

(and vice versa). In late Imperial Russia, the six decades between the end of the Crimean War in 1856 and the fall of the monarchy in 1917, science and scientists moved out of the shadow of the state and onto the public stage in an unprecedented fashion, in large part because civil society was becoming an increasingly significant social, economic, and political factor in Russian public life. The natural sciences, theoretical and applied, contributed both ideologically and practically to the educated public's desire for purposive action to achieve the goals of renewal and reform of society. And scientists increasingly pursued their own professional agendas and social duties within the networks and institutions of Russian civil society. These developments in turn had significant impacts on the development of scientific disciplines, career patterns, and the construction of scientists' professional and social identities.

This paper explores several aspects of the complex interactions between scientists, science, and civil society in the period before the 1905 revolution. The term "civil society" itself can be problematic when applied to late-nineteenth-century Russia, and the first section of the paper delves into the relevant historiographic debates and presents several of the Russian terms that better reflect the peculiarities of social development in that country. One striking characteristic of Russian civil society was its social ethic, which included a strong mandate to act in the service of society as a whole. That moral imperative became central to many scientists' conceptions of their identity and role in Russian society.

The growing importance of the interactions between scientists and society in late-nineteenth-century Russia was reflected in the emergence of a newly vibrant public culture of science, a development that is examined in the second section of the paper. As the practitioners, patrons, and audience for science expanded beyond the narrow confines of state circles, and the government allowed its citizens an increased measure of associational freedom, the public pursuit of science became an important element in the lives of educated Russians. This was reflected in the thickening social networks that bound together practitioners and the public, and contributed to the growth of the structures of civil society in this period. While faith in science as a tool for social reform and renewal became part of the worldview of most educated Russians, some found more extreme ideological implications in the pursuit of natural science, whether materialism and atheism or, more likely, an association between scientific progress and the growth of freedom and equality.

During this same period, a newly self-conscious scientific profession was engaged in the process of constructing professional identities based on expertise, membership in the cosmopolitan scientific community, and potential service to society. As discussed in the third section of the paper, that ethos of social responsibility allowed scientists to present their professional activities as civic activism aimed at ameliorating the problems of Russia and its people rather than just serving narrow professional interests or the state, as radical critics charged. But scientists varied in their estimation of what science could offer to the common good: some stressed the contribution to material progress made by applying science to the problems of Russia, while others argued that the pursuit of theoretical science was most important for inculcating a critical, tolerant, and antiauthoritarian worldview. In a time of limited civil liberties, professional gatherings such as the Congresses of Russian Naturalists and Physicians functioned not just as avenues for advancing professional interests, but also as spheres of autonomous public discourse, association, and initiative.

The programmatic claims of scientists concerning their positive contributions to society tells us little, however, about the actual interactions between scientists and the public, or about the impact of those interactions on scientific careers and disciplines. For that reason, the final section of the paper presents a case study focused on the early institutionalization of bacteriology in Russia in the 1880s. This was a new field of science that promised dramatic applications for medicine and agriculture. Scientists and the interested public interacted in the civic realm to exploit this new science for their own purposes in the name of building the material bases for a modern society. The resulting public patronage of bacteriology was crucial in shaping the development of that field in Russia and had an enormous impact on the careers of the scientists involved as well, even if their interests did not always coincide with those of their patrons. In this case, at least, the extension of civil society in the late nineteenth century had produced the civic institutions and sphere of critical discourse that allowed the varying interests to be mediated independent of the direct control of the state.

CIVIL SOCIETY IN RUSSIAN TERMS

> In Russia, the State was everything, civil society was primordial and gelatinous.
> Antonio Gramsci[1]

> Despotism, which by its nature is suspicious, sees in the separation among men the surest guarantee of its continuance, and it usually makes every effort to keep them separate. . . . [A] despot easily forgives his subjects for not loving him, provided they do not love one another.
> Alexis de Tocqueville[2]

Studies of civil society have become a growth industry within Russian history in the past decade and a half. While this trend owes a great deal to the contemporaneous events in the Soviet Union and its successor states, it only accelerated a preexisting movement to refocus attention on aspects of social and cultural history that had been long neglected in the study of the Russian past. In their search for Russian civil society and its associated values, historians of late Imperial Russia are now tracing the construction of new social identities and spheres of public activity independent of the state through studies of cultural patronage, Freemasons, philanthropy, professionals, voluntary associations, local self-government, sexual identity, consumer culture, and urban life—to name just a few examples.[3] In short, the concept of civil

[1] Quoted in John Ehrenberg, *Civil Society: The Critical History of an Idea* (New York: New York Univ. Press, 1999), p. 208.

[2] Alexis de Tocqueville, *Democracy in America*, ed. by Phillips Bradley, 2 vols. (New York: Alfred A. Knopf, 1945), vol. 2, p. 102.

[3] See, e.g., Edith W. Clowes, Samuel D. Kassow, and James L. West, eds., *Between Tsar and People: Educated Society and the Quest for Public Identity in Late Imperial Russia* (Princeton, N.J.: Princeton University Press, 1991); Harley D. Balzer, ed., *Russia's Missing Middle Class: The Professions in Russian History* (Armonk, N.Y.: M. E. Sharpe, 1996); David Wartenweiler, *Civil Society and Academic Debate in Russia, 1905–1914* (Oxford: Clarendon Press, 1999); Laura Engelstein, *The Keys to Happiness: Sex and the Search for Modernity in Fin-de-Siècle Russia* (Ithaca, N.Y.: Cornell Univ. Press, 1992); Jane Burbank and David L. Ransel, eds., *Imperial Russia: New Histories for the Empire* (Bloomington: Indiana Univ. Press, 1998); Ben Eklof, John Bushnell, and Larissa Zakharova, eds., *Russia's Great Reforms, 1855–1881* (Bloomington: Indiana Univ. Press, 1994); Catriona Kelly and David Shepherd, eds., *Constructing Russian Culture in the Age of Revolution: 1881–1940* (Oxford and New York: Oxford Univ. Press, 1998); Adele Lindenmeyr, *Poverty Is Not a Vice: Charity, Society, and the State in Russia* (Princeton, N.J.: Princeton Univ. Press, 1996). In the Russian lan-

society has been successfully deployed as a probe to explore that social and moral space that lay "between Tsar and people." This is not to say, however, that this trend has not engendered a great deal of controversy along the way. Much of the debate is prescriptive in nature, concerned with ascertaining the "success" of Russian civil society in light of Western models or the subsequent events of the early twentieth century. But some historians ask how the concept of civil society was defined and discussed in the Russian setting, especially given its unique social dynamics and structures. This issue is linguistic in part, but it also goes to the heart of understanding the peculiarities, and ambiguities, of civil society in the Russian context.

Like most of us, historians of Russia find it much easier to use the term "civil society" than to define it. One influential formulation, by the editors of a groundbreaking 1991 collection of essays titled *Between Tsar and People: Educated Society and the Quest for Public Identity in Late Imperial Russia*, describes the institutional, legal, and moral concomitants to civil society in broad terms:

> The elusive concept of civil society implies a critical mass of educated individuals, voluntary associations, journalistic media, professional societies, universities, patronage networks, cultural organizations, and other structures that establish intermediate identities between the family and the state. Central to the notion of civil society is the assumption of a basic legal framework that protects freedom of organization and choice. One might also assume values that protect the idea of a "civic sphere," a sense of a "public culture" that legitimizes social collaboration as an indispensable adjunct to family life, on the one hand, and the realm of state and administrative power, on the other. . . . The presence of civil society implies agreement on two things: the state should not and cannot do everything, and people are public as well as private creatures.[4]

Clearly, the writers of the above passage carefully eschew any implication of equating civil society with liberal democratic regimes or bourgeois societies. That caution is obviously warranted in the case of Russia, which was characterized in this period by its autocratic regime, relative paucity of civil freedoms, and lack of a politically powerful or cohesive middle class. For those reasons, many historians of Russia have been much influenced by the models of civil society developed by German historians as well as the Habermasian paradigm of the public sphere.[5] Also hinted at in the above passage is the importance of the moral or ethical dimensions of the concept of civil society, what historian David Wartenweiler terms the "dualism" inherent in the concept of civil society:

> There is civil society as a structural entity, identified by Hegel as a domain of social interaction nestled between the strictly private and the purely public sphere of state and

---

guage literature, the leading scholar in the study of voluntary associations is A. D. Stepanskii, who has published a series of works on public organizations—*Samoderzhavie i obshchestvennye organizatsii Rossii na rubezhe XIX–XX vv.* (Moscow: Moskovskii gosudarstvennyi istoriko-arkhivnyi institut, 1980); *Obshchestvennye organizatsii v Rossii na rubezhe XIX–XX vekov* (Moscow: Moskovskii gosudarstvennyi istoriko-arkhivnyi institut, 1982); *Istoriia nauchnykh uchrezhdenii i organizatsii dorevoliutsionnoi Rossii* (Moscow: Moskovskii gosudarstvennyi istoriko-arkhivnyi institut, 1987). Also see V. R. Leikina-Svirskaia, *Intelligentsiia v Rossii vo vtoroi polovine XIX veka* (Moscow: Mysl', 1971); and idem, *Russkaia intelligentsia v 1900–1917 godakh* (Moscow: Mysl', 1981).

[4] Samuel D. Kassow, James L. West, and Edith W. Clowes, "Introduction: The Problem of the Middle in Late Imperial Russian Society," in *Between Tsar and People* (cit. n. 3), pp. 3–14, on p. 6.

[5] See, e.g., ibid., pp. 9–11, as well as Joseph Bradley, "Voluntary Associations, Civic Culture, and Obshchestvennost' in Moscow," in *Between Tsar and People* (cit. n. 3), pp. 131–48.

> politics. But civil society represents a certain social ethic, defined by social responsibility, public benefit, and service to society, thus challenging Hegel's negative moral judgement and his focus on economic activity.[6]

For Russians, this social ethic was particularly important in defining civil society, Wartenweiler argues, since "from its very inception the theory of an emancipated society contained a critical ethical component."[7] It also fit well with the Russian intelligentsia's suspicion of and disdain for commercial activity and private gain. This was the public sphere without bourgeois values.

An even better sense of the intricacies of Russian conceptions of civil society can be gained through an analysis of how Russians themselves framed these issues in the nineteenth century, focusing particularly on the changing meanings of the terms used to denote the structures and social interactions of public life outside the sphere of the state.[8] Abbott Gleason, for example, points to the evolving definitions of one important term in the Russian vocabulary, *obshchestvo* (society), in order to capture the evolution of the Russian sense of civil society at that time.[9] He points out that in the eighteenth and early nineteenth centuries, *obshchestvo* was most often understood to denote the aristocratic elite of "high society," as well as connoting a potentially broader social distinction between the educated and Westernized members of "society" and the common people, the *narod*. Even as *obshchestvo* began to lose its aristocratic connotations by the mid-nineteenth century, this "binary opposition" between educated society and the common folk continued to be crucial for Russians. Even more important, a new set of opposing terms came into common usage: *obshchestvo* versus *gosudarstvo* (the state). By the post–Crimean War era then, according to Gleason's account, the meaning of *obshchestvo* had evolved toward "more modern and politically challenging notions of a social elite, devoted to changing the stagnant society around them."[10]

Even more redolent of the diverse meanings of civil society was the term *obshchestvennost'*, which entered Russian discourse at a later date, coming into wide usage relatively late in the nineteenth century.[11] It is often translated into English as "educated society" or "the public," but there was more involved. *Obshchestvennost'* implied a Westernized elite, but one based on education and expertise rather than birth, and drawn from diverse social groups and estates. In addition, it also implied a set of values—rational and secular—as well as a disposition to actively serve the

[6] Wartenweiler, *Civil Society and Academic Debate* (cit. n. 3), p. 6.

[7] Ibid.

[8] Other, more recent works, which will not be discussed here, focus on the importance of Russian legal theory and notions of republican citizenship in forming Russian analyses of civil society. See ibid., especially chap. 3, for an exposition of how notions of civil society were deployed in legal and political theory in turn-of-the-century Russia. For discussions of the elaboration of notions of citizenship in the context of imperial rule, see Dov Yaroshevski, "Empire and Citizenship," in *Russia's Orient: Imperial Borderlands and Peoples, 1700–1917*, ed. Daniel R. Brower and Edward J. Lazzerini (Bloomington: Indiana Univ. Press, 1997), pp. 58–79; and Austin Lee Jersild, "From Savagery to Citizenship: Caucasian Mountaineers and Muslims in the Russian Empire," in ibid., pp. 101–14.

[9] Abbott Gleason, "The Terms of Russian Social History," in *Between Tsar and People* (cit. n. 3), pp. 15–27.

[10] Ibid., pp. 18–9.

[11] There seems to be some disagreement about when and how this word was coined. Gleason suggests after the 1860s, but Vadim Volkov disagrees, putting its origins much earlier, in at least the 1840s, if not the late eighteenth century.

broader public good; possession of both was necessary for one to be considered a member of *obshchestvennost'*.[12] Joseph Bradley, in his work on Moscow voluntary associations, emphasizes the pragmatic civic-mindedness implicit in the word:

> In early twentieth-century Russia the term *obshchestvennost'* signified the public sphere, a sense of public duty and civic spirit, increasingly in an urban context, and the groups possessing these values. In the Russian tradition, *obshchestvennost'* promoted activity for the public good rather than for private gain. *Obshchestvennost'* suggests the practical and purposive activity of the citizen rather than the state service of the government official or the visionary service of the revolutionary.[13]

But while *obshchestvennost'* could serve to set off or distinguish a certain segment of educated society, this concept also had an integrative implication. As Vadim Volkhov points out, *obshchestvennost'* held out the promise of creating cohesiveness among otherwise diverse groups from across the spectrum of Russian society. "Potentially, then, the term supplied a common identity to people of different estates and professions (and varying political views) who committed themselves to social duties."[14] This argument points to the possibility that common conceptions of the public good could even bridge the gap between state and society. But it must be kept in mind that the most heated historiographic debates concerning the implications of the civil society paradigm have revolved around this very issue of the extent of social cohesion that was achieved or even possible given the political and social fragmentation of late Imperial Russian society.[15]

[12] Gleason, "Terms" (cit. n. 9), p. 22.

[13] Bradley, "Voluntary Associations" (cit. n. 5), pp. 131–48, on p. 131. Other commentators concur in pointing to the links between *obshchestvennost'* and both a rational and progressive worldview and commitment to a desire to work for change, social improvement, and "progress." For example, Vadim Volkov has recently argued that originally *obshchestvennost'* meant "social solidarity, the specific qualities developed by people living in society. A secondary meaning signified the 'progressive' or rationalistic sector of society. . . . That is, *obshchestvennost'* meant both the qualities of social engagement, and the sector of society most likely to manifest such qualities." Vadim Volkov and Catriona Kelly, "*Obshchestvennost', Sobornost'*: Collective Identities," in *Constructing Russian Culture* (cit. n. 3), pp. 26–7. That seems to be a consensus position judging by other works cited in n. 3.

[14] Volkov, "*Obshchestvennost', Sobornost'*" (cit. n. 13), p. 27.

[15] Once we situate Russian civil society more precisely in evolving notions of *obshchestvo* and especially *obshchestvennost'*, there still remains for many historians the fundamental question of whether Gramsci and Tocqueville weren't correct in characterizing Russian civil society as rudimentary, underdeveloped, ineffectual, and relatively powerless vis-à-vis the state. There is great skepticism among historians concerning the autonomy and cohesiveness of civil society in Russia, the extent to which *obshchestvennost'* truly became a scaffolding for cohesive new identities and public initiative, or rather remained "incomplete" or "unstable." One of the major issues dividing historians is the issue of social cohesion versus fragmentation in late imperial society. The basically integrative notion of a cohesive public sphere comes right up against a dominant strain in the historiography that emphasizes social fragmentation as the major feature of Russian social change in this period. See, e.g., Leopold Haimson, "The Problem of Social Stability in Urban Russia, 1905–1917," *Slavic Rev.* 23 (1964): 620–42, and (1965): 1–22; Alfred J. Rieber, "The Sedimentary Society," in *Between Tsar and People* (cit. n. 3), pp. 343–66. The notion of a public based on the conception of *obshchestvennost'* puts emphasis on cultural factors, broadly speaking. For many historians, the fact that there was limited development of *political* power for civil society is more decisive. For example, one historian of the Russian professions has asserted: "There were many elements of "civil society" . . . . But there was not a cohesive middle class. Most important, Russian society was not able to develop the network of intermediating institutions that allow social groups to play a role in the political system." Harley Balzer, introduction to *Russia's Missing Middle Class* (cit. n. 3), pp. 3–38, on p. 22. And no doubt there is a tendency toward Whiggish thinking among many historians of Imperial Russia, who know, after all, that it ended badly for civil society in Russia! Historical accounts of the strength of the public sphere counter what Jane Burbank has called the "narratives of failure" that

As the above discussion makes clear, the second half of the nineteenth century was a crucial period for the crystallization of the meanings of *obshchestvo* and *obshchestvennost'*, and, by extension, the emergence of civil society.[16] The mid nineteenth century was a time of dramatic change in Russia. The country's humiliating defeat in the Crimean War (1853–1856) convinced the new tsar, Alexander II, and his government that fundamental changes in key institutions of the empire were necessary for the state's continued existence as a European power of the first rank. The ensuing "era of great reforms" witnessed an array of state-initiated changes, most notably the abolition of serfdom in 1861, and included a revamping of rural and urban governance, education, the judicial system, and the military. Although in later years the government attempted to roll back some of these concessions in defense of the integrity of the autocratic system, the social, political, and economic consequences of the reform period proved impossible to reverse completely.

Along with these reforms, the government granted a carefully measured dose of greater freedoms designed to enlist educated society in the cause of strengthening the state and supporting its agenda. While far from offering the degree of civil liberty granted after the 1905 revolution, these were still important concessions. The government eased censorship and instituted a policy of *glasnost'* to encourage public debate on key issues such as emancipation of the serfs. It streamlined and somewhat liberalized the process of obtaining permission to form voluntary associations and professional organizations.[17] A new university statute in 1863 bestowed a degree of self-governance and academic freedom on the professorate. And the reforms of local administration in the 1860s and 1870s led to new structures for self-governance (the urban duma and the rural zemstvo) that could provide an outlet for public initiative at the local level.[18] But it was not just concessions from above that created a sense of renewal in Russian educated society. A palpable new spirit was abroad, one of energy and dedication to service and reform. *Samodeiatel'nost'* (initiative) and *samostoiatel'nost'* (independent action) became the bywords of the day. As Adele

---

have distinguished some of the late imperial historiography. See Jane Burbank, "In Place of a Conclusion," in *Imperial Russia* (cit. n. 3), pp. 333–45, on p. 334.

[16] This is not to deny that "the public" had a much longer history in Russia. See Douglas Smith, *Working the Rough Stone: Freemasonry and Society in Eighteenth-Century Russia* (Normal: Northern Illinois Univ. Press, 1999).

[17] Adele Lindenmyer, "The Rise of Voluntary Associations during the Great Reforms: The Case of Charity," in *Russia's Great Reforms* (cit. n. 3), pp. 264–79. For other important works on associations in late Imperial Russia, also see Bradley, "Voluntary Associations" (cit. n. 5); Stepanskii, *Samoderzhavie i obshchestvennye organizatsii Rossii* (cit. n. 3); and Alexander Vucinich, *Science in Russian Culture*, 2 vols., vol. 2: *1861–1917* (Stanford: Stanford Univ. Press, 1963, 1970), pp. 214–34. On professionals, see the articles on engineers, schoolteachers, university professors, and lawyers in *Russia's Missing Middle Class* (cit. n. 3); Harley Balzer, "The Problem of Professions in Imperial Russia," in *Between Tsar and People* (cit. n. 3), pp. 183–98; and Nancy Frieden, *Russian Physicians in an Era of Reform and Revolution, 1856–1905* (Princeton, N.J.: Princeton Univ. Press, 1981).

[18] For more on the relations between the zemstvos and the building of civil society, see Charles Timberlake, "The Zemstvo and the Development of a Russian Middle Class," in *Between Tsar and People* (cit. n. 3), pp. 164–82; Theodore Porter and William Gleason, "The Zemstvo and Public Initiative in Late Imperial Russia," *Russian History* 21 (1994): 419–37; and Thomas E. Porter, *The Zemstvo and the Emergence of Civil Society in Late Imperial Russia, 1864–1917* (San Francisco: Mellen Research Univ. Press, 1991). The urban city council, or duma, came into existence in 1870; the zemstvo was created in most provinces of European Russia in 1864, in the immediate aftermath of the emancipation of the serfs. While there is debate, now and in the past, over the question of whether such local governments were sufficiently autonomous to be considered "public" or functioned merely as an arm of the state, it seems clear that the most active of these pursued opportunities for independent public initiative to the extent that was possible in the Russian system.

Lindenmyer argues, it was this new public spirit that accounted for the explosion of associational activities in the era of great reforms and pushed the government to allow its citizens a greater degree of autonomy in their public lives.[19]

But any balanced view of the coalescing of Russian civil society must also take due note of the fact that public life continued to function within the strict limitations set by the autocratic state. It is for this reason that most historians take care to emphasize "process rather than result" when examining the dynamics of *obshchestvennost'*.[20] Imperial Russia was a statist society with an autocracy extremely reluctant to relinquish political or cultural power to its subjects. That fact strongly colored all public activities since in Russia "all independent public initiative was political, even when it was not overtly oppositional, because it challenged, indirectly if not directly, the autocracy's control over society."[21] The new freedoms granted to Russians were limited, and many of them were retracted within a decade. The members of educated society made up only a small portion of the population and were mostly urban in what was still an overwhelmingly rural society. Although professionals had become more visible and active members of society, their small numbers, growing internal divisions, and ambivalent, "Janus-faced" relationship with the state limited their autonomy.[22] Local self-government was only quasi-representative at best, as political power was unequally distributed among the rural estates (and later essentially restricted to the nobility). The regimes of Alexander III (1881–1894) and Nicholas II (1894–1917) were marked by further attempts to reduce local governments to mere instruments of state power and a series of "counterreforms" intended to limit or reverse many of the previously granted liberties of the reform era.

How have science and scientists figured into the historiography of Russian civil society? In the studies noted above, it is possible to find occasional mention of the role of science and its practitioners in the construction of Russian civil society. For example, Joseph Bradley in his study of Moscow voluntary associations notes the large numbers of scientific societies, museums, and exhibitions that dotted this urban landscape and their importance in creating Moscow *obshchestvennost'*.[23] And many of the numerous studies of Russian educated professionals have focused on applied and pure scientists in their occupational roles as teachers, professors, doctors, engineers, and zemstvo employees.[24] But to illuminate more clearly why and how science came to be so closely associated with *obshchestvennost'* and its central notions of purposive activity, civic improvement, rationality, and critical debate,we

[19] Lindenmyer explains the origins of this new public posture as a response to the sense of optimism created by the government reforms as well as the sense of duty felt by educated Russians to improve the conditions around them. See Lindenmyer, *Poverty Is Not a Vice* (cit. n. 3), pp. 120–3. Alexander Vucinich grapples with many of the same issues in *Science in Russian Culture* (cit. n. 17), vol. 2, pp. 3–12.

[20] The phrase belongs to Kassow et al., "Introduction" (cit. n. 4), on p. 10.

[21] Lindenmeyr, *Poverty Is Not a Vice* (cit. n. 3), p. 196.

[22] The phrase is Nancy Frieden's; her work on physicians has been particularly influential. See her *Russian Physicians* (cit. n. 17).

[23] Bradley, "Voluntary Associations" (cit. n. 5), p. 147. Thus he includes "belief in progress and in science" as one of the central values of *obshchestvennost'*.

[24] Besides the recent collection of essays entitled *Russia's Missing Middle Class* (cit. n. 3), see Frieden, *Russian Physicians* (cit. n. 17); John F. Hutchinson, *Politics and Public Health in Revolutionary Russia, 1890–1918* (Baltimore: Johns Hopkins Univ. Press, 1990); Samuel D. Kassow, *Students, Professors, and the State in Tsarist Russia* (Berkeley: Univ. of California Press, 1989); and James C. McClelland, *Autocrats and Academics: Education, Culture, and Society in Tsarist Russia* (Chicago: Univ. of Chicago Press, 1979).

now turn to an examination of the links between the new public culture of science and the changing social structure and dynamics of Russian society in late-nineteenth-century Russia.

RUSSIAN SCIENCE GOES PUBLIC

> I began to work in the decade of the sixties. Remember what forward movement there was then, what irresistible forward motion seized the whole of society, and with it, what passion for science was awakened.
>
> Professor of Medicine Sergei P. Botkin[25]

> In democratic countries, the science of association is the mother of science.
>
> Alexis de Tocqueville[26]

The second half of the nineteenth century was a period of pervasive change in Russian science, a time typically portrayed as representing the maturation of Russian science and its achievement of international standing in the scientific community.[27] But there was more. In this period after 1855 a significant *public* culture of science began to coalesce outside the narrow confines of state institutions and circles.[28] The creation of this public culture of science was at least partially a consequence of the state-initiated policies, although it also flowed from a myriad of unintended intellectual and cultural changes. This was evident in the transformation of the practitioners, audiences, and patrons of science as well as the emergence of new public spaces in which they interacted. (I use "public" here and throughout the paper in the Russian sense of *obshchestvennyi*, signifying the realm of society apart from the state and its sphere.) These developments were all associated closely with the structures and moral economy of Russia's coalescing civil society. At the same time, the enhanced cultural authority of science and the acknowledgment of the natural sciences as instruments of progress and modernization were closely tied to the evolving values of *obshchestvennost'*.

Since at least the reign of Peter the Great (1689–1725), the Russian state had been the chief patron and employer of scientists and other technically trained personnel.[29] For Peter and his successors on the Russian throne, scientific patronage promised to

[25] Remarks to the assembly honoring his career, *Trudy Obshchestva Russkikh Vrachei v Sankt-Peterburge* (St. Petersburg, 1886): 217.

[26] Tocqueville, *Democracy in America* (cit. n. 2), vol. 2, p. 114.

[27] See, e.g., Loren Graham, *Science in Russia and the Soviet Union: A Short History* (Cambridge: Cambridge Univ. Press, 1993), p. 32.

[28] I borrow the phrase "public culture of science" from historians of Enlightenment science such as Jan Golinski, who situate scientific practitioners, activities, and knowledge within the emerging public sphere of the eighteenth century. Jan Golinski, *Science as Public Culture: Chemistry and Enlightenment in Britain, 1760–1820* (Cambridge: Cambridge Univ. Press, 1992). The discussion that follows owes a great deal to the participants in two conferences on science and society held in St. Petersburg, Russia, in 1994 and 1995: "Images of Science for Patrons and the Public: Historical Legacy for the Baltics, Russia, and the Ukraine," and "Regionalism, Local Interests, and Science: The Historical Legacy for the Baltics, Russia, and the Ukraine." Several of the papers from the 1994 conference were published and will be referred to separately below. See, especially, D. A. Alexandrov, "Nauka dlia uchenykh, gosudarstva i obshchestva: Opyty sotsial'noi istorii nauki," *Vop. Ist. Est. Tekh.*, no. 1 (1995): 21–3.

[29] On the history of Imperial Russian science, see Vucinich, *Science in Russian Culture*, vol. 1: *A History to 1860* (cit. n. 17); ibid., vol. 2; and Graham, *Science in Russia and the Soviet Union* (cit. n. 27), pp. 9–75.

confer practical benefits on the empire and enhance national prestige. While this pattern of state-driven scientific activity was common on the continent of Europe, the relative weakness of the ties between scientists and the broader public was particularly striking in Russia. Peter introduced European science into his country from the top down, founding an Academy of Sciences in 1725 stocked with prominent foreign scholars long before his country had any universities or even a system of secondary schooling. Natural science remained an alien cultural import for decades; even in the early nineteenth century interest in this Western pursuit was concentrated within state-run institutions and small circles in the court and government. The number of practitioners was small, and their ranks were dominated by foreigners and Baltic Germans who often pursued theoretical, even esoteric, research interests in relative isolation from their surroundings.[30] By the 1840s, however, there were signs of intensified interactions between some elements of the scientific community and the still tiny but growing educated public.[31] But then came the European revolutions of 1848, and the repression of public life and restrictions on education that followed submerged such changes, as Nicholas I (1825–1855) attempted to seal Russia off from the contagion of liberalism.[32]

In the wake of the defeat in the Crimean War, and the accession of Alexander II (1855–1881) to the throne, the pendulum of state priorities swung in the opposite direction, back to a practical appreciation for the potential benefits of scientific activ-

[30] One important line of work in the academy was the task of exploring, cataloging, and classifying the empire and its natural resources—hardly a theoretical pursuit, but very much done in the interests of the state. Links between science, the state, and aristocratic culture in eighteenth-century Russia have recently come under renewed scrutiny. See A. V. Bekasova, "'Uchenye zaniatiia' russkogo aristokrata kak sposob samorealizatsii (na primere grafa N. P. Rumiantseva)," *Vop. Ist. Est. Tekh.*, no. 1 (1995): 24–39; and Michael Gordin, "The Importation of Being Earnest: The Early St. Petersburg Academy of Sciences, *Isis* 91 (March 2000): 1–31.

[31] For an overview of science in this period, see L. A. Tarasevich, "Nauchnoe dvizhenie v Rossii v pervoi polovine XIX veke: Estestvoznanie i meditsina," in *Istoriia Rossii v XIX veke*, 9 vols. (Granat: Moscow, n.d.), vol. 6, p. 285–308. In an attempt to foster economic development, the Imperial Free Economic Society was founded in 1765, followed between 1800 and 1858 by a dozen state-chartered regionally based agricultural societies where the taste for science among improving landowners could be satisfied. See Joan K. Pratt, "The Russian Free Economic Society, 1765–1915," Ph.D. diss., Univ. of Missouri–Columbia, 1983. The Russian Geographical Society, which would become a major force in the public culture of science later in the century, was founded in 1845 to serve state interests in imperial exploration. See Nathaniel Knight, "Science, Empire, and Nationality: Ethnography in the Russian Geographical Society, 1845–1855," in *Imperial Russia* (cit. n. 3), pp. 108–43; and Catherine Clay, "Russian Ethnographers in the Service of Empire, 1856–1862," *Slavic Rev.* 54 (Spring 1995): 45–62. The ranks of educated Russians broadened under the impact of the Enlightenment and greater access to schooling in the reigns of Catherine II and Alexander I. The number of universities in the Russian empire rose from one in the eighteenth century to six after 1803. At these universities, learned societies appeared in the first decade of the nineteenth century, and there are indications that by the 1840s at least, some university professors were becoming more a part of a burgeoning Russian intellectual life. Nathan Brooks, "Public Lectures in Chemistry in Russia: 1750–1870," *Ambix* 44 (1997): 1–10.

[32] In the words of one contemporary, Russia in that period took on "the quiet of a graveyard, rotting and stinking, both physically and morally." Quoted in Nicholas V. Riazanovsky, *Nicholas I and Official Nationality in Russia, 1825–1855* (Berkeley: Univ. of California Press, 1959), p. 219. While that may be an overstatement in light of the works cited in n. 31, the shift in mood was palpable. The tension between the autocracy's need for the practical benefits of science and its fear of science's pernicious ideological effects created an ongoing dilemma for the state in this period and indeed up until the fall of the monarchy. As Loren Graham has explained, "[T]he rulers of nineteenth-century Russia favored the development of science and technology if given the assurance that it would not undermine the existing political and social order." Graham, *Science in Russia and the Soviet Union* (cit. n. 27), p. 32.

ity. The urgent tasks for the state were suddenly economic and technological modernization, and cautious reforms were undertaken with the goal of maintaining Russia's great power status with a minimum of political or social change. The minister of Public Instruction described the government's new priority in 1856: "*Nauka*, citizens, has always been one of the most important requirements for us, but now it is the primary one. If our enemies have an advantage over us, it is only by virtue of their learning."[33] In the educational reforms that followed, opportunities for the teaching and study of natural sciences expanded significantly.[34] The number of students studying natural sciences in the university system rose from a few dozen before the Crimean War to almost nine hundred in 1872, reaching seventeen hundred in 1880.[35] The government again allowed scholars to travel abroad for advanced training, and relaxed censorship allowed the freer flow of Western ideas and their dissemination in Russia. Although many of the academic freedoms granted in this period were soon retracted, the pendulum never swung back to the repression of the pre-reform era; despite its dangers, the need for a scientifically and technically trained elite had just become too pressing for the state.

The state's education policies had both immediate and long-term consequences for the community of scientific practitioners in Russia. Not only were the sheer numbers of professional practitioners significantly larger than ever before, but this new generation of scientists was also much more representative, both ethnically and socially, of Russian society as a whole. Science graduates now had greater opportunities for employment, not only with the state but also in the public sector as local governments' need for agronomists, statisticians, engineers, physicians, and veterinarians began to rival the state's by the end of the century. These shifts in the size, composition, and social origins of the scientific community meant that the ties between scientists and educated society intensified and broadened in the late nineteenth century, thus contributing to the emergence of a public scientific culture.

Other changes in state policies in the reform era—including expanded educational

[33] A. S. Norov, quoted in Kliment A. Timiriazev, "Probuzhdenie estestvoznaniia v tret'ei chetverti veka," in *Istoriia Rossii* (cit. n. 31), vol. 7, p. 2. *Nauka*, while often appropriately translated as "science," also means "learning" or "scholarship" more broadly. It is quite comparable to the German *Wissenschaft*.

[34] One of the most important for our purposes was the new university charter of 1863, which reorganized the teaching of science in all the country's universities. Separate natural science faculties were created, with sixteen professors and three *dotsenty* (lecturers) divided among twelve departments. To expand career opportunities for younger scholars, the post of *privat-dotsent* was created. This charter also provided for a greater degree of university autonomy and academic freedom than had been granted before, although that provision proved to be short lived, especially due to student unrest and the emergence of the revolutionary movement. On the topic of higher education in this period, see the previously cited works by Graham (cit. n. 27), Kassow (cit. nn. 3, 24), and McClelland (cit. n. 24); and Daniel R. Brower, *Training the Nihilists: Education and Radicalism in Tsarist Russia* (Ithaca: Cornell Univ. Press, 1975), which is especially good on science education; and of course the magisterial study by Vucinich, *Science in Russian Culture* (cit. n. 17), vol. 2, pp. 35–65.

[35] Charles E. Timberlake, "Higher Learning, the State, and the Professions in Russia," in *The Transformation of Higher Learning, 1860–1930*, ed. Konrad H. Jarausch (Chicago: Univ. of Chicago Press, 1983), pp. 325–6. (See also the chaps. by Patrick Alston, James McClelland, and Daniel R. Brower in the same volume.) Medicine was by far the most popular field of study in the universities: approximately 3,700 students chose that field of study in 1880. Law, which prepared a student for a career in civil service as well as private practice, drew about as many students as the sciences in 1880. Training for engineers, military physicians, agronomists, and other specialized personnel took place in the system of technical secondary and higher schools outside the universities; enrollments in those schools increased even more rapidly than in the universities in the late nineteenth century.

opportunities at all levels and loosened restrictions on some kinds of associational activity—also encouraged the greater public dissemination of science in several other ways, especially by contributing to the striking growth in public spaces for the practice and promulgation of science and an expanded base of literate, interested Russians. To many observers, one of the hallmarks of mid-nineteenth-century Russian culture was the extent to which virtually the whole of educated society had suddenly been gripped by an unprecedented enthusiasm for science, an intellectual spasm reflected in the broadened audience for scientific ideas, practices, and potential applications. Evidence of these new consumers of science is not hard to find: they were the readers of Darwin and his popularizers, of Sechenov's *Reflexes of the Brain*, and of the "thick" journals with their hefty quotient of popular science. These consumers attended the lectures given by university professors and itinerant entrepreneurs, as well as the public meetings of scientific societies.[36] In social gatherings they debated the merits of Justus von Liebig's discoveries for improving soil fertility.[37] They visited the new natural history museums, as well as the increasingly common public exhibitions on technology, electricity, hygiene, and related subjects. Such public spaces and meeting grounds—long common in western European countries—had been quite rare in Russia before 1850 but dotted the urban landscape by the end of the century.[38]

Learned societies and professional organizations were particularly important settings for the pursuit of science in the public sphere. Before the Crimean War, Russia had boasted fewer than twenty such societies, from the St. Petersburg-based Free Economic Society and Russian Geographical Society to agricultural associations and university-affiliated scientific societies around the country. Although few in number, these societies were important all out of proportion to their numbers, as they formed a significant area of autonomous public activity in pre-reform Russia and thus served as seeds for the further generation of civil society.[39] In the decades after the Crimean War, the associational impulse that gripped educated society led to the creation of literally hundreds of learned and applied scientific societies; only charitable organizations were created in larger numbers in that period.[40] This apparent violation of Tocqueville's dictum that despots strive to separate their subjects reflected rather a careful balancing act by the relevant government ministries to allow communication and collaboration between experts in these fields (and often interested laypeople as well) as long as there was minimal political threat and demonstrable benefit to the state. Local and regional organizations were therefore fa-

[36] One enterprising German rented the Imperial Circus in St. Petersburg for presentation of his geology lectures, illustrated with magic-lantern slides and maps. Timiriazev, "Probuzhdenie estestvoznaniia" (cit. n. 33), p. 22.

[37] As related in a scene from Ivan Turgenev, *Fathers and Sons* [1862] (New York: Norton, 1989), p. 20.

[38] D. A. Ravikovich, "Muzei mestnogo kraia vo vtoroi polovine XIX–nachale XX veka (1861–1917 gg.)," in *Ocherki istorii muzeinogo dela v Rossii* (Moscow: Izd. "Sovetskaia Rossiia," 1960), vol. 2.

[39] Wartenweiler, *Civil Society and Academic Debate* (cit. n. 3), pp. 19–20.

[40] On scientific associations, see Vucinich, *Science in Russian Culture* (cit. n. 17), vol. 2, pp. 214–34; Stepanskii, *Istoriia nauchnykh uchrezhdenii* (cit. n. 3); and the very extensive entries on *Obshchestva* and *Sel'skokhoziaistvenniia obshchestva* in *Entsiklopedicheskii slovar' Brokgauz-Efron* (St. Petersburg, 1900). Bradley, in his essay on Moscow *obshchestvennost'* (cit. n. 5), demonstrates the prominence of associations with a scientific or technical focus in the civic life of that city and reminds us that Russians were not a "nation of joiners."

vored over national ones, and applied fields such as agriculture, medicine, and public health predominated. These associations functioned as important contributors to the promotion of the core values of *obshchestvennost'*: dedication to practical improvement, civic consciousness, and public rather than narrow class, ethnic, or private interests.[41] This was often true even of exclusively professional associations such as medical societies and university-based learned societies, which increasingly reached out to laypeople through publications, publicity, and educational projects.[42]

A peculiarly Russian setting for the public pursuit of science were the zemstvos (*zemstva*), the organs of rural self-government instrumental in creating the beginnings of the social and communications networks of civil society in the vast hinterlands of European Russia.[43] Provincial and district zemstvo meetings, extension activities, and publications all could serve to bring together landowners, peasant farmers, professional zemstvo employees (including physicians, agronomists, and statisticians), and sometimes interested outside experts to tackle agricultural and health problems in particular. This function flowed from the original enabling legislation of 1864 that created these administrative organs. The zemstvos were given responsibilities, in their local areas only, for controlling epizootic diseases, preventing crop losses, encouraging economic development, and improving public health. By 1880 one begins to see evidence that the more activist zemstvos at least were putting this directive into action, reflecting both rising tax revenues and a turning away from political action to local "small deeds" reform efforts.[44] And although the state proceeded in later years to restrict the autonomy, power, and financial

[41] Bradley, "Voluntary Associations" (cit. n. 5), pp. 146–8. While forming a public association of any kind in this period still required the approval of the central government, different ministries were involved in the process: the Ministry of the Interior oversaw charitable and medical societies, the Agricultural Ministry approved agricultural societies, etc. Universities, on the other hand, were given the power to create learned societies in 1863.

[42] See V. S. Savchuk's groundbreaking work on the natural history societies of southern Russia, which has provided the most thorough depiction of the workings of scientific associations and their importance for civil society. See his *Estestvennonauchnye obshchestva iuoga Rossiiskoi imperii vtoraia polovina XIX–nachalo XX v.* (Dnipropetrovs'k: Vyd-vo DDU, 1994). E. I. Lotova, in *Russkaia Intelligentsiia i voprosy obshchestvennoi gigieny* (Moscow: Medgiz, 1962), has examined the Russian Society for Public Health Protection, a national organization with independent local branches, making similar arguments. Note that one must be careful not to automatically ascribe the qualities of *obshchestvennost'* to all professional societies, as the example of the Odessa medical societies demonstrates. One of the city's three medical societies in the 1890s excluded Jews and other non-Orthodox, non-Russian practitioners; a second served as the trade organization for physicians associated with the spas that dotted the Black Sea coast around the city. While neither of those organizations would have been considered as contributing to *obshchestvennost'*, the third society was a different matter. The Society of Odessa Physicians was open to all medical practitioners, including Jews, and its members became involved in a number of public health, educational, and other activities aimed at improving the general welfare of the city's population. That society showed the requisite degree of social responsibility and civic activism necessary for a group to be considered part of *obshchestvennost'*. See E. Khekten, "Nauka v mestnom kontekste: interesy, identichnosti i znanie v postroenii Rossiiskoi bakteriologgii," *Vop. Ist. Est. Tekh.*, no. 3 (2001): 36–62.

[43] For an overview on the zemstvo and its significance in late Imperial Russia, it is still best to start with the collection of essays in Terence Emmons and Wayne S. Vucinich, eds., *The Zemstvo in Russia: An Experiment in Local Self-Government* (Cambridge: Cambridge Univ. Press, 1982). For more recent treatments of the zemstvo that examine its importance for civil society, see the works cited in n. 18.

[44] See, e.g., R. T. Manning, *The Crisis of the Old Order in Russia: Gentry and Government* (Princeton, N.J.: Princeton Univ. Press, 1982), especially pp. 3–64.

resources of the zemstvos, these moves did not significantly curtail applied scientific activities such as the agricultural and medical services.[45]

Some of the most important results of the activities of voluntary associations and local governments came in their roles as patrons of science. The appearance of new public sources of scientific patronage, along with a revitalization of state and private patronage of science, became an important component in the public culture of science.[46] Rural and urban governments, medical societies and naturalists' groups all functioned as sources of financial and institutional support for a wide range of scientific activities, and agricultural societies emerged as one of the most important types of public, nonstate patrons of science in the post-emancipation period.[47] This public patronage resulted in the creation of a myriad of scientific research and service institutions, including meteorological observatories, natural history surveys, agricultural research stations and fields, and medical laboratories. It was not coincidental that these projects fell for the most part into the realm of applied sciences such as agronomy, earth sciences, and public health, as patrons more than general audiences tended to be particularly shaped by estimations of the potential and actual benefits of science, particularly within the changed economic climate of post-reform Russia. And public patronage could be a powerful influence on scientific careers and disciplines, as will become evident below.

Clearly, the natural sciences and their pursuit had become a central component of the worldview of educated Russians in the mid and late nineteenth century. But why did natural science suddenly take on this aura of cultural authority, privileging it above many other forms of culture? What meanings were attached to the pursuit of scientific knowledge by this public? No one single answer to that question exists, and the variety of possible responses helps us to understand how science could function as a mediating element between diverse groups in Russian society. One long-standing aspect of the appeal of science in Russia had been its association with Western culture. Since the eighteenth century science had been yet another imported European cultural commodity (like French fashions or Italian opera troupes), a form of polite knowledge and rational entertainment that bestowed a certain cultural cachet on its devotees. Though these aristocratic associations still lingered after the mid-nineteenth century, more reflective of the spirit of the age of great reforms was the identification of science with a modern, rationalist, progressive worldview that valued practical activity, technical competence, individual initiative, and the improvement of self and society.[48] While the government-initiated reforms helped to

[45] In fact, just the opposite may be true. Kendall Bailes claims that zemstvos saw a 400% rise in spending on agricultural research, extension services, etc., between 1894 and 1904. See Bailes, *Science and Russian Culture in an Age of Revolutions: V. I. Vernadsky and His Scientific School, 1863–1945* (Bloomington: Indiana Univ. Press, 1990), pp. 86–91. Ol'ga Elina's findings on the growth in experimental fields and stations conform to the same pattern, as does the establishment of medical and sanitary services by the zemstvos. It should also be noted that many urban city councils (dumas, or *dumy*) followed a similar pattern of involvement in medical and scientific pursuits, as will be discussed below.

[46] On state patronage see, e.g., Iu. A. Laius, "Uchenye, promyshlenniki i rybaki: nauchno-promyslovye issledovaniia na Murmane, 1898–1933," *Vop. Ist. Est. Tekh.*, no. 1 (1995): 64–81; as well as the discussion in Alexandrov, "Nauka dlia uchenykh, gosudarstva i obshchestva" (cit. n. 28).

[47] O. Iu. Elina, "Nauka dlia sel'sko khoziaistva v Rossiiskoi Imperii: Formy patronazha," *Vop. Ist. Est. Tekh.*, no. 1 (1995): 40–63.

[48] See Vucinich, *Science in Russian Culture* (cit. n. 17), vol. 2, pp. 1–14; and Lindenmeyr, *Poverty Is Not a Vice* (cit. n. 3), pp. 120–4.

set the context for this national mood, the emphasis was on the possibilities for individual renewal (especially via education) and public initiative. The values and modes of expression associated with science fit with the palpable sense of optimism at this time and the faith in the possibility of improvement, progress, and modernization. And in a more concrete sense, too, the successes of applied science in this era demonstrated the importance of the natural sciences and technology as practical tools for the immediate modernization of the economy and society. As one young engineering student put it, science represented "indisputable knowledge primarily in the sense that it left the least room for doubt" and provided "something unequivocal to cope with the mass of problems which showered down on us."[49] This utilitarian attitude toward science as an instrument of modernization cut across many political and ideological lines in late Imperial Russia.

But for many Russians, the pursuit of the natural sciences could take on more politicized ideological implications, particularly in the decade of the sixties. For them, science represented a particular form of enlightenment, inculcating the values not just of rationality and utilitarianism, but also of materialism, atheism, and scientism. The works of materialist philosophers such as Karl Vogt, Ludwig von Feuerbach, Ludwig Büchner, and Jacob Moleschott, which were suddenly available in the period after the Crimean War, were clearly an influential source for these ideas.[50] And, of course, there were homegrown philosophers such as the nihilists Dimitri Pisarev and Nikolai Chernyshevsky, who elevated science above all other types of learning and saw individual scientific enlightenment as a "panacea for social ills."[51] Thus when the character of the nihilist medical student Evgenii Bazarov, in Ivan Turgenev's novel *Fathers and Sons* (1862), proclaimed that "a single decent German chemist is worth a dozen poets," he was expressing both a sense of science as purposive action and an extreme scientism that became stereotypical of the radical youth of that generation who wished to sweep away traditional authority of all kinds.[52] It is, however, important to not overstate the impact of materialism or nihilism; for most educated Russians, especially scientists, belief in evolution or physiological materialism did not translate into political radicalism.[53]

Positivism was yet another influential philosophical trend that both elevated the authority of science and pointed to the potential challenges to the established order inherent in its pursuit. It was undoubtedly a much more influential ideology among Russian scientists and the educated public than nihilism ever was. In Russia, positivism was above all associated with Henry Thomas Buckle's *History of Civilization in England*, the first Russian edition of which appeared in 1863 (before any translations of Auguste Comte). Buckle's theory of social evolution identified positive knowledge, skepticism, and rationality—as particularly evidenced by the scientific revolution and the growth of the inductive method in England—as the main engines of both social progress and the growth of freedom. Indeed, for Buckle, science and

[49] Quoted in Brower, *Training the Nihilists* (cit. n. 34), p. 158.

[50] Indeed, some young men, such as the future Nobel Prize winner Il'ia Mechnikov, learned German just to be able to read those works. Olga Metchnikoff, *The Life of Elie Metchnikoff, 1845–1916* (Boston: Houghton Mifflin, 1921), p. 31.

[51] Vucinich, *Science in Russian Culture* (cit. n. 17), vol. 2, p. 24.

[52] Turgenev, *Fathers and Sons* (cit. n. 37), p. 19.

[53] This argument is made by both David Joravsky, *Russian Psychology: A Critical History* (Oxford: Basil Blackwell, 1989); and Daniel Todes, *Darwin Without Malthus: The Struggle for Existence in Russian Evolutionary Thought* (New York: Oxford Univ. Press, 1989).

democracy were causally linked, as he demonstrated in his account of the background to the French Revolution. Buckle directly credited the flowering of the sciences in France in the second half of the eighteenth century to the growth of belief in equality and liberty:

> The intimate connexion between scientific progress and social rebellion, is evident from the fact that both are suggested by the same yearning after improvement, the same dissatisfaction with what has been previously done, the same restless, prying, insubordinate, and audacious spirit.[54]

He went on to assert that the passion for science that seized every corner of French society, from the high nobility to the scarcely literate, helped to create social solidarity among the progressive elements of French society and overcome previous divisions and inequalities:

> As long as the different classes confined themselves to pursuits peculiar to their own sphere, they were encouraged to preserve their separate habits; and the subordination, or, as it were, the hierarchy, of society was easily maintained. But when the members of the various orders met in the same place with the same object, they became knit together by a new sympathy. . . . [T]he pleasure caused by the perception of fresh truths, was now a great link, which banded together those social elements that were formerly wrapped up in the pride of their own isolation.[55]

This vision of science as a force for social cohesion would have carried unmistakable suggestions of the possible links between science, *obshchestvennost'*, and political reform for the Russian reading public. And when Buckle went on to conclude that "the hall of science is the temple of democracy," it is not difficult to imagine that at least a portion of the Russian public read into this historical account a blueprint for Russia's future development.[56] While the fashion for positivism (as for the other radical ideologies of the 1860s) soon passed, its influence was long lived. Populists, who represented the next wave in social criticism in Russia in the 1870s, were usually strongly critical of the extreme scientism of their predecessors. But as one populist critic of Buckle admitted, as a result of Buckle's ideas "the program of the people labeled 'conservatives' includes, among other things, a hatred for democratic ideals and a fear of the harmful influences of science. [On the other hand,] the program of people labeled 'liberals' includes an admiration for both the natural sciences and democratic ideals."[57]

Thus, in the middle decades of the nineteenth century, clear signs of an expanding public culture of science in Russia existed. This development in turn was part of a broader upsurge of public life that signaled the coalescing of Russia's civil society. As scientists and educated laypeople alike increasingly "assumed that the successes of science are able to best promote general progress," scientific enterprises moved to the center of the project of improvement and modernization that lay at the heart

[54] H. T. Buckle, *The History of Civilization in England*, 2 vols. (New York: D. Appleton and Co., 1875), vol. 1, p. 658.
[55] Ibid., pp. 661–2.
[56] Ibid., p. 662.
[57] N. K. Mikhailovskii, quoted in Vucinich, *Science in Russian Culture* (cit. n. 17), vol. 2, p. 28.

of *obshchestvennost'*.[58] While that project could serve to unite people with varying political views, for those imbued with positivism or more radical philosophies, science also marked the path toward more thoroughgoing social and political change.

BETWEEN THE IVORY TOWER AND THE WORKSHOP

> Nature is not a cathedral but a workshop, and man is the craftsman.
>
> Evgenii Bazarov in Turgenev's *Fathers and Sons* (1862)[59]

> . . . permanent inequality of conditions leads men to confine themselves to the arrogant and sterile search for abstract truths, while the social condition and the institutions of democracy prepare them to seek the immediate and useful practical results of the sciences. This tendency is natural and inevitable.
>
> Alexis de Tocqueville[60]

In the period of rapid social and cultural change in Russia after the mid-1850s, several factors pushed scientists to reevaluate their careers, identities, and professional values. On the one hand, Russian scientists were undergoing a process of professionalization similar to that seen among western European scientists in this period. On the other hand, scientists too were influenced by the emerging social ethic of *obshchestvennost'*, which centered around the importance of purposive activity for the public good rather than for individual profit or state interests. Could scientists as educated professionals serve the common good, and if so, how? As highly trained, relatively well-paid professionals in the world's first underdeveloped nation, how could they justify their privileges in a country where poverty, premature death, and illiteracy were the lot of the majority of people?[61] Would scientists continue to be defined by their traditional roles as servants of the state, especially in light of their high degree of dependence on state employment and the autocracy's ability to limit professional autonomy? Were they mere *chinovniki* (bureaucrats), or could they function as *obshchestvennye deiateli* (public activists) serving society's needs more broadly? These questions became more pressing by the very end of the nineteenth century as many educated Russians increasingly lost their faith in the state's willingness and competence to serve the needs of all Russians. So for the Russian scientific community as a whole, a key challenge in this period was how to link their professional activities to broader, public goals, to both serve society and achieve self-realization as autonomous professionals. Such concerns put the process of professionalization at the heart of the contemporaneous process of the construction of civil society in Russia. Whether scientific publicists emphasized the practical benefits of applied science in improving the material conditions of life or the necessity of cultivating a skeptical, rational scientific outlook as a prerequisite for creating a critical-thinking public, they were striving to place natural science at the center of the enterprise of reform and improvement.

[58] I. I. Mechnikov, "Rasskaz o tom, kak i pochemu ia poselilsia za granitsei" (1909), republished in *Stranitsy vospominanii: sbornik avtobiograficheskikh statei* (Moscow: Izd. Akademii Nauk SSSR, 1946), pp. 77–86, on p. 78.

[59] Turgenev, *Fathers and Sons* (cit. n. 37), p. 19.

[60] Tocqueville, *Democracy in America* (cit. n. 2), vol. 2, p. 46.

[61] See David Joravsky, "The Perpetual Province: 'Ever Climbing Up the Climbing Wave,'" *Russ. Rev.* 57 (Jan. 1998): 1–9. This was not a new question for the Russian intelligentsia, of course.

Although it sounds a bit paradoxical, at the same time the scientific community was becoming more reflective (socially and ethnically) of Russian society, the profession was becoming more Westernized. A great migration abroad for advanced training began within weeks of the death of Nicholas I, particularly to the nearby laboratories of Germany.[62] Those scientists returned as research-oriented, highly trained specialists who identified more closely with the international community, which set research problems and distributed status. Once reestablished in Russian institutions, the scientists tried to obtain from the authorities the same laboratories, equipment, time, and money for research enjoyed by their colleagues in the West, and they aspired to publish the results of their work in specialized journals abroad and at home. Another mark of the growing sense of professional identity was the desire of Russian scientists to express their increasing self-consciousness as a professional community. This associational impulse could be satisfied to only a limited extent by university learned societies or specialists' gatherings. In pursuing the goal of an interdisciplinary, national organization, Russians were again striving to emulate Western models such as the British and German scientific associations. The organizers of the First Congress of Russian Naturalists, which met in St. Petersburg in the winter of 1867–1868, explained their motivations in the following terms:

> Russia, more than any other country of Europe, needs to unite and strengthen its scientific strength: its huge distances and difficulty of communication, isolate each of our scientific centers and often surround the lone investigator with a veritable desert. Thus, a very familiar feeling among us is the desire for rapprochement with other scientists. It is this rapprochement that is the chief goal of the Congress of Russian Naturalists. Let the conviction settle on each of us who gives himself up to the study of nature that there exists a common bond, that there is no such remote corner that even a single isolated worker cannot expect to be helped by word or deed.[63]

That first congress attracted approximately 465 attendees, about one-quarter from outside the capital, to hear 153 papers in six sections (although published reports indicate that as many as 2,500 people were present at some of the general sessions). One of the major accomplishments of this first congress was the laying of the groundwork for an expanded network of university-based naturalists' societies that would help promote collegiality and collaboration between national meetings. The national congresses continued to be held at irregular intervals over the next fifty years at various universities all over the empire and reflected the growth of numbers and specialization in the Russian scientific community.[64] In 1889, when the congress

[62] The travels and study of the physiologist I. M. Sechenov were typical. See his *Autobiographical Notes*, ed. Donald B. Lindsley and trans. Kristan Hanes (Moscow: Academy of Medical Sciences, 1952), pp. 65–96.

[63] The announcement of the congress is reprinted in N. Kin., "Sovremennaia zapadnaia nauka na russkoi pochve," *Otechestvennye zapiski* (1870): otd. 2, 204–5. Note that subsequent meetings added physicians to the title of the meetings.

[64] For background on the congresses, see Vucinich, *Science in Russian Culture* (cit. n. 17), vol. 2, pp. 79–83; and A. V. Pogozhev, *Dvadtsatipiatiletie estestvenno-nauchnykh s'ezdov v Rossii, 1861–1886* (Moscow: Tip. V. M. Frisch, 1887). This latter work also contains a complete bibliography of the publications of the first eight congresses. In addition, V. S. Savchuk, *Estestvennonauchnye obshchestva iuoga Rossiiskoi imperii vtopaia polovina XIX – nachalo XX v.* (Vidavnitstvo, 1994) examines the congresses held in Odessa. Professor N. A. Grigor'ian of the Institute for the History of Science and Technology of the Russian Academy of Sciences has also studied these congresses; I am grateful to her for sharing her unpublished work on this subject with me.

met in St. Petersburg for the third time, there were more than 2,000 registered attendees (almost half from outside the city) listening to 400 papers in eleven separate sections; by the eve of World War I these congresses were regularly attracting some 5,000 scientists and physicians.[65]

But although Russian scientists were able to adopt some of the trappings of Western professionals, their autonomy was limited by suspicious state bureaucrats who exhibited profound ambivalence about allowing such initiative and self-organization. On the one hand, government officials, even the most conservative, understood that there were advantages to be gained from encouraging the pursuit of science, and they certainly wanted to reap the utilitarian benefits of scientific progress and educated technicians and specialists. On the other hand, state officials at that time were eternally vigilant about the dangers of national gatherings, which were viewed as potential seedbeds for political organizing. The Ministry of Public Instruction bureaucrats who oversaw the congresses made certain that speakers and their topics were preapproved. While anyone who "pursued the natural sciences in a scholarly manner" was allowed to attend the meetings, and guests were also permitted, actual participation (as a presenter or even in the discussions) was restricted to "those scientists who have published communications or research in natural science or other closely related disciplines, and also science instructors in higher and secondary schools."[66] These restrictions would presumably ensure a scholarly, politically innocuous, discourse. A worsening political situation in 1904 caused the ministry to cancel the twelfth congress in Odessa; the congress did not take place until 1909 (and then in a less politically volatile venue). Ironically, in an apparent oversight, zemstvo statisticians—a notoriously radical group—were allowed to meet as a subsection at several congresses. But overall the Congresses of Russian Naturalists and Physicians were relatively tranquil affairs, especially compared to the turmoil of university campuses at the end of the century (or to more radicalized professions such as zemstvo physicians).

Nonetheless, despite their evident docility, the university scientists who organized the congresses could not obtain permission to transform them from periodic meetings into a permanent, national organization. This would have been a particularly desirable step given that the lack of a permanent organizing committee hindered the planning of meetings; in addition, long-term projects (scientific expeditions, research stations, and the like) proved difficult to mount without some continuity between gatherings. The scientists no doubt felt that their claim to a national organization was at least as persuasive as that of physicians, who had been allowed to form a national society in the mid-1880s. At the eighth congress in 1890, proposals were made to create a permanent organizing committee as well as a Russian Association for the Advancement of Science. The Ministry of Public Instruction turned down the first proposal on the spot and spent years finding reasons to not approve the charter for the association.[67] These conflicts between scientists and bureaucrats paralleled the ongoing struggles between the liberal university professorate and the state over such issues as academic freedom, university governance, and control of the

[65] *Dnevnik 10-go s"ezda russkikh estestvoispytatelei i vrachei v Kieve*, no. 1, 21 Aug. 1898, p. 15.

[66] Ibid., pp. 4–7.

[67] *VIII S"ezd russkikh estestvoispytatelei i vrachei v S.-Peterburge ot 28 dekabria 1889 do 7 ianvaria 1890 g.* (St. Petersburg: Top. V. Demakova, 1890), pp. 49–50.

curriculum.[68] And they demonstrate how a struggle for increased professional autonomy could take on the shades of a political fight for greater freedom of association in the public realm.

While state bureaucrats erected significant barriers to the achievement of some of the scientists' professional ambitions, it was the radical intelligentsia who posed the greatest intellectual challenge to scientists' professional identity and self-image in the last third of the nineteenth century. A striking shift occurred among radical youth as the fad for science as an intellectual panacea in the 1850s and 1860s gave way to a renewed questioning of the utility of learning in light of the immediate material and political needs of the masses. As one erstwhile student of the natural sciences framed the dilemma:

> Work or learning; that is, should one devote oneself to scholarship in order later to lead a life of a privileged educated professional? Or remembering one's debt to the people, [should one] abandon the institutions of higher education to take up a trade [and work with the people]?[69]

Or, more pointedly: "If you are going to train for such a long time in chemistry, when will you begin to work for the revolution?"[70] For populist philosophers such as P. L. Lavrov and N. K. Mikhailovskii the issues were perhaps more complex, but the conclusions much the same. Lavrov and Mikhailovskii, themselves trained in the sciences, led the philosophical backlash against the extreme scientism, positivism, and utilitarianism of such writers as Buckle and the Russian nihilists. Their analyses of the role of science in modern society are complex and important; in this context it is their critique of the "ivory-towered" scholar that is relevant.[71] For Lavrov, specialization and dedication to "pure science" had the pernicious effect of isolating scientists from the moral and social implications of their work. They were in danger of becoming a new aristocracy, cut off from society and also alienated from the true nature of science itself. In Mikhailovskii's evocative phrase, the scientific specialist "is not a man but an organ, a part of a man."[72] In reviving Alexander Herzen's critique of "Buddhism in Science," Lavrov and Mikhailovskii were reminding scientists of their moral duty, as members of a privileged minority, to give back to the suffering masses on whose backs scientific progress had been achieved. Russian scientists faced a stark choice, according to historian David Joravsky: "Was the service of pure science appropriate for ideal *intelligenty* or only for self-seeking careerists? Would a life in science help to achieve truth and justice—*pravda*, 'rightness', combines both concepts—or would it be a retreat from struggle to the cloister (*zamknutost'*) of a new privileged class, a secular priesthood?"[73] Or, to put it another way, would scientists demonstrate the social ethic and activism implicit in *obshchestvennost'*? A concrete instance of the critique of scientists' narrowly professional

[68] See Samuel Kassow, "Professionalism among University Professors," in *Russia's Missing Middle Class* (cit. n. 3), pp. 197–221, especially pp. 198–201.

[69] Quoted in Brower, *Training the Nihilists* (cit. n. 34), p. 186.

[70] Ibid, p. 187.

[71] The following discussion is based on Alexander Vucinich's discussion in *Social Thought in Tsarist Russia: The Quest for a General Science of Society, 1861–1917* (Chicago: Univ. of Chicago Press, 1976), pp. 39–44.

[72] Vucinich, *Science in Russian Culture* (cit. n. 17), vol. 2, p. 43.

[73] Joravsky, *Russian Psychology* (cit. n. 53), p. 52.

concerns could be found on the pages of the influential journal *Otechestvennye zapiski* (Notes of the Fatherland). In a review of the published proceedings of the First Congress of Russian Naturalists, the writer roundly criticized the congress's stated goal of promoting collegiality. In a time when there were so many more crucial issues to be discussed—such as the government's recent decision to curtail the teaching of natural sciences in the classical *gymnasia* or the issue of women's higher education—that focus ensured that the results of the congress would be "null."[74]

Russian natural scientists wrestled with these criticisms at both a collective and a personal level. The congresses, public events covered by the general press as well as professional journals, became one forum where professional identity was shaped by everything from the choice of the keynote speakers to the specific disciplines represented in the subsections and the time devoted to each. One needs to be wary of making conclusions too sweeping about the evolution of Russian science from these meetings since each congress was organized by the science faculty of a different university, each of which seems to have had a unique method of slicing up the pie of knowledge. But some general trends did crystallize over time. At first glance, the congresses indeed seemed to confirm Lavrov's worst fears. Most papers were written by specialists, for a specialized audience. The sections themselves became more numerous, reflecting the disciplinary developments of the nineteenth century and the resulting fragmenting of knowledge. On the other hand, Lavrov might have approved the noticeable trend toward including more sections in applied science on the programs. Medicine and hygiene were the earliest of these applied sciences; later on fields such as agronomy, statistics, mechanics, and aviation made appearances.[75] Just how much applied science to include was often a most contentious issue; it appears that university scientists who dominated the organizational committees were often conscious of accusations that their enterprises were too removed from the concerns of everyday life. These scientists also felt pressure from congress attendees, a large number of whom were employed by public entities, especially zemstvos, in fields such as medicine, agronomy, and statistics.[76]

Even when applied sciences were not well represented on the program, the general sessions at each congress often became a forum to highlight the benefits of science to the broader society, underlining the ability of science to solve concrete problems

[74] Kin.,"Sovremennaia zapadnaia nauka" (cit. n. 63), p. 205. The presidium of the congress received a petition requesting that the congress discuss the issue of women's higher education; the response was to cautiously decline based on the fact that there was no pedagogical section to take up the issue. On the other hand, the same journal had earlier run a review of the published proceedings of the first congress by the young Il'ia I. Mechnikov, who savagely criticized the participants for failing to achieve the scientific standards of their Western counterparts. I. I. Mechnikov, "Retsenziia 'Trudy I s"ezda russkikh estestvoispytatelei v Sankt-Peterburge, proiskhodivshego s 28/XII 1867 po 4/I 1868,'" *Otechestvennye zapiski* 2 (April 1869): 244–55. These two very different critiques exemplified the difficulties that scientists faced in steering a course between the ethos of professional achievement and service to society.

[75] The eighth congress in 1890, for example, included sections on mineralogy, agronomy, and scientific medicine and hygiene, while the tenth congress—held in Kiev in 1898—added sections on aviation and mechanics, and a subsection on statistics.

[76] One early example of this pressure came at the fourth congress in Kiev in 1873, where the local medical society presented a strongly worded petition to include a section on *obshchestvennaia meditsina* (public or zemstvo medicine) in future congresses. F. M. Suvorov and N. O. Kovalevskii, eds., *Protokoly i rechi obshchikh sobranii chetvertago s'ezda Russkikh estestvoispytatelei* (Kazan, 1873), pp. 5–6. Also see Pogozhev, *Dvadtsatipiatiletie estestvenno-nauchnykh s'ezdov v Rossii* (cit. n. 64), pp. 112–33.

and improve the conditions of life. As we saw in the previous section, a utilitarian emphasis on science as a tool for achieving material progress was both uncontroversial and accorded well with the values of *obshchestvennost'*. But it was also the case that such speakers often explicitly cataloged the shortcomings of Russia (and often its government as well), thus raising controversial social or political issues that would have been difficult to voice in other public forums. In the early congresses, hygiene was often cited as a concrete example of the direct contributions that science could make to the quality of human life. With death rates declining in western Europe at a time when they were climbing in Russia, hygiene served as a salutary example of the need for more scientific expertise to be applied to the health problems of Russians. Such was the message at the Kazan congress of 1873, for example, when attendees heard back-to-back speeches on "Hygiene and Civilization" and "Concerning the Relations of Hygiene to Each Member of Society."[77] Another type of argument for the utility of science for Russian society focused on the need to study and catalog the flora, fauna, geology, meteorology, and ethnography of a vast, diverse, and, it was argued, exceptional empire. Many of the scientific projects undertaken by the congresses, such as the Crimean Committee and the Sevastopol Biological Station, reflected this emphasis. It is particularly revealing to note that both of these justifications for the value of the scientific enterprise in Russia had long been deployed in the context of the needs of the state and were now being transferred to justify the benefits that civil society would gain from the pursuit of science by professionals.

But not all scientists were so eager to defend their enterprise solely on the basis of its practical utility, preferring to argue that the pursuit of pure science also contributed to social progress. The comparative embryologist Il'ia Mechnikov made such an argument for the importance of theoretical science for Russian society to the seventh congress in Odessa, over which he presided in 1883. It was a dramatic moment: the assassination of Alexander II had taken place just two years earlier, and in the aftermath of that event Mechnikov himself had resigned from Novorossiia University in protest against the heightened politicalization and repression of academic freedom on that campus. In his speech, Mechnikov advised his audience not to be distressed by the fact that this particular congress program featured very little by way of applied science, being devoted instead almost completely to "so-called pure or theoretical, even university science."[78] While he acknowledged the good that applied science had accomplished for humankind, Mechnikov emphatically rejected the common view that "for us, theoretical science is an impermissible luxury." Instead, he argued, Russians needed above all to strengthen the pursuit of pure sci-

[77] N. A. Andreev, "Ob otnoshenii gigieny k kazhdomu iz chlenov obshchestva," in *Protokoly i rechi obshchikh sobranii chetvertago s'ezda Russkikh estestvoispytatelei* (cit. n. 76), pp. 69–82; and I. P. Skvortsov, "Gigiena i tsivilizatsiia," in ibid., pp. 83–92. Andreev's talk was a particularly good example of a political argument made in a scholarly guise as he used Rudolf Virchow's metaphor of the body politic to build an argument in favor of further state reforms. As the published abstract of the speech explained, "[D]eveloping this analogy between the conditions of the health of an organism and the health of a state, N. A. Andreev showed that the recent reforms in Russia, thanks to which Russia has moved with a quick pace down the path of progress, could not have corresponded better to the healthful functioning of society." Suvorov, *Protokoly i rechi obshchikh sobranii chetvertago s'ezda Russkikh estestvoispytatelei* (cit. n. 76), p. 16.

[78] I. I. Mechnikov, "Vstupitel'noe slovo predsedatelia s'ezda," in *Sorok let iskaniia ratsional'nogo mirovozzreniia* (Moscow: Nauchnoe slovo, 1913), pp. 201–5.

ence—though not because it was a weapon to tear down established traditions, as the Nihilists claimed, or even because a scientific worldview could provide certain answers to the moral or social problems of life, as Herbert Spencer argued. Rather than viewing contemporary science as the source of certain answers to issues (whether scientific or social), Mechnikov asserted that it was only via the "theoretical working out of questions of natural science" that an individual could develop the skeptical frame of mind and rational worldview that led one to the critical examination of one's own principles and toleration of divergent views. The study of natural science provided a method for carefully working out the best solution to any particular problem and for approaching as near as possible to the truth.[79] As evidence of the effectiveness of this scientific method, he cited the example of medicine, where great progress had been achieved merely through the critical questioning of the efficacy of therapeutic procedures such as bloodletting. Mechnikov went on to obliquely link his argument to recent political events, condemning the results of the lack of critical thinking that had led to extremism in Russian political life:

> We have already seen examples of the practical application of uncritically examined high principles, and these prior experiences are sufficient to encourage the very deepest skepticism. It [skepticism] should produce social toleration, train us to look truth squarely in the eye, not shying away from its consequences, and in any event, if possible, banish from the ethical pharmacopeia the use of "heroic medicine."[80]

Mechnikov ended by asserting that, to the extent that the congress promoted theoretical science, it was accomplishing an important task of promoting the development of a rational worldview among fellow Russians. "In the expectation of a better future, we should consider it our responsibility to defend the interests of theoretical knowledge in Russia, despite all the obstacles that might be encountered."

Mechnikov's views offer just one example of the ways in which late-nineteenth-century Russian scientists, particularly the professorate, elaborated upon "the mystique of pure *nauka*," often in defense of academic freedom and university autonomy. Such views could be seen as narrowly self-interested defense of privilege, implying no broader social or political purpose. Mechnikov, for one, claimed to be apolitical, and it is clear from other sources that he saw his mission as a professor to entice students away from politics to "true *nauka*."[81] But he was also a devoted reader of Buckle, and Mechnikov's preference for the pursuit of science as the most direct means to reform the ethical foundations of society reflected his faith that "pure" science could provide a vital social benefit to society since its study led to habits of the mind that were the prerequisites to building a free society.[82] This type of

[79] Ibid, p. 205.

[80] Ibid., p. 204. These comments are sufficiently vague that it is rather difficult to know whether Mechnikov intended to condemn the actions of radicals, such as the ones who assassinated Alexander II, or of the autocracy and its supporters, or all of them.

[81] "I tirelessly preached faith in *nauka*, by which I meant true *nauka*, which is often ridiculed as 'science for science's sake.' I showed that it is necessary to study it so that a person can act consciously, and determine the source of the greatest good. Imbued with these feelings, I tried to foster them in young people, who were being attracted more and more to the side of politics." I. I. Mechnikov, "Rasskaz o tom" (cit. n. 58), pp. 77–86, on p. 79.

[82] Mechnikov's reading notebooks from the 1860s and 1870s show that he dipped into Buckle several times. Arkhiv Rossiisskoi Akademii Nauk (hereafter cited as ARAN), f. 584, op. 1, d. 268 and d. 292.

argument in defense of the scientific enterprise was long lived in Russia; in the early twentieth century other Russian scientists, most notably K. A. Timiriazev and V. I. Vernadskii, developed their own variations on this theme, linking science and democratic reform even more explicitly in a way the Buckle might also have recognized.[83]

In their visions of the contributions of science to society, then, scientists were building directly from their privileged position as possessors of the specialized knowledge and technical skills they believed necessary for the uplift of Russia. This was a different conception of service to society than that held by the radical intelligentsia; this was not self-abnegation in service directly to the *narod*, with its attendant assumption of the masses' innate superiority or of the need to serve them directly as a zemstvo doctor or village teacher.[84] It is worth noting as well that this ethos of service became a key motif for Russians eager to distinguish their exceptional professional values from those of scientists in other countries. According to one Russian scientist, for example:

> The average German pursues science *[nauka]* as a profitable trade—profitable not only for himself personally, but also for the people and the state. Many Englishmen and Frenchmen pursue science as an interesting and noble sport, not giving a thought to its utility. But one often finds Russians, and Slavs in general, to be motivated by a sacred enthusiasm which regards the pursuit of science as the only way to achieve a tolerable if incomplete world view, and the search for truth as both an irresistible personal need and a moral duty before the fatherland and all of mankind.[85]

However one might question the accuracy of this depiction of German, French, or English science, it does convey the moral fervor that lay at the heart of this ethos of service to a common good, an ethos that fit squarely within the developing notions of *obshchestvennost'*. Last, but certainly not least, the notion of service to society provided a public identity removed from that of a servant of the state, a mere *chinovnik* (state bureaucrat). That was a priority for educated Russians in most professions, even as they remained dependent upon the state in a myriad of ways.[86]

So did Russian scientists live up to these ideals, or were they guilty of careerism and the meaningless pursuit of esoteric abstractions, as their radical critics charged? That question is too broad to answer with any precision. Across the whole community of Russian scientists there existed significant variations in career paths due to

[83] On Vernadskii, see Bailes, *Science and Russian Culture* (cit. n. 45); James McClelland discusses both Vernadskii and Timiriazev in his study *Autocrats and Academics* (cit. n. 24).

[84] See Kendall Bailes, "Reflections," in *Russia's Missing Middle Class* (cit. n. 3), pp. 39–54.

[85] V. A. Mikhel'son, "Rasshirenie i natsional'naia organizatsiia nauchnykh issledovanii v Rossii," *Priroda*, nos. 5–6 (1916): cols. 696–8, quoted in McClelland, *Autocrats and Academics* (cit. n. 24), p. 73.

[86] As Sidney Monas has observed, professors found themselves in a particularly tricky situation in this regard. "The intelligentsia, by no means all radicals and revolutionaries, became increasingly alienated from the state and state service. Publicistic writing tended strongly to distinguish between service to the state (*gosudarstvo*) and 'public' institutions, representative bodies like the zemstva, the 'free' professions, and publicistic and cultural activity (*obshchestvo*). (In this dichotomy, the universities were in an anomalous position: professors were technically civil servants [*chinovniki*], holding government rank; at the same time the universities enjoyed a certain chartered independence, and professors were generally considered *obshchestvennye deiateli* rather than *chinovniki*.)" Monas, "The Twilit Middle Class of Nineteenth-Century Russia," in *Between Tsar and People* (cit. n. 3), pp. 28–37, on p. 34.

disciplinary differences, geographical location, and employment. For example, Nathan Brooks has suggested that when chemists remade their discipline along German lines in the mid-nineteenth century, the more "professionalized" they became, the more they cut themselves off from formerly important local audiences, patrons, and interests.[87] That picture is certainly in accord with what Steven Shapin has referred to as the "canonical account of the historical relations between science and the public," where professionalized science is associated with the weakening of the public's power over scientists and a widening gap between the two camps. But was the experience of Russian chemists typical of other scientists? That is a question worth asking since the "canonical account" does not otherwise appear to work well in the Russian context.

Indeed, it can be argued that to a striking degree many Russian scientists, especially among the elite of the profession, moved closer to the public as they searched for ways to combine science with service to society. Some scientists could apply their specialized knowledge to specific economic or social problems. For example, the mineralogist V. V. Dokuchaev became the key figure in the creation of soil science in Russia, his work supported by the Free Economic Society, the St. Petersburg Society of Naturalists, and various zemstvos. Some Russian scientists demonstrated their commitment to contributing to the common good by pursuing a virtual second career in a field of applied science, one often far removed from their primary scientific interests. For example, A. M. Butlerov, a leading structural chemist, was also a pioneer in establishing modern techniques of beekeeping in Russia, while the comparative embryologist Alexander Kovalevskii became the leading Russian scientist in the fight against phylloxera, an economically devastating disease of grapevines. It is interesting to note that while all of these scientists published their primary research in western European languages as well as Russian, these secondary careers were conducted in Russian for an exclusively Russian audience and often at the instigation of a public organization. For scientists such as K. A. Timiriazev, who believed that education in science was a "school for democracy," the popularization of science became the key to serving society. And then there was V. I. Vernadskii, who seems to have combined all of the above-mentioned activities with zemstvo service as well. No wonder his wife complained that "in Russia, it is difficult to be a scientist. The surrounding life swallows one up."[88]

Thus professionalization and the extension of civil society were two intimately connected processes in late Imperial Russia. Professional meetings and organizations provided some of the first opportunities for associational life for many educated Russians, and the meetings functioned as avenues not just for enhancing professional identity, but for contributing to the extension of autonomy and free speech in the public sphere as well. Driven in part by a desire to separate from the state, many scientists developed an ethos of service that linked professional values with the moral world of *obshchestvennost'*, thus encouraging them to find ways to link their enterprises to the needs of society.

[87] See Nathan M. Brooks, "Alexander Butlerov and the Professionalization of Science in Russia," *Russ. Rev.* 57 (Jan. 1998): 10–24. Chemists may well have formed an exceptional group as they were also among the earliest to form a discipline-specific professional association.

[88] Bailes, *Science and Russian Culture* (cit. n. 45), p. 65.

THE PUBLIC CULTURING OF RUSSIAN BACTERIOLOGY

As regards our mutual nightmare, practical activity . . .
Lev Tsenkovskii to Il'ia Mechnikov (1886)[89]

And when Russian thought, in spite of the constraint to which it was subject, tried to manifest itself in free action and in conscious applications to practical life, it often met with obstacles which were not favorable to its development.
A. S. Lappo-Danielevsky (1917)[90]

As we have seen, after mid-century many Russian scientists were interested in linking their professional activities more closely to the needs and demands of society and in addressing the new nonspecialist audience for science. At the same time, society itself was becoming a more powerful force, one that could command scientists' attention and pay their bills. But what did this service to science and society look like in practice? How did scientists interact with their audience and potential patrons, and what can this tell us about the dynamics of Russian civil society? Did the emergence of public interest in and patronage of science change scientific career patterns or the ways in which scientific disciplines developed? In order to answer such questions we need a rather fine focus, which will be supplied here through a case study of scientists, the public, and the discipline of bacteriology. The history of bacteriology in Russia provides a particularly clear example of the importance of the interactions between science and society. The discoveries of the 1880s promised exciting, immediate benefits for agriculture and health—two key areas of interest for a broad swatch of Russian society. And the influence of public patrons on bacteriology, a new field, lacking any prior patterns of institutionalization, can be clearly seen. The scientists involved found themselves actively courted by public activists seeking to persuade them to apply their expertise to this promising field. These scientists interacted on the public stage with local politicians, landowners, physicians, and even some government officials, all of whom shared a vision of science as a means of achieving social and economic progress and tended to view scientific activity in utilitarian terms, not as enlightenment so much as a solution to local needs and problems. The associations and networks of local civil society had a major role in encouraging and shaping the enterprise of bacteriology in Russia.

Two scientists figure prominently in this story: Lev S. Tsenkovskii and Il'ia I. Mechnikov, commonly awarded the title of founders of Russian bacteriology by historians.[91] Both men came out of the natural history tradition, having pursued chiefly morphological studies of lower organisms in botany and zoology, respec-

[89] L. S. Tsenkovskii to I. I. Mechnikov, 22 Feb. 1886, reprinted in *Bor'ba za nauku v Tsarskoi Rossii* (Moscow: Gos. Sotsial'no-Ekonomicheskoe Izd., 1931), pp. 133–4, on p. 133.

[90] Alexander S. Lappo-Danielevsky, "The Development of Science and Learning in Russia" in *Russian Realities and Problems*, ed. J. D. Duff (Cambridge: Cambridge Univ. Press, 1917), pp. 153–229, on p. 228.

[91] The following discussion of the creation of Russian bacteriology is drawn from Elizabeth A. Hachten, "Science in the Service of Society: Bacteriology, Medicine, and Hygiene in Russia, 1855–1907," Ph.D. diss., Univ. of Wisconsin–Madison, 1991. I do not mean to imply that these two men were the only Russians involved in the emerging field of bacteriology; there were many others, particularly medical and veterinary pathologists. But these are two most commonly cited by Soviet historians as founders of the Russian schools of veterinary (Tsenkovskii) and medical (Mechnikov) bacteriology.

tively, and developed skills in microscopy and related techniques. They both were drawn to bacteriology in part by their long-standing research programs and theoretical allegiances. For Tsenkovskii, who already specialized in the study of bacteria and fungi, it was his allegiance to the doctrine of pleomorphism that helped spark his interest in vaccine development; for Mechnikov, it was his long-standing fascination with parasitical phenomena and evolutionary theory, as well as the desire to defend his controversial phagocytic theory of inflammation, that led him from comparative embryology to the study of pathogenic bacteria and immunity. But the shift from descriptive natural history to an experimental and applications-driven field such as bacteriology was hardly straightforward or easy, and it is difficult to see how this would have occurred without the intervention of other actors: an audience that demanded not just knowledge but the immediate benefits of new breakthroughs, and patrons willing to act to ensure them. Key to Tsenkovskii's and Mechnikov's transformations from priests of theoretical science to craftsmen of an applied field was the support and interest of the public. Of course, these scientists had their own intellectual and professional interests, which did not necessarily always coincide exactly with the interests and needs of society. And although Tsenkovskii and Mechnikov clearly found it expedient to pursue these connections and combine their scientific agendas with *obshchestvennost'*, that path was not always smooth or ultimately satisfying for all parties.

Before 1885 the most important pull on these scientists was the interest shown by agricultural societies and zemstvos in the possible applications of bacterial studies. The link between bacteriology and agriculture reflected the fact that many of the early advances in bacteriology pertained directly to problems involving animal and plant health. But the keen interest in these developments by groups such as the Free Economic Society, the Southern Russian Agricultural Society, and activist zemstvo boards was not just fortuitous. These groups were closely attuned to scientific advances and confident that science would provide solutions to Russian agricultural, medical, and economic problems. Frequent discussions appeared in their publications about the need for science-based innovations, such as experimental stations and modern agricultural education, as the surest routes for escaping Russia's backwardness. In particular, these landowners and scientists often looked to American efforts in this field, ascribing the greater success of American agricultural products on the world market to that country's scientific edge over Russia. Such views seem to have been particularly prevalent in southwestern Russia and Ukraine, an area of large-scale agriculture where landowners seemed especially willing to invest in scientific research as a direct way of improving productivity.[92] In short, these organizations reflected both the structural and moral components of Russian civil society: a Westernized, educated elite imbued with a rational, scientific worldview, organized and acting in the public realm outside the state in order to achieve socially useful goals.

If there was one figure who embodied for Russian agriculturalists the achievements and promises of Western science, it was the French scientist Louis Pasteur. Pasteur's work figured prominently in Russian agriculture journals from the

[92] These statements are based on my study of the publications of groups such as the Southern Russian Agricultural Society. For a more comprehensive study of the connections between science and agriculture see Elina, "Nauka dlia sel'sko khoziaistva" (cit. n. 47).

mid-1870s on. But it was in 1881, with his dramatic public demonstration of the anthrax vaccine at Pouilly-le-Fort, that Pasteur's research began to attract widespread attention in Russia.[93] Pasteur's success against anthrax (and a few years later with the rabies vaccine) was a crucial ingredient in creating a consensus among Russians for fostering bacteriological research in their own country. In a country where anthrax was a widespread and costly problem, it is not surprising that both state and society responded quickly to Pasteur's breakthrough. But instructive differences existed in their approaches. Following the lead of Austria-Hungary which immediately invited one of Pasteur's assistants to demonstrate the vaccine there, the Russian Ministry of State Domains (which functioned as the agricultural ministry it became a decade later) quickly announced plans to buy enough vaccine from Pasteur to test on ten thousand lambs the next spring.[94] However, the members of the Free Economic Society (Vol'noe Ekonomicheskoe Obshchestvo) (FES) had more ambitious plans for the anthrax vaccine. This St. Petersburg-based organization was one of the oldest voluntary associations in Russia, founded in 1765 by Catherine II, who granted its charter hoping to encourage civic involvement and an "improving spirit" among her subjects.[95] The society pursued its main goals of promoting scientific agriculture and ameliorating the conditions of rural life through a large number of programs, including economic research, beekeeping, agricultural and soil studies, the encouragement of animal husbandry, dairy production, agricultural education, and support of smallpox vaccination. With a membership composed primarily of state officials, members of the nobility, and the scientific elite of St. Petersburg, the FES defies easy categorization, functioning rather as a meeting ground between state and civil society.

It was a highly placed government official, V. I. Kovalevskii of the Ministry of Finance, who launched the group's anthrax vaccine project in a speech at an FES meeting on 31 October 1881. Kovalevskii pointed out that reliable vaccines to prevent epizootic diseases could potentially save millions of rubles for Russian agriculture (although he had to concede that sheep anthrax itself, the disease against which Pasteur's vaccine was designed, was not very important economically in Russia). Nonetheless, he argued that "Russia, more than any other country, needs to take an active posture in regard to this extremely important question . . . because this method of vaccination promises to bring in the future (perhaps in the near future) further important discoveries and new, even more fruitful practical applications."[96] Presumably it was Russia's very backwardness that necessitated a more aggressive approach to assure that in the future the vaccine could be produced in Russia by Russians. Promising that the anthrax vaccine would be "one more brilliant page in the activi-

[93] On Pasteur and the initial anthrax testing at Pouilly-le-Fort, see Bruno Latour, *The Pasteurization of France* (Cambridge, Mass.: Harvard Univ. Press, 1988), pp. 85–7; and Gerald Gieson, *The Private Science of Louis Pasteur* (Princeton, N.J.: Princeton Univ. Press, 1995), pp.145–76.

[94] The similar history of the anthrax vaccine in Hungary is mentioned in *Correspondence of Pasteur and Thuillier Concerning Anthrax and Swine Fever Vaccination*, trans. and ed. Robert Frank and Denise Wrotnowska (Birmingham: Alabama Univ. Press, 1968). These events regarding vaccination, both scientific and political, were followed very closely on the pages of Russian publications such as *Arkhiv Veterinarnykh Nauk*.

[95] On the Free Economic Society, see A. N. Beketov, comp., *Istoricheskii ocherk dvadtsatipiatiletnei deiatel'nosti Imperatorskogo Vol'nogo Ekonomicheskogo Obshchestva s 1865 do 1890 goda* (St. Petersburg: Tip. V. Demakova, 1890); and Pratt, "Russian Free Economic Society" (cit. n. 31).

[96] V. I. Kovalevskii, "O predokhranitel'nom privivanii sibirskoi iazvy i o zhelatel'nom sodeistvii etomu delu," *Trudy Imper. Vol'nogo Ekonomicheskogo Obshchestva*, no. 3, vyp. 4 (Dec. 1881): 425.

ties of our society," Kovalevskii proposed sending two scientists—a botanist and a veterinarian—to Paris to acquaint themselves with the new method.[97] The society's main role would come later, when the vaccine was ready for production and distribution throughout Russia.[98]

The society's choice for a botanist fell on Lev Tsenkovskii, at that time a professor in the natural sciences faculty at Kharkov University. Although Tsenkovskii was considered one of the leading Russian botanists of his day, his career to that point had not been marked by conspicuous activity in the public sphere. But that changed abruptly with the commission from the FES.[99] The mission to Paris itself proved unsuccessful as Pasteur intended to preserve a commercial monopoly on the vaccine. But Tsenkovskii had been bitten by the bacteriological bug and was determined to develop his own workable vaccine, motivated in part by his conviction that Pasteur's attenuation method supported the Russian's own theory of bacterial pleomorphism.[100] Returning to Kharkov University, Tsenkovskii proceeded to convert his botanical study into a bacteriological laboratory. Gathering assistants from the Kharkov Veterinary Institute, he began his work from scratch, transforming himself in the process from a descriptive morphologist to an experimental bacteriologist. (Tsenkovskii finally succeeded in producing a workable prophylactic inoculation, but not until shortly before his death in 1887. Hence he did not live to see his method become the preferred vaccine in Russia.)[101]

Aside from his theoretical and technical problems with the vaccine, Tsenkovskii struggled to find financial backing. Although the Free Economic Society granted him five hundred rubles for vaccine development upon his return from France, that was the end of the society's patronage of anthrax research. Besides the relative failure of the Paris mission, the society seems to have been discouraged by the disappointing results from the first Russian field trials of Pasteur's vaccine in 1882.[102] Tsenkovskii's fortunes began to turn around, however, as word of his project spread

[97] Ibid., pp. 425–6.

[98] While Kovalevskii's speech seems to have received an enthusiastic reception from the other laypeople present, many of the scientists in attendance that day were far more skeptical. Indeed, it is quite intriguing that Kovalevskii assumed the mantle of an expert on bacteriology in this setting. In particular, many of the physicians and veterinarians present expressed doubts not only about the theoretical basis and efficacy of Pasteur's vaccine, but also about Kovalevskii's proposal to send a botanist, along with a veterinarian, to Paris. Kovalevskii viewed bacteria as plants, albeit microscopic ones, which demanded special culturing in a laboratory setting, thus necessitating the services of a botanist. In the end, Kovalevskii's proposal prevailed—helped by the eagerness of wealthy members such as Grand Duke Nikolai Nikolaevich to contribute generously to the cause. "Zhurnal sobraniia I Otdeleniia I. V. E. Obshchestva 9 noiabria 1881 goda," *Trudy IVEO* no. 2, vyp. 1 (1882): 79–97.

[99] The two scientists from the FES were not the only Russians leaving for Paris in the winter of 1881–1882; two other veterinary pathologists were sent for the same purpose by various government bureaus (including the state stud farm), leading at least one medical journal to wonder why these missions were not better coordinated. *Vrach* 3 (1882): 94, 178.

[100] Pasteur's approach was premised on the theory that manipulating the conditions of life of the bacilli would result in drastic changes in bacterial physiology and morphology, an approach that assumed the existence of bacterial pleomorphism. Tsenkovskii did not seem to have any difficulty squaring the germ theory of disease with his notion of the nonspecificity of bacteria, attributing the differing physiological impact of bacteria to changing environments or stages of the life cycle. See, e.g., "Mikroorganizmy: Bakteriial'nye obrazovaniia," *Mir* 2 (1882): 45–70.

[101] For more on Tsenkovskii's career, see A. I. Metelkin, *L. S. Tsenkovskii: Osnovopolozhnik otechestvennoi shkoly mikrobiologov, 1822–1887* (Moscow: Medgiz, 1950).

[102] See P. Gordeev, "Pervye opyty Pasterovskoi privivki v Rossii," *Veterinarnyi vestnik* 1 (1882): 171–92.

through the agricultural and scientific circles in the southern region of Russia, aided by the efforts of a former student, a large landowner in nearby Kherson province who became personally involved in the project as one of Tsenkovskii's assistants. Soon Tsenkovskii started receiving financial support and donations of experimental animals from sources such as the Kharkov Agricultural Society, the Southern Russian Agricultural Society, and the Kharkov and Kherson zemstvos. With this support came the obligation to regularly solicit funds, report on his progress, and ultimately deliver a workable product; all tasks Tsenkovskii had not had to face in his life as a university-employed botanist.[103] Tsenkovskii clearly felt some ambivalence about his new social role, especially the necessity of constantly courting patrons. He expressed some of this frustration in a letter to his friend and former colleague Il'ia I. Mechnikov. But although Tsenkovskii considered practical activity to have been a "nightmare" at times, he nonetheless found that the opportunity to build closer ties between the scientific world and Russian society was in itself a positive sign of the increasing role of science for the Russian public.[104] Thus, at a celebration of his thirty-fifth year of scientific activity, Tsenkovskii could note with pleasure that "[s]ociety is beginning to be interested not only in public activists, writers, and artists, but also in scientific specialists. This is all the more precious since up to this time we have not been spoiled by the attention of the public."[105]

At the same time that Tsenkovskii was laboring over the anthrax vaccine, Il'ia I. Mechnikov was developing new research interests touching on microorganisms and their role in nature. Like Tsenkovskii, Mechnikov had been drawn to the study of microbes for a mixture of practical and theoretical reasons, and much of the early impetus and financial support for this work had came from agricultural associations in southern Russia. His first foray into finding practical applications for his scientific training had come in the late 1870s, when the Southern Russian Agricultural Society and the Odessa Entomological Commission had recruited him to develop a method for biological pest control against a highly destructive weevil then wreaking havoc in the wheat fields of Novorossiia. This biological pest control project quickly had taken on great fascination for Mechnikov. In the following years, he published several scholarly and popular works on the diseases of weevil larvae, plunged into debates with skeptical entomologists, and became a member of a special zemstvo commission on the matter. In short, Mechnikov followed a common pattern among Russian scientists of developing a second, applied scientific career.[106] Thus, Mechni-

[103] See L. S. Tsenkovskii, "O pasterovskikh privivkakh (doklad na V entomologicheskom oblastnom s"ezde v Khar'kove)," *Arkhiv veterinarnykh nauk* (1886): otd. 5, 301–26; and "Otchet o privivkakh antraksa v bol'shikh razmerakh," *Sbornik Khersonskogo zemstva*, no. 6 (1886): 1–19.

[104] In fact, most of Tsenkovskii's frustrations in this particular letter seem to have been focused on the belated offer of financial support from a state ministry.

[105] "Rech' professora L. S. Tsenkoskogo," *Iuzhnyi Krai*, no. 1772 (19 Feb. 1886), p. 2.

[106] In her dissertation on Mechnikov, A. V. Sorokina mentions finding a manuscript source in the Odessa University Library that allowed her to date his interest to this subject to 1875. "I. I. Mechnikov i razvitie otechestvennoi mikrobiologii," Kand. diss., Institut vaktsini i syrovotok im. I. I. Mechnikova, 1968, p. 23. On the biological pest control work, see I. I. Mechnikov, *Khlebnyi zhuk: Bolezni lichinok khlebnogo zhuka* (Odessa: Tip. P. Frantsova, 1879); idem, "Materialy k ucheniiu o vrednykh nasekomykh iuga Rossii," *Zapiski Novorossiisskogo universiteta estestvoispytatelei* 4 (1880): 1–10; idem, "Zamechaniia na sochinenie g. Lindemana," *Sel'skoe khoziaistvo i lesovodstvo*, no. 6 (1880): 131–49. Unfortunately, the execution of the idea proved far more difficult than Mechnikov had anticipated, and although he continued his research off and on for more than a decade, he never managed

kov's belief in "science for science's sake" did not prevent him from pursuing practical applications for his work in the civic realm.

In 1882, Mechnikov's circumstances changed significantly when he resigned his university position; for a time, he even considered pursuing a career as a zemstvo entomologist. But a new scientific passion had begun to consume Mechnikov: the elaboration and defense of his phagocytic theory of cellular immunity.[107] As the phenomenon of the "struggle for existence" between phagocytes and pathogenic microbes loomed ever larger in his work, Mechnikov established a small bacteriological laboratory in his Odessa apartment and began to search for financial support and an institutional base for his work (a desire shared by a number of other would-be bacteriologists—mainly young, underemployed physicians—in the city).[108]

As it turned out, another resident of Odessa was thinking along those same lines. Liudvig Adol'fovich Marovskii, a physician by training and an elected representative on the Odessa duma (city council), was a state official serving as an inspector for the Medical Board (Meditsinskoe Upravlenie) of the Ministry of the Interior. He was thus an actor in both city government and state administration and intimately acquainted with local medical and sanitary conditions. Marovskii thought that it would be highly desirable for prevention and control of epidemic disease to establish a bacteriological laboratory in Odessa.[109] And despite his position in the state medical bureaucracy, Marovskii intended from the beginning for his planned bacteriological facility to be supported by the city government. In the last weeks of 1885, Marovskii approached Mechnikov and Nikolai Gamaleia (a local physician and self-taught bacteriologist) with his proposal for a municipal bacteriological laboratory. The two bacteriologists embraced the project, although it should be noted that they saw it primarily as an opportunity to pursue their own research interests. The fact that this could be accomplished through affiliation with a publicly funded institution, and not the state, may have been an additional bonus. In the best tradition of the Russian intelligentsia, both men proclaimed their willingness to serve without pay.[110]

That this alliance between the civic-minded medical inspector and the bacteriologists produced results in less than six months was due only to a wholly unforeseen event: the unprecedented public excitement created by Louis Pasteur's development

to achieve any success in field trials. For the last known reference to this project in relation to Mechnikov, see L. Zdroevskii to I. I. Mechnikov, 10 June 1890, ARAN, f. 584, op. 4, d. 24, ll. 1–2.

[107] See Mechnikov's first three works on the phagocytic theory: "O tselebnykh silakh organizma" (1883), in *Akademicheskoe Sobranie Sochinenii*, 16 vols., ed. L. A. Zil'ber (Moscow: Izd. Akademii Meditsinskikh Nauk, 1950), vol. 6, pp. 22–9; "O gribkovom zabolevanii dafnii" (1884), in ibid., vol. 6, pp. 30–40; and "Issledovaniia o vnutrikletochnom pishchevarenii u bespozvonochnykh" (1884), in ibid., vol. 6, pp. 3–21. On Mechnikov's theory, see A. E. Gaisinovich, "100 let fagotsitarnoi teorii I. I. Mechnikova," *Priroda*, no. 8 (1983): 12–22; and Alfred I. Tauber and Leon Chernyak, *Metchnikoff and the Origins of Immunology: From Metaphor to Theory* (Oxford: Oxford Univ. Press, 1991).

[108] In 1885 the Odessa Entomological Commission funded a research project by Mechnikov on the bacteriology of rinderpest, with the goal of developing a vaccine against the disease. However, not only did the research quickly stall, but the commission could not provide the ongoing support and institutional base that Mechnikov desired.

[109] Nikolai F. Gamaleia, *Vospominaniia* (Leningrad: Medgiz, 1947), p. 44. Marovskii's initial conception of a municipal bacteriological laboratory was that it would function as a means of disease control, especially epidemic disease control, by providing physicians and public health officials with rapid and certain diagnoses of cholera, diphtheria, tuberculosis, and other diseases.

[110] Metchnikoff, *Life of Elie Metchnikoff* (cit. n. 50), p. 127.

of a rabies vaccine. This became a local issue for Odessa residents in January 1886 after the newspaper *Odesskii vestnik* (Odessa Herald) reported that an area woman and her four children, who had been bitten by an apparently rabid dog, were in the isolation ward of the local hospital waiting to see if they had been infected with the fatal disease. Responding to these reports, a local magistrate wrote to express his view that Odessa "might have the resources to do more for [the victims] than just isolate them, when about 200 rubles would be sufficient to send them to the genius Pasteur." (He did not fail to point out that several cities in the United States and Hungary had already successfully raised money to send rabies victims to Paris.)[111] Soon a public subscription campaign was launched on the pages of the newspaper to do exactly that.

But it was only after Marovskii and Mechnikov came forward to link the humanitarian effort to send bite victims to Paris to their proposal to establish a bacteriological laboratory in Odessa that the public began to enthusiastically support the newspaper's campaign—a reflection above all of Mechnikov's immense scientific authority. Marovskii hurried to unveil his plan to his fellow city councilmen, justifying the investment by promising public health and economic benefits. Marovskii reeled off a long list of diseases—smallpox, anthrax, swine erysipelas, glanders, rinderpest, and sheep smallpox—that could thus be eliminated as threats to Odessa and the surrounding area. (What he did not mention was that half of the vaccines on his list had yet to be developed anywhere!) The inclusion of animal diseases on this list was, however, no accident, as it allowed him to emphasize the direct economic benefits of bacteriology for Odessa and its surrounding region. A number of the duma members were themselves involved in the lucrative foreign trade in agricultural products, and as a whole, the duma had been most willing in the past to fund municipal projects that directly contributed to improving Odessa's economy.[112] The final appeal Marovskii made on behalf of his project was to the civic pride of his audience:

> The establishment of a bacteriological station in the city of Odessa will have other favorable consequences, apart from purely sanitary and economic benefits, since it will attract to the city the interest of the whole region, make it the center of the dissemination of knowledge of incalculable importance, and cover it with glory. . . . Odessa will be among the first cities in Russia to respond to the call of science.[113]

Perhaps Markovskii had in mind that the proposed station would be a fitting complement to the two other large public projects then underway, projects that symbolized the city's wealth, modernity, and progressive spirit: a full-floating sewerage system (the first in Russia) and an opera house modeled on Vienna's. At the duma meeting, the proposal found yet another important supporter: the head of the coun-

[111] *Odesskii vestnik*, no. 20, 22 Jan. 1886, p. 2.

[112] Patricia Herlihy, *Odessa: A History, 1794–1894* (Cambridge, Mass.: Harvard Univ. Press, 1986), pp. 457–74; and Frederick W. Skinner, "City Planning in Russia: The Development of Odessa, 1789–1892," (Ph.D. diss., Princeton Univ., 1973), chap. 6. Odessa's urban self-government had been considerably strengthened by government legislation in 1863 and 1870, which provided much enhanced scope for autonomous action. The Western orientation and cosmopolitan, commercial bent of the city's political elite, and the economic prosperity of the city, had produced a relatively activist *duma* in this period.

[113] L. A. Marovskii, [Speech to Odessa Duma], *Vedomosti Odesskogo Gorodskogo Obshchestvennogo Upravleniia*, no. 10, 1 Feb. 1886, pp. 1–2, on p. 1.

cil's executive board, N. K. Vel'koborskii, who was well situated to appreciate the promise of bacteriology as he was also a member of several agricultural associations and a representative to the Kherson provincial assembly.[114] As the proposal took shape under his direction, it included a provision to allow the involvement of the Kherson zemstvo and other interested area zemstvos as financial supporters of the new enterprise. Events moved quickly from there as yet other elements of Odessa's civil society began to contribute to the effort as well. Two days after Marovskii's speech to the duma, the Society of Odessa Physicians announced that an anonymous donor (an area landowner) proposed to donate a thousand rubles to fund a scientific mission to Paris to study Pasteur's rabies treatment. This philanthropist had requested that the medical society choose the most appropriate person for the job and be responsible for monitoring his progress in Paris.[115] Here we see a linking of private philanthropy, professional expertise, and public initiative to achieve civic goals in a way that would not have been possible a few decades earlier. At the urging of Mechnikov, the society chose Nikolai Gamaleia for the mission to Paris. He set off for Pasteur's laboratory in early February, accompanied by several of the bite victims. Several weeks later, they returned to Odessa after having undergone apparently successful courses of treatment with the rabies vaccine.

Mechnikov had one of the most visible roles in maintaining and encouraging public interest in a bacteriological station during this time. The high point of his campaign was a series of public lectures on the subject of bacteria and their significance, which he delivered in late March 1886. Organized as a charity event to raise money for the St. Petersburg Women's Medical Courses, the lectures attracted standing-room-only audiences to the hall of the Odessa Stock Exchange, drawn by the subject matter and the cause as well as Mechnikov's reputation as a spellbinding speaker.[116] Mechnikov emphasized his viewpoint that practically oriented laboratories were needed to further expand the scope of preventive vaccines and to develop proper sanitary-hygienic and disinfection measures against epidemic diseases. For Mechnikov, only independent, public institutions could accomplish this task:

> University and academy laboratories, which pursue scholarly and educational goals, cannot satisfy the requirements for bacteriological stations, which should serve as a connection between science and practical life. Above all, such stations should be on the alert for scientific achievements, adapt them to local conditions, and, when possible, tackle new problems as well.[117]

And this work, he declared, had to be pursued in Russia itself. The peculiar problems of the country demanded unique solutions that foreign scientists could not be

[114] For more on urban development and politics in Odessa, see Skinner, "City Planning," and Herlihy, *Odessa* (both cit. n. 112).

[115] The donor was apparently a landowner from the nearby province of Bessarabia, N. V. Stroevsko. Although not a particularly wealthy man, he was a well-known and generous philanthropist.

[116] The venue and especially the specific charitable purpose of these lectures strongly marked Mechnikov's lectures, and the whole bacteriological enterprise, as connected to liberal, progressive elements in Odessa's society. Two of these lectures, "Obshchii Ocherk Bakterii" (15 March 1886) and "Bor'ba s bakteriiami" (27 March 1886) were republished in full in the collection of Mechnikov's work *Akademicheskoe Sobranie Sochinenii*, 16 vols., ed. G. V. Vygodchikov (Moscow: Izd. Akademiia Meditsinskikh Nauk, 1955), vol. 9, pp. 9–34. The other lecture, delivered on 21 March 1886, did not survive in manuscript form, but the newspaper account is also published in ibid., vol. 9, pp. 357–61.

[117] Mechnikov, "Bor'ba s bakteriiami" (cit. n. 116), p. 33.

expected to address successfully. Echoing Kovalevskii of the FES, Mechnikov warned: "Russia finds itself in very distinctive conditions and therefore cannot be satisfied with certain prepared results obtained in the West."[118] Thus Mechnikov framed bacteriology as a pursuit peculiarly suitable for public patronage by local public institutions.

As these events unfolded in Odessa during the spring of 1886, the central government's response remained notably negative. The Medical Department of the Ministry of Internal Affairs attempted to discourage the activities of the Society of Odessa Physicians, for example, turning down its petitions to permit Gamaleia to travel to Paris (the refusal was received in Odessa two months after Gamaleia's departure).[119] Even when plans for a station in Odessa were already well advanced, the department was still expressing its disapproval to local authorities, noting that similar establishments were already being discussed in St. Petersburg, Moscow, and Kazan, and advising that any money gathered in Odessa would best be spent in supporting those efforts in "one of the large cities of our country."[120] What motivated this opposition is not clear. Perhaps it was a reflex action by a bureaucracy that preferred to maintain central control over such local initiatives. In the end, however, the ministry acquiesced to the city's plans, swayed perhaps by Pasteur's own support for the Odessa station or merely by the fact that the city had gone ahead with its plans despite the orders from St. Petersburg.[121]

Thus over the course of a few months in Odessa, the discovery and reception of Pasteur's rabies vaccine helped create a receptive public for bacteriological science and provided the impetus for the establishment of a publicly sponsored institution devoted to its exploitation. Once the station was framed as an economic and medical necessity for Odessa, a complex network of interacting individuals and groups were recruited to further the cause.[122] From this story the contours of civil society in Odessa and the surrounding Kherson province clearly emerge: a thick network of civic-minded individuals, linked through public institutions and pursuing goals that focused on material betterment, bearing out one historian's characterization of Odessa as "a bourgeois, middle-class city *par excellence*."[123] The rabies vaccine, and bacteriology more broadly, had different meanings for the scientists and public involved. For scientists such as Mechnikov and Gamaleia, local *obshchestvennost'* created opportunities to combine professional expertise with social duties and achieve

[118] Ibid., p. 33. Mechnikov also threw himself into the technical arrangements for the bacteriological station. See Iakov Bardakh, "Vospominaniia o Mechnikove," *Vrachebnoe Delo*, nos. 15–17 (1925): 1197–8. It is interesting to note that the picture that emerges of Mechnikov's activities on behalf of the station before its opening, and indeed after as well, runs counter to the impression cultivated in later years by Mechnikov and his wife that he was relatively uninvolved in these events.

[119] Gosudarstvennyi Arkhiv Odesskoi Oblasti (hereafter cited as GAOO), f. 5, op. 1, d. 1527 (1886–1887), ll. 19–39; and *Protokol Zasedaniia Obshchestva Odesskikh Vrachei*, no. 14, 17 May 1886, p. 126.

[120] Medical Department to the Vremennyi Odesskii General-Gubernator, 24 April 1886, GAOO, f. 5, op. 1, d. 1527 (1886–1887), ll. 21–2. Of course, Odessa *was* one of the largest cities in Russia—fourth in population after St. Petersburg, Moscow, and Kiev.

[121] In a letter to the Russian ambassador in Paris, Pasteur had recommended that "if the Russian government and the city of Odessa wish to organize an anti-rabies vaccination laboratory, there is no one better to appoint at this time than Dr. Gamaleia." *Protokol Zasedaniia Obshchestva Odesskikh Vrachei*, no. 14, 17 May 1886, p. 128.

[122] In particular, the overlapping memberships in various medical, scientific, and agricultural societies in Odessa seem to have helped facilitate the work of the supporters.

[123] Skinner, "City Planning" (cit. n. 112), p. 174.

their personal goals of establishing a comfortable base for their researches. The patrons looked for immediate practical benefits, preferably economic, from their investment in bacteriology, as well as the civic glory that had been promised. And, most broadly, Pasteur's discovery was interpreted as a symbol of modernization and progress, of the benefits that science could bring to Odessa and the whole of Russia. Whatever the specific public attitudes, the proponents of the station—especially Marovskii, Mechnikov, and Vel'koborskii—successfully translated this receptivity into institutional support for bacteriology.

The whole affair had political ramifications as well, despite the fact that vaccines and bacteriological stations would seem to lack any overt ideological implications. Representatives of conservative and state interests were consistently arrayed against the actions undertaken by Odessa society and scientists, whether that involved sending bite victims to Paris or opening a local bacteriological station.[124] The reason for this is probably not found in the content of the science but rather in the fact that a publicly initiated and executed undertaking was inherently threatening to the autocratic order. Mikhailovskii's insight that "the program of the people labeled 'conservatives' includes, among other things, a hatred for democratic ideals and a fear of the harmful influences of science" seems to be have been borne out here.[125]

The Odessa experience was not unique. No fewer than eight other "Pasteur" stations sprang up in Russia between 1886 and 1889, in Kharkov, Moscow (two), St. Petersburg, Kiev, Warsaw, Samara, and Tbilisi (more in fact than in any other single country). And with the exception of the ones located in the capitals, these stations were all publicly initiated and funded, generally either by local governments or by medical societies, to serve real or perceived immediate local needs.[126] Even in cities such as Kharkov and Kiev, where medical professors were part of the effort to establish these stations, there seems to have been no desire to link the project to the state-run universities. Rather, these stations reflected local initiative, civic consciousness, and the desire to apply the latest scientific advances to solve urgent local problems. And the prevalence of public sponsorship of bacteriological institutions continued after 1890 as voluntary associations and urban and rural governments alike continued to show a willingness to tackle problems—such as public health and agricultural improvement—that the central state had ignored or inadequately handled.

But this is not to say that the central state played no role in encouraging and supporting this field. As might be expected, in St. Petersburg (and to a certain extent

[124] For example, besides the state opposition detailed above, the politically conservative newspaper *Novorossiiskii telegram* publicly opposed the subscription drive undertaken by the liberal newspaper *Odesskii vestnik*, while the most consistent and vocal critics of the rabies vaccine and the bacteriological station were state-employed, overtly anti-Semitic local medical inspectors. (Their anti-Semitism was pertinent because one of the first employees of the bacteriological station was a Jewish physician, Ia. Iu. Bardakh.)

[125] N. K. Mikhailovskii, quoted in Vucinich, *Science in Russian Culture* (cit. n. 17), vol. 2, p. 28.

[126] For example, the rabies station in Samara, which began offering preventive treatments in July 1886, was funded by the provincial zemstvo and operated out of its hospital. The Warsaw rabies station, the second to begin operation in the Russian empire, received part of its financial support from the Warsaw Council of Public Charity, which also permitted bite victims the use of its hospital facilities. The station proper, however, was located in the private bacteriological laboratory of the local physician who founded it. In Kharkov and Kiev the local medical societies took on the task of organizing and financing bacteriological stations (with the help of private donors and local governments' support). See Hachten, "Bacteriology, Medicine, and Hygiene in Russia," (cit. n. 91), pp. 137–46.

in Moscow as well) the medical bureaucracy, military officials, and court members were prominent initiators and patrons of bacteriological institutions.[127] Interestingly, the university medical faculties tended not to be important centers for research or even specialized training in bacteriology.[128] One reason was the highly restrictive university charter of 1884, which limited the ability of medical faculties to found new chairs. (In comparison, the Women's Medical Courses, established primarily through public and private initiatives, were able to introduce bacteriology into their programs with no difficulty.) Nonetheless, the continued prominence of state-funded institutions is a useful reminder of the continued importance of the central government in support of science, including the new discipline of bacteriology. But there was an important difference from earlier periods in Russian history: other elements of Russian society now also had the desire and means to support scientific activities, and there were alternative social and political structures through which these projects could be carried out.

The fate of Il'ia Mechnikov as director of the Odessa Bacteriological Station can help us to better understand the challenges and contradictions that emerged when the public sphere became the common ground on which professional, social, and state interests were pursued by a variety of actors. After playing such a large role in the station's creation, Mechnikov found his tenure as its director surprisingly short. In a memoir written twenty years later, he blamed interference by state officials, unreasonable demands from patrons, hostility from the medical profession, and incompetence and infighting among his junior colleagues as reasons for his departure from Odessa in 1888.[129] The reality was a bit more complex. While it was true that some state medical bureaucrats in Odessa tried to interfere in the station's operations, other state officials came to the defense of the bacteriologists. Public patrons did put demands on the station for practical results, but those results had been promised by Mechnikov and the other initiators of the project themselves in their drive to obtain public funds. In a bid to demonstrate their institution's utility, Mechnikov and Gamaleia signed a contract with a French company to produce and distribute Pasteur's anthrax vaccine in southern Russia. That action by itself brought down a storm of protest on the heads of the bacteriologists, whose venture into commercialism was criticized as a violation of scientific ethics and social responsibility. Finally, an early attempt to use Pasteur's vaccine resulted in the death of thousands of sheep belonging to a local landowner, bringing the threat of lawsuits against the station personnel. In short, the station did not become the refuge for research that Mechnikov had desired, and he therefore abandoned it to its fate and fled, finding refuge in the newly founded Pasteur Institute. He remained in Paris, Mechnikov explained, because there "the goal of scientific work could be achieved outside of politics or any sort of public activities. In Russia the obstacles come from above, from below,

[127] In Moscow, one rabies station was organized in a military hospital and another in a state-supported hospital. In St. Petersburg, the major backer of bacteriology was Prince Peter A. Ol'denburg, a notable medical philanthropist who organized the first rabies station (importing Pasteur's nephew to set it up) and, in 1890, expanded it into the Imperial Institute of Experimental Medicine, a freestanding multidisciplinary research center.

[128] Again, the exception was Moscow, where the establishment of a city-funded bacteriological laboratory at the university provided the basis for the development of a bacteriological institute. See Hachten, "Bacteriology, Medicine, and Hygiene in Russia" (cit. n. 91), pp. 328–38.

[129] I. I. Mechnikov, "Rasskaz o tom" (cit. n. 58), pp. 77–86.

from the side, making that dream impossible to fulfill."[130] Thus, Mechnikov became one Russian scientist who found a comfortable ivory tower in which to pursue his scientific interests. But most Russian scientists, of course, were left to pursue their scientific work amid the realities of Russian life, pulled between service to state and to society and conscious of the social duties of a highly trained expert living in a backward country.

CONCLUSION

Both science and civil society took on increased importance in the culture of late-nineteenth-century Imperial Russia at the same time that these phenomena became increasingly intertwined in complex and unexpected ways. This paper has traced this deepening relationship through an examination of the expansion of a public culture of science, the reconfiguration of scientists' professional identities and values, and the shaping of scientific careers and disciplines in late Imperial Russia. Analyzing the history of Russian science and civil society in tandem helps to enrich our understanding of both phenomena. Certainly there is room in the historiography of Russian civil society for more consideration of the place of science and scientists in the construction of the moral and structural aspects of *obshchestvennost'*. Scientific organizations and projects formed an important element of Russian associational life, while the pursuit of science epitomized the rational and secular temper of that portion of educated society devoted to purposive action for the public good and affecting positive change in society. And if civil society is conceived as the space where the interactions of individuals with divergent agendas can be mediated through civic institutions, shared values, and common social goals, then the natural sciences very often served as both the symbol and goal of progress in the name of the common good.

By the same token, examining Russian science through the lens of civil society serves to reveal aspects that have been previously overlooked or discounted by historians working in the dominant historiographic paradigm. The view of Russian science as state driven and outstanding primarily for achievements in the theoretical realm may be balanced by a due appreciation for the importance of applied scientific fields and activities pursued outside the sphere of the state. In this regard, one may recall Tocqueville's argument that science in hierarchical, despotic countries is characterized by the "the arrogant and sterile search for abstract truths," whereas people living in democracies "seek the immediate and useful practical results of the sciences."[131] Judging by the case of decidedly undemocratic Russia, Tocqueville overstated this dichotomy. Autocratic Russia did exhibit to some degree the associational life and sphere of public initiative that Tocqueville connected to the pursuit of science for its utilitarian benefits.

The expanding domain of public life had an identifiable impact on scientists and their activities as well. Many scientists became much more interested in linking their professional activities to the needs and demands of *obshchestvo*, and in addressing

[130] Ibid., on p. 86. One should add that Mechnikov was able to pursue his dream due in large part to the financial support derived from the income from his wife's family's estates.

[131] See n. 60.

the new public audience for science. These developments reflected a pragmatic understanding of the need to search out expanded employment opportunities as well as the traditional ethic of service to the people, society, and the common good that was so strongly embedded in the consciousness of the Russian intelligentsia. Indeed, it was within the interstices of *obshchestvennost'* that issues of professional ambitions and social duties were sorted out, since, as David Wartenweiler points out, "[c]ivil society allows the mediation of personal interests and social needs, individual freedom and social responsibility."[132] But the impact of these developments on individual scientists was highly variable, depending on discipline and location, among other factors. And as the case study of bacteriology indicated, scientists often only uneasily balanced the competing demands of *obshchestvennost'* and professional interests. Scientists responded to opportunities created by the Russian public, but they had their own intellectual and professional agendas, which frequently took precedence over the common good. Employment by public entities was no panacea either, and scientists in the employ of zemstvos, for example, had complaints similar to those in state service. The ivory tower was an attractive alternative for many.

Scientists maneuvered within a social space defined not just by public, professional, and moral obligations, but also by the state. Although Russians often thought in terms of a binary opposition between *obshchestvo* and *gosudarstvo*—society and the central state—in reality the division was not always so stark. Scientists themselves frequently straddled that divide. So too did state officials from V. I. Kovalevskii in the Finance Ministry to L. A. Marovskii, the medical inspector in Odessa. So while certain agents and policies of the central government could throw up obstacles to scientific activities of all kinds, there were also "improving" state officials whose goals of modernization and utilitarian conception of science provided common links to the public sphere, and who often found it expedient to work through public channels to achieve these goals. If *obshchestvennost'* potentially provided an integrating factor in Russian society, the pursuit of science (or at least certain scientific projects) sometimes served to bring together elements of both state and society. But one way or another, it is almost impossible to discuss civil society and the public culture of science without considering the influence of the state as well. Anthropologist Chris Hann has recently made a similar point in discussing the development of post-Soviet civil society in Eastern Europe, warning: "[T]he assumption of an overriding antagonism between state and society is futile. If these terms can serve at all, the task must be to investigate their complex and continuous interactions."[133] This point is equally well taken for late Imperial Russian society.

Finally, it is well to remember that the focus here has been on the late nineteenth century and on a generation of Russians profoundly affected by the changes of the post–Crimean War era. It is possible to talk of an identifiable ethos that bound together the members of the relatively small scientific community at that time. The lack of legal outlets for political activity helped to elevate the importance of the pursuit of science and the "small deeds" reform efforts associated with the public activists. The early twentieth century, however, brought profound changes in these

[132] Wartenweiler, *Civil Society and Academic Debate* (cit. n. 3), p. 6.

[133] Chris Hann, "Introduction: Political Society and Civil Anthropology," in *Civil Society: Challenging Western Models*, ed. Chris Hann and Elizabeth Dunn (New York: Routledge, 1996), pp. 1–26, on p. 9.

patterns. The Russian Revolution of 1905 and its aftermath ushered in a dynamic period of political and social change. The revolutionary potential of *obshchestvennost'* and autonomous public activity was fulfilled in 1905 when professionals and their associations became important political actors. At the same time, the expanded civil liberties and opportunities for legal political activities opened up new opportunities for public activism, and liberal scientists such as Vladimir I. Vernadskii and Kliment A. Timiriazev could clearly make their cases in the public sphere for the links between science and democracy. Another wave of associational activity after 1905 significantly broadened the scope of the public culture of science, while a new spurt of public patronage of science and education significantly expanded the opportunities for scientific work under nonstate auspices. At the same time, the first years of the twentieth century brought noticeable social and political fragmentation, creating new fault lines across Russian educated society and within specific professions.[134] But while both science and civil society were more fragmented and diverse after 1905, their complex relationship continued to unfold on the basis of the developments of the late nineteenth century that have been traced here.

[134] John F. Hutchinson has examined this process in detail for the Russian medical profession in *Politics and Public Health in Revolutionary Russia, 1890–1918* (Baltimore: Johns Hopkins Univ. Press, 1990), pp. 38–43.

Statistical Utopianism in an Age of Aristocratic Efficiency

*Theodore Porter**

ABSTRACT

The modern history of science is commonly associated with an inexorable move toward increasing specialization and, perhaps, a proliferation of expert discourses at the expense of public discourse. This paper concerns the standing of science as a basis for public authority in late-Victorian and Edwardian Britain, and suggests that, in relation to the political order, this standing remained tenuous. These themes are exemplified by the career of Karl Pearson, founder of the modern school of mathematical statistics and something of a social visionary. Like Huxley and other scientific naturalists, Pearson wished to incorporate science into a reinvigorated "general culture" and in this way to reshape an elite. Statistics, seemingly the archetypal form of specialist expertise, was conceived as an almost utopian program to advance intelligence and morality in what he sometimes referred to as a new aristocracy.

INTRODUCTION

THE DEEDS OF STATISTICIANS do not appear in epic guise, as historical reason on horseback, and so it is fitting that they take place only once. In Marx's sardonic account of the reinvented past, Louis Napoleon reenacted scenes from his uncle's heroic career so as to disguise the bourgeois banality of 1848 as it actually unrolled and thereby to hold off the proletarian millennium. The triumphs of bureaucracy appear less glorious and less personal. They have been attained quietly, through the cunning of reason, and could never have arisen as the outcome of a human plan. Reversing, in the name of scientific modesty, the emperor's narrative of inherited grandeur, the epic ambitions of the founders of statistics were reduced by their heirs to a record of mathematical discoveries, leading inexorably to our modern forms of knowledge and administration. Lost or ignored were not only the ambiguities and uncertainties, but above all the social ambition and historical vision that gave meaning to the technicality.

Quantitative technologies such as statistics and accounting seem to rank among the dullest and most inevitable tools of a modern civic order. The English biometricians who created modern statistics, however, were not merely accommodating social and bureaucratic pressures. They advanced a bold, almost utopian, program that

* Department of History, University of California, Los Angeles, 6265 Bunche Hall, Los Angeles, CA 90095-1473; Tporter@history.ucla.edu.

I gratefully acknowledge the Manuscripts Division of the Science Library, University College, London, for permission to quote from the Karl Pearson Papers.

 0369-7827/02/1701-0008$10.00

generated bitter and persistent controversy. Karl Pearson, who really founded the field, understood the statistical enterprise as something vast and magnificent. His was a vision that looked backward to the Middle Ages as well as forward to the future and that drew from the values of aristocracy no less than from those of scientific expertise. He devoted himself to statistics as the straight and narrow way to comprehend the great issues of biological evolution and, with evolution, of social science. With no more than a hint of irony, he used terms such as "gospel" for the statistical methods he taught.

When Pearson was discouraged or depressed, his career appeared to him in the guise of a full-blown tragedy, not least because he believed it would be reduced by the next generation to a few discoveries, isolated from the great struggle that had given them meaning:

> As I look back on it now I think life has been largely a failure. I have sacrificed everything . . . to the idea of establishing a new tool of science which would give certainty where all was obscurity and hypothesis before. I have made many enemies and few friends in the process, for I was upsetting old idols and endeavouring to replace them by new gods whom scientists of the old training would not accept. Now that I am old, the younger men, especially in America, Russia and Japan are rushing in and devastating all the paths I have tried to hew through the jungle. Twenty years hence a curve or a symbol will be called "Pearson's" & nothing more remembered of the toil of the years.[1]

The campaign for statistics was no inexorable process of rationalization, advanced by legions of little men exercising their modest bureaucratic or scientific functions, but a titanic struggle. If statistical sciences were to have an important function in society and state, Pearson and his allies had to create that role. In attempting to do so, Pearson had not been able to simply appeal to some incipient public rationality. That, too, he had undertaken to create.

Despite Pearson's efforts, mathematical statistics has since become almost invisible except to its practitioners. "Pearson," as predicted, now names a coefficient and is associated with a test. Thus do the sciences reduce their complicated ancestors to a single dimension. In social histories of scientific and professional life, a wider field of implications is made visible, but often only by narrowing severely the range of possibility, contingency, and complexity. In the most elemental narrative of modernizing professionalism, knowledge—with the passing of time—becomes specialized, and specialists gain authority. Sometimes this rise of expertise is portrayed as a response to the needs of an increasingly complex society, sometimes as a usurpation of what by right should be debated in a vigorous public sphere. Expertise is, in any case, a defining feature of "modernity," the teleological destination of our time.

EXPERTS AND THE LIBERAL ORDER

The rise of expert authority, often at the cost of open democratic debate, is a central theme of cultural histories of science and the professions. The process is, in a sense, almost timeless, having been identified in periods—and nations—as widely

[1] Karl Pearson to Elisabeth Cobb, 2 April 1927, Pearson Papers 9/6, University College, London. He wrote this as his wife, Maria (Cobb's sister), lay dying. He marked the letter "Please destroy" at the top and reiterated the point in very strong language in the opening and closing sentences. Scrawled at the top of the last page we find, in Pearson's now somewhat shakier hand: "Never sent K.P."

scattered as those once believed to have nurtured a rising bourgeoisie. Yet the period of incipient modernism, roughly 1880 to 1920, has been most closely identified with the triumphs of science-based expertise. A distinguished historiography defines the social and political circumstances of professionalization. These include the economic changes of what is often called the second industrial revolution, involving a shift from craft to science in such decisive enterprises as electric power and organic chemicals. They include also the heightening of liberal anxieties associated with the politics of nationalism, irrationalism, and intolerance, and with an often rebellious and sometimes revolutionary labor movement. In Britain, working-class radicalism was less threatening than in the leading nations of continental Europe, but the greatly expanded franchise of 1867 and 1884 created troubling uncertainties. By 1914 "liberal England" had perhaps already succumbed to a "strange death." Professionalism involved, as one of its dimensions, an effort to reconstruct authority in the face of unruly democracy.

Harold Perkin called his influential survey of British social history from the 1880s to the 1980s *The Rise of Professional Society.*[2] He supplies, however, disappointingly little by way of social history of the professionals themselves: their family backgrounds, educational requirements, career patterns, and clients or employers, and the legal structures of professional practice. Nor does he discuss their work practices, still less their ideas and ambitions. He does point to the crucial element of cultural conservatism that informed the new professional ideals of the *fin de siècle.* The simplified trajectory presumed in many allusions to the significance of the period fails to recognize even this factor. Professionalism continues often to be interpreted in functionalist terms, as a necessary element, and hence a progressive one, in a complex modern economy and society. Here I will explore an alternative view, to see how far we can comprehend the scientism of the turn of the twentieth century as utopian rather than simply functional, as aristocratic rather than administrative, and as nostalgic as well as forward looking.[3] The new regime that invests great authority in expert specialists—if indeed we have created one yet—was scarcely even inchoate in early modernist Britain. I address these issues through a study focused on the career of a single individual: Karl Pearson, the dominant figure in the biometric group that, in many ways, created the modern field of statistics.

Pearson's work in this field began in the 1890s, when statistics as a mode of study and form of reasoning was very much on the upswing.[4] The *fin de siècle* offered new institutional possibilities for scientists who knew how to seize them. Discipline builders such as Pearson and the economist Alfred Marshall took advantage of a growing university system to create new academic fields that could be brought to bear on fundamental issues of political society. But their programs were in no way

[2] Harold Perkin, *The Rise of Professional Society: England since 1880* (London: Routledge, 1989).

[3] I draw inspiration from T. W. Jackson Lears, *No Place of Grace: Antimodernism and the Transformation of American Culture, 1880–1920* (Chicago and London: Univ. of Chicago Press, 1981).

[4] See, e.g., Alain Desrosières, *La Politique des grands nombres: Histoire de la raison statistique* (Paris: Editions La Découverte, 1993); J. Rosser Matthews, *Mathematics and the Quest for Medical Certainty* (Princeton, N.J.: Princeton Univ. Press, 1995); Lorenz Krüger et al., eds., *The Probabilistic Revolution*, 2 vols. (Cambridge, Mass.: MIT Press, 1987); Stephen M. Stigler, *The History of Statistics: The Measurement of Uncertainty to 1900* (Cambridge, Mass.: Harvard Univ. Press, 1986); Theodore M. Porter, *The Rise of Statistical Thinking, 1820–1900* (Princeton, N.J.: Princeton Univ. Press, 1986).

responses to a widely felt consciousness of some pressing need. For his part, Marshall wanted not only to build a discipline, but to reshape the economic common sense of an incipient business elite as well. These leaders Marshall hoped to train and advise, however, were unlikely to be moved by recondite or specialist arguments. Hence there is cultural as well as biographical significance in Marshall's well-known injunction that economists might deploy analytical tools to solve a problem but should subsequently "burn the mathematics" and give explanations in plain English. Pearson, more radically, insisted on the mathematics and brought a succession of postgraduate students into his biometric laboratory to teach them how to wield it.[5] He did not, however, succeed in his own time in winning for the new statistics an important role in bureaucracy or the professions. British politics and administration in the late nineteenth and early twentieth centuries offered at most bit parts to specialized academic discourses. Pearson aspired to something more exalted.

From about 1892, when Pearson wrote his philosophical *Grammar of Science* and turned to statistics, that exaltation involved an unfailing advocacy of science—of counting or measuring and statistical analysis—as the proper basis for knowledge on almost every practical topic. For social questions, as for biological ones, he demanded Darwinian evolutionary explanations, which he combined somehow with a positivist insistence that we never know causes, but can only process sensory information. His doubts about our access to reality did not diminish his assertiveness. The mature style of Pearson's public lectures and popular writings was uncompromising and aggressive, even militant. He left, in these writings, some notable hostages to fortune. Bernard Semmel cites him for particularly extreme statements of imperialistic aggression. Darwinian struggle among humans, in Pearson's view, was no longer a matter of individual survival, but rather of the race. No thoughtful socialist, Pearson announced, would hesitate "to cultivate Uganda *at the expense of its present occupiers* if Lancashire were starving. Only he would have done this directly and consciously, and not by way of missionaries and exploiting companies."[6]

A PROFESSIONAL CLASS

Donald MacKenzie and Bernard Norton emphasized Pearson's identification with a new professional class and deployed this element of social history to help explain his eugenic and statistical ambitions. MacKenzie interpreted his historical materials using more recent writings on the "new class" and on scientific controversies in relation to interests, in writing one of the classic works of Edinburgh sociology of science, *Statistics of Britain, 1865–1930* (1981). In it he articulated a clear program for relating science to social structures. The program drew from a neo-Marxist understanding of class, while permitting each individual a comparatively subtle and nuanced class position. MacKenzie did not seek to deduce scientific choices from class interests but pursued the more modest and reasonable goal of interpreting the

[5] On Pearson's laboratories, see M. Eileen Magnello, "The Non-Correlation of Biometrics and Eugenics: Rival Forms of Laboratory Work in Karl Pearson's Career at University College, London," *Hist. Sci.* 37 (1999): 79–106, 123–50.

[6] Karl Pearson, "Socialism and Natural Selection" (1894), reprinted in *The Chances of Death and Other Studies in Evolution*, 2 vols. (London: Edward Arnold, 1897), vol. 1, pp. 103–39, on p. 111, quoted in Bernard Semmel, *Imperialism and Social Reform: English Social-Imperial Thought, 1895–1914* (Cambridge, Mass.: Harvard Univ. Press, 1960), p. 42.

science retrospectively as an expression of such interests. He recognized an inescapable element of individual distinctiveness, and he allowed a serious cognitive dimension. Individual reasoning he understood not as an alternative to the pursuit of interests, but, in part, as its vehicle. He portrayed Pearson, justly, as a highly sophisticated scientific thinker, one who did not assume philosophical positions passively but read widely and made choices. For this reason, MacKenzie argued, Pearson figured as a singularly excellent specimen of the type: his ideas "reflected professional middle-class interests uncomplicated by particularistic commitments."[7]

This last point is difficult to interpret. MacKenzie discussed a few of the many alternatives to Pearson's program, and the controversies to which they gave rise. These alternatives were not dead ends, but genuine possibilities, and indeed many of them played a positive role in the reshaping of this far-flung field of statistics. I follow MacKenzie, and the tradition of sociology of knowledge on which he drew, in seeking to situate science in social history by locating the scientist in relation to larger social groupings. But it is no longer possible to suppose that "society" is sufficiently monolithic to determine uniquely the possible practices or beliefs of a science, and there is no reason to think that an uncomplicated class interest would have the advantage over more heterogeneous interests. Indeed, "class" has lost much of its persuasiveness in recent decades, in part because the standard Marxian categories seem too confining. The very concept of a "new" professional class was cultivated as part of a rebellion against orthodox Marxism and was worked out by Yugoslav intellectuals, notably Milovan Djilas, to account for the sad bureaucratization and stultification of Soviet Communism. Without abandoning the crucial economic dimension, we need to go further in incorporating the elements of a serious cultural history in the relations of science to civil and political society.

Pearson's advocacy of an important social role for professionals and scientists was uncompromising. He expressed it in biological terms and seemed to delight in emphasizing the contradictions between the harsh demands of science and the teachings of a soft, Christian morality. Still, he never lost sight of the ethical dimension of his arguments, even, or indeed especially, when they were most pitiless. He became the harshest of social Darwinists, a champion of neo-Darwinian socialism. The "very fact," he wrote in 1887, "that Socialism is a morality in the first place, and a polity only in the second," must not tempt us to endorse soft, misguided conceptions such as Christian socialism.[8] In place of Christian otherworldliness, Pearson insisted on a socialistic attention to life in this world. And in contrast to the reverence for all of humanity demanded by Comtean positivism, he endorsed the

> more practical faith . . . that the first duty of man [is] . . . to the group of humans to which he belongs, and that man's veneration is due to the State which personifies that social group. . . . Corporate Society—the State, not the personified Humanity of Positivism—becomes the centre of the Socialist's faith. The polity of the Socialist is thus his morality, and his reasoned morality may, in the old sense of the word, be termed his religion.[9]

[7] Donald MacKenzie, *Statistics in Britain, 1865–1930: The Social Construction of Scientific Knowledge* (Edinburgh: Edinburgh Univ. Press, 1981), p. 92; also Bernard J. Norton, "Karl Pearson and Statistics: The Social Origins of Scientific Innovation," *Soc. Stud. Sci.* 8 (1975): 3–34.

[8] Karl Pearson, "The Moral Basis of Socialism" (1887), in *The Ethic of Freethought* (London: T. Fisher Unwin, 1888), pp. 317–45, on p. 318.

[9] Ibid., p. 319.

Pearson grounded socialist morality in Darwinian struggle. With the rise of civilization, he explained, the struggle for existence among men had ceased to operate at the level of the individual. Natural selection had not on this account lost its force, but it had been displaced to a different level, the competition among human societies, or states. A state, to be effective, required the loyalty and devotion of its citizens, who should work for the collective good rather than personal interests. They must, indeed, abandon the prerogatives of individualism, including the right to reproduce in the haphazard manner of the liberal order. Since collective survival required efficiency at the level of individuals, reproduction had become inevitably a question of social welfare rather than personal choice. Pearson typified, perhaps in an exaggerated way, the alliance of eugenics to the welfare state. Thus it is misleading to speak of social Darwinism here. He regarded public education, state support of child rearing, and rational organization of the economy as allied to eugenics, forming collectively the basic elements of an effective social state. A similar pattern of commitments was common among Fabians and others with advanced political views.[10]

Throughout his career, Pearson assigned to science a fundamental role in administration and politics. The irrationalities of liberalism, of a society and economy based on the individualistic pursuit of wealth, he rejected forcefully. The corrupt aristocratic forms of his own day, which permitted hereditary elites to inherit political influence as well as great wealth, seemed only to compound the irrationality. Such elites had a parasitical relation to society, in his view; they had done nothing to earn their status, and he doubted that many could ever have achieved it by their intellectual merits. In 1887, still nurturing some of the working-class sympathies of his youth, he argued that a society would benefit by recruiting its leaders, "the best heads and the best hands," from all ranks of society.[11] However, as his eugenics became more extreme, and his democratic fervor subsided, he ceased to identify so strongly with the presumed interests of working people. He shifted to the view that hereditary talent was concentrated in certain classes, though certainly not in the titled aristocracy or inheritors of business wealth. He wrote in 1902:

> With rough practical efficiency a man's work in life is settled by his caste or class. This is not so undesirable as it might at first appear; it is a largely unconscious differentiation of the nation into workers of different types, who marry within their caste, and—if we remember how few generations are needed for a special human group to breed true—thus preserve to a large extent their special usefulness.

Too many workers who move up the social ladder, he continued, fail at those higher levels. We must remember, too, "that the middle class in England, which stands there for intellectual culture and brain-work, is the product of generations of selection from other classes and of in-marriage."[12]

That class had become indispensable. He expressed alarm that its failure to reproduce at the rate of inferior classes, and especially of the most callow and irresponsible, was sapping the English nation of its vigor. At the same time, the continued failure of his own society to place the most capable men in positions of power, and

[10] Diane B. Paul, "Eugenics and the Left," *J. Hist. Ideas* 45 (1984): 567–90.
[11] Ibid., p. 328.
[12] Karl Pearson, *The Function of Science in the Modern State* (1902; reprinted, Cambridge: Cambridge Univ. Press, 1917), pp. 9–10.

to give them the authority to organize industrial and political life, put Britain at a disadvantage. No wonder, he remarked, that the Germans are beating us. Pearson offered these failures of national organization as explanation for British military failures during the Boer War in the first years of the new century, an indignity that at last drew wide public attention to calls by Pearson, the Fabian Sidney Webb, and others for urgent measures to advance national efficiency.[13] Pearson called for selective reproduction and brain-stretching scientific education for the middle classes, so they could give form and direction to national life. Economic competition, he declared, is "not the correct attitude from the standpoint of science." Britain has "the flesh, blood, and sinews of a nation, but to make it foremost in the struggle, to make it a homogeneous, highly-organized whole, you must have a complex nervous system."[14]

AN ARISTOCRACY OF SCIENCE

The nervous substance to which Pearson referred was, of course, science, a term he used in a broad sense (even if he was aggressive in privileging the quantitative forms to which he devoted his career). Social and economic problems were no less susceptible to scientific treatment than were questions in mechanics or biology. He described science not only or mainly as a set of technical tools, but also as the foundation of social morality. In his *Grammar of Science*, Pearson defended science as the proper core of education, above all on account of its moral dimension. Scientific method forms citizens by teaching them to give up the merely personal and to accept as valid only what is true for everyone, irrespective of prejudices and selfish interests. Thus for Pearson, science stood for renunciation and self-denial. Yet self-denial here was also self-aggrandizing, at least for the scientific elite itself. "We must aristocratise government at the same time as we democratise it; the ultimate appeal to the many is hopeless, unless the many have foresight enough to place power in the hands of the fittest."[15]

Who were these aristocratic leaders, these outstanding specimens of a scientific or professional class? Pearson did not champion special privileges beyond what his favored elite deserved as a result of their personal merits or their genetic endowment. He became, however, a keen admirer of the English constitution and the processes of gradual adaptation for which it stood. In 1910, drawing inspiration from Francis Galton, he put forward a eugenic scheme for reforming the House of Lords. That body, he explained in a letter to the *Times*, "has too often been recruited by mere plutocrats, by political failures, or by men who have not taken the pains necessary to found or pre-

[13] Simon Szreter, *Fertility, Class, and Gender in Britain, 1860–1940* (Cambridge: Cambridge Univ. Press, 1996), p. 184.

[14] Karl Pearson, *National Life from the Standpoint of Science* (London: A and C Black, 1905), pp. 33, 54, on p. 13.

[15] Quote from Pearson, "Moral Basis of Socialism" (cit. n. 8), p. 306; see Theodore Porter, "The Death of the Object: Fin de Siècle Philosophy of Physics," in *Modernist Impulses in the Human Sciences, 1870–1930*, ed. Dorothy Ross (Baltimore and London: Johns Hopkins Univ. Press, 1994), pp. 128–51; idem, *Trust in Numbers: The Pursuit of Objectivity in Science and Public Life* (Princeton, N.J.: Princeton Univ. Press, 1995); idem, "Reason, Faith, and Alienation in the Victorian Fin-de-Siècle," in *Wissenschaft als kulturelle Praxis, 1750–1900*, ed. Hans Erich Bödeker, Peter H. Reill, and Jürgen Schlumbohm (Göttingen: Vandenhoeck & Ruprecht, 1999), pp. 401–14. On the opposition of Victorian moralists to selfishness: Stefan Collini, *Public Moralists: Political Thought and Intellectual Life in Britain, 1850–1930* (Oxford: Oxford Univ. Press, 1991), p. 65.

serve an able stock." It "wants rather more than less of the hereditary principle—where I understand by 'principle' the application of the truth drawn from the observation that, for good or bad, children, in a certain marked and measurable degree, resemble their parents."[16] Later, in an uncharacteristic spirit of social inclusion, he evoked "a hereditary nobility, an aristocracy of worth, and it is not confined to any social class; it is a caste which is scattered throughout all classes; let us awaken it."[17]

Can such proposals be squared with his unswerving faith in statistics? They can: his ideal was to breed up a superior population, and then to educate it to the highest level. He believed that science, especially statistics, had to be a vital element in the intellectual and ethical formation of this elite. It would not consist of mere technical specialists, however. We should be careful not to confuse it with Max Weber's idealized bureaucracy, neatly subdivided into little domains of professional competence. For one thing, Pearson emphasized—in a way Weber also could have sanctioned—the moral as well as the intellectual qualifications of his elite. More than that, Pearson identified its superiority with a mastery not of recondite knowledge, but of "that classified experience which we term wisdom." He called for a new aristocracy of active leaders. "The training of an oligarchic class in statecraft is the first and perhaps hardest task of the modern state."[18]

He wrote this last line around 1900, by which time his professed socialism had become distinctly conservative in some respects. A eugenic meritocracy defined for him the proper course "between the Charybdis of Democracy and the Scylla of an Hereditary Peerage."[19] This view involved a considerable evolution from his radical democracy—and even Marxist enthusiasm—of the 1880s. But he required that inherited abilities prove themselves in every generation, that leaders demonstrate their intellectual attainments through high-level service to society. Thomas Carlyle's ideal of the hero and hero-worship, which appealed so strongly to "men of science" in Galton's generation, remained attractive to Pearson. He believed firmly in meritocracy, an aristocracy of talent, which in a eugenic regime must come once again to be defined in terms of lineage.

Pearson's ideal of the Carlylean hero involved, of course, much more of scientific or intellectual achievement than the original. Yet the hero would transcend all specialism. Writing in the 1920s, Pearson regretted that science had, since the war, become "almost entirely one of the professional roads to a living," insisting that "science as a pursuit must always stand higher than science as a profession."[20] Not so much specialist expertise as a supple, disciplined mind, qualified the ruling classes. Pearson envisioned the higher statistics as an indispensable part of their mental equipment.

This notion of a scientifically literate elite was, in his time, more plausible as a strategy for bringing together knowledge and power than was a tyranny of petty experts. The period has too often been characterized in intellectual histories of the

[16] Karl Pearson, "Primogeniture and Heredity," letter to the *Times*, 31 March 1910, p. 8.

[17] Karl Pearson, *The Life, Letters, and Labours of Francis Galton*, 3 vols. in 4, vol. 3A: *Correlation, Personal Identification, and Eugenics* (Cambridge: Cambridge Univ. Press, 1914, 1924, 1930), p. 353.

[18] Pearson, *National Life* (cit. n. 14), p. 14; idem, *Function of Science* (cit. n. 12), p. 16.

[19] Pearson, *The Life, Letters, and Labours of Francis Galton* (cit. n. 17), vol. 2: *Researches of Middle Life, p. 122.*

[20] Ibid., pp. 154–5.

late-Victorian and Edwardian periods, especially when they turn to "social Darwinism" and eugenics, as straightforwardly an "age of science." Such a characterization is not wholly false. Science was widely discussed, and in the aftermath of the Boer War, many believed it urgent to choose rulers who comprehended the scientific method. In actuality, however, the political elite continued to come up through Eton and Oxford, equipped with a classical education rather than a scientific one. In 1902 Winston Churchill wrote to H. G. Wells, in response to his *Anticipations of the Reaction of Mechanical Progress upon Human Life and Thought*: "Nothing would be more fatal than for the Government of States to get in the hands of experts. Expert knowledge is limited knowledge, and the unlimited ignorance of the plain man who knows where it hurts is a safer guide than any rigorous direction of a specialized character."[21] Harold Perkin, who quotes this line, dismisses Churchill's protest as the vain effusion of a man now out of his time. Many contemporaries were not so sure. Even Wells did not imagine that experts had as yet acquired much influence. "The modern democracy, or democratic quasi-monarchy, conducts its affairs as though there was no such thing as special knowledge or practical education. The utmost recognition it affords to the man who has taken the pains to know, and specifically to do, is occasionally to consult him upon specific points and override his counsels in its ampler wisdom."[22]

Churchill's disdain for mere experts confirmed Wells's assessment of the standing of the man of science in relation to higher affairs of state. Both recall the charge that so troubled T. H. Huxley after he had assumed the tenuous role of statesman of science:

> How often have we not been told that the study of physical science is incompetent to confer culture; that it touches none of the higher problems of life; and, what is worse, that the continual devotion to scientific studies tends to generate a narrow and bigoted belief in the applicability of scientific methods to the search after truth of all kinds? How frequently one has reason to observe that no reply to a troublesome argument tells so well as calling its author a "mere scientific specialist."[23]

He countered this criticism by saying that science is vindicated by its utility in practical life. Mainly, however, he undertook to refute the charge by claiming a greater dignity for science. There was once a time, he explained, centuries ago, when textual editing and interpretation, the rediscovery of ancient learning, represented the best of intellectual life:

> The representatives of the Humanists, in the nineteenth century, take their stand upon classical education as the sole avenue to culture, as firmly as if we were still in the age of the Renascence. Yet, surely, the present intellectual relations of the modern and the ancient worlds are profoundly different from those which obtained three centuries ago. . . . This distinctive character of our own times lies in the vast and constantly increasing part which is played by natural knowledge. . . . Thus I venture to think that the pretensions of our modern Humanists to the possession of the monopoly of culture and to the exclusive inheritance of the spirit of antiquity must be abated, if not abandoned.[24]

[21] Churchill to Wells, 17 Nov. 1902, quoted in Perkin, *Professional Society* (cit. n. 2), p. 169.

[22] H. G. Wells, *Anticipations of the Reaction of Mechanical and Scientific Progress upon Human Life and Thought* (New York and London: Harper & Brothers, 1902), p. 167.

[23] Thomas H. Huxley, "Science and Culture" (1880), reprinted in *Science and Education: Essays* (1893; reprinted, New York: Greenwood Press, 1968), 134–59, on pp. 140–1.

[24] Ibid., pp. 149, 152.

Huxley thus answered critics such as Matthew Arnold not by contesting the importance of culture, but by asserting the claim of science to an important part of it. And it was not specialist science that he defended; it was science as an indispensable tool of understanding. Recognizing implicitly that such a defense was, by itself, inadequate, he replied also with the very form of his essays. He cultivated an elegant and learned style. He wrote not only of physiology and morphology, but also of education, religion, ethics, and the history of philosophy. While proud of his scientific knowledge, he did what he could to refute the charge of being a mere specialist. He refused, for example, the label "scientist." He wrote in 1894, "To any one who respects the English language, I think 'Scientist' must be about as pleasing a word as 'Electrocution.'" Perhaps he supposed, as did many British "men of science" of his generation, that this was another of those execrable Americanisms.[25] In a cultivated country, on the eastern shore of the Atlantic, science could be refined as well as practical.

Expertise among the Victorians was, of course, indispensable, but also, as Wells noted, rather lowly. The word "expertise" is related etymologically to "experience," and in the nineteenth century an "expert" was more likely to be a person who did things than someone who had formally studied them. In educational terms, expertise was acquired principally by practical training rather than by scientific schooling. Expertise was the domain of specialists and thus suggested narrowness and the distortion of personality, the fashioning of students into useful tools rather than educated persons. Wells expressed the point with brutal sarcasm: "The man of special equipment is treated always as if he were some sort of curious performing animal."[26] Ironically, although science often bore the brunt of this negative view of expertise, it did not necessarily reap the credit for the positive, utilitarian dimension of expertise. Huxley, who repeatedly avowed his respect for technical knowledge, felt obliged to answer objections by practical engineers that formal instruction in science was useless as well as narrow, that it was ill suited to inculcate a practical skill. As a teacher of engineers at University College, Pearson faced this skepticism every day.

While he defended the utility of science, Pearson resolutely opposed the idea that science education should be distorted by the industrial needs of the moment. He referred approvingly to the university as a medieval corporation, a union of teachers, guiding students across "the passage from receptivity to self-production, from apprenticeship and journeymanship to mastership and the full freedom of the guild." Huxley, as it happens, proved an unreliable ally in this campaign, being less inclined to remake collegiate education in London after the German model. He worried that while the professoriate included "broad-minded, practical men" as well as specialists, it contained also "a fair sprinkling of one-idead fanatics."[27] Pearson, recognizing that the epithet was meant for him, countered that this was merely "the usual name for those who attempt to carry out in practice what they preach in theory."[28] As he had said earlier, "My own view is simply that we should endeavour to obtain for London a University on the model of Berlin—a University in which the highest research, ordinary academic teaching, and even popular lecturing have been

[25] Sydney Ross, "Scientist: The Story of a Word," *Ann. Sci.* 18 (1962): 65–85, on p. 78.
[26] Wells, *Anticipations* (cit. n. 22), p. 168.
[27] T. H. Huxley, "A Professorial University for London," letter to the *Times*, 6 Dec. 1892, p. 11.
[28] Karl Pearson, "A Professorial University for London," letter to the *Times*, 8 Dec. 1892, p. 10.

repeatedly found associated with the same Professor. Witness such names as those of Helmholtz, Du Bois Reymond, and Virchow."[29] Both Pearson and Huxley wished to raise specialist knowledge to a higher level through science, and so to erase the boundary between culture and practice. The university's highest mission, wrote Pearson, is "the education and training of those citizens whose knowledge and thought are to leaven the community, . . . the teaching of the teachers, . . . the preparation of that staff of scientists, specialists, leaders of industry, and representatives of culture in and outside the learned professions, upon whom the welfare of the nation so largely depends."[30]

Pearson's combination of modernist urgency with medievalist sympathies, typified by his representation of the ideal university as a guild of scholars and a fount of efficiency, is altogether exemplary, and these themes by no means lacked appeal at the time. Yet we must understand that his mission was boldly ambitious and that it largely failed. Mere expertise was perhaps never more strictly subordinated to general culture than in the decades around 1900. Mid-nineteenth-century British government had been comparatively open to experts. Men such as Edwin Chadwick, William Farr, and John Snow, whose positions were justified by experience and specialist knowledge, were able to act vigorously and independently within new bureaucratic agencies. This had depended on the relative inactivity of their formal superiors, many of whom held office because of hereditary titles.[31] But the Northcote-Trevelyan reform of the British Civil Service, implemented by Prime Minister William Gladstone in 1870, established an examination system for government officials that gave all the advantages to Oxbridge. Mere experts, in the new system, were examined on knowledge relevant to their posts; they subsequently worked their way up through the lower ranks on the basis of practical experience. Further advancement into the higher ranks for these specialists was not contemplated. The upper levels were filled by men of general culture, as displayed by their demonstrated competence in elite academic subjects, especially classical languages, but including modern languages and mathematics.[32] The new upper civil servants regarded public service as a calling, and no longer behaved in the manner of absentee landlords or clerics. Rather than advancing within a single department, they followed a career track that moved from ministry to ministry.

Scientific education did not fit well into this system at any level. Even when the government was compelled to look outside its own ranks for expertise, as it did most systematically during the First World War, it preferred the practical experience of

[29] Karl Pearson, "A Professorial University for London," letter to the *Times*, 3 Dec. 1892, p. 4.

[30] Karl Pearson, *The New University for London: A Guide to Its History and a Criticism of Its Defects* (London: Unwin, 1892), pp. 8, 26, 58.

[31] Lawrence Goldman, "Experts, Investigators, and the State in 1860: British Social Scientists through American Eyes," in *The State and Social Investigation in Britain and the United States*, ed. Michael J. Lacey and Mary O. Furner (Cambridge: Cambridge Univ. Press, 1993), pp. 95–126, p. 108. On science and expertise, see also Roy MacLeod, *Public Science and Public Policy in Victorian England* (Aldershot, Hampshire: Variorum, 1996). Powerful expert administrators did not disappear from Britain after 1870 but were more likely to be found in municipal administration than in the civil service; see Szreter, *Fertility, Class, and Gender* (cit. n. 13).

[32] John Roach, *Public Examinations in England, 1850–1900* (Cambridge: Cambridge Univ. Press, 1971); R. K. Kelsall, *Higher Civil Servants in Britain from 1870 to the Present Day* (London: Routledge & Kegan Paul, 1955), pp. 60–2; Peter Gowan, "The Origins of the Administrative Elite," *New Left Review* 61 (March-April 1987): 4–34.

businesspeople and industrialists to the more formalized knowledge of scientists.[33] Pearson at that time tried to save his biometric laboratory by enlisting it in the war effort, which meant converting it to a humble factory for preparing graphs and computing tables.[34] This was useful work for his mostly female "computers," but did not draw on the higher wisdom of statistics. The wartime mobilization did not create an exalted political or administrative role for science in Britain.

In science, as in administration, the late nineteenth century in Britain was a time of professionalization based on meritocracy. The rise of scientific professionalism refers in this context to several interrelated developments: among them the growth of universities and an increasing number of university posts, and the restriction of scientific societies, most notably the Royal Society, to a more selective membership. Scientific careers and honors should henceforth be within the control of one's academic peers, whose judgments would involve an assessment of competence and originality in the particular field.

The growth of a scientific profession is by now a standard theme of historical scholarship. It has been offered as an explanatory context for the effort to dissociate science from Christianity in the post-Darwinian controversies and for the turn to a more technical, mathematical style (the "marginal revolution") in British economics.[35] Yet this professionalization cannot be understood simply or mainly as a move to technique, to specialism, or to a closed disciplinary discourse. Sheldon Rothblatt has remarked that John Stuart Mill and Matthew Arnold, however divided on the deference due tradition, agreed that education should convey culture and nurture an intellectual aristocracy.[36] Pearson clearly endorsed this conception, too, in both his writings and his teaching efforts. To be sure, his commitment to the cultivation of specialist knowledge was unmistakable. Mastery of a special field at a very high level was for him, as for Huxley, the minimal requirement for a university professor and a qualification far too scarce on the island of leisured dilettantes and practical men. At the same time, he believed deeply that a proper education in a special subject could be elevating, rather than narrowing. It should awaken the capacity to think and analyze for oneself, and to recognize the special competences of others, that true citizenship demanded. In its highest incarnation, science was to be for Pearson what culture was for Arnold—a way of thinking and living that would take one beyond self and class. The scientist should, like the Carlylean hero, stand above mere consideration of career. He had at his command not only a body of information, but also method. Scientific method defined a general intellectual standard that, *ipso facto*, was simultaneously a moral one.

[33] Roy MacLeod, ed., *Government and Expertise: Specialists, Administrators, and Professionals, 1860–1919* (Cambridge: Cambridge Univ. Press, 1988).

[34] Pearson Papers, files 600–606 (cit. n. 1).

[35] Frank M. Turner, "The Victorian Conflict between Science and Religion: A Professional Dimension" (1978), in *Contesting Cultural Authority: Essays in Victorian Intellectual Life* (New York: Cambridge University Press, 1993), chap. 7; A. W. Coats, "Sociological Aspects of British Economic Thought," *Journal of Political Economy* 175 (1967): 706–29; George J. Stigler, "The Adoption of the Marginal Utility Theory," *Hist. Polit. Econ.* 4 (1972): 571–86.

[36] Sheldon Rothblatt, *The Revolution of the Dons: Cambridge and Society in Victorian England* (London: Faber & Faber, 1968), pp. 128–32; see also James Moore, "Geologists and Interpreters of Genesis in the Nineteenth Century," in *God and Nature: Historical Essays on the Encounter between Christianity and Science*, ed. David C. Lindberg and Ronald L. Numbers (Berkeley: Univ. of California Press, 1986), pp. 322–50.

The professional ideal in Victorian science derived in part from admiration for the German university. Pearson, who came out of King's College, Cambridge, a thorough Germanophile and proceeded to undertake a year of study in Heidelberg and Berlin, knew the German academic scene. In Germany, as in Britain, there were serious tensions between natural science and humanistic scholarship. But the mandarins of German science, men such as Hermann von Helmholtz and Emil Du Bois-Reymond, aspired to the high standing of *Kulturträger*, or "culture bearer." They balanced the achievements of mathematics and the laboratory with elegant, accessible public lectures on philosophical and artistic topics—and, perhaps necessarily, on Goethe.[37] The point was not only to display refinement and wide learning, though this was important enough, but also to demonstrate the pertinence of natural science to philosophical and aesthetic questions. Huxley and William Kingdon Clifford exemplified for Pearson a similar ideal in Britain, one to which he also paid tribute with brilliant lectures and essays on a variety of historical, philosophical, social, and scientific topics in the 1880s and 1890s.[38] The ideal of cultivated refinement was less manifest in his dogmatic and abrasive eugenic lectures after about 1900. But a commitment to general learning remained there, as vigorous as ever. In the new century, he insisted, the vital questions of social life had become the business of science, which he defended as applicable to every aspect of life. Yet science remained, for him, a calling, one that seemed now to be defended most effectively in Britain and France. He began to criticize German science as plodding, systematic, and routine, the product of research factories rather than of inspired truth seekers.

LOOKING BACKWARD

I have shown already how Pearson framed his professional ambitions by appealing to images of aristocracy. Donald MacKenzie calls attention to Noel (subsequently Lord) Annan's argument that the Huxleys, Stephens, Darwins, Trevelyans, and their ilk formed a kind of "intellectual aristocracy" in this period.[39] Pearson was perhaps a more marginal, because less sociable, figure than his contemporaries in the Darwin and Huxley lines, yet there are reasons to take this idea—of his participation in an intellectual aristocracy—seriously. In many ways it provides a more persuasive context for Pearson's ambitions than does the "new class" or the rootless "intermediate class" introduced by Eric Hobsbawm to account for the ineffectiveness of the Fabians.[40] We should recall that the Edwardian movement for "National Efficiency" was led not by some northern manufacturer, but by the titled Archibald Philip Primrose, earl of Rosebery. A cohort of landed aristocrats and privileged children of the middle class, educated at Eton or Rugby and then Oxford or Cambridge, supported the movement.[41] This was not the rebellion of a new class, but the self-conscious reformation of an old one.

[37] See Timothy Lenoir, "The Politics of Vision: Optics, Painting, and Ideology in Germany, 1845–95," in *Instituting Science: The Cultural Production of Scientific Disciplines* (Stanford, Calif.: Stanford Univ. Press, 1997), 131–78.

[38] See essays in his *Ethic of Freethought* (cit. n. 8); and *Chances of Death* (cit. n. 6).

[39] Noel Annan, "The Intellectual Aristocracy," in *Studies in Social History: A Tribute to G. M. Trevelyan*, ed. J. H. Plumb (London: Longmans Green, 1955), pp. 241–87.

[40] Eric Hobsbawm, "The Fabians Re-Considered," in *Labouring Men: Studies in the History of Labour* (London: Weidenfeld & Nicolson, 1968), 250–71.

[41] G. R. Searle, *The Quest for National Efficiency: A Study in British Politics and Political Thought, 1899–1914*, 2d ed. (London: Ashfield Press, 1990), pp. xviii-xix.

To be sure, few if any of these men were invested in technical knowledge and scientific methods to the same extent as Pearson. But, as we have seen, there are reasons to doubt whether a language of technicality is adequate to express even Pearson's view of the role of statistics or of scientific method. While he emphasized that there must be a differentiation of function within science, and also among the leaders of society, his faith did not call for humble specialists to inherit the earth. His exaltation of self-effacement idealized not the little man at his desk performing calculations, but the highly educated, morally vigorous product of good breeding and rigorous education, who was thereby fitted for a position of real leadership. If he did not portray statistics as outright successor to high culture and historical learning, it was because he continued to value these, even if he largely abandoned his historical and literary inquiries when he discovered the promise of quantification. He envisioned statistical knowledge as an endowment of elites, not of mere technicians. Yet he was impatient with real aristocrats because he believed most who were not idle were uninformed or misguided about science. He operated, nevertheless, in the administrative culture of Northcote-Trevelyan: he would preserve an elite civil service of generalists but educate them in scientific method rather than dead languages—not because the old system of education had no value, but because its advantages could now be attained more efficiently. He intended his *Grammar of Science* to offer the same educational benefits as had once been supplied by the study of Greek grammar, a rigorous mental training for future leaders.[42] Statistics, which by 1895 he regarded as central to a proper modern education, offered something more: specific analytic tools to investigate scientific problems, which for him included social ones. It was not, however, to be the mere tool of expert specialists, but part of a more effective intellectual formation, which should be complemented by historical learning and literary culture. The formal methods of statistics would not negate aristocracy, but redefine it.

Pearson's ambitions for science were very much the product of his age, yet his was not the realistic ideology of a rising professional class. He pursued, in a way, a modern utopia, one preserving more than might at first be suspected of the medieval sympathies of his early career. It is instructive to consider some themes of that work. In 1882, a decade before he turned to statistics, he had poured his soul into a "nineteenth-century passion play," because, as he explained, shared rituals of this kind were important for every society. For most of a decade he wrote sympathetic essays about pre-Reformation Germany. He took a dim view of radical utopianism, on the grounds that transformations of society must take place gradually and that revolution can bring only misery and destruction. He made vivid its dangers in his youthful essays on the fanatical revolutionism of the Reformation period. Meister Eckhart and Jan van Leyden had undertaken to create all at once, and on the basis of mystical illumination, what must be achieved slowly, in the fullness of time, by science. Yet Pearson appreciated their fervor, their *Schwärmerei*. The Anabaptist Jan Mathys, for example, "may have been a fanatic, his idea may have been false; still, he fought and died for a *spiritual* notion—his grace the bishop fought and triumphed

[42] Karl Pearson, *The Grammar of Science* (London: Walter Scott, 1892), p. 8n, where he recalled the intellectual benefits of studying Greek grammar, even after he had entirely forgotten the language. Helmholtz, also, depicted scientific education as providing the same benefits as grammatical instruction, though science is superior, he added, because its laws have no exceptions.

for *himself*."[43] In one of his essays, Pearson contrasted the "enthusiasm of the market-place" with the broader view of science. The unreflective fervor of the masses was dangerous, indeed tragic. The point was not to suppress it, but to guide it—by science, the "enthusiasm of the study." It was indeed enthusiasm, which for him was always to be preferred over complacency and selfishness. Working-class radicalism, however, required guidance, and "brainworkers" were its natural partners in the struggle against the parasitic classes living from inherited capital.

This talk of mental labor was the language of the Comtean religion of humanity, whose English leaders called for an alliance of "brains and numbers."[44] There is a higher morality than wealth, Pearson preached, and it is imperative that the "wealth-owning" classes learn this:

> They must be taught a *new morality*. Here, again, is a point on which we see the need of a union between the educative and hand-working classes. The labourers with the head must come to the assistance of the labourers with the hand by educating the wealthy. Do not think this is a visionary project; at least two characteristic Englishmen, John Ruskin and William Morris, are labouring at this task; they are endeavouring to teach the capitalistic classes that the morality of a society based upon wealth is a mere immorality.[45]

Only if the laborers of the hand rose in revolt would those of the head desert them, because they knew from history that revolution never leads to permanent benefit.

Pearson's invocation of Ruskin and Morris as proof positive that he was being realistic, not visionary, is not altogether convincing in retrospect. Especially as a young man, he shared much of their viewpoint, not least their admiration for folk culture and their nostalgia for the medieval world. Like Morris in Iceland, Pearson traveled among peasants in Norway in hopes of recovering an era long past.[46] Even as an old man, he remembered with admiration and longing his youthful experience of a passion play in the Austrian Tyrol, a play he described as fully medieval. And, as for Morris in his *News from Nowhere* (1890), this deep past contained within it a good deal of Pearson's imagined future. Thus, in his admiring study of German passion plays, he concluded that while they cannot be resuscitated, they might, with the aid of "cultured men," be reinvented in a form appropriate for a new age:

> They may help to guide new labour organisations to a sense of their social responsibility; they may assist in converting trades-processions into civic pageants, and mass-meetings into folk-festivals. They may aid the tendencies of the time to level down in wealth and to level up in knowledge. . . . Then perhaps it may come about that those social instincts, which are in truth more intense to-day than in Athens, Jerusalem, or Nürnberg of old, will cease to be so diverse and confused in expression as they are now; they will find one watchword to arouse all classes of the community; then and not till then will anything worthy of the name of a folk-religion be possible, then and not till then can a great religious festival be again a reality.[47]

[43] Karl Pearson, "The Kingdom of God in Münster" (1884), reprinted in *Ethic of Freethought* (cit. n. 8), pp. 263–313, on p. 312.
[44] Christopher Kent, *Brains and Numbers: Elitism, Comtism, and Democracy in Mid-Victorian England* (Toronto: Univ. of Toronto Press, 1978).
[45] Karl Pearson, "Socialism: In Theory and Practice" (1884), reprinted in *Ethic of Freethought* (cit. n. 8), pp. 346–69, on p. 362.
[46] Carl Schorske, "The Quest for the Grail: Wagner and Morris," in *Thinking with History: Explorations in the Passage to Modernism* (Princeton, N.J.: Princeton Univ. Press, 1998), pp. 90–104.
[47] Karl Pearson, "The German Passion-Play," in *Chances of Death* (cit. n. 6), pp. 246–406, on p. 406.

No more than Comte or Marx (and rather less than Morris) did Pearson think the medieval world could be resuscitated. From his earliest writings, he argued that science led to a different future. Yet much in the past deserved to be recovered. As he explained in some early lectures of 1882 on "German Social Life & Thought in the 16th Century," the Reformation had been the birth of individualism. Looking beyond it to the sixteenth century we discover an "old socialistic age," when the town regulated every form of expression and activity. In those days, there was little or no poverty, and the churches took care of what there was. The medieval tradesman "performed not for the sake of profit, but under the notion of a moral duty towards his fellows, a duty created by the advantages the community had conferred upon him." This was a prosperous age, yet with no proletariat. The handicraftsman was "full of energy artistic and intellectual." Alas, forces within already threatened that golden age, which soon succumbed entirely to the individualism of the Reformation. The anarchical nineteenth century, as Pearson called it, was the product of this "Deformation age." The spirit of individualism led modern discontents to declare the equality of all, or communism, a system "quite unlike true socialism." This history was thus a tragedy, one that had continued to Pearson's own time. Yet the most desirable features of that time, which had existed under the aegis of regulation and religion, could be recovered in a new era of knowledge and freedom. The history of mankind might even be defined "*as the development of man towards an ideal human freedom*." The great German philosopher Hegel, Pearson explained, had come to the same conclusion on somewhat different grounds.[48]

Pearson's own utopian tendencies—his open admiration for the implicit faith, communitarian harmony, spontaneous artistic labor, and anti-individualism of the medieval period—no longer supplied the framework of his political viewpoint after 1890. He buried this socialism that looked longingly to folk traditions and church beneath an aggressive rhetoric of racial struggle and scientific method. Yet his opposition to the wasteful, competitive order of economic individualism remained intact. He hoped, too, that evolutionary biology and statistics might frame a new intellectual order, as the Catholic Church had provided a unified intellectual order in the age of Erasmus. Or, in a different modality, Carlylean hero-worship might be joined to science within institutions resembling the reformed civil service introduced in the later Victorian period. However radical he may have been, Pearson imagined the future in terms of the cultural forms of his own time and in relation to the past as he understood it, especially the era before the Reformation. Statistical reason he conceived not as a tool of bureaucratic routinization, but as a union of reason and personal renunciation, suited to form the leaders of a modern socialist moral economy.

THE CUNNING OF HISTORY

The term "civil society" signified little in English in 1900 and even now circulates mainly in academic discussions. The Victorians expressed their discontent with the state (read: "government") politically, and the "unpolitical" had very little standing as a middle-class ideal. Pearson's view of political action was rather unusual for an

[48] Course of lectures, 1882, on "German Social Life & Thought in the 16th Century," Pearson Papers 47/3 (cit. n. 1).

engaged Englishman of his time. He resolved at the beginning of his professional life, often against strong temptations, to keep clear of day-to-day politics. This meant, for him, following the high road of science—not a disengagement from political life but an effort to assess it and critique it on the basis of principle. His vision of scientific method had everything do to with the organization of economic and social life, but not in isolation from matters of state. For British intellectuals, to reshape the social order was self-evidently to reconfigure the political order as well.

Pearson, like Galton before him, was moved by a deep sense of the inadequacy of contemporary politics and religion in the face of monumental intellectual and social changes. The ills of industrial society, made plain with the emergence around 1880 of a British labor movement, were very much on his mind. The obsolescence of the old order was also indicated by its halfhearted response to new forms of knowledge, especially Darwinian evolution, to which Pearson also looked for solutions to social problems. These were precisely the years when the legitimacy of government by patrician landowners began to be seriously doubted and when their economic position seriously weakened.[49] The deep discontent of Pearson as a young radical in the early 1880s changed form over the ensuing decades, particularly as his sympathy for working people gave way to a preoccupation with efficiency; yet his commitment to the formation of an elite of intelligence and merit remained in place. From the beginning he saw science as the proper foundation of this new aristocracy, and his discovery of the promise of statistics only sharpened his sense of mission.

It is this dimension of the statistical faith that I have called utopian. Pearson's increasing use of "aristocracy" to characterize an elite to be formed through eugenic breeding and scientific education points to the limits of his social radicalism and is perfectly consistent with his view that neither in nature nor in history does progress take place through revolutionary leaps. Pearson devoted his life to statistics not to promote administration by narrow experts, but in quest of something more glorious, and hence tragic, involving statistics as a rational basis for public credibility. It would provide authority not for specialized experts, but for men of broad scientific culture. They would govern wisely and efficiently in the interest of the collective, which demanded sacrifice from individuals. Such a vision of political order, joining statistical technique to the values of personal cultivation in a scientific aristocracy, soon became unimaginable, even, it seems, to historians. It does not fit with a narrative of monotonic technical advance nor with one of ever advancing bureaucratic rationalization.

It is, of course, not irrelevant that, however successful as a statistician, Pearson failed as a social visionary. Aristocrats and Oxford-educated political leaders adopted the languages of efficiency and eugenics when these became popular, without acceding to the program for science that underlay it. Pearson, by no means a radical opponent of democracy, appealed to the public through articles in the highbrow press, in hope that voters might replace outdated landed elites with a genuine scientific aristocracy of talent. Meanwhile, his program of research on the analysis of quantitative data developed slowly into a common language and body of methods for researchers in many fields. In this respect, his scientific success was extraordi-

[49] David Cannadine, *The Decline and Fall of the British Aristocracy* (New Haven, Conn.: Yale Univ. Press, 1990).

nary, even if some of his followers, such as R. A. Fisher, were also his sharpest critics. The growing prestige of their versions of statistics weighed heavily on his mind as he looked back on the tragedy of his career in the despairing letter of 1927 from which I quoted in the introduction. Yet, from a wider perspective, Pearson's followers were advancing and not combating his statistical program.

Pearson could not, however, reshape the relations of science to society. A succession of scientific visionaries—to say nothing of the unreflective claims of scientific "myopiaries"—have imagined that science might become the basis for a superior education, inculcating a higher level of general culture, rather than functioning as a rubric for technical training and teeming specialization. Pearson's *Grammar of Science* was a notable effort in this tradition and an inspiration to many. The project for general education through science, however, has largely failed. Because of its failure, schemes such as Pearson's for a new elite grounded in science have become almost unthinkable, and, for just this reason, should not be forgotten by history.

The Civic Uses of Science: Ethnology and Civil Society in Imperial Germany

*By H. Glenn Penny**

ABSTRACT

In an effort to move us toward a better understanding of the relationship between science and civil society in imperial Germany, this essay explores the creation and development of ethnographic museums in Hamburg, Berlin, Leipzig, and Munich. German ethnology became a locus of institution building during the imperial period, and an analysis of the actions and rhetoric surrounding these museums shows that the central developments in each institution were closely tied to the changes underway in its city. In each case, there was a mutual transformation of city and science that followed a general shift in the topography of public domains. Experts and professionals began colonizing and dominating areas of public space that autodidacts and local associations initially had carved out and occupied. This move reduced the importance of local associations in the science of ethnology at the same time that such civic associations were beginning to wane in general. The impact of this shift on these museums can tell us much about the ways in which changes and variations in civil society can affect the form and function of scientific institutions; it may also provide us with new answers to old questions about the "rise of the German sciences."

INTRODUCTION

ETHNOLOGY RAPIDLY ROSE to prominence as a public science in late-nineteenth-century Germany. Its subject—the exploration of humanity in all its many variations—appealed to the Humboldtian and cosmopolitan interests of many autodidacts and educated elites. The new science gave them a means to connect with the non-European world, and the artifacts they acquired from across the globe functioned as concrete evidence of their scientific authority and worldliness. Indeed, the newness of this science, combined with its focus on material culture, offered aspiring ethnologists and their supporters a new kind of intellectual authority—one that stemmed from experience rather than formal education. Moreover,

* Department of History, University of Missouri, Kansas City, 205 Cockefair Hall, 5100 Rockhill Rd., Kansas City, MO 64110–2499

I would like to thank Tom Broman, Lynn Nyhart, Kathy Olesko, and the two anonymous referees for *Osiris*. Ian McNeely deserves a special word of thanks for helping me gain a better understanding of civil society in nineteenth-century Germany. I also benefited from comments by participants at the conference on science and civil society at the University of Wisconsin in Madison. Funding for this research was provided by the Social Science Research Council—Berlin Program for Advanced German and European Studies. I am also grateful to the Center for European Studies at Harvard University for providing me with a congenial place to write the initial draft.

 0369-7827/02/1701-0009$10.00

ethnology's link to museums placed it outside the universities, in a site that these newcomers could occupy and use as a base to gain considerable municipal support. Because of these appeals, German ethnology became a locus of civic institution building during the imperial period (1871–1918), and German ethnologists experienced great success.

When the Kaiserreich was founded in 1871, it possessed only a scattering of ethnographic collections; ethnology was just beginning to emerge as a field of research and had no significant institutional base. But during the imperial period, museums became the institutional center for ethnological inquiry. Independent organizations in a number of German cities quickly began spearheading the movement to build these institutions and fill them with collections from throughout the world. In 1887 Berlin opened the world's largest freestanding ethnographic museum, and by the first decade of the twentieth century, an array of German cities possessed internationally acclaimed ethnographic institutions that quickly became the envy of scientists across Europe and the United States.[1]

As ethnology moved into these new institutions and became a professional discipline, its development, especially the fate of ethnographic museums, was closely linked to the transformation of German civil society and changes at large in the public sphere. Civil society and civic culture were tightly intertwined in imperial Germany. Indeed, if we conceive of civil society as that "realm of associations—the layer of supposedly non-coercive organizations located between the family and the state"[2]—we have to acknowledge, as James Sheehan pointed out three decades ago, that civil society in nineteenth-century Germany stemmed largely from, and was located in, German civic culture.[3] So much of German organizational life, scientific or otherwise, was polycentric, organized around German cities and regions that were often engaged in avid competition in the arts, the sciences, industry, and trade.[4] "German science" benefited from the polycentric character of this new state because clusters of scientists could draw on that staunch intra-German competition to gain impressive local support for their own institutions and projects.[5] The support these

[1] See, e.g., O. M. Dalton, *Report on Ethnographic Museums in Germany* (London: Her Majesty's Stationery Office, 1898); and Northcote W. Thomas, introduction to *The Natives of British Central Africa*, by Alice Werner (London: Archibald Constable, 1906). Cf. Annie E. Coombes, *Reinventing Africa: Museums, Material Culture, and Popular Imagination in Late Victorian and Edwardian England* (New Haven, Conn.: Yale Univ. Press, 1994); and H. Glenn Penny III, "Cosmopolitan Visions and Municipal Displays: Museums, Markets, and the Ethnographic Project in Germany, 1868–1914," Ph.D. diss., Univ. of Illinois at Urbana-Champaign, 1999.

[2] Charles S. Maier, preface to *Paradoxes of Civil Society: New Perspectives on Modern German and British History*, ed. Frank Trentmann (New York: Berghahn Books, 2000), p. ix.

[3] James J. Sheehan, "Liberalism and the City in Nineteenth-Century Germany," *Past Present* 51 (1971):116–37. I am grateful to Ian McNeely for this reference and his insight.

[4] Brian Ladd, *Urban Planning and Civic Order in Germany, 1860–1914* (Cambridge Mass: Harvard Univ. Press, 1990); Robin Lenman, *Artists and Society in Germany, 1850–1914* (Manchester: Univ. of Manchester Press, 1997); Glenn Penny, "Fashioning Local Identities in an Age of Nation-Building: Museums, Cosmopolitan Traditions, and Intra-German Competition," *Germ. Hist.* 17 (1999): 488–504.

[5] For the importance of intra-German competition in the rise of the German sciences, particularly with regard to the universities, see Joseph Ben-David, *The Scientist's Role in Society: A Comparative Study* (Englewood Cliffs, N.J.: Prentice Hall, 1971). This essay builds on Ben-David's argument even as it diverges from it. Ben-David focused heavily on Berlin, stressed the importance of leading professors and the concepts of *Bildung* (education or self-improvement) and *Wissenschaft* (science) and argued that German science was pushed forward by the needs of individual states competing with each other for the best scholars in a kind of free-market system in which productivity was linked

scientists received from local elites not only provided them with many opportunities, but also allowed public interest and discourse to have a strong influence over new (and thus more malleable) sciences such as ethnology.

In this essay, I propose to contribute to our understanding of the ways civic culture shaped the German sciences by comparing and contrasting the development of ethnographic museums in four cities: Hamburg, Berlin, Leipzig, and Munich. I argue that a combination of cosmopolitan interests, civic self-promotion, and staunch intra-German competition for status drove the rise of German ethnology during the late nineteenth century. I stress that an eager desire to refashion local identities on a thoroughly international stage provided the central motivation for the support ethnology received across Germany. I argue as well that the fates of these institutions were closely tied to the interests of their respective cities and that their history illustrates the dynamic and interactive relationships between science and civil society in imperial Germany. Ethnology, I contend, flourished during this period because a range of different cities provided this science with particularly fertile ground for its development. It emerged as a science just as Germany's cities were beginning a phase of unprecedented growth. In a number of places, organizations of citizens embraced this new international science and museums that displayed it because ethnology offered cities a means for exhibiting their worldliness and thus refashioning themselves.

The largest and fastest growing cities produced the grandest and best-known museums, but these institutions did not all develop in the same way. Indeed, I argue that the central developments in each institution were closely tied to the specific changes underway in its city as well as more general changes in the public sphere. In many ways, we can see a mutual transformation of city and science taking place in these museums—reflecting the consistencies and differences in the local contexts that fostered the institutions' growth, and stemming in large part from the striking shift in the topography of public domains during the last decades of the nineteenth century. On the one hand, German ethnology and many of the organizational bodies that structured everyday life in German cities became more professionalized, more closely tied to municipal and state institutions, and thus less reliant on the kinds of independent associations we regard as the basis of civil society. On the other hand, as the populations of German cities radically expanded and their demographics diversified during these decades, so too did the variety of players who helped shape public discourse.

The exclusivity commanded by the *Bürgertum*—those educated and financially better-off elements of the middle classes who dominated many of the associations that constituted German civil society, and who continue to dominate the historiography today—was challenged by the rising importance of the lower classes and the expansion of commercial consumer culture. As Frank Trentmann has argued, while the *Bürgertum* played a critical role in the creation of civil society in Germany, civil society became much more than the *Bürgertum*'s extension.[6] New groups gained

---

to expansion. In contrast, I move away from a Prusso-centric model and argue that while intra-German competition was critical to German ethnologists' success, civic self-promotion and an eager desire to refashion local identities on an international stage were the critical elements of this competition.

[6] This view has gained considerable support recently, and a number of historians and other scholars have focused on pointing out the disparate elements that participated in civil society and helped

increasing political consequence in the wake of the demographic revolution in Germany's largest cities around the turn of the century, and they demanded a more inclusive public sphere. The museums were recast from places that had been created for the use of local associations, international scientists, and a few educated elites to places that were often meant to accommodate a broad, socially diverse public, including schoolchildren, members of the working classes, and people from the surrounding countryside.

Such changes had a noticeable impact on both science and civil society, forcing the educated elites who directed German ethnographic museums to make their institutions more socially inclusive and to remake their science into something that this broader and more diverse public would find useful and pleasing. Through the 1880s and 1890s, the museums presented vast, complicated, geographic arrangements of material culture from all over the globe, which were intended to provide the raw material for ongoing scientific research as well as for each visitor's own studies or self-improvement. After the turn of the century, however, visitors increasingly encountered tightly organized, didactic displays that made single, well-illustrated points. One display might, for example, compare the kinds of animals people used for transportation in different areas of the world, or compare the different kinds of bridges and vessels they built with materials at hand, in order to illustrate the relationship between geography and technical innovations.[7] By 1910 the museums could no longer simply present their extensive research collections to the public. They had to illustrate elements of the collections' application in a clear, didactic way.

The impetus for these changes—both the ethnologists' move toward embracing a more inclusive public and the municipal authorities' efforts to place the museums in the hands of professionals—was part and parcel of a more general trend in civil society and civic government. Experts and professionals began colonizing and dominating areas of public space that had been initially carved out and occupied by autodidacts and local associations.[8] This was much more than a parallel movement in the professionalization of ethnology and city government. It was dialogic. As the structure of city life shifted, and the role of associations in it changed, so too did the character of ethnographic institutions and the associations based on science that were located in German cities.[9] This transformation, I shall suggest, can tell us much about the ways changes in the character of civil society can affect the form and

shape discourse in the public sphere. As a result, historians of Germany have begun to move beyond their focus on the *Bürgertum* and civil society as an intellectual ideal and toward exploring the complexity of civil society "on the ground." See *inter alia*, Trentmann, *Paradoxes of Civil Society* (cit. n. 2).

[7] Indeed, two of Karl Weule's displays in Leipzig in 1910 did just that: *Führer durch die Sonderausstellung über die Wirtschaft der Naturvölker* (Leipzig, June 1909); *Illustrierter Führer durch die Sonderausstellung über Transport- und Verkehrsmittel der Naturvölker und der außereuropäischen Kulturvölker* (Leipzig, summer 1910).

[8] See, *inter alia*, Ladd, *Urban Planning* (cit. n. 4); Lenman, *Artists and Society in Germany* (cit. n. 4); and Jan Palmowski, *Urban Liberalism in Imperial Germany: Frankfurt am Main, 1866–1914* (Oxford: Oxford Univ. Press, 1999).

[9] Although this paper is based on the history of German ethnology and German ethnographic museums, it does not include extensive discussion of the museum displays. Nor does it attempt to explain the ways in which the disputes about the ordering and display of artifacts contributed to changes in the method and content of the discipline over time. I have written about this elsewhere, but I did not revisit that argument here because it did not directly affect the shift I analyze in this essay. For information on these topics, see H. Glenn Penny, "Bastian's Museum: On the Limits of Empiricism and the Transformation of German Ethnology," in *Worldly Provincialism: German*

function of scientific institutions, as well as provide us with new answers to old questions about the "rise of the German sciences" during the nineteenth century by illustrating the role civic culture played in that process.[10]

TIMING: ETHNOLOGY AND CIVIL SOCIETY IN IMPERIAL GERMANY

Timing was a critical component of ethnology's success in imperial Germany. Ethnology, which would begin to emerge as a scientific discipline in Europe in the late 1860s, benefited from the wave of civic associations that began forming after the Napoleonic Wars, especially the increasing number of natural scientific associations founded following the 1848 revolutions.[11] Many of the people involved in these associations, and many who would later champion ethnographic museums, were inspired by the travel literature that began flooding Europe at this time. Alexander von Humboldt's cosmopolitan vision enchanted them, and many eagerly took up his challenge to pursue total histories of humanity and the world.[12] Indeed, Humboldt's efforts to fashion a total empirical and harmonic picture of the world formed the backdrop to the ethnographic project envisioned by Adolf Bastian, director of Berlin's ethnographic museum from 1873 to 1905 and perhaps the leading ethnologist of the day. Humboldt inspired Bastian's efforts to gather all knowledge of human history—ethnological, philosophical, psychological, anthropological, and historical—into a huge synthesis[13] He also motivated an array of other scientists, such as Mortiz

---

Anthropology in the Age of Empire, ed. H. Glenn Penny and Matti Bunzl (Ann Arbor: Univ. of Michigan Press, forthcoming); and idem, "Cosmopolitan Visions and Municipal Displays" (cit. n. 1).

[10] On the rise of the German sciences, see especially Ben-David, *Scientist's Role in Society* (cit. n. 5); and Friedrich Paulsen, *Die deutschen Universitäten und das Universitätsstudium* (Berlin: Ascher, 1902). For overviews of the historiography on the German sciences, see Kathryn Olesko, ed., *Science in Germany: The Intersection of Institutional and Intellectual Issues, Osiris*, 2d ser., 5 (1989); and more recently, Arleen M. Tuchman, "Institutions and Disciplines: Recent Work in the History of German Science," *J. Mod. Hist.* 69 (1997): 298–319. See also Lynn K. Nyhart, "Civic and Economic Zoology in Nineteenth-Century Germany: The 'Living Communities' of Karl Möbius," *Isis* 89 (1998): 605–30; and the works cited by Olesko, Tuchman, and Nyhart.

[11] The classic account of these associations is Thomas Nipperdey, "Verein als soziale Struktur in Deutschland im späten 18 und frühen 19. Jahrhundert," in *Gesellschaft, Kultur, Theorie: Gesammelte Aufsätze zur neueren Geschichte* (Göttingen: Vandenhoeck & Ruprecht, 1976), pp. 174–205. More comprehensive is Otto Dann, *Vereinswesen und bürgerliche Gesellschaft in Deutschland* (Munich: R. Oldenbourg, 1984). On the natural science associations, see Andreas Daum, "Naturwissenschaften und Öffentlichkeit in der bürgerlichen Gesellschaft: Zu den Anfängen einer 'Populärwissenschaft' nach der Revolution von 1848," *Hist. Z.* 257 (Aug. 1998): 57–90. For a concise account in English, see David Blackbourn, *The Long Nineteenth Century: A History of Germany, 1780–1918* (New York: Oxford Univ. Press, 1998), pp. 278–80.

[12] On Humboldt's impact on Germans' popular imagination, see Andreas Daum, *Wissenschaftspopularisierung im 19 Jahrhundert: Bürgerliche Kultur, naturwissenschaftliche Bildung und die deutsche Öffentlichkeit, 1848–1914* (Munich: R. Oldenbourg, 1998). See also Michael Dettelbach, introduction to *Cosmos: A Sketch of a Physical Description of the Universe*, by Alexander von Humboldt, trans. E. C. Otté (1858; reprinted, Baltimore: Johns Hopkins Univ. Press, 1997), vol. 2, pp. vi–xlvii.

[13] Bastian's *Mensch in der Geschichte*, for example, was dedicated to Humboldt and sketched out the beginning of the project the museum director pursued until his death. See Adolf Bastian, *Der Mensch in der Geschichte: Zur Begründung einer psychologischen Weltanschauung*, 3 vols. (Leipzig: O. Wigand, 1860). See also idem, *Alexander von Humboldt. Festrede* (Berlin: Wiegend und Hempel, 1869). Cf. Annemarie Fiedermutz-Laun, *Der kulturhistorische Gedanke bei Adolf Bastian: Systematisierung und Darstellung der Theorie und Methode mit dem Versuch einer Bewertung des Kulturhistorischen Gehaltes auf dieser Grundlage* (Wiesbaden: Franz Steiner Verlag, 1970).

Wagner, who became director of Munich's ethnographic museum in the 1860s, to travel and collect the material traces of humanity's many variations.[14]

Moreover, as the population of Germany dramatically increased and the size of its cities grew at an unprecedented rate, Germans were caught up in a process of continual renewal. Germany's largest cities were gripped by a wave of municipal self-fashioning, led by future-oriented citizens who shared cosmopolitan outlooks, a strong civic pride, and an unbridled optimism, and who were driven by a desire to transform their homes from provincial centers into *Weltstädte*, or "world cities." They were keenly aware of the cultural capital[15] that came with participating in scientific endeavors and possessing internationally recognized institutions of art and science.[16] These conditions created the opportunity for a range of individuals outside the universities to establish alternative institutions for the study of humanity and to participate in the production of knowledge. Indeed, in late-nineteenth-century Germany, ethnology was an emerging discipline led by intellectual newcomers in cities caught up in the throes of change.

Of course, European imperialism provided an important context for the pursuit of ethnology as well. There is no question that, as George W. Stocking Jr. has argued, colonialism was the "*sine qua non* of ethnographic fieldwork."[17] Colonial expansion expedited the very act of going into "the field" and provided many of the basic structures and conditions for ethnologists' experiences. But while the colonial context would prove increasingly important for German ethnology as the century drew to a close, we cannot simply superimpose on German history the British experiences that inform so much of our understanding about the relationship between colonialism and science. German interests in ethnology, and Germans' efforts to create ethnographic museums, preceded by decades Otto von Bismarck's move to gain colonies in 1884. Even if we were to accept recent efforts to push Germans' colonial interests back into the eighteenth century,[18] the actions and rhetoric surrounding the creation of German ethnographic museums make it clear that cosmopolitan visions and the dramatic transformation of German cities proved much more critical to ethnology's development and success than interests in empire.[19]

[14] Hanno Beck, "Moritz Wagner in der Geschichte der Geographie," Ph.D. diss. Univ. of Marburg, 1951; cf. Sigrid Gareis, *Exotik in München: Museumsethnologische Konzeptionen im historischen Wandel* (Munich: Anacon, 1990).

[15] By cultural capital I mean the social and cultural power that individuals and groups (i.e., associations, cities, regions, and nation-states) gleaned by supporting scientists and scientific institutions. This is a softer form of power than political or economic strength, but it was often harnessed by these supporters for building up or re-creating their own identities and reputations. For further explication, see Pierre Bourdieu, *Distinction: A Social Critique of the Judgement of Taste*, trans. Richard Nice (Cambridge, Mass.: Harvard Univ. Press, 1984).

[16] For a good example of the range of cultural and scientific activities that gained support in one such city, Frankfurt am Main, see Lothar Gall, "Zur politischen und gesellschaftlichen Rolle der Wissenschaften in Deutschland um 1900," in *Wissenschaftsgeschichte seit 1900*, ed. Helmut Coing et al. (Frankfurt am Main: Suhrkamp Taschenbuch Verlag, 1992), pp. 9–28. For a more general assessment, see Blackbourn, *Long Nineteenth Century* (cit. n. 11), p. 274.

[17] George W. Stocking Jr., "Maclay, Kubary, Malinowski: Archetypes from the Dreamtime of Anthropology," in *Colonial Situations: Essays on the Contextualization of Ethnographic Knowledge*, ed. George W. Stocking Jr. (Madison: Univ. of Wisconsin Press, 1991), p. 10.

[18] Susanne Zantop, *Colonial Fantasies: Conquest, Family, and Nation in Precolonial Germany, 1770–1870* (Durham, N.C.: Duke Univ. Press, 1997).

[19] The character of the science was affected by imperialistic ambitions and connections by the end of the century, but this was not the context that was most responsible for either the museums' creation or growth. See Penny, "Fashioning Local Identities in an Age of Nation-Building" (cit. n. 4).

The rise of local associations and the corresponding expansion of civil society in Germany created a context that disposed many of Germany's leading citizens to lend financial and political support to cultural and scientific endeavors. Scholars have calculated that by 1870 "one German citizen in two belonged to an association,"[20] and Andreas Daum has argued that between 1840 and 1870 up to eighty new associations devoted to the natural sciences alone were founded.[21] In most cases these endeavors were dominated by Germany's growing middle classes, the *Bürgertum*, which became increasingly well organized on the local level and began moving into areas of society and culture traditionally dominated by royalty.[22] Within the cities they often involved themselves in associations devoted to renewal, such as projects focused on social and moral improvement (for example, hygiene, welfare for the working classes, and public health care), and they eagerly engaged in fashioning signs of their own cultural progress, such as art museums, opera houses, and botanical and zoological gardens. The zoos are a particularly important example. During the same period in which Daum has noted the rapid growth of natural science associations, fourteen new zoological gardens were founded across Germany, not just in university towns, but also in a number of free cities and the capitals of different states. In most cases their creation was strongly tied to civic pride, municipal development, and a desire by leading citizens to exhibit their worldliness.[23]

Germany's largest and best-known ethnographic museums were founded in the late 1860s and early 1870s in four very different cities.[24] Hamburg was a free city and mercantile harbor, Leipzig a landlocked commercial center, Berlin the national capital and an important industrial hub, and Munich the capital of Bavaria, which the historian Veit Valentine termed "the 'classic state' of anti-national reaction."[25] Despite their differences, these cities shared several characteristics. They were the largest cities in imperial Germany,[26] and during the period 1850–1910, they all experienced fantastic growth: Berlin grew from a provincial capital to a city of more than two million people, the populations in Hamburg and Munich increased fivefold, and Leipzig's population of 63,000 in 1850 grew to nearly eleven times that size by 1910.[27]

[20] Blackbourn, *Long Nineteenth Century* (cit. n. 12), p. 278.

[21] Daum, "Naturwissenschaften und Öffentlichkeit" (cit. n. 11), p. 67.

[22] Dann, *Vereinswesen* (cit. n. 11).

[23] Annelore Rieke-Müller and Lothar Dittrich, *Der Löwe brüllt nebenan: die Gründung zoologischer Gärten im deutschsprachigen Raum, 1833–1869* (Cologne: Böhlau Verlag, 1998), pp. 7, 264.

[24] A number of smaller museums were also founded in smaller cities: Bremen, Kassel, Darmstadt, Karlsruhe, Lübeck, Stuttgart, Freiburg im Breisgau, as well as more medium-size cities such as Dresden, Cologne, and Frankfurt. The smaller museums, however, were generally regarded as less scientific and those in the medium cities were founded much later. Dresden was the exception: the royal collections were refashioned into a zoological and anthropological-ethnological museum in 1879. Sierra Ann Bruckner, "The Tingle-Tangle of Modernity: Popular Anthropology and the Cultural Politics of Identity in Imperial Germany," Ph.D. diss., Univ. of Iowa, 1999, p. 97; Michael Hog, *Ziele und Konzeptionen der Völkerkundemuseen in ihrer historischen Entwicklung* (Frankfurt: Rit G. Fischer Verlag, 1981), pp. 92–106.

[25] Veit Valentine, *Geschichte der deutschen Revolution von 1848–1849*, 2 vols. (Berlin, 1931), vol. 2, p. 565, cited in Celia Applegate, *A Nation of Provincials: The German Idea of Heimat* (Berkeley: Univ. of California Press, 1990), pp. 26–7.

[26] Although Dresden, Cologne, and Breslau were larger than Leipzig in 1871, Leipzig had outgrown the other three by the turn of the century.

[27] Such growth rates were characteristic of Germany at this time. See Jurgen Reulecke, *Geschichte der Urbanisierung in Deutschland* (Frankfurt am Main: Suhrkamp, 1985), p. 203.

The rapid demographic growth combined with the escalating technological advances of the second industrial revolution placed these cities in a state of almost perpetual flux, which many of their citizens optimistically embraced as something to be channeled and negotiated, rather than resisted. In their attitudes toward civic reform, the possibilities of science, even their own national and local pasts, Germans evidenced a future-oriented optimism.[28] Monuments built by local associations to celebrate national identity gradually shifted away from themes that connected Germany to older empires and toward celebrations of the nation's potential.[29] The preservation of older buildings by civic societies, a popular endeavor at this time, was focused as much on the future as on the past.[30] Scientists inside and outside the universities who argued for the replacement, or at least the augmentation, of philology by natural and ethnographic sciences—including archaeology, ethnology, prehistory, and that branch of art history focused on material culture—introduced "a new set of cultural symbols that focused on the future rather than the past."[31] Moreover, the wave of civic pride that swept Germany during the imperial period captured precisely these attitudes, inducing city planners not only to confront and control the changes in their cities, but to harness the opportunities that accompanied those changes in order to make the cities' futures.

As local elites, municipal governments, and civic associations set out to reshape their cities and themselves, they did so with an eye toward international developments. In Hamburg, Berlin, Leipzig, and Munich, they focused increasingly on becoming *Weltstädte*. During their efforts, the citizens of these and other German cities looked outside as well as inside their new nation to discover what this might entail. It was not just about size. The conventional definition of a *Großstadt*, or "large city," was one that contained at least 100,000 people.[32] But this term also connoted "comfort, sophistication, and cultural amenities,"[33] while the term *Weltstadt* suggested a certain cosmopolitan character that might set a city such as Hamburg or Berlin on a plane with London, New York, and Paris. The high numbers of foreign students in German universities, art schools, and academies certainly helped increase the cosmopolitan character of the cities hosting them.[34] But so too did the possession of scientific institutions that could claim to be setting the pace in an international science, especially a science that showed a connection with the wider world.[35] Such institutions, particularly those that could be opened to the public, functioned

[28] Gall, "Zur politischen" (cit. n. 16), p. 23.

[29] Wolfgang Hardtwig, "Bürgertum, Staatsymbolik und Staatsbewußtsein im deutschen Kaiserreich, 1871–1914," *Geschichte und Gesellschaft* 16(3) (1990): 269–95.

[30] Rudy Koshar, *Germany's Transient Pasts: Preservation and National Memory in the Twentieth Century* (Chapel Hill: Univ. of North Carolina Press, 1998), pp. 9–10.

[31] Kathryn Olesko, "Civic Culture and Calling in the Königberg Period," in *Universalgenie Helmholtz: Rückblick nach 100 Jahren*, ed. Lorenz Krüger (Berlin: Akademie, 1994), pp. 22–42. See also Walter Hochreiter, *Vom Museumtempel zum Lernort: Zur Sozialgeschichte deutscher Museen, 1800–1914* (Darmstadt: Wissenschaftliche Buchgesellschaft, 1994), pp. 58–86; Suzanne L. Marchand, *Down from Olympus: Archaeology and Philhellenism in Germany, 1750–1970* (Princeton, N. J.: Princeton Univ. Press, 1996); and idem, "The Rhetoric of Artifacts and the Decline of Classical Humanism: The Case of Josef Strzygowski," *Hist. Theory* 33 (1994): 106–30, on p. 109.

[32] Ladd, *Urban Planning* (cit. n. 4), p. 14.

[33] Ibid., p. 240.

[34] Lenman, *Artists and Society* (cit. n. 4), p. 125.

[35] For an intriguing look at the potential for natural history to help groups of people cultivate a certain cosmopolitanism, see Nyhart, "Civic and Economic Zoology" (cit. n. 10), p. 615.

as exceptional municipal displays that allowed a city to exhibit its sophistication and raise its cultural value.[36] This was precisely the appeal of ethnology and ethnographic museums: ethnology was a new, international science that took in the world, a science in which the *Weltstädte* of the United States and Europe were also actively engaged. It was an unrestricted, international field ready to be captured. And as a science that was not yet established in the universities, and which many people argued was best pursued in museums since its knowledge base was largely derived from material artifacts, ethnology was open to an array of individuals with an eye on the future rather than a focus on the past.

The rising interest in ethnology and the creation of German ethnographic museums gave a number of individuals a chance to move beyond the intellectual fringes in German society and into institutional positions that would enable them to participate in the production of knowledge. Moritz Wagner, for example, who produced a veritable mountain of journalistic and scientific publications based on his extensive travels and who ultimately influenced Friedrich Ratzel's theories of *Anthropogeographie* as well as the field of natural history in Germany,[37] continued to be regarded as a mere journalist and a dilettante by German educated elites. (He complained until his death in 1887 that the "armchair scholars" in the universities never took him seriously.)[38] Because he lacked the proper academic credentials, Wagner was unable to obtain a regular university position.[39] But with his appointment as curator of Munich's ethnological collections in 1862 and as the first director of the ethnographic museum in 1868, Wagner gained a certain intellectual credibility and an institutional base from which he could participate more actively in academic discussions about the multiplicity of humanity.[40] Similarly, Hermann Obst, a physician from Leipzig who dreamed of going abroad but was denied the chance to take part in expeditions to the North Pole and East Asia in the 1860s, found his intellectual niche and gained social stature and institutional power by directing the Leipzig ethnographic museum, which quickly became the second largest of its kind in Germany. He was a founding member of the association dedicated to maintaining this museum, a group that consisted of businessmen, journalists, and others who were considered an intellectual grade below Germany's educated elites. They shared Obst's enthusiasm for ethnology, recognized the cultural capital of this international science, and used their museum to carve out a new cultural territory among the educated elites that could be

[36] This cultural benefit was recognized not only in Germany, but in other countries, such as the United States, as well. See, e.g., Douglas Cole, *Franz Boas: The Early Years, 1858–1906* (Seattle: Univ. of Washington Press, 1999), p. 189; Sally Gregory Kohlstedt, "Museums: Revisiting Sites in the History of the Natural Sciences," *J. Hist. Biol.* 28 (1995): 151–66; Mary P. Winsor, *Reading the Shape of Nature: Comparative Zoology at the Agassiz Museum* (Chicago: Univ. of Chicago Press, 1991).

[37] Beck, "Moritz Wagner in der Geschichte der Geographie" (cit. n. 14); Helmut Ganslmayr, "Moritz Wagner und seine Bedeutung für die Ethnologie," *Verhandlungen des 38. Internat. Amerikanistenkongresses* 4 (1972): 459–70.

[38] Gareis, *Exotik in München* (cit. n. 14), pp. 46–7.

[39] Wagner's only titles were an honorary doctorate from Erlangen university, which he received following his 1838 publications on Africa, and an honorary professor title at the University of Munich, which he received in 1862. The museum provided him with the means for gaining a foothold in the Munich university, yet his honorary titles were not well respected, and he consistently felt himself on the fringes. Wolfgang J. Smolka, *Völkerkunde in München: Voraussetzungen, Möglichkeiten und Entwicklungslinien ihrer Institutionalisierung, c. 1850–1933* (Berlin: Dunker & Humboldt, 1994), especially pp. 54, 77–8, and 80–2.

[40] Ibid., p. 49.

occupied without university degrees. Within German ethnology, in other words, a new set of scientists created new institutions in rapidly developing cities and participated in a process of conscious self-fashioning on individual, institutional, and municipal levels.

ETHNOLOGY IN A POLYCENTRIC NATION

Imperial Germany's polycentric nature was a boon for ethnologists. Even if scholars disagree about what constitutes "good science," there is little question that the arts and sciences thrived across Germany largely because the Reich constitution left cultural affairs in the hands of the federal states.[41] This precipitated a heightened, and ultimately productive, competition among regional centers that complemented the already existing Prussian laws granting municipal autonomy. It also stimulated interest in local affairs that went well beyond zoning ordinances, municipal taxes, and the workings of government institutions. Indeed, a conscious effort at civic self-promotion was endemic in imperial Germany. Cologne and Düsseldorf, for example, were staunch rivals, and publicists and civic leaders in Düsseldorf argued that as a center of commerce, efficient government, and culture their city was at least as good as, if not better than, its larger neighbor to the south, and they eagerly listed their individual achievements to drive their point home.[42] As a result of such competition, museums, theaters, art schools, universities, and many scientific institutions in these and other German cities retained a decidedly local character during the imperial period—one they retain to this day.

Germany's polycentric nature and the degree to which cultural institutions were tied to local interests and needs make it difficult to generalize about the development of German ethnology, particularly with regard to German ethnographic museums. But the rhetoric and actions of the museum supporters in Hamburg, Berlin, Leipzig, and Munich also make it clear that they did share certain motivations, even if the "social lives" of their ethnographic institutions sometimes played out in markedly different ways.[43] In each case, local politicians, businessmen, and educated elites used their museums as an extension of their identity politics. They supported these institutions in order to enhance or reshape their cities' reputations and self-images, showing that the cities were centers of culture and learning, that the citizens recognized the importance of science, and that they were willing and able to make sacrifices to support it.

The commotion raised by the Danish ethnologist Kristian Bahnson in Hamburg provides a particularly instructive example. When Hamburg's ethnographic museum was founded in 1877–1878, the movement to create significant scientific institutions in Hamburg had not really begun yet. Even when it did, in the late 1880s, the process moved forward sporadically. Thus few of the leading members of Hamburg's govern-

[41] On the Reich constitution, see Volker R. Berghahn, *Imperial Germany, 1871–1914* (Providence R.I.: Berghahn Books, 1994), pp. 196–201. For a comparison with the arts, see Lenman, *Artists and Society* (cit. n. 4), p. 24.

[42] Ladd, *Urban Planning and Civic Order in Germany* (cit. n. 4), p. 12.

[43] In the growing literature on collecting, many scholars have shown the ways in which following objects through time and sketching out the social transactions that take place around them can shed much light on the societies engaged in these transactions. Something similar can be done with institutions. See, especially, Arjun Appadurai, ed., *The Social Life of Things: Commodities in Cultural Perspective* (Cambridge: Cambridge Univ. Press, 1986).

ment seemed to understand the need for this particular type of display, and the museum languished during its early years. In 1887, when Bahnson published his well-known article on European ethnographic museums, which surveyed the growth of ethnology in Europe over the past half-century, he strongly criticized the state of Hamburg's museum. He lamented the limited space given to the arrangements, the disorganization of the collection, and the fact that "to a certain degree the ethnographic principle, which initially structured the ordering, was not consistently followed." Geographic organization, he noted, had fallen to the wayside, and chaos had intervened. Most important, he chastised Hamburg's city fathers while noting that the German city with perhaps the greatest opportunity for producing a major international ethnographic institution had not even attempted to reach its potential.[44]

The reaction in Hamburg was quick and strong. Newspapers reprinted Bahnson's evaluations and comments about the sorry state of Hamburg's museum. Some papers took his criticisms further by portraying the museum as one in a state of chronic disrepair. Interestingly, none of the newspapers reproached the museum's director; they all blamed the city. They complained about the stain on Hamburg's image, and one critic pointedly warned that if the city's leading citizens continued to behave like "philistines," placing their business ahead of everything else, then Hamburg would "perhaps remain an important trade and manufacturing city," but it would "also sink to the level of a provincial city, a fate from which Hamburg should remain protected!"[45]

This public accusation that Hamburg's citizens, despite their economic success, remained short-sighted provincials quickly gained the attention of the city government. Governmental reports in the next week termed the author of these criticisms a "competent judge," and the *Senat* agreed that something had to be done. As a result, they quickly reevaluated the museum's budget, moved for the museum's reorganization, and transferred the museum to a location that would accommodate a wholesale rearrangement.[46] This expeditious action by Hamburg's government reflected an entirely new attitude toward ethnology and the city's ethnographic museum, an attitude that would lead to the institution's explosive growth around the turn of the century. The public outburst, in other words, forced the city council to realize that the fate of this scientific institution was closely tied to the members' own reputations.

Over the next three decades, concerns about Hamburg's honor continued to play a fundamental role in the development of the ethnographic museum; at the same time, the museum continued to be regarded as a means for refashioning the city's image. With this in mind, Senator Werner von Melle began a campaign to revamp

[44] Kristian Bahnson, "Ueber ethnographischen Museen mit besonderer Berücksichtigung der Sammlungen in Deutschland, Oesterreich und Italien," *Mittheilungen der Anthropologischen Gesellschaft in Wien* 18 (1888): 109–64, on p. 117.

[45] *Hamburg Correspondent*, 1 Nov. 1887. Other articles appeared in the *Hamburg Fremdenblatt* on 17 Nov. 1887.

[46] As a result of this decision, the ethnographic museum was given the top floor in a new building designed for the natural history museum, with the expectation that this was a temporary accommodation until an even better location could be found. "Mittheilungen der Bürgerschaft an den Senat," 29 Feb. 1888, in Hamburg Staatsarchiv (HSA) 361–5, Hochschulwesen I, C II, b 10; and "Auszug aus dem Protocolle der Ersten Section of the Oberschulbehörde," 8 Nov. 1887, in HSA 361–5, Hochschulwesen I, C II, b 10.

all of Hamburg's scientific institutions in the last decade of the nineteenth century.[47] Major decisions governing large acquisitions or changes in the museum were consistently based on their importance for the city's image.[48] Moreover, competition and constant comparisons with other German cities often guided the local government's choices. For example, during a critical meeting of Hamburg's city parliament in 1903, in which the pastor of the St. Pauli-Kirche brought the question of creating a monumental new building for their ethnographic museum into open discussion, comparison, emulation, and Hamburg's particular virtues provided the cornerstones of debate. Noting the vast numbers of ethnographic objects from all corners of the globe that frequently arrived in the local harbor only to be unloaded and whisked away by rail to other cities, the pastor pointed to the successes of Hamburg's counterparts as a reflection of the city's own failures and argued strongly that the citizens of Hamburg should be securing these objects for themselves:

> It often happens that valuable collections, which could be acquired for a few thousand marks, pass through here to Leipzig and other cities. It has in fact come to the point that Leipzig, which in all seriousness cannot be described as a maritime city (laughter), has already outdistanced our ethnographic museum. Indeed, Bremen is currently in a good position to overtake us in this respect. And it is high time that we bring to life the public interest for these things in an exceptional manner, through a worthy accommodation for [our] ethnographic museum. We in Hamburg must, and even those with little connection to these things have to agree, possess the best ethnographic museum, because nowhere else are the conditions for the propagation of such a museum so advantageous as in the continent's first maritime city. Gentleman, much has already been lost, let us now ensure that not everything will be.[49]

Similar examples abound in the public and private discussions of why Hamburg should create and continue to support such a museum, and many similar examples can be found in the other cities as well. In each case the residents' rationale for supporting their own museum combined emulation and a general discussion of the virtues of science with a form of particularism and a desire to exceed their counterparts' achievements.

Berlin's ethnographic museum was also linked to that city's image—as the capital of both Prussia and the new nation state. The Prussian Cultural Ministry was motivated to exceed any efforts in the Austrian, British, and French capitals, while carefully ensuring that Berlin stayed ahead of any efforts in other German cities as well—reacting to any notable expansion or acquisition in either Hamburg or Leipzig. In 1879, for instance, when the ethnologist Feodor Jagor complained to the

[47] Von Melle began lobbying for the introduction of science on a serious level in 1886, arguing for changes in the direction of the art museum and for better general education in the city. He especially condemned the fact that so many of Hamburg's children were forced to go elsewhere for their education. Werner von Melle, *Dreizig Jahre Hamburger Wissenschaft, 1891–1921* (Hamburg: Hamburger Wissenschaft Stiftung, 1924), p. 5.

[48] See, e.g., the debate over the purchase of the Thomannschen collection from Burma, gathered by a Hamburg resident and offered for sale to other European and American cities as well: "Mitteilung des Senats an die Bürgerschaft," 29 July 1907, in HSA 361–5 I, Hochschulwesen I, C II, b 15, Band III: 1904–1908. Cf. Glenn Penny, "Municipal Displays: Civic Self-Promotion and the Development of German Ethnographic Museums, 1870–1914," *Social Anthropology* 6 (1998): 157–68, on p. 162.

[49] Von Melle, *Dreizig Jahre* (cit. n. 47), p. 272.

Prussian Cultural Ministry that the new museum building Berliners had been promised six years earlier had not yet been built, he not only stressed that the international movement was leaving the German capital behind, but also warned that even lesser cities, not to mention the capitals of other nations, had already recognized and acted on the need for such institutions:

> Everywhere ethnographic museums are being founded, existing collections completed, and the currently accepted scientific standards fittingly established. In France they have founded an ethnological museum of the greatest scale, one equipped with everything current scientific standards require; in Vienna, the most famous architects of the land are erecting a splendid building . . . ; in a short time Leipzig, and probably also Hamburg will build their own museums for this purpose.[50]

The actions of these cities alone, he argued, "should suffice to show that it is high time for us to take practical steps as well to join in this movement and develop the Berlin collection in a manner fitting to the capital of the German Reich." And, he added, "if the collections of the Royal Museums are not imperial collections," then "it is a point of pride for the Prussian state." Such arguments proved effective, and the Berlin ethnologists gained the world's first freestanding ethnographic museum. Yet even after Berlin opened this new building to the public in 1887, and it became the leading institution of its kind, arguments such as Jagor's continued to be made about the importance of maintaining the city's position, of continuing to lead, rather than follow, the scientific stream.[51] Even in Berlin, in other words, where the connections between museum and university were closest, the support the museum received had little to do with the science itself, and everything to do with honor, image, and prestige.

In Leipzig as well, discussions of the museum often focused on questions of image and honor and the particular character of that city. Indeed, the fact that Leipzig's museum had been created by a local association rather than through governmental decree was championed as its great advantage.[52] The citizens regarded their museum's exceptional status in the international community of science as an accurate measure of Leipzig's importance as a leading university city, a site of international trade, and a place crafted by self-made men. Consequently, the museum's ethnologists and supporters not only made concerted efforts to maintain this reputation by pursuing ever bigger and more coveted acquisitions, but repeatedly attempted to usurp Berlin's leading position while playing on Leipzig's own exceptional status.

[50] Feodor Jagor, W. Reih, Simon to Königlichen Staatsminister der geistlichen, Unterrichts- und Medizinal-Angelegenheiten (KSG) Robert von Puttkamer, 11 Dec. 1879, in Geheimes Staatsarchiv Preußischer Kulturbesitz (GSA), I. HA, Rep. 76, Ve. Sekt. 15, Abt. III, Nr. 2, Bd. 2.

[51] Ibid. Such discussions of the city's honor and the need for Berlin to maintain its position as a world leader were not limited to the correspondence between government offices; they arose in public venues such as Berlin's newspapers as well. For example, during the public debates on the museum's future that took place in a local paper in 1900, one writer proclaimed: "The world reputation of our collection, to which members of all German states, indeed the entire scientific world have contributed, and which has been attained through the outstanding directors and able departmental staffs, cannot be permitted to fade, but must be further increased." *Vossische Zeitung*, 3 July 1900. Similar articles appeared in *Berliner Tageblatt*, 17 Feb. 1901, and *Tägliche Rundschau*, 18 Feb. 1901, among others.

[52] The failed attempts by Otto Georgi to create a *Handelsmuseum* (trade museum) illustrate how an institution offering a practical, business utility did not stir up support in the merchant city, while the ethnographic museum, which offered a different kind of utility—promulgating a higher image for the city—did.

In a newspaper article from 1874, for example, one Leipzig writer made comparisons similar to those stressed by Jagor in Berlin but changed the nature of his evaluation to favor the particularities of this Saxon city. Stressing that extensive ethnological collections already existed in "Berlin, Vienna, London, Paris, St. Petersburg, Copenhagen, and other world cities," he noted the exceptional potential of Leipzig's institution, which was not dependent on a national government, but had been generated by a locally based "international organization" that was in a particularly advantageous position to bring "together an overview of the nature and products of all of humanity from all of time."[53] Leipzig, he argued, rather than Berlin or one of these other world cities, should be leading ethnology, because the cosmopolitan character of this science could best be realized in a city "lying in the heart of Germany, yes, of civilized Europe," with its "ever-growing connections, with its position in worldwide trade," its "rich and extensive transportation networks and its multifarious intellectual and material power," and most important, its university. Leipzig's cosmopolitan character, he argued, made it the natural place for such an important international institution.

Just as ethnography could be made to conform to the goals of Leipzig's educated elites or the aspirations of the businessmen in Hamburg's *Senat*, so too could it be appropriated to help ordain Munich, Germany's self-proclaimed cultural center. Here, too, the fate of the museum was closely tied to civil society, international trends, and the city's self image. Yet in Munich these factors came together in a much different way. Munich's museum was one of Germany's oldest; but because it did not initially meet the interests or needs of a city engaged since the national unification in an aggressive struggle with Berlin over artistic reputations, it was essentially ignored during the last three decades of the nineteenth century.[54] This all changed abruptly, however, with the growing international interest in non-European and exotic art during the first decades of the twentieth century. As a result of this movement, Munich's powerful artistic community developed a sudden interest in ethnology and the city's ethnographic museum. The museum became an asset for the city almost overnight, and its fate quickly changed. Responding to this interest, the local government immediately hired Lucien Sherman, a university-trained expert on Asian and Oriental arts, as its new director; he gave the entire museum a complete makeover, and its budgets and acquisitions grew exponentially.[55] In one year, 1907, Sherman was able to procure more artifacts for the museum than his predecessor had obtained in the previous decade, and during his first ten years as director, he acquired more than ten times as much as the Bavarian royal family had collected in

[53] *Dresdener Zeitung*, 5 Sept. 1874. This sort of rhetoric appeared again and again when discussing the particularity of Leipzig's municipal landscape. Cf. B. S., "Unser Museum für Völkerkunde," *Leipziger Tageblatt*, 5 June 1878; and Otto Georgi, *Vortrag das Grassi-Museums betreffend*, 11 April 1884, in Stadtarchiv Leipzig (LSA), Kap. 31, no. 14: 40–63.

[54] Smolka, *Völkerkunde in München* (cit. n. 39), p. 153. For a nice discussion of this competition, see Lehman, *Artists and Society* (cit. n. 4), pp. 102–6.

[55] Despite his focus on the high art of India and Asia, Sherman was not an artist, but a dedicated scientist who specialized in an area of ethnography particularly suitable for Munich. Sherman was the first professional museum man to take over this collection. He had a Ph.D. and had a *Habilitation* (second doctoral degree), was an excellent organizer, and was a sound, if not brilliant, scientist. Scientists in and outside of Munich respected his work, and his qualifications as a scientist enabled him to become the first director to gain a university professorship. All these factors contributed to his and the museum's success, but it was the link to the art complex of the *Kulturstadt* (cultural city) that paved the way for that success.

the previous four hundred years. Munich's museum also gained a special place among ethnographic institutions because of its distinctive concentration on non-European art. Because the emphasis lay on the possession of exotic art rather than international science, the intersection of international trends in promoting ethnology with intra-German competition and local interests was different here than in other German cities, but the reason for the museum's explosive growth was the same: once the museum was firmly linked to the city's image, it began to boom.[56]

THE TRANSFORMATION OF CIVIL SOCIETY AND GERMAN ETHNOGRAPHIC MUSEUMS

The civic uses of ethnographic museums in Germany shifted over the course of the imperial period from a broadly cosmopolitan ideal embraced by autodidacts and their closest supporters to educational institutions meant to reach broader audiences. During the 1870s and 1880s, the cosmopolitan character of ethnology and the ambitious breadth of ethnographic museums—which their creators envisioned as sites for the accumulation of material culture from all peoples across the globe and throughout time—appealed to a generation inspired by Alexander von Humboldt's cosmopolitan vision.[57] Humboldt's penchant for total histories and massive empirical projects not only provided the critical intellectual backdrop for leading ethnologists such as Berlin's Adolf Bastian, but also influenced the physicians, businessmen, professionals, and others who helped those such as Hermann Obst in Leipzig create their own institutions. Humboldt's efforts to fashion a total empirical and harmonic project had captured the attention of Germany's middle classes during the era of reaction that followed the 1848 revolutions, and those efforts continued to hold the *Bürgertum*'s interest through at least the 1880s. Indeed, when one considers that Franz Boas was wont to label himself a "cosmographer" long after immigrating to the United States in 1887 and counted his bust of Humboldt as one of his most prized possessions, it is clear that the Humboldtian vision had a rather long intellectual legacy among German ethnologists.[58]

Yet as humanism and science "went disciplinary" during the imperial period, splitting into ever more refined slices of institutionally supported knowledge, Humboldt's penchant for sweeping theories fell from favor, and despite his continued importance as a cultural icon, professional scientists quickly began distancing themselves from him.[59] Similarly, within ethnology, one of the sciences that clung to Humboldt's vision the longest, the professionalization of the discipline saw a with-

[56] Once they entered into this competition, the Munich officials, like their counterparts in Hamburg, Leipzig, and Berlin, were driven forward by the constant comparisons and rivalries between the cities. See Professor Karl Theodore von Heigel to Königlich Bayerischen Staats-Ministerium des Innern für Kirchen- und Schulengelegenheiten (K&S), 24 March 1910, in Bayerische Hauptstaatsarchiv (BHA), MK 19455.

[57] On the widespread impact of Humboldt on the German middle classes, see Daum, "Naturwissenschaften und Öffentlichkeit" (cit. n. 11); and idem, *Wissenschaftspopularisierung* (cit. n. 12), especially pp. 138, 167, 274–80.

[58] Cole, *Franz Boas* (cit. n. 36), pp. 26, 109, 123; for a nice portrait of this intellectual trajectory, see Matti Bunzl, "Franz Boas and the Humboldtian Tradition: From *Volksgeist* and *Nationalcharakter* to an Anthropological Concept of Culture," in *Volksgeist as Method and Ethic: Essays on Boasian Ethnography and the German Anthropological Tradition*, ed. George W. Stocking Jr. (Madison: Univ. of Wisconsin Press, 1996), pp. 17–78.

[59] Daum, "Naturwissenschaften und Öffentlichkeit" (cit. n. 11), p. 86.

drawal from early efforts to fashion total empirical projects as a younger generation of ethnologists responded to a range of internal and external pressures. Armed with multiple university degrees, practical training in the largest museums, experience in the field, and in many cases university positions, the generation of ethnologists that took over the museums from 1904 to 1910 strove to create more systematic forms of collection, documentation, organization, and display.[60] They sought to move beyond what they regarded as the haphazard character of earlier collecting, to bring the disorganization of the museums' catalogs and displays under control, and to set more stringent standards for the operation of the institutions and for the practice of ethnology abroad. One of the strongest pressures for change within the museums themselves came from the cities, which were caught up in their own wave of professionalization.[61] The introduction of specialists and "experts" into various municipal offices left city councils with little interest in autodidacts and dilettantes running local museums and an increased willingness to foster a new generation of professionals in a range of institutions.[62]

Moreover, by the first decade of the twentieth century, participation in a cosmopolitan project was no longer enough to legitimate the escalating costs of maintaining such scientific institutions, and as these institutions were professionalized, they became more closely linked to local governments, which in turn began calling for more attention to the interests of those classes outside the *Bürgertum*. In exchange for their financial support, local governments began demanding that the museums be refurbished to conform to the needs of a changing public sphere. The rising importance of the lower classes in local and national politics (not to mention the sheer growth of their numbers within the cities) made local governments increasingly interested in addressing the needs of those citizens and perhaps even redirecting them through public education. Ethnographic museums, like other museums, needed to be accessible to nonscientists, serve general, educational purposes, and respond to the interests in their respective cities. These changes helped to fundamentally remake the field. There was, in essence, a transformation of German ethnology and German ethnographic museums that directly corresponded to the more general changes in civic culture and civil society taking place around the turn of the century, as they became more recognizably inclusive.

In the 1870s and 1880s, both the museums and the publics who visited them were still relatively small. The educated elites who supported and frequented these museums generally shared or at least accepted ethnologists' convictions that ethnographic museums first and foremost should be research institutions. Thus before the

[60] Germany's leading ethnological museums gained new directors shortly after the turn of the century—Hamburg in 1904, Berlin in 1905, and Leipzig and Munich in 1907. Younger men who had completed both a doctoral degree and a *Habilitation* replaced the old directors. All of the newcomers had close connections to the local university (or in Hamburg, the Colonial Institute once it was opened, and then the university when it was founded in 1919). None of the earlier directors had gained a *Habilitation,* only two had honorary Ph.D.'s, and consequently only Bastian and some of his assistants were connected in any significant way to a university. See Penny, "Cosmopolitan Visions and Municipal Displays" (cit. n. 1).

[61] Ladd, *Urban Planning* (cit. n. 4), pp. 20, 33, 238; Palmowski, *Urban Liberalism in Imperial Germany* (cit. n. 8), p. 3. As Rudy Koshar's work on preservation shows, even the more informal tasks centered on maintaining municipalities were professionalized at this time. Koshar, *Germany's Transient Pasts* (cit. n. 30), p. 67.

[62] Indeed, art historians had long since replaced the older artists who had initially directed Germany's art museums. Lenman, *Artists and Society* (cit. n. 4), p. 56.

turn of the century, the museums in Hamburg, Berlin, Leipzig, and Munich received very few letters of criticism or discontent from their local governments or visitors, most of whom followed the ethnologists' lead and attempted to take in the totality of the exhibits laid out before them according to geographic principles. That is not to say that these ethnologists operated free of public influence during these decades; their supporters had always demanded a useful science. But in the 1870s and 1880s, the utility of such scientific institutions lay in their very existence, which allowed residents to fashion their cities into *Weltstädte*. By the turn of the century, however, the municipal bodies that supported these museums faced a rapidly changing public sphere. Thus when museums in general shifted toward educational tasks, ethnographic museums were expected to conform, and ethnologists were expected to create didactic displays that were, if not enlightening, then at least accessible and pleasing.[63] Rather than cabinets and tables filled with comprehensive collections of material culture from across the globe, the exhibits should be limited, focusing on easily understandable comparative displays based on representative objects. Because of the differences between the cities, the process of professionalization and the turn toward public interests took place in strikingly different ways. Yet equally striking is the force of the more general movement, which required the transformation of even the most renowned and financially secure of these museums, illustrating the formative power that public discourse had over this science.

The Munich museum experienced the quickest and most dramatic transformation. Munich's was the oldest of the four museums examined in this essay, and its connection to the Bavarian government both ensured and limited its existence. The museum, founded in 1868 as one of the Royal Collections, received its budget from the state, and its directors were responsible only to the Bavarian monarchy and its ministries. Moritz Wagner, the first director, paid careful attention to scientific developments among his counterparts in northern Germany and expressed the desire to create an important scientific institution.[64] Wagner, however, was unable to connect with the people of a city in which, as one newspaper put it, "ethnological literature is probably nowhere else so little respected, read, and purchased."[65] Similarly, Max Buchner, the seasoned Africanist who took over the museum after Wagner's death in 1887, was "unable to bring his vision of an ethnographic museum into harmony with that of the artistically oriented city of Munich."[66] Like Wagner, Buchner "suffered under the strong lack of interest of the population."[67] His attempts to expand the collections and make changes in the museum along lines pursued by ethnologists in the north were hindered by the museum's meager funding, while his efforts to improve the museum's financial state were curtailed by the fact that, as Wolfgang Smolka has noted, "the misery of the ethnographic collection was only a reflection of the torturous existence under which all the scientific collections of the artistic

[63] This was an international movement toward education that stemmed to a large degree from the United States. The movement encompassed all kinds of museums but did not affect them simultaneously. For a general discussion of the "New Museum Idea" and its implications, see Tony Bennett, *Birth of the Museum: History, Theory, Politics* (London: Routledge, 1995).

[64] Smolka, *Völkerkunde in München* (cit. n. 39), p. 89; Gareis, *Exotik in München* (cit. n. 14), pp. 56–61.

[65] "Das neue ethnographische Museum in München," in *Beilage zur Allgemeine Zeitung*, Nr. 8–11, 8–11 Jan. 1868, cited in Gareis, *Exotik in München* (cit. n. 14), p. 56.

[66] Smolka, *Völkerkunde in München* (cit. n. 39), p. 154.

[67] Ibid., p. 154.

city of Munich existed."[68] Consequently, both Wagner and Buchner worked in a small museum that served an equally small public, and what audience their museum did have was limited to local and visiting scientists, a modest number of patrons, and the state.[69]

When changes did arrive, they seemingly did so in the hands of Lucien Sherman, who replaced Buchner as director of the museum in 1907, and who, as noted in the section "Ethnology in a Polycentric Nation," quickly began increasing the museum's collections at a dramatic rate and completely refashioning its displays. He thinned out the otherwise cramped exhibits by storing artifacts in different locations around the city. He used this extra space to create *Schausammlungen*, or didactic exhibits, that prominently displayed particular objects, placing them on different colored backgrounds or creating freestanding exhibits that would increase the aesthetic appeal of the displays. He used mirrors to light the backs or interior of larger pieces and transferred the collections from overcrowded tables to the vertical space of cabinets that would provide better viewing.[70] To help him conceptualize the new arrangements for permanent displays and temporary exhibits, Sherman called on Munich's local artists.[71] The local press discussed his efforts, which were well received by the general public, and ever greater numbers of visitors soon began frequenting the once-isolated museum.[72]

These changes were so abrupt that it would be easy to give sole credit to Sherman, portraying the museum as a place neglected early on by two somewhat interested, but not particularly motivated, directors and later reinvigorated by an enthusiastic individual who reshaped it and brought it up to international standards.[73] Without taking anything away from Sherman's abilities, however, we should bear in mind the degree to which these changes, and in fact Sherman's very presence in the museum, were a response to the demands and interests of Munich's artistic and scientific associations and civic leaders. In fact, it was no accident that Sherman became director in 1907. He was carefully chosen from a field of nine candidates because his academic background in Indian and East Asian art and cultures made him the most likely to bring the museum into line with artistic interests in the city.[74]

Thus by the time Sherman took charge, the general attitude in Munich toward the science of ethnology was already shifting, and the museum was well on its way to gaining a much broader audience. The new audience comprised influential patrons, local and visiting artists, and art critics such as Wilhelm Hausenstein, who was closely connected to the artist Paul Klee and the well-known group of modernists

[68] Ibid., p. 124, on p. 141.

[69] Gareis, *Exotik in München* (cit. n. 14), p. 86. Smolka, *Völkerkunde in München* (cit. n. 39), p. 93.

[70] This had actually been initiated by Buchner. Gareis, *Exotik in München* (cit. n. 14), p. 97.

[71] Ibid.

[72] Between 1908 and 1909 alone, the figures grew from 9,101 to 15,518. Walter Lehmann, "Bericht des K. ethnographischen Museums in Müchen III (1910)," *Sonderausdruck aus dem München Jahrbuch der bildenden Kunst*, 1911, I. Halbband, pp. 143–61, cited in Gareis, *Exotik in München* (cit. n. 14), p. 98.

[73] This, in fact, is ultimately Smolka's argument. Smolka, *Völkerkunde in München* (cit. n. 39), pp. 292–6.

[74] These nine candidates, who ranged from self-educated travelers with limited experience in museums to well-known museum men such as Willy Foy and well-credentialed scientists such as Sherman—who had both a doctorate and a *Habilitation* and was the editor of the *Orientalische Bibliographie*—are described in detail in Smolka, *Völkerkunde in München* (cit. n. 39), pp. 170–4.

that made up the Blaue Reiter. While Sherman's own creative energies were certainly responsible for much of the momentum with which these changes took place, he followed a direction already sketched out by patrons and members of Munich's local elites who were interested in seeing the museum flourish. Like his counterparts in northern Germany, Sherman created displays meant for public consumption. But he did so in an attempt to raise the museum's aesthetic appeal more than its educational potential, serving a civil society in which artistic interests were much more prominent, and the general, "uneducated public" played much smaller roles, than in the north. Thus Sherman's presence in the museum as well as his efforts as director were ultimately a response to his particular city's traditional interests in high art and its newly developed interest in the art of "exotic" peoples.

In stark contrast to the situation in Munich, the directors of the Leipzig ethnographic museum had to pay greater attention to general public interests and demands from the outset because their museum lacked any strong state support and suffered from tenuous finances. The association that founded the museum in 1869 relied on contributions from its members, entry fees paid by visitors, and donations from an array of patrons. From the beginning, the directors and members sought out a broad public base of support both inside and outside the city. The museum's creation was followed by a series of promotional lectures, ranging from presentations on "fire and its role in societies" to the "great roads of world transportation," which were all reviewed in local papers.[75] Once the museum was opened, detailed discussions of donations, acquisitions, memberships, and visitors remained a central part of the museum's yearly reports until the city took over the institution in 1904.[76] Moreover, almost every significant action taken by the museum's supporters—large and small acquisitions, new displays, the opening of new rooms, movement to new buildings, visits by the king and queen of Saxony—was discussed in local newspapers.[77] And from the outset, the "uneducated public" maintained an important presence in the planning of its directors and activities of the museum. In short, from the first days of its existence, the Leipzig museum maintained an uncommonly consistent and even exceptional public profile; its directors and supporters went to great lengths to sell their endeavor to an array of individuals; and they committed themselves to serving a broad stratum of Leipzigers with varied interests and desires.

Gradually, however, the Leipzig museum became a city-run institution. It moved from having no governmental support at its creation, to having partial support in 1884, to being an institution under the auspices of the city council in 1904. As its relationship to the local government changed, so too did its public focus. As in Mu-

[75] The lectures for 1872, for example, are listed in *Leipziger Zeitung*, 14 Dec. 1871, and the individual lectures were covered in a series of articles in both this paper and the *Leipziger Tageblatt* under the title "Vorlesungen zum Besten des Deutschen Centralmuseums für Völkerkunde." These ran from Jan. through March 1872.

[76] This remained consistent through the entire run of the museum's reports from 1873 to 1900. See Museum für Völkerkunde zu Leipzig, *Bericht des Museum für Völkerkunde zu Leipzig*, 24 vols. (Leipzig, 1874–1900). These yearly reports changed notably after the city took over the museum. Membership was no longer highlighted in the same manner in the second series. *Jahrbuch des Städtischen Museums für Völkerkunde*, vols. 1–6 (Leipzig, 1906–1914).

[77] The fact that the Museum für Völkerkunde zu Leipzig archive has six large volumes of news clippings for this period of the museum's history is the best indication of the breadth of coverage the institution received in local papers.

nich, international scientists provided the Leipzig museum with legitimacy from 1868 to 1914, and wealthy patrons were important for gaining new acquisitions and maintaining the museum even after it became a city institution. Until the city took over, local elites had been the museum's most regular visitors and had dominated the membership of the association that founded and supported the institution. But as the population of the city grew and the museum's audience expanded in size, Leipzig's city fathers became increasingly interested in having the institution play a bigger role in public education, changing the dynamics between the museum and its audience and pushing ethnologists toward popular, didactic displays.

Despite the somewhat different focus, these changes were not unlike those initiated by Sherman in Munich. Like Sherman, Karl Weule, the new director, was a professional with both his Ph.D. and his *Habilitation*. He had come to Leipzig from the Berlin museum in 1899 and had become the director of the Leipzig museum following Hermann Obst's death in 1906. He was the first ethnologist to gain a position at Leipzig's university.[78] While Sherman sought more aesthetic arrangements that would appeal to patrons and local elites, Weule (in addition to cataloging the museum's extensive collections and endorsing ever larger collecting expeditions) sought a new kind of comparative display that would allow him to carefully balance his museum's commitment to international science with his desire to communicate with a broader public—two seemingly different goals. Despite their different approaches, both directors were largely motivated by a desire to secure their museums' positions in their respective cities. Sherman sought to reshape his institution to better serve his city's artistic needs; Weule—once his museum was firmly established as a civic institution—worked to maintain his commitment to broader education. Both directors, in other words, took part in a more general movement toward creating new kinds of ethnographic displays in an effort to make their institutions more "useful" by meeting particular, local needs.

Like the Leipzig museum, the Hamburg museum was initially set up by a group of local citizens with an interest in ethnology.[79] But once it was firmly established in 1878, its development was in many ways closer to that of Munich's museum. Much like the initial directors of the Leipzig and Munich institutions, the first director of Hamburg's ethnographic museum, Carl Wilhelm Lüders, was not a trained scientist, but an autodidact who, like Munich's Wagner and Buchner, had largely gained his authority through his travels.[80] He, too, found himself in charge of a scientific institution in a city where scientific endeavors stirred little public interest and, in this case, a city that did not even have a university. Like Hermann Obst, the first director in Leipzig, Lüders had to rely on donations and patronage networks for acquisitions, though his salary and his institution's existence, like those of his

[78] Indeed, with Weule, Leipzig saw the creation of the first Universitäts-Ordinariat für Völkerkunde in 1920.

[79] Hamburg's Natural History Association. Ferdinand Worlée and Adolph Oberdörfer were elected in 1866 to take charge of the city's loose collection of ethnographica, which they cleaned, ordered, and published in a modest guide to the 332 pieces in 1867. This is generally regarded as the genesis of the museum, but it was not actually founded as an ethnographic museum and given its own location until 1877–1878. Jürgen Zwernemann, *Hundert Jahre Hamburgisches Museum für Völkerkunde* (Hamburg: HMfV, 1980), pp. 3–4.

[80] Lüders, in fact, essentially gained his position because he contributed a substantial collection from his travels to the museum. Ibid., pp. 10–1.

Munich counterparts, were secured by the city. After a short period of time, he also withdrew into his own humble projects, even more content than Wagner and Buchner with his small museum and equally limited audience.

As in the case of Munich, the most profound changes appeared to have arrived in the hands of a new director, Georg Thilenius, who took charge of the museum in 1904. But just as in Munich, the director was chosen for his professional credentials;[81] he set about reforming the institution in a manner that had already been largely sketched out by Senator Werner von Melle and his associates, who hoped that they could use the museum to re-create the city's image. Just as Sherman revamped his museum to serve Munich's artistic sensibilities and respond to new interests in non-European and "exotic" art, so too did Thilenius refashion Hamburg's museum into a leading scientific establishment, one that would not only serve the growing interests of the general public, but also become the stepping-off point, as von Melle had hoped, for creating Hamburg's Colonial Institute, and later, its university.[82]

While the process of transforming Hamburg's museum resembled what took place in Munich, the motives of the local government and others who initiated this process were much closer to those of their counterparts in Leipzig. In Hamburg, too, civic reform was under way by 1900,[83] and from the outset of the discussions about creating a new, monumental museum, members of the museum commission, the local government, and the scientists agreed that the institution's "primary goal" would be to serve the "education of the general population." As one senator defined it, this "broad public" would include not only the local elites and foreign visitors, but also people from "all of Hamburg's classes, . . . from the area around the city, the people from the countryside, small villages, and the suburbs."[84] No one ever challenged this point during the discussions, and even Thilenius, despite his own professional ambitions and commitment to creating a first-rate research institution, agreed wholeheartedly with these convictions and set out to design a museum that could serve multiple purposes and publics.[85] Here, too, ethnologists and their supporters held didactic displays at a premium.

Sherman and Thilenius both used their particular abilities to respond to the equally particular needs of their cities. For the art city of Munich, Sherman moved quickly to fill his museum with collections primarily from *Kulturvölker*, or "civilized people," in Asia and India. For the trade center of Hamburg, Thilenius moved even faster to obtain artifacts from an array of rapidly vanishing *Naturvölker*, or

[81] Thilenius, too, had both university degrees, had taken three major research trips abroad, and had gained a position at the university of Breslau in 1900 as *außer-etatsmäßiger Extraordinarius für Anthropologie und Ethnologie* and as *Kustos* of the anatomical collection, which had given him considerable experience with collections and displays.

[82] And like Sherman, Thilenius was given the funds. The total budget of the museum rose from 21,870 Marks in 1904 to 112,150 marks in 1910. Von Melle, *Dreizig Jahre Hamburger Wissenschaft* (cit. n. 47), p. 509.

[83] For discussions of professionals, the introduction of a civil service in Hamburg, and the reformism of Social Democrats there, see Richard Evans, *Death in Hamburg: Society and Politics in the Cholera Years, 1830–1910* (New York: Penguin Books, 1987), especially pp. 539–56.

[84] "Mitteilungen des Senats an die Bürgerschaft," no. 109, in HSA, CIIa, Nr. 16, Bd. I. See also "Bericht des von der Bürgerschaft am 22 June 1904," no. 38, in HSA, CIIa, Nr. 16, Bd. I, and "42 Sitzung der Bürgerschaft," 14 Dec. 1904, in HSA, CIIa, Nr. 16, Bd. I.

[85] For a precise rendition of what this entailed, see Georg Thilenius, *Das Hamburgische Museum für Völkerkunde, Museumskunde* 14 (1916).

"natural peoples," living in Africa, South America, and most notably, the Pacific. As a result, the people who visited these museums entered two fundamentally different worlds represented by two different kinds of *Schausammlungen*: one filled primarily with porcelains, silks, Buddhist and Hindu carvings, and steel weapons; the other dominated by hand-hewn Micronesian canoes, African masks, and weapons tipped with shark teeth, bone, stone, and shells.[86] The social and cultural contexts in each of these cities dramatically affected what ethnologists could do, and the ethnologists worked with city officials and others to channel and shape the international science of ethnology to serve local interests and needs.

In this cast of German institutions, the well-known Berlin museum was ultimately the anomaly. Its financial security and its prestige in the international world of science allowed its ethnologists to focus on their scientific projects for decades while ignoring all the layman's demands. Unlike the other museums, this institution had a large and consistent budget from the moment of its creation in 1873. It also had a range of extremely wealthy supporters, who provided its ethnologists with surplus funds for major acquisitions, and the best-known ethnologist in Germany as its director. Adolf Bastian and many of his assistants had advanced degrees and taught classes at the Berlin university; none of the museum directors in the other three cities had similar credentials or were closely affiliated with their cities' universities until after 1904. (In Hamburg, Thilenius had to wait for his position to be created along with the university in 1919.) After 1889 the Berlin museum had a federally decreed monopoly on all collections originating in German colonial territories, a restriction that made such collections noticeably scarce in other German museums. Ethnologists at the Berlin museum had a much closer working relationship with the German foreign office, the military, and the admiralty, and the prestige of being the largest museum of its kind gave it a special position within the international community of science.

The Berlin museum's secure funding, ample resources, prestige, and elite support allowed its ethnologists to continue keeping their objects in magazine-style arrangements like library stacks, eschew the creation of *Schausammlungen*, remain focused on "elite science," and essentially ignore a broader public. Thus while Thilenius was busily creating an institution that would serve "all of Hamburg's classes" and Weule was conceiving of new types of displays, the Berlin museum's ethnologists seldom considered the "uneducated public" and were much less willing to implicate them in reshaping the displays. As a result, Germany's largest ethnographic museum in Germany's largest city served a combination of visitors that was less socially diverse than in Hamburg and Leipzig.

Yet even in Berlin, where Bastian stood adamantly opposed to rearranging his museum, the virtues of adopting didactic displays aimed at a more general audience were being hotly debated among his supporters and assistants by the turn of the century, and the pressure to follow this controversial international trend continued to rise. In the late 1890s, pressing space limitations initiated discussions among Berlin's ethnologists about dividing the collection into two separate displays—one for public consumption and the other for scientific research—as many natural

[86] These differences, of course, are ones of emphasis. I do not mean to imply that either museum was completely void of the kinds of collections most coveted by the other, only that because of each museum's particular focus, its audiences had necessarily different visual experiences.

history museums had already done and Thilenius and Weule soon would do.[87] Initially, Bastian and most of the museum's ethnologists argued strongly against any division of the collection. But only a decade later, the relentless overcrowding combined with the fact that other German museums had begun following this international trend toward didactic displays made the change seem inevitable. American museums led this transformation, and many American ethnologists were also up in arms about the changes taking place in some of their largest museums.[88] Felix von Luschan, who became director of the Berlin museum following Bastian's death in 1905, harnessed these testaments from Franz Boas and others to dissuade Berlin's museum administration from endorsing radical measures, from following actions so extreme that American scientists had washed their hands of the results. But the movement toward didactic displays in general, and the new focus on the "uneducated public" in particular, had arrived in Berlin to stay. Luschan, in fact, began publicly endorsing such arrangements as early as 1905 after carefully rethinking his position, and he was only prevented from implementing a wholesale rearrangement of the museum after Bastian's death because he was engaged in designing an entirely new museum building—a project that was not completed until after World War I.[89] Despite the Berlin ethnologists' initial convictions and their commitment to "scientific arrangements," any strong position against displays aimed at an "uneducated public" became essentially untenable by 1914; after the war, when the Socialists came into power in Berlin, a return to the more "scientific" collections was unthinkable.[90] By this time the social utility of ethnographic museums had shifted, German and non-German cities had embraced the change, and Berlin, too, was forced to conform.

CONCLUSIONS: A USEFUL SCIENCE

As institutions that became increasingly geared toward public consumption, German ethnographic museums during the imperial period provide us with a window into the shifts in the civic utility of scientific institutions and insight into the roles that civic culture and civil society can play in these transformations. In this particular case, the associations and individuals who created these museums promoted a useful

[87] An exchange of correspondence and recommendations by the directors of all sections of the museum are located in the archive of the Museum für Völkerkunde in Berlin (MfVB), "Erweiterungsbau des Königlichen Museums für Völkerkunde," vol 1. In 1900 only Luschan spoke positively about *Schausammlungen* but was still highly critical of attempts in London and Paris. By the end of the decade, however, everyone agreed that these changes had to be made. This debate was reproduced in part, but with some misrepresentation of Felix von Luschan's statements, by Sigrid Westphal-Hellbusch, "Zur Geschichte des Museums," in *Hundert Jahre Museum für Völkerkunde Berlin*, ed. K. Krieger and G. Koch (Berlin: Reimer, 1973), pp. 1–100.

[88] Franz Boas, "Some Principles of Museum Administration," *Science* 25 (1907): 921–33; George A. Dorsey, "The Anthropological Exhibits at the American Museum of Natural History," *Science* 25 (1907): 584–9. Their private evaluations were even stronger than their published condemnations. Dorsey characterized the attempts by the new director of the American Museum in New York City to "popularize the science of ethnology" as a "vulgariz[ation]" of the science, and Franz Boas stated acidly to von Luschan that the New York museum had "sacrificed . . . all attempts at scientific accuracy, truthfulness, and efficiency to the popular clamor for striking exhibits." Boas to Luschan, 18 Oct. 1909; Dorsey to Luschan, 14 Nov. 1909; and Luschan to the General Administration of the Royal Museums (GVKM), 24 Nov. 1909, in MfVB "Umzug," vol. 10, 2124/09.

[89] Indeed, the displays were not actually reorganized until 1924. See "Die Neuordnung des Museums für Völkerkunde," *Berliner Tageblatt*, 26 Oct. 1924.

[90] Following the 1918 revolutions, the "elite" focus of Prussian museums came under heavy attack. See Karl Scheffler, *Berliner Museumskrieg* (Berlin: Cassirer, 1921), p. 22.

science, one that appealed to two varieties of cosmopolitanism: the intellectual interests of the *Bürgertum* that were captured in the celebration of Alexander von Humboldt and his cosmopolitan vision, and the desire of many members of this class to refashion their rapidly growing cities into *Weltstädte* (world cities).

However, the use value of these institutions, and indeed the science they grounded, shifted over time as the interests of their supporters changed. These changes followed in the wake of a transformation in the public sphere that, particularly in the northern cities, witnessed the growing importance of the lower classes in civic culture. They also dovetailed with similar changes taking place in municipal governments in some of these cities. Jan Palmowski has argued that as municipal governments became increasingly politicized with the rise of the Socialists around the turn of the century, they also grew in complexity, and cities were increasingly run by experts: "Whereas before the turn of the century cities were governed by the local notables, after 1900 a degree in jurisprudence and specialization in particular branches of municipal government became prerequisites for holding office in the all-powerful *Magistrat* [municipal authorities]."[91] At the same time that local notables were replaced by professional administrators in city governments, self-educated directors of ethnographic museums were replaced by professional ethnologists who had all the requisite university degrees and therefore were deemed better suited to adjust these museums to their cities' new interests and needs. This shift reduced the importance of local associations in the science of ethnology just as such associations were beginning to wane in significance in civic culture in general. Ethnological inquiry gained a foothold in the universities, which soon provided the young professionals with chairs, thus helping to finish off the legacy of Humboldt's cosmopolitan vision in both this science and its public articulation (the museums). In short, the professionalization of ethnology and the popularization of these scientific institutions were part of a mutual transformation of this science and civil society.

Attempts to make science useful for the state, be it national governments or local city councils, often have proven fruitful for the production of knowledge. Timothy Lenoir has even argued that in some cases the willingness of scientists to adopt rhetorical strategies that sell their sciences to the proper authorities has been "essential to the growth of knowledge."[92] There is little question that the growth enjoyed by German ethnographic museums during the imperial period and even the form they assumed were directly connected to ethnologists' success in pitching their science as useful to local and regional governments. Indeed, there is ample evidence to show that the smaller museums with the more tenuous bases of funding were the ones that moved first and most quickly toward the use of entertaining and didactic displays.[93] Those with the weakest financial bases had to adjust quickly to the winds of change, while ethnologists in Berlin, with the largest, and most financially secure of these museums, could resist this transformation the longest. But even the directors of the Berlin museum—with their unequaled financial support, their close connection to the Berlin university, the array of scientists on their staff with the highest

[91] Palmowski, *Urban Liberalism in Imperial Germany* (cit. n. 8), p. 3. Cf. Sheehan, "Liberalism and the City" (cit. n. 3), p. 123.

[92] Timothy Lenoir, "A Magic Bullet: Research for Profit and the Growth of Knowledge in Germany around 1900," *Minerva* 26 (1998): 66–88, on p. 66.

[93] This is further substantiated when one looks at the even smaller museums that were not included in this paper See Bruckner, "The Tingle-Tangle of Modernity" (cit. n. 24), p. 125.

academic degrees, and the institution's international reputation—could only temporarily resist the trends set in motion by the transformation of civil society and the changes in the public sphere. This, I think, helps to underscore the dynamic and interactive relationship between science and civil society in imperial Germany in ways that give us insight not only into the critical role civic culture played in the rise of ethnology and perhaps other German sciences but also into the degree to which social and cultural forces can reshape the form and function of scientific institutions in both productive and unproductive ways.

MODERN FORMULATIONS: THE TWENTIETH CENTURY

Civil Society, Science, and Empire in Late Republican France:
The Foundation of Paris's Museum of Man

*By Alice L. Conklin**

ABSTRACT

In 1938 an important civic institution for the support of science opened in Paris: the Musée de l'homme, a state-of-the-art anthropological museum under the direction of Paul Rivet. From the outset, there was more to the museum than met the eye. A socialist and humanist, Rivet used both bones and objects to try to convince the public of the equality of all peoples and cultures. Behind these displays lay research facilities designed to lure practitioners from all branches and schools of French anthropology under a single roof, that humanity might be studied there in its totality. Ironically, at this most "democratic" of museums, artifacts and funds from the empire were essential. Therefore colonialism at the Musée de l'homme went unchallenged.

INTRODUCTION

> That everywhere, throughout our continental and overseas territory, the museum might become what it should be: the home of the masses, where they will come to know themselves and to become conscious of their civilization.[1]
>
> Jacques Soustelle, 1936

> I think France has at present the most beautiful museum and center for ethnological studies in the world. This seems to me . . . particularly opportune at the moment, when the idea of the empire is finally penetrating the minds of the masses. I am convinced that a visit to our public galleries contributes greatly to

* Department of History, University of Rochester, 458 Rush Rhees Library, Rochester, NY 14627-0055; ackn@mail.rochester.edu.

I would like to thank Lynn Nyhart and Tom Broman for the invitation to join in the Civil Society and Science Workshop, in which this paper originated, and for all their patience, support, and incisive comments in seeing the article through to completion. I'm also grateful to the feedback from other workshop members. Special thanks go to Glenn Penny for sharing his expertise on ethnographic museums with a neophyte, to Florence Bernault and Jean Pedersen for their early critical reading, and to the anonymous reader and Kathy Olesko for their wonderful suggestions for revisions. Research support was provided by grants from the Fulbright Commission and the John Simon Guggenheim Foundation. Archives of the Collège de France cited with permission.

[1] Jacques Soustelle, "Musées Vivants," *Vendredi*, June 26, 1936, p. 1 ["Que partout, dans tout le territoire continental et d'outre-mer, le musée devienne ce qu'il doit être: la maison des masses, où elles viendront se connaître elles-mêmes et prendre conscience de leur civilisation."]

 0369-7827/02/1701-0010$10.00

developing this essential notion, which is so prevalent among the English for example, among the indifferent or ignorant.[2]
Paul Rivet, 1939

In 1938 an important civic institution for the support of science opened in Paris: the Musée de l'homme, an ethnographic museum administered by the Ministry of National Education with partial funding from the French colonies.[3] In an era marked by traveling shows of exotica, world fairs, colonial exhibitions—in which "primitives" from the colonies were still physically placed on display for the masses—and the advent of Nazi racism, the Musée de l'homme claimed to enact for all French a new, more progressive science of the world's cultures and races. Its principal founders, Paul Rivet and Marcel Mauss and their students,[4] were not only maverick anthropologists seeking to renovate the study of man in France, but also active socialists. The very title they adopted for their new museum was meant to evoke the French left's commitment to a politics of tolerance and respect for non-Western peoples in the face of Hitler's misuse of the race concept. Dedicated to bringing artifacts from living and fossil man to the attention of scholars and laypeople alike, the Musée de l'homme could and would, it was argued, convince the wider public of the full humanity of the peoples of Africa, Asia, and Oceania, many of whom inhabited the French empire.

The Musée de l'homme thus qualifies as a critical and self-conscious outpost of a longstanding civil society tradition in 1930s France, one associated with the system of higher education in particular and with intellectuals in general.[5] During the

[2] Paul Rivet à Jean Zay, Ministre de l'Éducation Nationale, 23 Jan. 1939, no. 133, Archives du Musée de l'homme (hereafter cited as AMH), 2 AM 1 A 12. ["Je pense qu'actuellement la France a le plus beau musée et centre d'études ethnologiques du monde. Ceci m'apparaît et vous apparaîtra aussi singulièrement opportun au moment où l'idée d'empire pénètre enfin dans l'esprit des masses populaires. J'ai la conviction, en effet, qu'une visite de nos salles publiques contribue puissamment à développer chez les indifférents ou non-avertis cette notion essentielle, qui est si générale chez le peuple anglais, par exemple."]

[3] It was unusual for a national museum to fall under the authority of the Ministry of National Education, rather than the Ministry of Fine Arts. But the Musée de l'homme's predecessor, the Musée d'ethnographie at the Trocadéro, had already been attached to an institution of higher learning—the Muséum national d'histoire naturelle—and the new museum inherited this particular linkage.

[4] The best known of these students included Georges-Henri Rivière, Charles Lecoeur, Marcel Griaule, Michel Leiris, Deborah Lifchitz, Therese Rivière, Germaine Tillion, Jacques Soustelle, Georgette Soustelle, Anatole Lewitsky, Boris Vildé, Henri Lehmann, André Schaeffner, Denise Paulme, Henri Lhote, Harper Kelley, Robert Gessain, Paul Victor, Jacques Faublée, Marcelle Bouteiller, Jeanne Cuisinier, and André Leroi-Gourhan.

[5] The museum was affiliated in particular with two institutions (which in turn fell under the control of the Ministry of National Education): the Muséum national d'histoire naturelle (Muséum), and the "New Sorbonne" as the University of Paris came to be called after the reforms of 1880 and 1896. Both had strong republican traditions of defending their autonomy, particularly under the Second Empire. On the history of the Muséum, see Camille Limoges, "The Development of the Muséum d'histoire naturelle of Paris, c. 1800–1914," in *The Organization of Science and Technology in France (1808–1914)*, ed. Robert Fox and George Weisz (Cambridge: Cambridge Univ. Press, 1980), pp. 211–40; Camille Limoges, "Une 'République de savants' sous l'épreuve du regard administratif: le Muséum national d'histoire naturelle, 1849–1863," in *Le Muséum au premier siècle de son histoire*, ed. Claude Blanckaert, Claudine Cohen, Pietro Corsi et al. (Paris: Éditions du Muséum national d'histoire naturelle, 1997), pp. 65–85; and Christophe Bonneuil, "Le Muséum national d'histoire naturelle et l'expansion coloniale de la Troisième République," *Revue française d'histoire d'outre-mer* 86(322–323) (1999): 143–69; for the history of republicanism in the university and the importance of the institution to the development of civil society under France's Third Republic, see Philip Nord, *The Republican Moment: Struggles for Democracy in Nineteenth-Century France* (Cambridge, Mass.: Harvard Univ. Press, 1997), especially chap. 1. Modern historians have underestimated the vitality of French associational and institutional life in the face of a strong centralized state, but this

early part of the Third Republic (1871–1940) the professoriate of France's institutions of higher learning had won the right to self-government, in the name of values that typically denote a vital public sphere: openness, independent thinking, and universal manhood suffrage.[6] In the wake of the Dreyfus Affair, academic and nonacademic intellectuals alike also had come to take it for granted that they had a particular civic responsibility to lead public debate on major political controversies. From the turn of the century onward, an increasing number of university professors across the disciplines and across the political spectrum could and did freely engage the government, other members of *la cléricature* (as France's intellectual "class" liked to call itself), and the larger public on such varied issues as military preparedness, pacifism, communism, fascism, imperialism, and appeasement.[7] In the waning years of the Third Republic, a group of these politicized *universitaires* founded the Musée de l'homme. Created in the *universitaires*' own image, this public institution incarnated a particularly democratic facet of French civil society.

Yet for all their independence, the Musée de l'homme and the many other civic institutions of the Third Republic existed as much in tandem with the state as outside it or against it. This was especially true of the French system of higher education to which the Musée de l'homme scientists were formally attached. As Terry Clark has argued, many academics and political leaders during the Third Republic shared the same ideology, which in turn led to university-generated efforts to further national cultural and political goals.[8] Part of this ideology, along with the cult of liberty, was the cult of science—particularly the new science of man in his moral as well as physical aspects. As products and beneficiaries of the republicanized university, the Musée de l'homme anthropologists certainly embodied its most liberal and pro-science tendencies in the 1930s. They supported the Popular Front, rejected a

---

neglect is beginning to change. The pioneering works are Maurice Agulhon's trilogy—*La République au village* (Paris: Plon, 1970); *1848 ou l'apprentissage de la République* (Paris: Seuil, 1973); *Le cercle dans la France bourgeoise* (Paris: A. Colin, 1977)—and Christophe Charle, *Les élites de la République, 1880–1900* (Paris: Fayard, 1997).

[6] The July Monarchy had first liberalized the French university by creating a body of representatives of the university's multiple faculties, the Conseil de l'instruction publique, with whom the minister of public instruction had worked closely. Napoleon III's education minister then curtailed the autonomy that the *conseil* represented, by packing it with his own appointees. But in the 1860s republican *universitaires* fought back, organizationally and intellectually. This vigorous reaction of intellectuals against the state was duly recompensed when the republicans came to power. The Third Republic made the Conseil de l'instruction publique of the university once again an elective body, and granted authority over appointments and curriculum to elected faculty councils. Nord, *The Republican Moment* (cit. n. 5), pp. 33–47; also see Georges Weisz, *The Emergence of the Modern Universities in France* (Princeton, N.J.: Princeton Univ. Press, 1983). A similar battle was successfully waged at the Muséum against Napoleon III; founded by a revolutionary decree in 1793, the republican Muséum defended its autonomy with even more vigor than the *universitaires*. See Limoges, "Une république" (cit. n. 5); and Jean Claude Schnitter, "Le développment du Muséum national d'histoire naturelle au cours de la deuxième moitié du XIX siècle: Se transformer ou mourir," *Rev. Hist. Sci.* 49 (1996): 53–97.

[7] On the public role of intellectuals in twentieth-century France, see Jean-François Sirinelli, *Génération intellectuelle: Khâgneux et Normaliens dans l'entre-deux-guerres* (Paris: PUF, 1994); idem, *Intellectuels et passions françaises* (Paris: Fayard, 1990); Jean-François Sirinelli and Pascal Ory, *Les intellectuels en France, de l'Affaire Dreyfus à nos jours* (Paris: A. Colin, 1986); Christophe Prochasson, *Les intellectuels, le socialisme et la guerre (1900–1938)* (Paris: Seuil, 1993); Christophe Charle, *La République des universitaires (1870–1940)* (Paris: Seuil, 1994); Gisèle Sapiro also gives important background on writers in the interwar years in her *La guerre des écrivains* (Paris: Librairie Arthème Fayard, 1999).

[8] Terry Clark, *Prophets and Patrons: The French University System and the Emergence of the Social Sciences* (Cambridge, Mass.: Harvard Univ. Press, 1974), pp.19–20.

hierarchical view of the world's races, and opted for as great a dissemination of their science as possible. But they also failed to question some of the more repressive policies and practices of the hand that fed them. This tendency was especially evident in two domains: the impulse to impose, on the basis of their "infallible" scientific authority, a specific representation of other cultures on the public, whom they wished to educate; and an acceptance of empire, whose legitimacy these scholars did not yet openly question, in part perhaps because colonialism served their science so well.

With its close ties to France's centralized education system and the imperial nation-state, as well as its pedagogical project of fashioning scientifically enlightened citizens and subjects, the Musée de l'homme is a perfect example of the kind of semi-autonomous civic institution that the French Third Republic spawned repeatedly over the course of its seventy-year existence. The impulse to produce and popularize a nonracist science was a consciously humanist and internationalist one in an age of extreme nationalism and xenophobia. But the same impulse was imperialist in the way it defined and engaged its public and in the parts of the world it chose to represent. All citizens and subjects would be welcome in the Museum of Man, but only cultures and races outside of France would be included. Among the latter, those found in the French empire would be disproportionately represented. And there would be no dialogue yet between the ordinary peoples from whom the objects were collected and the scientists assembling and readying them for public consumption in the name of universal reason, mutual respect, and progress. Whether the "self-consciousness" Jacques Soustelle called for was achieved anyway, as the many different publics imagined by him forged their own gazes independently of the science on display, is an intriguing possibility—one that usefully reminds us that civil society, once its existence is guaranteed by certain democratic institutions, is never fully contained or controlled by them.

I. ANTHROPOLOGICAL MUSEUMS IN INTERNATIONAL PERSPECTIVE

Any serious assessment of the Musée de l'homme's potential as a civic institution practicing a democratic science must first consider the timing of the museum's creation. The Musée de l'homme was the end product of a century of French civic activism in the realm of science, during which competing groups of self-styled anthropologists pioneered their new discipline. As in other countries, the collection and display of objects was critical to this process, particularly in its early stages; what is striking about France is that anthropologists there remained wedded to the public display and the study of their objects much later than anthropologists elsewhere. As William Sturdevant argued in his now classic article on the subject, by the 1920s the "museum moment" in modern anthropology had already passed in the field's two uncontested leaders—Germany and the United States—and was on the wane in Great Britain.[9] Yet it is the very lateness of the French project that accounts for its explicit democratizing ambitions and its semicolonial status.

[9] See, especially, William Sturdevant, "Does Anthropology Need Museums?" *Proceedings of the Biological Society* 82 (1969): 619–49; Ira Jacknis, "Franz Boas and Exhibits: On the Limitations of the Museum Method of Anthropology," in *Objects and Others: Essays on Museums and Material Culture*, ed. George W. Stocking Jr. (Madison: Univ. of Wisconsin Press, 1985), pp. 75–111; George W. Stocking Jr., "Philanthropoids and Vanishing Cultures: Rockefeller Funding and the End of the Museum Era," in ibid., pp. 112–45; and Michael M. Ames, *Cannibal Tours and Glass Boxes: The Anthropology of Museums* (Vancouver: Univ. of British Columbia Press: 1992).

The reasons for the differences between France and other countries are complex and worth investigating with a brief detour through the history of ethnographic museums. Beginning in 1840, the first generation of anthropologists in Germany, the United States, and Great Britain helped to sponsor the founding of new, or the refurbishing of existing, ethnographic museums as a critical step in institutionalizing their fledgling discipline. A second impetus for ethnographic collecting in this period was the recent assembly of large systematic natural history collections, important sites for innovation and research in nineteenth-century biology. As an outgrowth of natural history, early anthropology was influenced by its parent discipline, and thus anthropologists began their own collections, in which the emphasis was on classification, typologies, and geographical distributions.[10] By 1890 state and municipal authorities, civic associations, and philanthropists, caught up in the imperialist rivalries and scientific controversies of the day, seriously began to support this museum movement. In the United States, Germany, and Great Britain, universities, too, began investing in ethnographic museums on school premises; at the same time they began hiring the first professors of anthropology for their faculties. The results of this combined support were impressive. The new, or expanded, ethnographic museums that emerged at the dawn of the twentieth century embodied the ambitions, wealth, and power of their home cities and nations. In an age of social reform and mass democracy, the museums also reflected the desire among elites to educate (and entertain) their newly enfranchised citizens scientifically. Finally, the specific research needs of the increasingly university-trained anthropologists hired to classify, display, and build up the collections further drove museum expansion in the decades leading up to World War I. These collections remained important for academic research and were often used as evidence for supporting the theoretical developments of the day, including the emergence of the concept of culture-areas.[11]

Between 1890 and 1920, new principles of museum arrangement—themselves a reflection of a burgeoning relativism among scholars—began to be debated. In this realm, however, the concrete results proved much less impressive. Museums were either unable, or refused, to adapt to the more culturalist ways of thinking about other peoples that were starting to be formulated; this was hardly surprising, given the unwieldy structures of museums, along with the time, energy, and expense that the mounting of any exhibit represented. The best-known example of growing tensions between "museum and gown" was the experience of Franz Boas. Boas worked

[10] Sturdevant, "Does Anthropology Need Museums?" (cit. n. 9), p. 622; Anita Herle, "The Life-Histories of Objects: Collections of the Cambridge Anthropological Expedition to the Torres Strait," in *Cambridge and the Torres Strait: Centenary Essays on the 1898 Anthropological Expedition*, ed. Anita Herle and Sandra Rouse (Cambridge: Cambridge Univ. Press, 1998) pp. 77–105.

[11] On these developments internationally, see Sturdevant, "Does Anthropology Need Museums?" (cit. n. 9); H. Glenn Penny III, "Cosmopolitan Visions and Municipal Displays: Museums, Markets, and the Ethnographic Project in Germany, 1868–1914," Ph.D. diss., Univ. of Illinois at Urbana-Champaign, 1999; idem, article in this volume titled "The Civic Uses of Science: Ethnology and Civil Society in Imperial Germany"; Annie Coombes, *Reinventing Africa: Museums, Material Culture and Popular Imagination in Late Victorian and Edwardian England* (New Haven, Conn.: Yale Univ. Press, 1994) ; Nélia Dias, *Le Musée d'ethnographie du Trocadéro (1878–1908): Anthropologie et Muséologie en France* (Paris: Editions du CNRS, 1991); Dominique Poulot, "Le Louvre imaginaire: Essai sur le statut du musée en France des Lumières à la République," *Hist. Reflect.* 17(2) (1991): 172–204; Daniel J. Sherman, *Worthy Monuments: Art Museums and the Politics of Culture in Nineteenth-Century France* (Cambridge, Mass.: Harvard Univ. Press, 1989); Tony Bennet, *The Birth of the Museum: History, Theory, Politics* (London: Routledge, 1995).

as a curator at the American Museum of National History in New York between 1895 and 1905, while also holding a position at Columbia University. Boas had always rejected the evolutionary and typological schemes that tended to dominate in American museums, as well as the geographic organization favored in Germany and England. His preference was for groupings according to ethnic similarities, which he meticulously worked out in his displays. His superiors, in contrast, remained wedded to the more easily transmitted dogma of evolutionism. They believed that the public could respond only to simple messages and refused to seriously undertake the education of a lay audience.[12] The major innovation in the presentation of objects during this period of museum overhaul, at least in the United States, was the diorama (an innovation pioneered by the French, ironically). The major ethos of the big museums was amassing as many artifacts as possible—certainly more than their counterparts in other countries—as well as "completing" existing collections. By 1906 Boas had determined that it was simply impossible to realize his anthropological vision within the confines of a public institution beholden to wealthy patrons, who rejected his more erudite presentations in favor of displays that first and foremost entertained the public. He also complained of being held hostage to curators too mired in administration to follow the most recent anthropological scholarship.

If 1890 marked a certain museum takeoff then the interwar years, not surprisingly, were a period of stagnation and even decline. World War I had a major impact upon museum collecting; everywhere it disrupted established networks of exchange, decimated the ranks of anthropologists in Europe, and challenged complacent western views of themselves and others. Even before the war, the intellectual ground had been shifting from debating evolutionary theory to exploring the historical diffusion of peoples and/or accepting cultures on their own terms. In a further development of these trends, professional anthropologists began to turn to "participant-observation" fieldwork and to collect oral testimonies rather than artifacts.[13] The object of anthropology was now to study and valorize the psychological as well as the historical relations of cultures, rather than assert the universality of Western civilization. This meant students had to learn from the "natives" themselves, instead of relying on museum collections. Within the museums, curators realized how difficult it was to reconcile satisfactorily the three vocations that they had originally set for themselves—entertainment, education, and the promotion of research. Once the rush to collect was on, moreover, the storage and display of ethnographic objects became increasingly problematic. According to Glenn Penny, by 1920 German curators were

[12] Jacknis, "Franz Boas" (cit. n. 9), pp. 82–3; David Cole, *Franz Boas: The Early Years, 1858–1906* (Vancouver: Univ. of Washington Press, 1999), chap. 14.

[13] In the nineteenth century, anthropologists internationally had tended to be armchair theorists; they compared, classified, and ranked data (physical and social) about "primitive" peoples whom they believed to be at an earlier evolutionary stage. Amateurs collected this data for them. These theorists hoped to discover universally valid laws about the development of human society toward ever higher standards of rationality, morality, and quality of life. By the twentieth-century Boas, W. H. R. Rivers, and Bronislaw Malinowski in particular had begun to pioneer the pluralistic study of the variability of discreet cultures through direct personal contact with non-Western societies, in place of grand theorizing about entire races. For these developments in Anglo-American anthropology, see George W. Stocking Jr., *Victorian Anthropology* (New York: Free Press, 1987); idem, *After Tylor: British Social Anthropology, 1881–1947* (London: Athlone, 1995); Elazar Barkan, *The Retreat of Scientific Racism: Changing Concepts of Race in Britain and the United States between the World Wars* (Cambridge: Cambridge Univ. Press, 1992).

literally running out of space.[14] In addition, anthropologists recognized that even as museums were overrun with objects, displays of specimens from specific cultures could never be complete. In the Smithsonian's Museum of Natural History, New York's American Museum of National History, Chicago's Field Museum (as well as the joint museum-university programs at Pennsylvania, Columbia, Chicago, and Berkeley), and apparently in museums and universities throughout Great Britain and Germany as well, the great collections of an earlier era were not substantially altered between 1920 and 1940.[15] With their underlying evolutionist or geographic frameworks and, in some places, popular appeal to the public, these spaces became ones the anthropologist-in-training visited only briefly in order to prepare for the more important fieldwork overseas.[16]

France followed a slightly different trajectory, and a very different timetable, from that outlined above. In 1925 three prominent intellectuals associated with the study of non-Western cultures—Marcel Mauss, Lucien Lévy-Bruhl, and Paul Rivet—were able at last to establish a toehold for anthropology at the University of Paris (Sorbonne). That year they founded an Institut d'ethnologie, which made university certificates and diplomas in the discipline available for the first time.[17] Mauss and especially Rivet, however, had still larger ambitions. Publicly acknowledging France's backwardness in the discipline relative to the other great powers, they quickly turned their efforts towards creating a state-of-the-art ethnographic museum as well. In other words, French anthropologists were investing deeply in the display of material culture at a moment when their counterparts internationally were, if not actually marginalizing the ethnographic museum, at least questioning its relevance for their science. Why were French intellectuals and politicians not following suit? To answer this question we must turn to the longer history of anthropology in

[14] Penny, "Cosmopolitan Visions" (cit. n. 11), chap. 5.

[15] My information on the situation internationally in the interwar years remains very incomplete at this point. Ira Jacknis has pointed out the decline of the joint university-museum programs in the United States. Jacknis, "Franz Boas" (cit. n. 9), p. 108. Annie Coombes suggests that the critical era for the organization of ethnographic museums in Great Britain (principally the Horniman, the Pitt Rivers Museum, the ethnography hall of the British Museum, and the Cambridge Museum of Archaeology and Anthropology) was pre–World War I, although her focus is on African collections in particular. Coombes, *Reinventing Africa* (cit. n. 11). A. C. Haddon finally completed the cataloging and displaying of the Torres Straits collections in the early 1920s, when he was appointed deputy curator of the Cambridge Museum. Herle, "Life-Histories" (cit. n. 10), pp. 101–2. Mary Jo Arnoldi has shown that the African collections were installed in the Smithsonian at the beginning of the twentieth century, added to in 1922, but then underwent no changes until the 1960s. Arnoldi, "From the Diorama to the Dialogic: A Century of Exhibiting Africa at the Smithsonian's Museum of Natural History," *Cahiers d'etudes africaines* 39(3–4) [nos. 155–156] (1999): 701–26.

[16] Here I have a bit of suggestive andecdotal evidence: Margaret Mead and Reo Fortune made a tour "of the great collections of Oceanic art in the museums of Germany" in 1927, after Fortune completed his degree in anthropology at Cambridge. Jane Howard, *Margaret Mead: A Life* (New York: Fawcett Columbine/Ballantine, 1984), p. 102. What part museums played in the training of anthropologists in the interwar years in Europe and England has not yet been adequately explored. We cannot assume that because collections did not grow or change, scholars abandoned them. In Germany curators of the major museums acquired university positions at about the same time in the early 1900s and apparently continued to incorporate museum-based research into their teaching all the way up until 1960. See Penny, "Cosmopolitan Visions" (cit. n. 11).

[17] The University of Paris included all the faculties in the Paris Académie after 1896. The *académie* was the basic administrative unit of the highly centralized French education system; under the Third Republic faculties in each *académie* were given greater autonomy, but also centralized into a single "university" with its own elected council. Clark, *Prophets* (cit. n. 8), chap. 1.

modern France, a discipline very much at the juncture of civil society, politics, and the state in the nineteenth century.

II. FRENCH ETHNOLOGY IN THE MAKING

The conventional answer to the question of why French anthropologists came to a major museum project so late is that they developed disciplinary and organizational coherence later than elsewhere and thus were playing catch-up in the late 1920s and early 1930s.[18] This backwardness in turn was partially due to the early emergence in France of two rival approaches to the study of man—physical anthropology, deemed properly scientific because it dealt with measurable facts, and the more "amateurish" ethnography, or description of cultures. Resistance by the French university system to recognizing the new social sciences more generally also helps account for the museum delay.[19] Exclusion from the university meant that as the study of man developed in France from the early nineteenth century on, aspiring anthropologists had to found other kinds of institutions to acquire and dispense their methods and new knowledge. These included non-degree granting public schools and learned societies, as well as, after 1871, private schools and museums. Yet without both secure public funding and the right to confer degrees, none of these dispersed and rival associations was able to develop the science of man beyond a certain point. The tale of French anthropology and its attendant museums over the long nineteenth century is one of a dense proliferation of overlapping and often antagonistic institutions (whose history is just beginning to be written). This network of "anthropological" sociability flourished in part due to the liberalism of the French state, which simultaneously protected citizens' right of association, respected the university's autonomy, and subsidized (minimally) the science of man in its extra-university locations. The Musée de l'homme ultimately emanated from this same liberal tradition; its founders

[18] The classic views of French backwardness have been put forward by Elizabeth A Williams, "Anthropological Institutions in Nineteenth-Century France," *Isis* 76 (1985): 331–48; and Donald Bender, "The Development of French Anthropology," *J. Hist. Behav. Sci.* 2 (1965): 135–54. Other perspectives on the origin and development of French anthropology include Jean Jamin, "L'histoire de l'ethnologie est-elle une histoire comme les autres?" *Rev. Syn.* 3–4 (1988): 469–83; Victor Karady, "Le problème de la légitimité dans l'organisation historique de l'ethnologie française," *Rev. Fr. Sociol.* 23(1) (1982): 17–35; and Claude Blanckaert, "On the Origins of French Ethnology: William Edwards and the Doctrine of Race," in *Bones, Bodies and Behavior: Essays in Biological Anthropology*, ed. George W. Stocking (Madison: Univ. of Wisconsin Press, 1983), pp. 18–55; two useful reference works are Gerald Gaillard, *Dictionnaire des ethnologues et des anthropologues* (Paris: A. Colin, 1997); and the older but more reliable Pierre Bonté and Michel Izard, eds., *Dictionnaire de l'ethnologie et de l'anthropologie* (Paris: PUF, 1991).

[19] The great success story in the implantation of the social sciences in the French university at the turn of the twentieth century, of course, was Durkheimian sociology. Indeed, this singular success was one of the many factors retarding anthropology's establishment in France as a discipline with its own special content and approach, and therefore worthy of a university chair and separate course of study. Durkheim and his school initially showed little interest in—and indeed outright hostility towards—physical anthropology as then defined in France; nor did they have much respect for France's other and more comprehensive anthropological tradition, ethnography, which they claimed lacked theoretical rigor. Rather than encourage the founding of an independent field, the Durkheimians simply assumed that the study of other cultures fell under the definition of sociology anyway. Clark, *Prophets* (cit. n. 8), chap. 6; Victor Karady, "Durkheim et les débuts de l'ethnologie universitaire," *Actes de la recherche en sciences sociales* 74 (Sept. 1988): 17–35; and Laurent Mucchielli, "Sociologie *versus* anthropologie raciale: l'engagement décisif des durkheimiens dans le contexte 'fin de siècle' (1885–1914)," *Gradhiva* 21 (1997): 77–95.

would nevertheless attempt to orient the relationship between their science, the public, and the state in new directions.

The cases of Paul Broca's École d'anthropologie and the Muséum national d'histoire naturelle (Muséum) are instructive examples of anthropologists' early vibrancy in France, as well as of their subsequent divisions and marginality by the end of the nineteenth century. Founded in 1876 to construct a scientific history of man, the École d'anthropologie stood at the center of a network of affiliated institutions, which included the Société d'anthropologie, Broca's Laboratoire d'anthropologie, a museum, and a library; initially funded exclusively from private funds, the École began receiving state aid in 1878. Its association with the École de médecine (where Broca had an appointment) ensured that an overwhelming percentage of its students and professors were doctors, thus securing the École d'anthropologie social prominence as well as a pronounced materialist and polygenist orientation.[20] Thanks to Broca's organizational acumen, the École d'anthropologie became an international leader in the classification and comparison of races on the basis of visible differences in anatomy, although it offered pioneering courses in sociology and linguistics as well. Only a minority of its researchers, however, was convinced that the differences being discovered between genders, races, and civilizations were fixed and genetic. Broca and most of his followers were neo-Lamarckians who enthusiastically supported the programs of physical and moral hygiene adopted by the early Third Republic. The government in return endowed the École and extended patronage to Broca's learned society, the prestigious Société d'anthropologie. Nevertheless, by 1900 the school and the society had lost much of their earlier dynamism due to factional divisions over anthropology's proper role in society and future direction and to external attacks questioning the use of the cephalic index to classify races and determine intelligence.[21] The École d'anthropologie nevertheless remained in existence, with many of its faculty members devoting themselves almost exclusively to physical anthropology and clinging stubbornly to increasingly contested methods of racial classification. Unfortunately, the development of blood typing and eugenics in the context of renewed xenophobia in the interwar years would give the study of race at the École a new and sinister lease on life.[22]

[20] For a history of Broca's laboratory, see Denise Férembach, *Le laboratoire d'anthropologie à l'Ecole pratique des hautes études (Laboratoire Broca)* (Paris: n.p., 1980).

[21] The literature on Broca and the École d'anthropologie is vast. For a synthesis of much of this literature, see Claude Blanckaert, "'L'anthropologie personnifiée': Paul Broca et la biologie du genre humain," preface to *Mémoires d'anthropologie*, by Paul Broca (Paris: Jean-Michel Place, 1989), pp. i–xiii; and idem, "Méthode des moyennes et notion de 'série suffisante' en anthropologie physique," in *Moyenne, milieu, centre: Histoires et usages*, ed. J. Feldman, G. Lagneau, and B. Matalon (Paris: Éditions de l'ÉHÉSS, 1991), pp. 213–43. Also helpful on the political commitments of Broca are Joy Harvey, "Races Specified, Evolution Transformed: The Social Context of Scientific Debates Originating in the Société d'Anthropologie de Paris (1859–1902)," Ph.D. diss., Harvard Univ., 1983; Nord, *Republican Moment* (cit. n. 5), chap. 1; and Michael Hammond, "Anthropology as a Weapon of Social Combat in Late Nineteenth-Century France," *J. Hist. Behav. Sci.* 16 (1980): 118–32. On the waning scientific credibility of craniometry, see William H. Schneider, *Quality and Quantity: The Quest for Biological Regeneration in Twentieth-Century France* (Cambridge: Cambridge Univ. Press, 1990), p. 218.

[22] No general history of the École d'anthropologie after 1900 exists, and the ideas and institutions of the physical anthropology community in the pre- and interwar period are just beginning to be studied. Elements of the evolution of racial theory among anthropologists in these years can be gleaned from Filippo Zerilli, "Il Questionnaire sur les métis della 'Société d'anthropologie de Paris (1908)," *Ric. Folklor.* 32 (1995): 95–104; idem, "Il dibattito sul meticciato Biologico e sociale nell'-antropologia francese del primo Novecento," *Archivio per l'Anthropologia e la Etnologia* 125 (1995):

The Muséum national d'histoire naturelle was one of France's most prestigious *grand corps scientifiques* in the nineteenth century. It was a state-funded research and teaching center, with courses, much like those of the Collège de France, open to the public at large as well as to traditional students preparing degrees elsewhere. The Muséum comprised a library as well as specialized museums and laboratories for the different branches of natural history, and its faculty was the most highly paid in France. A chair of anthropology had been added to the faculty in 1855, and the first incumbent, Armand de Quatrefages, became Broca's principal rival for leadership among French anthropologists. Quatrefages's teaching and orientation differed dramatically from Broca's. He defended a monogenist theory of the origin of the races and persistently criticized Darwin. He was also more interested than Broca in France's other, more comprehensive anthropological tradition—ethnography, the description of cultures, peoples, and races. In 1859 Edmond de Rosny had organized French ethnographers into their own society, the Société d'ethnographie. Twenty years later these ethnographers had achieved sufficient prestige to convince the minister of education to open France's first ethnographic museum and link it informally with the Muséum's anthropology chair,[23] thus giving Quatrefages a foot in both camps. Quatrefages's works circulated widely, and there is no question that he was as well known and respected as Broca, inside and outside France.

Yet despite this impressive intellectual capital, the development of anthropology at the Muséum national d'histoire naturelle, much like at its rival at the École d'anthropologie, was increasingly hampered by the institution's inability to confer diplomas on students taking the courses offered in the new social sciences.[24] As Elizabeth Williams has pointed out, the power to confer degrees formed the core of French higher education; in all academic fields the sequence of degrees from the *baccalauréat* to the *doctorat* was the essential prerequisite to success.[25] Thus by the turn of the century, Muséum anthropology courses were not attracting as many serious students as in the past, and those students who did persevere had difficulty establishing themselves professionally in the conservative university world of the Third Republic.

237–73; Jean-Pierre Bocquet-Appel, "L'anthropologie physique en France et ses origines institutionnelles," *Gradhiva* 6 (1989): 23–34; Schneider, *Quality and Quantity* (cit. n. 21); Marc Knobel, "L'ethnologie à la dérive: George Montandon et l'ethnoracisme," *Ethnologie française* 2 (1988): 107–13; and the following articles by Pierre-André Taguieff—"Face à l'immigration: Mixophobie, xénophobie ou sélection. Un débat français dans l'entre-deux-guerres," *Vingtième siècle* 4 (July-Sept. 1995): 103–31; "Théorie des races et biopolitique sélectionniste en France," *Sexe et race* 3 (1989): 12–60, and 4 (1990): 3–33; and "Immigrés, métis, Juifs: Les Raisons de l'inassimilabilité: Opinions et doctrines du Dr. Martial," in *Mélanges en l'honneur de Rita Thalmann* (Frankfurt: Peter Lang, 1994), pp. 177–221. Also helpful, but more superficial, are Régis Meyran, "Races et racisme: Les ambiguités chez les anthropologues de l'entre-deux-guerres," *Gradhiva* 27 (2000): 63–76; and Herman Lebovics, "Le conservatisme en anthropologie et la fin de la Troisième République," *Gradhiva* 4 (1988): 3–17.

[23] On the early history of ethnography in France, see George Stocking, "Qu'est-ce-qui est en jeu dans un nom? La 'Société d'Ethnographie' et l'historiographie de 'l'anthropologie' en France," in *Histoires de l'anthropologie*, ed. Britta Rupp-Eisenreich (Paris: Klincksieck, 1984), pp. 421–31; and Emmanuelle Sibeud's excellent dissertation, "La construction des savoirs africanistes en France, 1878–1930," Diss., Paris, ÉHÉSS, 1999, chap. 3.

[24] Williams, "Anthropological Institutions" (cit. n. 18), pp. 340–3; and Dias, *Le Musée d'ethnographie* (cit. n. 11), pp. 70–2; for the founding of the anthropology chair at the Muséum, see Claude Blanckaert, "La Création de la chaire d'anthropologie du Muséum dans son contexte institutionnel et intellectuel (1832–1855)," in *Le Muséum* (cit. n. 5), pp. 85–124.

[25] Williams, "Anthropological Institutions" (cit. n. 18), p. 342.

As the above description makes clear, anthropological (i.e., osteological) and ethnographic (i.e., material culture) museums were an integral part of the research and teaching at the École d'anthropologie and the Muséum national d'histoire naturelle at the end of nineteenth century. Not surprisingly, the museums, like anthropology itself, suffered from the slow and fractured development of the field in France. The three most important collections among the many small late-nineteenth-century museums—the Muséum's collection of skeletons, the Musée d'ethnographie's collection of cultural artifacts housed at the Trocadéro Palace, and Broca's collection of skulls—remained underadministered and underfunded (although not necessarily underused). This neglect was particularly true of the Musée d'ethnographie's collection, the only one of the three destined for a broad public (it would form the nucleus of the future Musée de l'homme). The Musée d'ethnographie had actually been at the forefront of the ethnographic museum movement internationally at its opening after the 1878 Universal Exposition in Paris (under the Americanist Ernest-Théodore Hamy, Quatrefages's student and Muséum assistant). It had boasted not just an important collection of artifacts inherited from the Paris World's Fair and a superb pre-Columbian American collection, but also a library devoted to ethnography. Hamy initially had grand plans to centralize all existing ethnographic collections in the city.[26] Unfortunately the money had soon dried up, and lack of space and personnel quickly reduced the Musée d'ethnographie to, in James Clifford's phrase, a dusty "jumble of exotica."[27] The principal supporter of ethnography outside the Muséum, the Société d'ethnographie, suffered a comparable decline. Its membership diminished from 470 in 1878 to 165 in 1903.[28]

On the eve of World War I, ethnographic studies began to revive in a way that physical anthropology did not. In an era marked by rapid colonial expansion and a growing demand that society be studied with scientific methods, a vanguard of French scholars interested in the study of other cultures claimed that ethnography, too, was a science and had to be organized and approached as such. No fewer than three new societies devoted principally to ethnography were founded between 1910 and 1914.[29] One impetus for this renaissance came from the Muséum anthropologists Réné Verneau (who succeeded to Hamy's chair in 1909) and the young Paul Rivet, who recognized that the study of man necessarily included his cultural as well as his physical attributes. Equally influential, according to Emmanuelle Sibeud, was the one group who had been regularly in contact with "exotic" or "primitive" societies since the 1880s: colonial administrators of the stature of Maurice Delafosse—France's pioneering Africanist—although the iconoclastic founder of the study of

[26] Other ethnographic collections could be found at the Louvre, the Bibliothèque nationale, the Musée de la Marine et des colonies, the Musée des antiquités nationales de Saint-Germain-en-Laye and the Musée de l'artillerie; Elise Dubuc, "Le futur antérieur du musée de l'homme," *Gradhiva* 24 (1998): 64–96, especially p. 74. There were many more in private hands; important collections also existed in the provinces.

[27] While little has been written on the history of the Musée de l'homme, Nélia Dias has fully reconstructed that of the Trocadéro and its connections to anthropology from 1878 to the early 1900s in *Le Musée d'ethnographie* (cit. n. 11). On the decline of the Trocadéro museum, see James Clifford, *The Predicament of Culture: Twentieth-Century Ethnography, Literature, and Art* (Cambridge, Mass.: Harvard Univ. Press, 1988), pp.135–48; and, especially, Réné Verneau, "Le Musée d'ethnographie du Trocadéro," *L'Anthropologie* 19 (1918–1919): 547–60.

[28] Sibeud, "La construction" (cit. n. 23), p. 136.

[29] The new societies were the Institut d'ethnographique international de Paris, the Institut français d'anthropologie, and the Société des Amis du Musée d'ethnographie du Trocadéro.

French folklore, Arnold Van Gennep, played a key role as well.[30] Also contributing to this intellectual realignment (by rejecting the notion that race was ever a determinant of civilization) and working now from within the university were the Durkeheimians: Émile Durkheim's nephew and close collaborator Marcel Mauss, who taught at the École pratique des hautes études, Sorbonne philosopher Lucien Lévy-Bruhl, and the linguist Marcel Cohen, who was based at the École nationale des langues orientales vivantes, to name the most important figures for the future of French cultural anthropology.[31]

However, differences regarding ethnography's content and methods, personal rivalries, and different professional affiliations prevented this eclectic community of ethnographers from developing a common purpose in the prewar years.[32] As Marcel Mauss noted in 1913 (in what was both a compliment to anthropologists in other countries and a disparagement of the French in this realm), the breakthrough represented by Durhkeimian as well as recent Anglo-American and German sociology had depended on the innovative use of ethnographic documents: "ethnographic facts, borrowed from inferior societies, now are an integral part of the totality of facts that the most classical disciplines consider." Yet, he lamented, France was doing nothing to study such facts or make them known.[33] That same year, in two articles surveying "ethnography in France and abroad," he concluded bitterly that even Switzerland and Sweden had done better than France in this domain, despite the latter's superior status as a great "scientific power and colonial power."[34] He then added:

> The cause and the also the consequence of the stagnation of ethnography in France is the absence or the inadequacy of institutions that might concern themselves with it. We

[30] Delafosse and Van Gennep are the best-known ethnographers in the period, but they were part of a larger group of administrators and folklorists studying the cultural and social development of peoples inside and outside France and Europe—even before Durkheimian sociology began to make such subjects respectable. By the end of the century, colonial administrators in particular had taken the lead in this public sphere, publishing in France's preeminent scientific journals articles based on direct contact with the peoples about which they were writing. Sibeud has argued that it was Van Gennep and Delafosse who effected the "intellectual revolution" at the turn of the century that made it now possible to talk of civilization in the plural. See in particular, Claude Blanckaert, "Fondements disciplinaires de l'anthropologie au XIXe siècle: Perspectives historiques," in *Politix* 29 (1995): 31–54; and Emmanuelle Sibeud, "La naissance de l'ethnographie africaniste en France avant 1914," *Cahiers d'Etudes Africaines* 34(4) [no. 136] (1994): 639–58; and Emmanuelle Sibeud, "Les étapes d'un négrologue," in *Maurice Delafosse: Entre orientalisme et ethnographie. L'itinéraire d'un africaniste (1870–1926)*, ed. Jean-Loup Amselle and Emmanuelle Sibeud (Paris: Maisonneuve & Larose, 1998), pp. 166–90; for the intellectual shift to cultural relativism in France generally, see Laurent Muchielli, *La découverte du social: naissance de la sociologie en France* (Paris: Éditions La Découverte, 1999).

[31] Durkheim himself led the way here in the wake of the Dreyfus Affair. Mucchielli, "Sociologie" (cit. n. 19), pp. 81–3. The French term *ethnographie* designated what would come to be called "cultural anthropology" in the Anglo-American academy, and *ethnologie* in French. The development of *ethnologie* is discussed in section 3 of this article.

[32] In particular, different groups could not agree on ethnography's relationship to physical anthropology as well as on its proper object and methods of study. See Sibeud, "La construction" (cit. n. 23), chaps. 3, 5–7; Filippo M. Zerilli, "Etnographia e etnologia al congresso di Arnold van Gennep (1–5 giugno 1914)," *Ric. Folklor.* 37 (1998): 143–52.

[33] Collège de France, Archives Marcel Mauss (hereafter cited as CFAMM), Marcel Mauss à Ministre de l'Instruction Publique, typescript, n.d. (1913), 2 pp., on p. 2. ["Les faits ethnographiques, empruntés à des sociétés inférieures, font désormais partie intégrante de l'ensemble des faits que considèrent les disciplines les plus classiques."]

[34] Marcel Mauss, "L'ethnographie en France et à l'étranger I," *Revue de Paris* (1913): 537–60, on p. 549. ["La France grande puissance scientifique et grande puissance coloniale."]

> have no teaching programs, no good museums, no centers of ethnographic research because we are not interested in ethnography. And, conversely, we do not interest ourselves in this science because there is no one amongst us who is particularly interested in its success. A science does not live only from verbal promises, it needs material and personnel. It needs permanent organizations, durable institutions that create and nurture it.[35]

Mauss—in what was clearly a self-interested bid to establish himself as an intellectual leader of this divided but resurgent scientific community—went on to outline what remained to be done if France wanted to rank on a par at the very least with London and Vienna. As an "observational science" like zoology, botany, geology, and physical geography, ethnography demanded three different orders of work: fieldwork, during which "documents" were collected; museum or archival work, where objects were classified, studied, and published; and teaching, where knowledge produced was communicated to specialists and "even the public at large."[36] That same year, Mauss actually drew up a proposal for a bureau of ethnography along these same lines; its explicit objective would be to "organize, encourage, and activate ethnographic studies in France, and particularly in the French colonies."[37]

Nothing came of the project as France mobilized for war, but Mauss's description of what France lacked in 1913 turned out to be almost a blueprint for the plans drawn up by Paul Rivet in 1928 for the future Musée de l'homme. The principal difference is that Rivet would prove even more ambitious than Mauss, seeking not just to organize ethnographic studies in France but to reform and consolidate the science of man more generally, with the museum as its institutional base. Exactly as Mauss had recommended, Rivet's new museum would be tripartite in vocation and organization. Because of the peculiarly late timing of the Musée de l'homme's birth, moreover, this "tripartism" would reflect a new symbiosis between the republican state, democratic politics, and a particular community of anthropologists in France. On the one hand, the rebirth of France's existing ethnographic museum culminated with the eventual coming to power of the socialists, which helped consciously cast Rivet's new anthropology in the image of the Popular Front. On the other hand, from the early 1930s the creators of the Musée de l'homme sought to capitalize on colonies but also—in theory at least—to humanize French rule, at a moment when the empire was reaching its apogee. As Mauss's words quoted above suggest, they accepted unquestionably that the colonies would yield the most ethnographic new facts and objects, and that as the ruler of so many "primitive" peoples around the globe, France had an obligation to science to collect these objects before they disappeared. Rivet and his collaborators always insisted quite sincerely that their museum was a

[35] Marcel Mauss, "L'ethnographie en France et à l'étranger II," *Revue de Paris* (1913): 815–37, on pp. 820–1. ["La cause et aussi la conséquence de la stagnation de l'ethnographie en France est l'absence où l'insuffisance des institutions qui pourraient s'en occuper. Nous n'avons ni enseignements, ni bons musées, ni offices de recherches ethnographiques parce que nous ne nous intéressons pas à l'ethnographie. Et, inversement, nous ne nous intéressons pas à cette science parce qu'il n'y a chez nous personne qui soit particulièrement intéressé à son succès. Une science ne vit pas que de beau langage, il lui faut un matériel et un personnel. Il lui faut des organes permanents, des institutions durables qui la créent et l'entretiennent."]

[36] Ibid., p. 821. [". . . une science d'observation . . . les documents . . . même du grand public."]

[37] CFAMM, Projet de création d'un Bureau ou Institut d'ethnologie, typescript, n.d. (1913), 7 pp., on p. 1. [". . . organiser, encourager et activer les études ethnographiques en France, et en particulier dans les colonies françaises."]

state institution and therefore by definition apolitical: "Politics play no role in our house, where everyone works together fervently in our common project, and leaves their political ideas at the door."[38] Yet in an era when civil liberties appeared threatened and colonialism still capable of reform, a committed democrat such as Rivet had a difficult time keeping his science and his politics separate.

III. THE FIRST POSTWAR STEPS: *ETHNOLOGIE* AT THE SORBONNE

We are well informed about the genealogy of the Musée de l'homme, thanks to the many accounts that its primary architects, Paul Rivet and his assistant director, Georges-Henri Rivière, gave of its development.[39] These accounts suggest that there were two stages in the museum project. During the first stage, which occurred in the1920s, a group of leading intellectuals sought to bring together the two strands of anthropology that existed institutionally and intellectually in France (physical anthropology and ethnography) into a unified discipline of "ethnology," and to implant ethnology in the university.[40] Refurbishing France's ethnographic museum, the Trocadéro, which would now be formally linked to the Institut d'ethnologie, was an integral part of the plan, as was the centralizing of the many different ethnographic and anthropological societies and libraries under a single umbrella institution. These years were also characterized by an attempt to "vulgarize" a more positive view of African civilizations in particular through books destined for a broader public.[41] But

[38] AMH, 2 AM 1 A 8, Georges-Henri Rivière à Rédacteur en chef, *Gringoire*, 28 May 1935, no. 2052. ["La politique ne joue d'ailleurs aucun rôle dans cette maison où tous collaborent avec ferveur à l'oeuvre commune et laissent au vestiaire leurs idées politiques, s'ils en ont."]; also see Soustelle's letter in which he criticized a journalist for confusing "the Musée de l'homme, a public institution and rigourously impartial, and a private association . . ." ["Votre correspondant a cherché à créer une confusion entre le Musée de l'homme, institution d'État et rigoureusement impartiale, et une association privée . . ."] 2 AM 1 A11, Jacques Soustelle à Rédacteur, *Revue des Lectures*, 22 Oct. 1938, no. 1680.

[39] See in particular the following articles by Paul Rivet—"L'étude des civilisations matérielles; ethnographie, archéologie, histoire," *Documents* 3 (1929):130–4; "L'ethnologie," in *La science française*, 2 vols. (Paris: Larousse, 1933), vol. 2, pp. 5–12; "Ce qu'est l'ethnologie," in *Encyclopédie française*, vol. 7: *L'espèce humaine: peuples et races* (Paris: Larousse, 1936), pp. 7'06–1-7'08–16; and "L'Ethnologie en France," *Bulletin du Muséum*, 2d ser. (Jan. 1940): 38–52. Equally important are Paul Rivet and Georges-Henri Rivière, "La réorganisation du Musée d'ethnographie du Trocadéro," *Outremer* (1930): 138–49; and Paul Rivet, Paul Lester, and Georges-Henri Rivière, "Le Laboratoire d'anthropologie du Muséum," *Nouvelles archives du Muséum d'histoire naturelle*, 62 ser., no. 12 (1935): 507–31. These articles tended to repeat the same information, as Rivet sought to publicize his reforms as widely as possible.

[40] In choosing the term "ethnology," Rivet was deliberately returning to French anthropology's original nomenclature. The first anthropological society in France was William Frederick Edwards's Société ethnologique, founded in 1839. Rivet was also clearly ignoring the meaning that the term had acquired at the École d'anthropologie, where the ethnology chair had always been a chair in raciology exclusively. Rivet's reforms of ethnology are also discussed by Benoît de l'Estoile, "Africanisme et *Africanism*: Esquisse de comparaison franco-britannique," in *L'Africanisme en questions*, ed. Anne Piriou and Emmanuelle Sibeud (Paris: EHESS, Dossiers Africains, 1997), pp. 19–42; Gerald Gaillard, "Chronique de la recherche ethnologique dans son rapport au CNRS, 1925–1980," *Cahiers pour une histoire du CNRS* 3 (1980): 85–106; and idem, "L'ethnologie avant le CNRS," ibid. 3 (1995): 85–129.

[41] Filippo M. Zerilli, "Maurice Delafosse, entre science et action," in *Maurice Delafosse* (cit. n. 30), pp. 157–8. Zerilli notes that Rivet in particular wished to "popularize" ethnology. He discussed the role that Flammarion had played in making astronomy popular with a larger public and hoped that ethnology too would find its "*vulgarisateur*." Maurice Delafosse briefly became that popularizer, at least as far as African ethnology was concerned, until his premature death in 1926. After leaving the colonial service in the early 1920s, he spent the last years of his life writing a series of books on Africa destined for a lay audience, which carried the same humanitarian message about the equality

it was only in the mid-1930s that the possibility of an entirely new museum emerged and with it a more definitive stance on the "race question" and a carefully considered plan on how to bring "ethnology to the masses" through the display of artifacts. France's imperial project, museographic trends internationally, and the escalating crisis in Europe all left their marks on the new institution's organization and content during these critical two decades.

In the end, the driving force behind the renovation of the Musée d'ethnographie in 1920s Paris was not Mauss but his fellow socialist and social scientist, and more gifted administrator, the anthropologist Paul Rivet (1876–1958).[42] Rivet, although a navy doctor trained in the old anthropometric methods of Broca, was one of the first anthropologists to question the use of visible physical properties to classify races right before and after World War I.[43] His interest was primarily in the prehistory of the Americas. Five years (1901–1906) spent in Ecuador on a scientific expedition transformed the young anthropologist into an accomplished linguist and ethnographer as well. The experience awakened in him what proved to be a life-long fascination with the migration patterns of Meso-America, and Rivet never gave up his passion for traveling to the field. A staunch Dreyfusard and internationalist, Rivet served brilliantly as a medical officer in World War I, then worked tirelessly with Franz Boas after the war to reestablish contacts between German and other Western scientists. He joined the socialist party and later, in the wake of the riots organized by the far right in 1934, would become a founding member of the Comité de vigilance des intellectuels anti-fascistes; in 1935 he would run for the Paris municipal council, where he became the first member of the future Popular Front to be elected to public office. Associated with the resistance group that formed at the Musée de l'homme immediately after the defeat by Germany, he fled France in February 1941 for South America.[44]

In the mid-1920s, however, Rivet's political career still lay in the future. During this more quiescent postwar decade, Rivet consolidated his standing as France's leading Americanist and his reputation as an innovative linguist, tireless organizer of congresses and missions, and lecturer at home and abroad. In 1928 Rivet's efforts paid off, when he inherited one of France's two chairs in anthropology: that of the

---

and plurality of all human cultures. These books included *Les Noirs de l'Afrique* (Paris: Payot, 1922); *L'âme nègre* (Paris: Payot, 1922); *Les civilisations négro-africaines* (Paris: Stock, 1925); and *Les Nègres* (Paris: Editions Reider, 1927).

[42] Mauss, whom Rivet saw regularly at both socialist and scientific meetings before and after World War I, remained the classic armchair anthropologist. Yet, though he relied on the fieldwork of others in his own work, he was as adamant as Rivet about the need to train a new generation to "encounter" their subjects and "*le terrain*" directly. The standard reference work on Marcel Mauss is Marcel Fournier, *Marcel Mauss* (Paris: Fayard, 1994); see also B. Karsenti, *L'homme total: Sociologie, anthropologie et philosophie chez Marcel Mauss* (Paris: PUF, 1997); and Camille Tarot, *De Durkheim à Mauss: L'invention du symbolique, sociologie et science des religions* (Paris: La Découverte/ M.A.U.S.S., 1999).

[43] Schneider, *Quality and Quantity* (cit. n. 21), pp. 218–9. Rivet was the founder in 1911 of the Institut français d'anthropologie, which broke with the physical anthropologists over issues of racial classification and promoted the study of culture.

[44] For an intellectual biography of Paul Rivet, see Fillipo Zerilli, *Il lato oscuro dell'etnologia* (Rome: CISU, 1998); see also Christine Laurière, "Paul Rivet, vie et oeuvre," *Gradhiva* 26 (1999): 109–28. Jean Jamin explores Rivet's political and scientific careers together in "Le savant et le scientifique: Paul Rivet (1876–1958)," *Bull. Mém. Soc. Anthropol. Paris* 1(3–4) (1989): 277–94; on his career as an Americanist, see Paul Edison, "Latinizing America: The French Scientific Study of America, 1830–1930," Ph.D. diss., Columbia Univ., 1999, especially chap. 11.

Muséum national d'histoire naturelle, where he had already been teaching for several years as assistant to Réné Verneau. By tradition since 1892, the chair's incumbent had also headed up the Trocadéro Musée d'ethnographie, although Verneau had neglected the museum in the absence of adequate support. In 1928 Rivet inherited that mantle as well. Unlike his predecessor he immediately formalized the link between the two institutions, securing for himself the museum's directorship and additional funding to hire an assistant director and two staff members. The Muséum chair now had a research center (*laboratoire*) for each of anthropology's two branches: physical and cultural. By then, however, Rivet, in conjunction with Mauss and Lévy-Bruhl, had already taken the first postwar steps for what they hoped would be a definitive reorientation and institutional reconfiguration of French anthropology—one that would bring it more in line with developments internationally and infuse it with their own particular brand of republican engagement.[45] They would not succeed completely: members of the École d'anthropologie in particular, as well as certain natural history professors at the Muséum, would resist both Rivet's attempt to meld physical and cultural anthropology into a common enterprise beyond their control and his questioning of their methods. They suspected, rightly as it turned out, that the study of man's physical traits would be the junior partner in the new ethnology. But the ethnographic vanguard did score enough institutional gains in the interwar decades to worry their competitors, some of whom would seek to recoup for themselves the Musée de l'homme under Vichy.[46]

If the prewar years had been marked by a sense of either stagnation or renewal but continued division on every French anthropological front, by the early 1920s a propitious combination of circumstances made a renovation and consolidation of the discipline appear possible and desirable. These circumstances included a new interest in humanizing colonial governance and a greater popular and official interest in the empire generally, an openness on the part of Durkheim's heirs to the constitution of "ethnology" as a university-based field in its own right, the loss of a whole generation of (sometimes feuding) scholars in the social sciences generally, and the coming of the left to power in 1924.[47] In 1925 Mauss, Lévy Bruhl, and Rivet presented the minister of public instruction and the minister of colonies with a specific plan to provide ethnology with an institutional and definitional coherence that it had lacked since at least the beginning of the century.[48] Each brought with him a particu-

[45] Rivet was especially influenced by Boas. He owed to Boas, he claimed, "my own understanding of the complexity of ethnology and the interdependence of its different branches. . . . [I]t is thanks to him that I imagined what a true museum of humanity should be, that is to say an immense diorama, where the visitor would find a complete tableau of world's races, civilizations and languages." ["C'est en lisant [Boas] que j'ai compris la complexité de l'ethnologie et l'interdépendence de ses différentes branches . . . c'est grâce à lui que j'ai imaginé ce que devait être un véritable musée de l'humanité, c'est à dire un immense diorama, où le visiteur trouverait le tableau complet des races, des civilisations, des langues du monde."] AMH, 2 AM 1 B 9 c/Rivet (1876–1958)/Nécrologie, Paul Rivet, "Tribute to Franz Boas," in *International Journal of American Linguistics* 24(4) (1958): 251–2; see also Filippo M. Zerilli, "La conscienza delle scienze dell'uomo. Il carteggio Boas-Rivet (1919–1941)," *Il Mundo 3*, anno 2, no. 1 (1995): 390–404.

[46] Henri Vallois, a physical anthropologist teaching at the faculté de médecine in Bordeaux, would become director of the museum in 1941. See Bocquet-Appel, "L'Anthropologie physique" (cit. n. 22), pp. 30–2.

[47] Sibeud, "Les étapes" (cit. n. 30), pp. 185–6; see also Victor Karady, "French Ethnology and the Durkheimian Breakthrough," *J. Anthropol. Soc. Oxford* 12 (1981): 165–76.

[48] CFAMM, "Note sur l'Institut d'ethnologie de l'Université de Paris," 20 Dec. 1925, 3 pp.; idem, "Institut d'ethnologie de l'Université de Paris," typescript, 1925, 6 pp.

lar community of support. Rivet had behind him the resources and prestige of a premier scientific institution, the Muséum national d'histoire naturelle. Classically trained Durkheimians, Mauss and Lévy-Bruhl helped open the doors of the University of Paris to the discipline of ethnology and helped ensure that in its new guise the science of man would also be sociological in orientation. All three men remained involved in socialist politics in the interwar years.[49] They also associated certain "colonial" scholars in their initiative: in particular Maurice Delafosse, who had left West Africa after the war and who was now teaching at the École nationale des langues orientales vivantes, and Louis Finot, an orientalist at the Collège de France, who in 1926 headed up the prestigious École française d'extrême Orient in Hanoi. These two men in turn could tap into the overseas networks of ethnographers studying the peoples of the empire.[50] With the help of these collaborators and other *camarades*, Rivet, Lévy-Bruhl, and Mauss would "relativize," professionalize, and popularize (but never monopolize) anthropological discourse and practice in France in the interwar years—and in the process give birth to a new scientific "Temple of Man."[51]

Perhaps the most important step in this campaign was the founding of the Institut d'ethnologie in 1925–1926 at the University of Paris. The new *institut* had no faculty of its own and was modestly subsidized by the Ministry of Colonies. Édouard Daladier, minister of colonies under the Cartel des Gauches (1924–1926), was a personal friend of Lévy-Bruhl's; without the financing provided by Daladier's ministry, the *institut* would not have gone forward. Although located within the Sorbonne, the Institut d'ethnologie was immediately linked to the Muséum national d'histoire naturelle, and the holder of the anthropology chair (Paul Rivet) automatically became one of the two secretaries general of the *institut*. Mauss was made the second secretary general, while Lévy-Bruhl became president. The Institut d'ethnologie had a permanent advisory council of twenty-eight members, half of whom represented other Paris faculties and institutions of higher learning; the École d'anthropologie was conspicuously absent from this list. Representatives from either the Ministry of Colonies or the colonies themselves composed the other half. In addition to publishing a new series of monographs in anthropology, the Institut d'ethnologie would sponsor missions overseas, as well as offer courses that, for the first time, were formally integrated into the degree cycle of the university. Mandatory core courses included descriptive ethnography by Mauss, physical anthropology by Rivet, linguistics by Marcel Cohen, and the ethnography of Africa and Asia. Students would

[49] On the founding of the Institut d'ethnologie, see in addition to Karady, "French ethnology" (cit. n. 47); Lucien Lévy-Bruhl, "L'Institut d'ethnologie de l'Université de Paris," *Revue d'ethnographie et des traditions populaires* 23–24 (1925): 233–6; and Clark, *Prophets* (cit. n. 8), pp. 188, 200, 214, 227, 232–3.

[50] On the links between Delafosse, postwar ethnography in France, and the Institut d'ethnologie, see Sibeud, "La construction" (cit. n. 23), pp. 634–42. Louis Finot had taught Mauss Sanskrit at the Ecole pratique des hautes études, and they had stayed close personally and professionally. On Finot's career see Fournier, *Marcel Mauss* (cit. n. 42), p. 566; Pierre Singaravélou, *L'école française d'Extrême-Orient, ou l'Institution des marges (1898–1956): Essai d'histoire sociale et politique de la science coloniale* (Paris: Harmattan, 1999).

[51] The term "Temple of Man" comes from Rivet, "Le Laboratoire d'anthropologie" (cit. n. 39), p. 529. ["temple de l'Homme."] Neither the Société d'ethnographie nor the École d'anthropologie chose to link up with the Musée de l'homme or its new methods, although relations with at least one of their chief spokesmen, the politically conservative Louis Marin, remained cordial. For one interpretation of the ideas of Louis Marin, see Herman Lebovics *True France: The Wars over Cultural Identity, 1900–1945* (Ithaca, N.Y.: Cornell Univ. Press, 1992).

supplement these with related courses of their choice already being taught in other Paris institutions of higher learning; these courses ran the gamut from zoological and biological anthropology to linguistics, from comparative physiology of the human races to descriptive ethnography, from archaeology to folklore.[52] Such coursework would lead to either a *diplôme d'études supérieures* in ethnology and/or a certificate in ethnology to apply toward a *licence* in sciences or letters (each of these *licences* required four such certificates in different subjects for a student to graduate). More ambitious students could use their training to prepare for fieldwork, and then go on and write doctoral theses; with no chair in anthropology yet established in any French university, such a choice was necessarily risky but also alluring. Thanks to the contacts of Rivet, Mauss, and Lévy-Bruhl, funds for fellowships to support such fieldwork became available to their best students on a regular basis. Students from the École coloniale, who were encouraged to opt for the certificate in ethnology, could likewise advance their careers in the colonial administration if they passed the certificate requirements.[53] The colonial subsidy was justified in another way as well: the *institut* was to be available to colonial governments at all times for the study of races and social facts (*les faits sociaux*) and to give advice on the preservation of collections and monuments.[54] In all these ways the Institut d'ethnologie created the conditions for the emergence of a critical cluster of young scholars and fieldworkers who viewed themselves as the only legitimate practictioners of a new and properly scientific anthropology in France.[55]

The creation and particular configuration of the Institut d'ethnologie immediately made clear Rivet, Mauss, and Lévy-Bruhl's intellectual and professional objectives. Only by successfully introducing ethnology into France's existing degree cycle could they be sure of reproducing anthropology as they defined it. The variety of courses they chose to list and the institutions they associated themselves with indicated, moreover, that their definition of the field was a capacious one—but with a culturalist bias. Rivet insisted that modern French ethnology studied man in all his characteristics and not just his physiological traits. Such a definition, he repeatedly maintained, was not innovative. It simply returned French anthropology to the roots from which, Rivet clearly implied, a certain school in particular had erroneously diverged. The first three holders of the Muséum's chair—Quatrefages, Hamy, and Verneau—and the first two holders of the chaire d'anthropologie at the École d'anthropologie—Broca and Léonce Manouvrier—had all always accepted that the study of man must be, in Rivet's words, "synthetic." "Their constant preoccupation was to

[52] Archives du Rectorat de Paris, Institut d'ethnologie, Carton 26, "Annexe du décret du 1er aôut 1925, Règlement"; idem, Carton 25, Séance du Conseil d'administration, 25 Nov. 1925 and 27 May 1927; CFAMM, "Institut d'ethnologie de l'Université de Paris" (cit. n. 48), p. 5.

[53] Rivet, "L'ethnologie" (cit. n. 39), pp. 39–40.

[54] CFAMM, "Note sur Institut d'ethnologie de l'Université de Paris," 30 Dec., 1925, 3 pp., p. 2.

[55] Paul Rivet made clear that he viewed ethnology under his leadership at the Institut d'ethnologie as the only legitimate anthropological science in France, when he wrote the minister to stop subsidizing the École d'anthropologie, a school that was now "finished." ["Suppression de la subvention à l'École d'anthropologie, école dont le rôle est fini depuis la création de l'Institut d'ethnologie de l'Université de Paris".] AMH, 2 AM 1 A 5, Paul Rivet à M. de Monzie, Ministre de l'Éducation Nationale, Jan. 21, 1933, no. 82. He was equally dismissive of the venerable old Société d'ethnographie de Paris, "a moribund and lusterless society, in which ethnology is still treated the old-fashioned way." ["une société sans rayonnement et sans activité où l'ethnologie est encore traitée à la vieille mode."] AMH, 2 AM 1 A 12, Paul Rivet à M. Deschamps, Gouverneur de la Côte française des Somalis, April 18, 1939, no. 657.

study human groups in their multiple aspects, that is to say to attempt to simultaneously define their physical and cultural characteristics, always using both the idea of race and the idea of civilization in their research."[56] Any anthropologist worth his or her salt, Rivet concluded, must always study all three aspects of a given human population: physical features, language, and cultures. But Rivet also went on to assert that, of the three types of information gathered, physical traces were the least reliable clues to a people's past, and thus present linguistic and cultural patterns were apt to reveal more.[57] Where better, moreover, for French ethnologists to begin this scientific work than in their own empire, close to hand, and so in need of their labor?

Yet if the Institut d'ethnologie was to constitute an essential teaching and publishing vector of the new ethnology, neither Mauss nor especially Rivet saw it as a standalone initiative. This was all the truer given that, materially speaking, the Institut d'ethnologie consisted of two rooms rented from the Institut de géographie, had no faculty of its own, and subsisted on a very modest budget that could be withdrawn at any moment. Mauss, as noted in section 2, had long identified the need for a serious ethnographic museum as part of any proper renovation of the science of man in France. Rivet shared this sentiment, and as heir apparent to the Muséum chair in anthropology was institutionally placed to take over the existing Musée d'ethnographie. When he did so in 1928, *ethnologie* in France now had a second institution from which to organize itself. Despite the *musée*'s delapidated state, Rivet surely sensed from the outset the possibilities it offered himself and like-minded reformers for the *rayonnement* of the ethnology they were trying to launch. Here was a state-funded institution, with its own library and its own building in the heart of Paris—a building sufficiently distant from its "mother institution," the Muséum national d'histoire naturelle, to give Rivet some freedom if not, alas, any real administrative autonomy. The collections, once expanded and properly classified, could give new visibility to the diverse cultures that the Institut d'ethnologie proposed to study. Serious ethnographers could be trained in contact with these artifacts and also be given positions at the *institut* as they prepared for their fieldwork, and then again when they returned to write up their results. Perhaps most important for Rivet, the Musée d'ethnographie also offered exciting possibilities for initiating the public into the new science, for bringing ethnology to the peoples of France and its colonies. While Mauss now shouldered the main burden of teaching at the Institut d'ethnologie and carried on his work of reviving *L'année sociologique*, Paul Rivet from 1928 on threw himself tirelessly into museum overhaul, very much along the lines sketched out above.

IV. A MUSEUM REBORN

It is difficult to know how much of Rivet's ten-year reform of the Musée d'ethnographie was thought out ahead of time. His initial plan appears to have been limited to rebuilding and reorganizing the existing museum's ethnographic collections and library, and to creating new research facilities on the premises. A commitment to

[56] Ibid., p. 507. [". . . synthétique. . . leur constante préoccupation à été d'étudier les groupements humains sous leurs multiples aspects, c'est à dire de chercher à définir simultanément leurs caractères physiques et leurs caractères culturels, associant toujours l'idée de race et l'idée de civilisation dans leurs recherches."]

[57] Rivet, "Le Laboratoire d'anthropologie" (cit. n. 39), p. 513.

reaching out to several publics at once was nevertheless also present from the outset, as Rivet's bold choice of the young Georges-Henri Rivière as his new assistant director of the Musée d'ethnographie demonstrated. Rivière was not an ethnologist, but a gifted graduate of the École du Louvre, a well-known and well-connected personality in Paris's glamorous and avant-garde art circles. Rivet apparently met Rivière when the latter helped to organize the first important exhibit of pre-Columbian art in Paris, in May of 1928—not at the Musée d'ethnographie, but at the Musée des arts décoratifs.[58] The exhibit was very successful with the Paris *mécénat* and suggested that private resources might now be effectively tapped for the public collection of ethnographic objects. In asking Rivière to copilot a renovation of the Musée d'ethnographie, Rivet secured the support of such art patrons as David-Weill and the vicomte de Noailles, without whose generosity none of his reforms could have been carried out. It was also because of Rivière that the writer Michel Leiris and the musicologist André Schaeffner joined the museum staff. The Paris *beau monde* subsidized the Musée d'ethnographie and donated their time and the surrealists patronized its exhibits as the museum became one of the favorite meeting places of the aristocratic, intellectual, and artistic elite as well as—for a while—the Parisian educated public.[59] In short, Rivière brought superb administrative, design, and "people" skills to the *Musée d'ethnographie*, to which he would devote his every waking hour until his resignation in 1937. It was nevertheless Rivet's socialist and scientific ethos that prevailed when it came to determining what would be displayed and for whom, and how ethnological research was to be carried out. The Rivet-Rivière combination proved to be a felicitous and somewhat surprising marriage of opposites, one that left a very particular stamp on the future evolution of French ethnography.

From their arrival in 1928, Rivet and Rivière made it clear that they intended to make of the Musée d'ethnographie both a research center and an institution of popular education. In keeping with the first goal, they immediately set about transforming the existing library into a more modern and comprehensive one modeled along the lines of those found in American universities. For the first time, France would have a research facility specializing in all aspects of anthropology and in both French and international publications. The library was to have a more specific function as well within the museum proper: it would serve as an "indispensable complement to the collections."[60] Regardless of trends internationally, Rivet continued to believe that theoretical initiation into the discipline of ethnology could only "be acquired on the basis of contact with collections."[61] The current collections, he felt, left a great deal to be desired. Much of what Rivet inherited was beyond repair, poorly displayed, and lacking adequate information concerning provenance. Never endowed with an acquisitions or conservation budget, the Musée d'ethnographie had depended since its opening on the largesse of individual donors for most of its objects. Collections

[58] On this influential exhibit, see Elizabeth Williams, "Art and Artifact at the Trocadéro," in *Objects* (cit. n. 9), pp. 145–66.

[59] Rivière still awaits his biographer. A good introduction to his early career can be found in Jean-François Leroux-Dhuys, "Georges-Henri Rivière, un homme dans le siècle," in *La Muséologie selon Georges-Henri Rivière: Cours de muséologie, textes et témoinages* (Paris: Dunod 1989), pp. 11–32.

[60] ["Indispensable complément des collections. . ."] AMH, 2 AM 1 A 2, Paul Rivet and Georges-Henri Rivière à Haut Commissaire, High Commissioner for India, 8 Sept. 1931, no. 1572.

[61] Rivet, "Le Laboratoire d'Anthropologie" (cit. n. 39), pp. 528–9. [". . . l'enseignement de cette science ne pouvant se concevoir d'une fason théorique et requérant une initiation technique qui ne peut s'acquérir qu'au contacte des collections."]

bequeathed in this fashion necessarily reflected the eclectic tastes of the individuals amassing them, not a carefully considered plan for how to represent the cultures of the world completely and methodically.

An equally important goal for Rivet in the late 1920s, then, was to expand and reorganize the Musée d'ethnographie's collections into an appropriate teaching tool for Institut d'ethnologie students. Yet he and Rivière also always insisted that their objects speak as much to the nonspecialist as to the budding scholar. Meeting both objectives required proceeding on several different fronts simultaneously. In their first years Rivet and Rivière embarked on a tour of all the major ethnographic museums of Europe and America, including those in the Soviet Union, to collect information on the most modern museographic trends. They also began encouraging travelers and residents abroad to collect for the museum, to try and fill its many gaps. At the same time, in order to control the quality of what was shipped, Rivet commissioned and distributed freely a thirty-page manual on what to collect and how to do so scientifically, complete with a sample *fiche* for cataloging. The collector would give the object's exact site and date of "harvest," name (or names) in the local languages, and function. Only pieces properly documented in this way would figure in Rivet's revamped edifice. These pieces, moreover, should not be exceptional or beautiful ones: what was needed were the most typical and practical objects of a given culture—its tools, weapons, utensils, and clothes—as well as its pottery, musical instruments, and games so that the daily lives of ordinary peoples might be adequately represented.[62] This emphasis on typicality and everyday life was one of several areas in which Rivet's voice appeared decisive. Such objects were of course cheap, since not in demand on the new and booming "primitive art" market. Since they were not rare, their status as scientific objects could more easily be asserted. Finally, the choice of the everyday reflected the other public, beyond *institut* students and researchers worldwide, Rivet hoped to attract into his museum: the ordinary French *citoyen* who presumably might recognize himself or herself in the techniques and products of daily life the world over.

Collection by amateurs was all well and good, but what was really needed were funds to send trained ethnologists overseas to carry out some of this work themselves. It would, moreover, serve no use to acquire additional objects if they could not be properly stored, studied, and displayed—which meant rethinking not just the library, but the museum's other facilities and functions. Thus between 1928 and 1931 Rivet and Rivière also worked tirelessly to increase the Trocadéro's public and private funding, without which no further reforms were possible. In the early 1930s they scored two important successes. In 1930 Parliament allocated Fr 750,000 for a two-year mission to Africa by a group of young ethnologists associated with the Musée d'ethnographie—the renowned Dakar-Djbouti mission—to collect as many ordinary artifacts as possible. This collection would constitute one of the cores of the renovated museum and provide a potent symbol of its renewal. Then, in 1931, parliament further authorized a significant rebuilding of the museum, including the acquisition of the best display cases available, at a cost of Fr 5,800,000 over five years. According to Rivet, before proceeding with "spectacular displays [*manifestations*] essential for drawing crowds" he and his staff had "to create from scratch the

[62] AMH, 2 AM 1 B 10, Michel Leiris, "Instructions sommaires pour les collecteurs d'objets ethnographiques," May 1931.

organization and interior modifications that allow for ordered work and scientific classification."[63] These supplementary funds helped to endow the old Trocadéro building with new heating, lighting, offices, a laboratory for disinfection and preservation of objects, storage space, and workspace *(salle de travail).* Meanwhile, volunteers—students training at the Institut d'ethnologie and society ladies—as well as the four professional staff members worked in the new spaces to catalog and repair what they could of the museum's existing 150,000 objects, along with the new ones that kept arriving. By 1935 Rivet and Rivière could proudly point to a collection of "cleaned and copiously labeled" objects, the most representative of which had now found their way into the new airtight cabinets that revealed the objects' full "spectacular and pedagogical value."[64] This same team of mostly young volunteers also oversaw the renovation of a series of new galleries, organized by continent; finally, they contributed endless hours during these years to organizing a succession of temporary exhibits designed to awaken and maintain the public's interest in the *musée* as it proceeded with its internal reorganization. This "collective labor"—critical to getting so much done so quickly—was perhaps, Rivet concluded, the most moving aspect of the regeneration of the Trocadéro in these years.[65]

By 1935 the Institut d'ethnologie, the refurbished Trocadéro museum, and the anthropological laboratory and chair at the Muséum national d'histoire naturelle—all now formally linked—appeared to constitute a coherent institutional nexus for ethnology to emerge in Paris. In 1936, in keeping with these trends, Rivet secured approval from the Muséum faculty for renaming his chair of anthropology a chair of ethnology of living and fossil man (*Ethnologie des hommes vivants et des hommes fossiles*). Yet he was still not satisfied; for true integration of the discipline and its maximum visibility, he sought further centralization under one roof. What was needed was a single site to fuse the libraries, collections of artifacts, work spaces, and seminar rooms of these three member institutions. In addition, Rivet dared to hope that he might convince the various private associations devoted to anthropology to house their libraries and hold their meetings at the Musée d'ethnographie. Because no single scholar could keep abreast of developments in all the fields related to the study of man in his totality, ethnologists would have to continue to specialize geographically by continent—hardly an ideal category, Rivet fully admitted. Africanists, Asianists, and Americanists would thus continue to have separate learned societies, just as the collections themselves would remain grouped according to continent. Nevertheless representatives from all fields should pool their knowledge and synthesize it for the public through collaborative work.[66] This sharing and production of knowledge would occur so much more easily if ethnologists could congregate in a single spot instead of constantly being obliged to cross the city for meetings or to duplicate their resources unnecessarily. Here we discover yet another democratic facet of Rivet's scientific project, hinted at above. Rivet was a staunch believer in

[63] Rivet, "Le Laboratoire d'anthropologie" (cit. n. 39), p. 517. ["Il convenait, en effet, avant de penser aux manifestations spectaculaires indispensables pour attirer la foule des visiteurs, de créer de toutes pièces les organisations et les modifications intérieures qui permettaient un travail ordonné et une oeuvre scientifique de classement."]

[64] Ibid., p. 519. ["nettoyés, remis en état, copieusement étiquetés et commentés, prenaient toute leur valeur éducative."]

[65] Ibid., p. 522. [". . . travail collectif. . ."]

[66] Ibid., p. 516; Rivet, "L'ethnologie" (cit. n. 39), pp. 45–6.

"le travail en équipes, la coopération intellectuelle,"[67] whether in the form of teams of students going on mission or working to refurbish the museum, collective works by scholars in related branches of the same discipline, or joint meetings by different kinds of specialists.

From Rivet's perspective, then, a single structure was essential for the kind of synthetic and cooperative knowledge he wished to produce. The opportunity to create one would come sooner than he thought and from an unexpected quarter. When plans for the Universal Exposition of 1937 got under way in 1935, officials decided to raze the old Trocadéro Palace and erect a bigger and more modern building on the spot, part of which would be used to rehouse the Musée d'ethnographie. With triple the space of the old Musée d'ethnographie, it would be possible to not only further expand the display halls, collections, and storage space, but also accommodate the many libraries, teaching facilities, and scientific collections that Rivet wished to bring together in one place—assuming that the societies concerned would agree to relocate.[68] This centralization of research facilities was partially achieved in the new building. Significant other facilities were added: a *photothèque*, a *phonothèque*, and a *salle de cinéma*, all equipped with the latest technologies available. Here even newer kinds of ethnographic objects could be stored or displayed: photographs, sound recordings, and movies.

Perhaps the most noticeable difference between the 1935–1938 renovation and the just completed refurbishing in the old Trocadéro Palace was the increased importance of the museum's pedagogical and popularizing vocation.[69] The very name given to the new installation—Musée de l'homme—was chosen to make the institution more palatable to the man in the street and to draw him spontaneously onto the premises. In one sense, Rivet and Rivière were merely introducing to France a development that museum curators and their governing boards had already embraced in other countries. But the political context of the mid- to late 1930s also influenced Rivet's decision in particular to proceed in this manner. As a socialist municipal councilor, Rivet supported the cultural policies of the new Popular Front, which swept into power in 1936. As an ethnologist, he also felt impelled to inoculate the public against the rising tide of fascist raciology unleashed by the election of the Popular Front by countering it with truthful science. The appointment in 1937 of one of the new graduates of the Institut d'ethnologie, Jacques Soustelle, to succeed the apolitical Rivière, further politicized the institution in these same directions. Soustelle was, if possible, an even more ardent believer than Rivet in the need for a democratization of all museums. In a revealing 1936 editorial titled "Living Museums," he argued that museums were the perfect instrument of popular education because of their "mode of direct and concrete presentation." People did not come "to be initiated into the multiple aspects of human nature through their immediate viewing of objects. . . ." But it was not enough merely to entice the masses into the

[67] Rivet, "Le Laboratoire d'anthropologie" (cit. n. 39), p. 516.

[68] Rivet, "L'ethnologie" (cit. n. 39), pp. 44–5. Rivière and Rivet managed to acquire for the Musée de l'homme the libraries of the Laboratoire d'anthropologie, Institut français d'anthropologie, Société préhistorique française, Société des Américanistes, and the Société des Africanistes; AMH, 2 AM 1 B 10/Coupures de presse/communiqués/Musée de l'homme, "Le Musée de l'homme," June 1937. The Société d'Anthropologie did not, however, agree to move its osteological collections; AMH, 2 AM 1 B 2/Coupures de presse, Henri Bouquet, *Le Monde Médical*, July 1, 1938.

[69] Paul Rivet, "Ce que sera le Musée de l'homme," *L'Oeuvre*, June 14, 1936.

museum as spectators: they had to become collaborators, acting on the museum in their turn. If many more museums were built, if they were open after work hours and displayed local cultures and art, people would come to know themselves and others better, and come to control the museum's content thanks to their presence and suggestions. Last but not least, once in the museum, the people would discover science there, and the best and the brightest—in true democratic fashion—would be recruited into its ranks.[70]

Vague, as many Popular Front proposals were, about how this elite-masses collaboration would actually work in practice, Soustelle's socialist dream of a living, popular museum nevertheless helped shape the Musée de l'homme that opened two years later.[71] From the outset, the museum was open in the evening. Its displays were spectacular but responsibly so, in the opinion of Soustelle. According to him, a tour of the museum (and there was a single, proper way to visit the museum) began with the hall devoted to the races of man. In keeping with Rivet's wish to consolidate osteological and ethnographic displays in one place, the collection of bones, skulls, and full skeletons formerly housed at the Muséum national d'histoire naturelle had been moved to the Musée de l'homme and formed the core of this gallery. The physical evidence underpinning classification of the principal human types had been chosen to reflect the most recent research. This *mise au point* was particularly necessary, given that "the notion of race is used . . . in the most confusing fashion."[72] There was in particular no such thing as a pure race: all races were the product of *metissage.* In addition, "deformed" skulls once considered "barbaric" lost their strangeness when juxtaposed with comparable ones from the French provinces.[73] The master narrative established from the very first gallery, then, was that of the unity of man.

The remaining displays were devoted to man-made objects, organized geographically or comparatively. As before, each continent had its own gallery and research

[70] Soustelle, "Musées Vivants" (cit. n. 1). [". . . par le mode de présentation direct et concret. . . pour s'initier aux multiples aspects de la culture humaine par la vue immédiate des objets. . . ."] .

[71] On the Popular Front cultural policy, see Pascal Ory, *La belle illusion: Culture et politique sous le Front populaire* (Paris: Plon, 1994). For background on the Musée de l'homme, see, in addition to Clifford, *Predicament of Culture* (cit. n. 27), the following articles by Jean Jamin—"L'ethnographie mode d'emploi: De quelques rapports de l'ethnologie avec le malaise dans la civilisation," in *Le mal et la douleur*, ed. Jacques Hainard and Roland Kaehr (Neuchâtel, Switz.: Musée d'ethnographie, 1986), pp. 45–79; "Le Musée d'ethnographie en 1930: L'Ethnologie comme science et comme politique," in *La Muséologie* (cit. n. 59), pp. 110–1; "Les Objets ethnographiques sont-ils des choses perdues?" in *Temps perdu, temps retrouvé*, ed. Jacques Hainard and Roland Kaehr (Neuchâtel, Switz.: Musée d'ethnographie, 1985), 51–74. Also useful are Fabrice Grognet, "D'un Trocadéro à l'autre, Histoire de métamorphoses," DEA de muséologie, Muséum nationale d'histoire naturelle, 1998; James Herbert, "Gods in the Machine at the Palais de Chaillot," *Museum Anthropology* 18(2) (1994): 16–36; Dubuc, "Le futur antérieur" (cit. n. 26), 71–92; and Nathalie Duparc, "Musée de l'homme, Musée national des Arts Africains et Océaniens, Musée Municipal d'Angoulême: Trois partis pris muséologiques différents" (Mémoire de maîtrise, Univ. of Paris, 1986).

[72] Jacques Soustelle, "Le Musée de l'homme," *Renaissance* 21 (1938): 17–21, on p. 19. ["Cette galérie représente une mise au point de nos connaissances particulièrement nécessaires aujourd'hui, alors que la notion de race est employée à tort et à travers de la façon la plus confuse."]

[73] M. V. Fleury, "La Science des races au Musée de l'homme," *Races et Racisme* 16-17-18 (1939): 1–5, on p. 3. This review appeared in the final newsletter of the Parisian group Races et Racisme, which had formed in 1936 to fight against German rearmament and the idea of Nordic superiority. Rivet, Lévy-Bruhl and Rivière were all founding members. The final issue (Dec. 1939) was dedicated exclusively to the Musée de l'homme, whose "true" science of mankind, they argued, represented the best possible response to the pseudo-scientific racism of the Nazis. For a preliminary analysis of the newsletter, see Meyran, "Races et racisme" (cit. n. 22).

department. But now each gallery was divided into two viewing circuits. Synthetic displays lined one side, summarizing the cultural traits of major ethnic groups, for the visitor in a rush, as well as for the schoolchildren now invited to come. On the other side was a series of cases offering in-depth contextualization for the more informed "amateur" or "initiated" visitor. In these instances a single custom of a particular group might be displayed through a specific object—a set of masks, for example, from a Dogon secret society. A bibliography was included, and all visitors were encouraged to go to the fourth floor and consult the appropriate works in the library. In addition, an entire gallery for arts and techniques, devoted to comparable objects from different culture areas, was being prepared.[74] From a museographic perspective the diorama was spurned as outdated (and too expensive) and was replaced by maps, explanatory panels, and photographs.[75] The entryway of each gallery featured an enormous map of the appropriate continent, with the names of ethnic groups filled in where known (colonial boundaries were also indicated). Panels everywhere contextualized in clear and simple language the functions of objects on display and gave information on the ethnic groups from which they came. Photographs showed how an individual item was produced and/or used. Rather than cramming each case full, the curators chose a select few objects from their massive reserves to highlight. In theory, objects would change regularly, to keep the collections "alive."[76]

Access to knowledge was not merely to be democratized: within the museum proper, the staff was to work democratically, in keeping with Rivet's long-held ideal of cooperation. The anthropologists, including Rivet himself, would take turns conducting groups around the museum, explaining the collections' organization. They also would divide administrative, research, and curatorial tasks equally.[77] In these organizational years, weekly meetings of the museum department heads (mostly Mauss and Rivet's students) were held not only to ensure uniformity of presentation, but also to make decisions collectively about how best to disseminate the science of man to those encountering it for the first time.[78] In his review explaining these guiding concepts, Soustelle anticipated charges of "vulgarization." The word, he insisted, was fundamentally wrong. The aim of the museum was to "diffuse":

> We hope that the individual from here [*l'homme d'ici*] . . . leaves the museum knowing and understanding better humankind elsewhere, whether the black from Africa or the Mexican Indian. We show that there were and that there exist different civilizations, more or less perfect in some respect or other, but all equally capable of practical and aesthetic invention; the museum that we wished to build is nothing more than a tableau of the collective efforts of humanity, under all climates, on all continents, an effort which has everywhere produced works worthy of respect.[79]

[74] Soustelle, "Le Musée de l'homme" (cit. n. 72), pp. 19–20

[75] AMH, 2 AM 1 A 8, Georges-Henri Rivière à M. Louis Taverne, Oasis, 23 Oct. 1935, no. 2080.

[76] Soustelle, "Le Musée de l'homme" (cit. n. 72), p. 18. See also the review by Jean Millot, "Le Musée de l'Homme," *Revue de Paris* 45 (Aug. 1938): 687–94.

[77] AMH, 2 AM 1 B 10 d/Bulletin mensuel d'informations/MH, 1(4) (July-Aug. 1939), "Le Laboratoire d'ethnologie: son organisation."

[78] Minutes from these weekly meetings, which began in 1931, can be found in AMH, 2 AM 1 D 14, "rapports hebdomadaires." The 1937 "rapports hebdomadaires" contain discussions of how the maps should be drawn, the size and color of the print to be used on the notices surrounding the objects, and how individual cases should be arranged for maximum clarity.

[79] Soustelle, "Le Musée de l'homme" (cit. n. 72), p. 21. [". . . nous ne vulgarisons pas . . . nous diffusons. Nous voulons que l'homme d'ici, le Parisien, le Français, sorte de ce Musée connaissant

The Musée de l'homme team, then, took seriously their mission to diffuse science to the masses. They were equally convinced of the novelty of their "synthetic" and "cooperative" approach. No other ethnographic museum in the world had sought to bring under one roof all facets of human history, along with the research and teaching facilities central to the discipline. Their museum, in the words of a staff member in 1939, was indeed a scientific organism, two-thirds of which remained invisible to the public—library, laboratory, vast card catalog, and extensive photography collection. But the museum was designed to immediately translate and communicate the most up-to-date research going on "upstairs" (the Institut d'ethnologie and some department *salles de travail* were located on the third floor, the library on the fourth) and "downstairs" (the basement housed the remaining departments) to the greater public welcomed in the galleries. Yes, there were labels and panels, but they were designed not so much "to explain" as "to make comprehensible." Conclusions were not asserted but proven. This was a "*musée raisonné*," which believed that a public not only could be educated, but wanted to be educated.[80] As late as December 1939, thus after France had declared war on Germany, the museum staff made the decision to stay open in "a symbolic gesture of defiance of the obscure forces menacing civilization." They had just opened a new exhibit on their most recent African missions. Confronted with the racist wave that was threatening to engulf the world, the museum was an "arsenal where any intelligent visitor will find impeccable arms . . . with which to answer—with the complete serenity provided by the revelations of an impassive science—the hateful attacks of the enemies of humanity."[81]

These remain stirring words, in many ways made more compelling by the well-known fact that the first resistance group operating in Paris began on the premises of the Musée de l'homme.[82] What greater proof, one might ask, of a democratically organized science, one that was institutionally and intellectually left-leaning from the moment Mauss, Lévy-Bruhl, and Rivet set out to consolidate but also recast the study of man in France, socialist in its message by the time of the Popular Front? Under the Popular Front, it is tempting to conclude, a ninety-year history of association between the republican state and various anthropologists transformed briefly into something more intimate, as like-minded politicians in power and a specific group of scientists came together around a particularly generous vision of the study of man.

Such a conclusion would not be wrong, but to emphasize only the "democratic" face of the Musée de l'homme–state nexus here would be misleading. The French

---

et comprenant mieux l'homme d'ailleurs, que ce soit le noir d'Afrique ou l'Indien mexicain. Nous montrons qu'il y a eu et qu'il existe des civilisations différentes, plus ou moins perfectionnées sur tel ou tel point, mais également capables d'invention pratique ou esthétique. Le Musée que nous avons voulu réaliser n'est pas autre chose qu'un tableau de l'effort collectif de l'humanité, sous tous les climats, sur tous les continents, effort qui s'est partout traduit par des oeuvres dignes de respect."]

[80] Fleury, "La Science" (cit. n. 73), p. 5. [". . . On ne tente pas d'expliquer . . . mais de faire comprendre."]

[81] Ibid., p. 5. [". . . c'est un défi symbolique aux forces obscures qui menacent la civilisation . . . le Musée de l'homme est un arsenal qui porte l'empreinte où tout visiteur réflechi trouvera des armes impeccables, des arguments impartiaux, avec la calme sérenité que confèrent les révélations d'une science impassible . . . aux attaques haineuses des ennemis de l'humanité."]

[82] Martin Blumenson, *The Vildé Affair: The Beginning of the French Resistance* (Boston: Viking, 1977); J. Blanc, "Le réseau du Musée de l'homme," *Esprit* (Feb. 1999): 89–105; and D. Fabre, "L'Ethnologie française à la croisée des engagements, 1940–1945," in *Résistants et Résistance*, ed. Jean-Yves Boursier (Paris: l'Harmattan, 1997), pp. 319–400.

state in the interwar years, and the Popular Front years were no exception, was as colonial as it was socialist—although significant reforms were attempted overseas by the socialist government.[83] A civic institution dependent on the state during this period thus ran a good chance of also being implicated in its imperial forms of governance. In the case of France's emerging community of ethnologists, the turn to empire in the 1920s and 1930s was as marked as their democratic sentiments, although less publicly proclaimed. The renovated Musée d'ethnographie, Rivet insisted in his 1931 plea to the government for more funds, would continue the educational work of the Colonial Exposition closing that same year.[84] On the eve of the opening of the Musée de l'homme, Rivet could still write Édouard Daladier, then minister of war, that his new museum was a "colonial museum" and request the presence of colonial troops at its inauguration.[85] No term better captures the embeddedness of empire even on the margins of the republican academy in the 1930s.

V. THE IMPERIUM OF SCIENCE AND THE SCIENCE OF EMPIRE

In late 1924 the eminent colonial administrator-ethnographer (and recently turned publicist) Maurice Delafosse wrote two articles for the leading newspaper of French colonial interests, *La Dépêche Coloniale*; they were titled, "When Will France Have a Colonial Museum?" and "Colonial Museums: What France Has and What It Should Have."[86] The first was written after a recent visit to Belgium's Congo Museum at Tervueren, whose imposing neoclassical building and rich, well-organized and brightly lit displays devoted to the Congo's "mineral and vegetal resources" and "flora and fauna," and "the anthropology of its peoples, and their material, social and religious civilizations" deeply impressed Delafosse. In his follow-up article, Delafosse went through the dispersed and impoverished holdings of France's different "colonial" collections, in Paris and the provinces, and argued for a single national museum that would centralize these holdings for all French to see, as well as coordinate and sponsor much needed scientific research on the colonies. When, he asked, would the French Parliament prove as generous as that of Belgium and as enlightened when it came to the administration of its empire?

Although Delafosse played an important role in renewing anthropology in France, he would not live to see his dream museum realized. Certainly, museum overhaul was on the agenda of scholars interested in Africa and Asia by the time of his death in the mid-1920s; but because social scientists rather than government or colonial

[83] Tony Chafer and Amanda Sackur, *French Colonial Empire and the Popular Front: Hope and Disillusion* (New York: St. Martin's Press, 1999); Fred Cooper, *Decolonization and African Society: The Labor Question in French and British Africa* (Cambridge: Cambridge Univ. Press, 1996).

[84] AMH, 2 AM 1 A 2, Paul Rivet à Maurice Foulon, Sous-secrétaire d'état au travail, 24 Oct. 1931, no. 1905. "At the moment when the Colonial Exposition is about to close its doors, it is indispensable that a decent permanent organization continue its educational work, all the more so since the bulk of the [ethnographic] collections which were assembled for it are going to be given to us." ["Au moment où l'Exposition coloniale va fermer ses portes, il est indispensable qu'un organisme permanent décent en continue l'oeuvre éducative, d'autant plus que la plupart des collections qui y ont été réunies vont nous être transmises."] The same argument is made in AMH, 2 AM 1 A 2, Paul Rivet à M. le Sénateur [sent to all senators], 5 Dec. 1931, no. 2210 bis.

[85] AMH, 2 AM 1 A 11, Paul Rivet à Edouard Daladier, Président du Conseil, 5 May 1938, no. 825. ["Mais notre Musée est avant tout un musée colonial."]

[86] "Quand la France aura-t-elle un musée colonial?" *La Dépêche Coloniale*, 28 Nov. 1924; "Musées coloniaux: Ce que la France a et ce qu'il lui faut avoir," *La Dépêche Coloniale*, 10 Dec. 1924.

elites took the lead in this initiative, the new institution was always designated officially "enthnographic" or "ethnological" rather than "colonial." Indeed, a specifically designated colonial museum, le Musée des Colonies, opened in the Bois de Vincennes after the 1931 Colonial Exposition, with the mandate of educating the public about the history of French colonization; a citizen wishing to know more about what France was accomplishing overseas, or about the resources available and the peoples inhabiting the colonies, was to turn to this institution for guidance. In contrast, the aim of the renovated Musée d'ethnographie and its successor, the Musée de l'homme, was more global and strictly ethnological. There would be no catalog here of the "mineral and vegetal" resources of French sub-Saharan Africa or Indochina or Algeria, only traces of material cultures worked by man and his physical remains through the ages.

What then did Rivet mean, when he called his museum a colonial one? Although it is tempting to dismiss the reference as purely rhetorical, designed to flatter a minister at a moment when the empire seemed more than ever essential to the grandeur and economic recovery of France, there appears to have been no irony or subterfuge intended.[87] Rivet's and Rivière's correspondence during the critical years of reorganization (1930–1938) makes clear that while they were not interested in colonial development per se, they never doubted that the empire was part of France and should remain so, or that the museum should contribute to the colonial cause. The empire's resources, consequently, should be made available to the museum particularly in the absence of metropolitan ones. This was all the truer if the colonies could expedite the modernization of *les sciences anthropologiques et ethnologiques* and consolidate France's international standing in these fields—and in the process salvage the vestiges of a human history that was disappearing before their very eyes, as the inevitable price to be paid for the march of progress. The attitude of Rivet and his young team toward the colonies was neither romantic nor paternalistic, but instrumentalist and vaguely humanist.[88]

The instrumentalisation of the empire in the Musée de l'homme's bid to unify anthropology in France manifested itself in a number of different but overlapping ways. I have touched on some already. Critical, in any listing, was the colonial subsidy. It contributed one half of the Musée d'ethnographie's budget, or Fr 150,000, in the early 1930s, although by 1934 this subsidy had decreased to Fr 50,000, and in

[87] Jacques Marseille, *Empire colonial et capitalisme français: Histoire d'un divorce* (Paris: Albin Michel, 1984).

[88] Jean Copans (*Critiques et politiques de l'anthropologie* [Paris: F. Maspéro, 1974]) and Daniel Nordman and Jean-Pierre Raison (eds., *Sciences de l'homme et conquête coloniale: Constitution et usage des sciences humaines en Afrique, XIXe-XXe siècles* [Paris: Presses de l'École normale supérieure, 1980]) were among the first to explore the relationship between colonialism and anthropology in the French context. New research is focusing on French colonial administrators' growing interest in physical anthropology and ethnology as instruments of colonial rule. See, e.g., Patricia Lorcin, *Imperial Identities: Stereotyping, Prejudice, and Race in Colonial Algeria* (London: Taurus, 1995); Michael Osborne, *Nature, the Exotic, and the Science of French Colonialism* (Bloomington: Indiana Univ. Press, 1994); Alice L. Conklin, *A Mission to Civilize: The Republican Idea of Empire in France and West Africa* (Stanford, Calif.: Stanford Univ. Press, 1997), chap. 6; Benoît de l'Estoile, "Science de l'homme et 'domination rationnelle': Savoir ethnologique et politique indigène en Afrique Occidentale Française," *Rev. Syn.* 4(3–4) (2000): 291–323; Sibeud, "La construction" (cit. n. 23); Gary Wilder, "Greater France in the Interwar Years: Negritude, Colonial Humanism, and the Imperial Nation-State," Ph.D. diss., Univ. of Chicago, 1999.

1938, it was Fr 48,000, out of an anticipated total budget of Fr 740,880.[89] Here it is important to have the larger picture of the Musée de l'homme's finances in view. The history of the Musée de l'homme sketched in section 4 may have sounded triumphal: a marshaling of key humanist scholars, a breakthrough into the Sorbonne, critical public support, funding at key junctures, and finally the socialists' inauguration in the late 1930s of, in Rivet's words, "the most beautiful museum and center for ethnological studies in the world."[90] At no point after certain initial outlays in the early 1930s, however, did the state endow either the Musée d'ethnographie or its successor institution with more than a minimal operating budget, or titularize the scientific personnel the museum desperately wanted. Rivet himself and half the department heads still took no salary in 1938.[91] Rivière used his personal funds to rebuild the library, and even when the museum tripled in size, no new guards or scientific personnel were authorized.[92] Given these straitened circumstances, exacerbated by the Depression, even a diminished subsidy from the colonies was critical to the museum's continued credibility, while at the outset the colonial contribution virtually guaranteed that the museum could remain open and pay its bills.

But the colonial connection went even further. As Rivière put it in a letter to the Rockefeller Foundation, to which he turned in his endless search for renovation funds in the early years: "And can we not foresee an immense and imminent extension of our ethnographic collections, if, as we hope, we succeed in assembling in our colonies the [kind of] collections that the scientific world expects of us?"[93] The colonies, in other words, would yield the bulk of the new artifacts of the renovated museum because an imperial power now apparently owed to the world the ethnological *mise en valeur* of its subjects. Fortunately, what was in the best interests of science internationally was also in the best interests of the museum, whose staff members had no other choice except to focus the bulk of their collecting in France's overseas territories—if they wished quick results. Part of Rivet's bid to reorganize

[89] AMH, 2 AM 1 A 3, "Note justificative," 6 Jan. 1931, pour la subvention de Fr 150,000 des Colonies; "Plan de travail pour l'exercise 1931 et les prochaines années," Jan. 1932. In these early years, museum personnel salaries (Fr 49,000) were paid by the Société des Amis du Musée de l'homme, a nonprofit organization, because expenses routinely outran receipts. Between 1929 and 1935 they contributed over half a million francs to the museum. AMH, 2 AM 1 A 11, Jacques Soustelle à M. Cogniot, Rapporteur du Budget de l'Éducation Nationale, 23 Nov. 1938, no. 2044. According to this report, Fr 542,000 would come from the Ministry of National Education, and 150,000 from museum entry fees.

[90] AMH, 2 AM 1 A 12, Paul Rivet à Jean Zay, Ministre de l'Éducation Nationale, 23 Jan. 1939, no. 13 ["Je crois qu'actuellement la France a le plus beau musée et centre d'études ethnologiques du monde . . ."]

[91] Jacques Soustelle à M. Cogniot (cit. n. 89). Out of the total operating budget of Fr 740,880, only Fr 133,115 were allocated for personnel; the museum had fourteen departments or services, only five of them headed by *fonctionnaires*.

[92] In 1938 there were still only six guards, the same as in 1929. Ibid. In his inaugural speech for the new library, Paul Rivet referred to Rivière's contribution of Fr 100,000. AMH, 2 AM 1 A 2, Paul Rivet à Louis Mangin, Directeur du Muséum national d'histoire naturelle, 19 June 1931; see the attached *projet de discours* by Rivet for inauguration of the library. Rivière himself put the figure at Fr 130,000 a few months later and claimed he had exhausted his resources. AMH, 2 AM 1 A 2, Georges-Henri Rivière à MM. Gutzwiller and Bungener, 30 July 1931, no. 1231.

[93] AMH, 2 AM 1 A 1, George-Henri Rivière à M. Selskar M.Gunn, vice president, Rockefeller Foundation, 11 July 1930, no. 875. ["Et devons nous pas prévoir une immense et prochaine extension de nos collections ethnographiques si, comme nous l'espérons, nous parvenons à faire dans nos colonies des collections que le monde scientifique nous réclame?"]

and consolidate ethnology in France was to win acclaim for innovating museographically, particularly by scrupulously documenting the provenance of each artifact and cataloging the museum's entire collection. To do so successfully meant practically starting from scratch; yet with no acquisition budget, the museum had little choice but to continue to rely on donations from private individuals as well as organize their own missions. Either way, the colonies could clearly provide rich harvests because conditions for collecting by any French citizen were most bountiful where the French flag flew. The empire was full, after all, of Frenchwomen and Frenchmen on site either as administrators, missionaries, or travelers, who could be tapped to keep an eye out for the ordinary objects, untouched by European influence, that the *musée* desired. While professional training helped in identifying and acquiring such objects, it was not a prerequisite. A collector could follow the instructions published by the museum, and then defer to the experts by handing over the objects.[94] Once the museum had entered into a relationship with such a person, French transit companies, or local colonial governments, could be counted on to ship the haul to the *métropole* at a discount. When the museum staff itself went on mission, the same rules would apply: they would collect objects, noting their origin and purpose, and rely on the French administration for hospitality, security, transportation, and invaluable local knowledge.

That the Musée de l'homme proposed to operate in this "colonial" manner became clear from the moment it sponsored the famous Dakar-Djbouti mission, whose timing coincided with the first round of museum renovation. From 1931 to 1933 Marcel Griaule led a team of six other young associates from the Institut d'ethnologie across West Africa and into Ethiopia, collecting several thousand objects and thus founding the African section of the Musée d'ethnographie and future Musée de l'homme.[95] The mission was considered a model for all subsequent museum expeditions thanks to its rich and cheap "haul," although no subsequent one would have such extensive funding. As Rivet put it in his annual report that year:

> Buying ethnographic objects on the art market is a solution as expensive as it is unscientific. . . . How much more preferable, as long as there is still time, to proceed with a direct harvest on the ground, either by individuals on mission or by those on the spot who are familiar with our ethnographic methods. Only in this way will the provenances be reasonably sure and the pieces are documented in a manner which increases their worth ten times. This is why, unlike art museums, which have significant acquisition funds, ethnographic museums require above all sufficient resources to send scientific missions, of which the Mission Dakar-Djbouti is the most complete to date.[96]

[94] AMH, 2 AM 1 A 3, Paul Rivet et George-Henri Rivière à Albert Sarraut, Ministre des Colonies, 29 Nov. 1932, no. 2561. In this letter Rivet and Rivière thank the minister for his circular of 21 July 1932 addressed to all governors general and governors of the colonies, "requesting that they cooperate with the Musée d'ethnographie du Trocadéro, the Institut d'ethnologie, and the learned societies attached to them." [". . . les engageant à entrer en relations avec le Musée d'ethnographie du Trocadéro, l'Institut d'ethnologie et les sociétés savantes qui s'y rattachent."] Rivet added that a very fruitful collaboration now existed between colonial administrators and the museum.

[95] Paul Rivet et Georges-Henri Rivière, "Mission ethnographique et linguistique Dakar-Djbouti," *Minotaure* 2 (1933): 3–5; Jean Jamin, "La Mission d'ethnographie Dakar-Djibouti 1931–1933," *Cahiers ethnologiques* 5 (1984): 1–179.

[96] AMH, 2 AM 1 A 3, Paul Rivet, "L'Activité du Musée d'ethnographie au cours de l'exercise 1931," Dec. 1931 ["acheter sur le marché d'art les objets ethnographiques est une solution aussi coûteuse que peu scientifique. . . . Combien préférable, pendant qu'il en est temps encore, de faire procéder à une récolte directe sur le terrain, soit par des chargés de mission, soit par des sédentaires rompus aux disciplines ethnographiques. Ainsi seulement les provenances présentent des caractères

The mission spent most of its time, needless to say, in the French colonies. This not only ensured easy access to artifacts, but also gave the team time to cultivate relations with local officials, whom both Griaule and Rivet in Paris were careful to thank in writing.[97] In addition, Rivière publicized the mission intensively in art, scientific, and colonial circles. Thanking cooperative officials and writing to various colonial propaganda agencies in France were as much a part of the strategy of investing in the empire as the actual collecting of objects. As they embarked upon their plans for expansion of the museum, Rivet and Rivière had to confront the fact that they had very few contacts, especially in the African colonies. Delafosse had been their point man there, and his premature death, in 1926, meant that Rivet had to forge his own imperial networks more or less from scratch. The mission to Africa was one means of establishing such networks, and not surprisingly Griaule would lead several teams back to the same parts of France's African empire throughout the 1930s.

The museum's involvement in the 1931 Colonial Exposition in Paris, taking place in the Bois de Vincennes, was also symptomatic of the move to invest ethnographically in the empire. The exposition, by bringing to Paris the governor general of every colony, offered Rivet and his staff a unique opportunity to start building colonial relationships, this time in France. Equally important, each colony provided the exposition with ethnographic displays, for which proper documentation, if not sent along at the time, was easily gained.[98] Rivet wrote to the head of every colony requesting permission to choose the best ethnograpic pieces from its display for deposit within the museum and gained full satisfaction.[99] In the same spirit, Rivière hastily put together the renovated museum's first temporary exhibit, devoted to the ethnography of the colonies and timed to coincide with the exposition. The museum then encouraged the many scientific congresses being held in conjunction with the exposition to come visit its new library and see its ongoing renovations.[100] In all these ways the museum sought to advertise the colonial dimension of its science and

---

de certitudes et les pièces sont accompagnées d'une documentation qui en décuple le prix. C'est pourquoi, à l'inverse des musées d'art qui disposent de crédits d'achat importants, les musées d'ethnographie ont surtout besoin de quoi organiser les expéditions scientifiques dont la mission DD est le type le plus accompli"]. 2 AM 1 A 3, "Rapport sur la Mission Dakar-Djbouti," 8 Jan. 1932.

[97] See, e.g., AMH, 2 AM 1 A 4, Georges-Henri Rivière à Colombani, Administrateur des colonies, 21 April 1932, no. 765.

[98] Just as Rivet thought of the Musée de l'homme as colonial, the exposition organizers aspired to be scientific—hence the ethnographic displays. On this aspect of the exposition, see Benoît de L'Estoile, "'Des races non pas inférieures, mais différentes': De l'Exposition coloniale au Musée de l'Homme," in *Les politiques de l'anthropologie: Discours et pratiques en France (1860–1900)*, ed. Claude Blanckaert (Paris: L'Harmattan, 2001), pp. 391–473. Regarding contact between the Musée d'ethnographie and the Exposition, see AMH, 2 AM 1 A 2, Georges-Henri Rivière à Michel Leiris, 21 Aug. 1931, no. 1435. Rivière wrote Leiris, who was on the Dakar-Djbouti mission, that he was way behind in museum affairs because the Colonial Exposition required "permanent contact" on his part ["nécessité de contacte permanent"]. His efforts were paying off, however, because he had already been promised the Indochinese and West Africa collections, most of which were excellent and identified.

[99] AMH, 2 AM 1 A 3, Georges-Henri Rivière à Michel Leiris, 18 Dec 1931, no. 2279 cf. bis. The only colony that failed to transfer its collections to the Musée de l'homme was Cameroon.

[100] Five congresses visited the museum in all—Congrès international des arts décoratifs, Congrès international d'anthropologie et d'archéologie préhistorique, Congrès des recherches scientifiques coloniales, Congrès des langues coloniales, Congrès de la société indigène—the last three of which had a distinctive colonial dimension. These congresses also offered an excellent opportunity to establish contacts with possible future collectors in the colonies. For example, Rivet and Rivière wroteto a certain Mr. Tiaeb Belkahiria, a Cäid in Tunisia, who visited the museum with the

to associate itself more closely with the empire, while still maintaining its distinctly scientific identity. For example, Rivet also used this opportunity to encourage the various governors general to found their own ethnographic museums in the colonies, to which he would happily send a student of his to help. Ethnology was not a vocation, he insisted, that could be improvised on the spot; it was a science that required the training only the Institut d'ethnologie could provide. Any overseas museum, Rivet insisted, had to have oversight from a scientifically recognized body in order to be properly organized.[101]

Both the Dakar-Djbouti mission and the Colonial Exposition helped to launch the Musée de l'homme on the colonial career that its imperial subsidy had initially legitimized. As the museum proceeded with its renovations between 1931 and 1935, and then again during 1936 and 1937, it fully exploited these early links to the empire. Of the many temporary exhibits mounted in these years, the two most important ones, in terms of preparation and success, were colonial: one on New Caledonia in January 1934—the first truly ethnological exhibit in France, according to Rivière, "in which the primitive societies . . . would be studied methodically from their natural milieu to their beliefs to their customs and techniques";[102] and an exhibit on the Sahara, May-December 1934, the museum's longest running exhibit in the 1930s.[103] Those among Rivet's collaborators at the museum who were also preparing theses chose colonial sites for their fieldwork. Administrators with an aptitude for ethnography were identified and encouraged, and some even recruited into the ranks of the new ethnology.[104] Museum personnel offered courses at the École coloniale, and École coloniale students took courses at the Institut d'ethnologie.[105]

Congrès international d'anthropologie et de préhistoire, to ask him if he would be willing to send them objects. "Everything interests us, except if it has been touched by a recent European influence. Not only objects from everyday life, their fabrication and documentation showing their fabrication and use." ["Tout nous intéresse, sous réserve bien entendu de ce qui représente une influence européene récente! Non seulement les objets de la vie courante, leur fabrication et la documentation en révélant la fabrication et l'usage."] AMH, 2 AM 1 A 5, Georges-Henri Rivière et Paul Rivet à Mr. Taieb Belkahiria, Caïd, Tunisia, 27 Oct. 1931, no. 1934.

[101] AMH, 2 AM 1 A1, Paul Rivet à Gouverneur Général de l'AOF, 30 June 1930, no. 1006; AMH 2 AM 1 A 3, Paul Rivet à Gouverneur Général de l'Indochine, 24 May 1932, no. 1074; Paul Rivet à Gouverneur Général de Madagascar, 18 July 1932, no. 1554.

[102] [". . . pour la première fois en France . . . les sociétés primitives de la Nlle. Calédonie seront méthodiquement étudiées du milieu naturel aux croyances en passant par les techniques et les usages."] AMH 2 AM 1 A 1, Georges-Henri Rivière à Ministre des Beaux-Arts, 15 May 1931, no. 740.

[103] See AMH, 2 AM 1 B 5 and 2 AM 1 B 6 for press clippings on the exhibit, 2 AM 1 A 4, 5, and 6 for the massive correspondence the exhibit generated. Although not exclusively colonial, the bulk of the objects came from French West and North Africa.

[104] Usually administrators were simply used as collectors of objects, which the scientists in the Musée de l'homme would classify and eventually analyze. For a fairly typical letter, see AMH, 2 AM 1 A 11, Thérèse Rivière à M. l'Administrateur, commune mixte de Barika, 27 Sept. 1938, no. 1399. Rivière, herself in charge of the museum's department on "white Africa" and at work on her doctorate, advised this official and former student of the Institut d'ethnologie on how to collect objects but nothing more. This approach can be contrasted with Paul Rivet's encouragement of another official, who both collected and wrote up his results. Rivet said he would publish the resulting monograph and encouraged the administrator to undertake a thesis. AMH, 2 AM 1 A 11, Paul Rivet à M. R. David, Administrateur adjoint des Colonies, 4 Oct. 1938, no. 1443. See also the list of Institut d'ethnologie students working at the Musée ethnographique du Trocadero, in AMH, 2 AM 1 A 8, Georges-Henri Rivière à Mlle. Rivet, Institut d'ethnologie, 31 May 1935, no. 2090.

[105] Jacques Soustelle taught a course in ethnology at the École coloniale in 1938. See AMH, 2 AM 1 A 11, Jacques Soustelle à Robert Delavignette, Directeur de École nationale de la France d'outremer [formerly École coloniale], 21 May 1938, no. 767. A portion of the Institut d'ethnologie students continued to come from the École coloniale throughout the 1930s.

Yet for all these links, the Musée de l'homme team remained resolutely detached from the actual work of colonizing. Although Rivet, Mauss, and Lévy-Bruhl had initially argued that the scientific knowledge produced by the Musée de l'homme's sponsors would make French rule overseas more humane, they did not subsequently make this argument very often. Nor did they, like their British counterparts, seek to find positions for their students in colonial service as government anthropologists (at least not in the 1930s).[106] Such reluctance to personally place themselves and their knowledge at the service of empire may have been due to the fact that all three men saw scientific fieldwork as a neutral and apolitical enterprise in which the anthropologist's primary task was to record ways of life untouched by colonialism. This did not mean that they were in any way opposed to the "civilizing" of native peoples or unconcerned with how such "development" took place. Rivet, for example, was entirely for the *revolution agraire* currently going on in Mexico to redistribute land and under the Popular Front argued that French officials in Algeria would have to contemplate a comparable policy there if they wished to retain the colonies.[107] Still, he insisted, the whole ethnographic project would be condemned if he and others decided to include in their presentation of other cultures "les aspects européens des civilisations d'outremer."[108] Such a position did not encourage Rivet to reflect upon colonial policy or attempt to advise officials, since the latter's job was to accelerate social and cultural change.

Equally absent, over the course of the 1930s, was a public celebration in the museum itself of the colonial connection. In the early 1930s the colonial press called attention to the importance of the founding of the Institut d'ethnologie and the reforms at the Musée d'ethnographie for improving imperial governance.[109] When the Musée de l'homme finally opened in 1938, however, colonial newspapers paid no particular attention, and press coverage in general left out all mention of a possible colonial vocation of the institution. In article after article, flattering accounts were given of the museum's extraordinary modernity (it even had a café bar), rational layout, and accessible scientific displays, as well as of its global representation of the world's exotic races and cultures. "Faire le tour du monde en deux heures" was a typical title in the press coverage of the opening, not "faire le tour de l'empire."[110] Cadets from the École coloniale, not colonial troops, were present at the opening; yet no one apparently recognized their uniform, probably because they were unfamiliar, but also because colonial representatives were not expected.[111] There thus appeared to be another part of the museum that was even more "hidden" from the eyes of the public than its research facilities: the colonial, and therefore French,

[106] Henrika Kuklick, *The Savage Within: The Social History of British Anthropology, 1885–1945* (Cambridge: Cambridge Univ. Press, 1991), chap. 5.

[107] AMH, 2 AM 1 A 3, Paul Rivet à Fernand Pila, Chef du Service des oeuvres françaises à l'étranger, 17 Nov. 1932, no. 2080; AMH, 2 AP 1 B 5, Paul Rivet, "Nous ne pourrons conserver l'Afrique du Nord que si nous faisons une grande réforme agraire," 8 May 1937 (press clipping, newspaper name not given).

[108] Paul Rivet à Fernand Pila (cit. n. 107).

[109] AMH, 2 AM 1 B 1, Georges Hardy, "La Politique d'expansion au Trocadéro, le musée d'ethnographie est organisé pour initier aux choses coloniales," *La Dépêche Coloniale*, June 6, 1931; AMH, Henri Paul Eydoux, "Un grand centre d'études coloniales, le Musée d'ethnographie du Trocadéro," *La Dépêche Coloniale*, 11 March 1932; AMH, "Le Musée d'ethnographie du Trocadéro est un centre actif de propagande coloniale," *La Dépêche Coloniale*, 24 Nov. 1933.

[110] See AMH, 2 AM 1 B 2 for the lack of colonial press coverage.

[111] AMH, 2 AM 1 B 2, "La Cape et l'Epée," *Paris-Midi*, 29 June 1938.

status of many of the cultures on display. Rivet's ethnology at the Musée de l'homme, then, was indeed colonial, but not in any overtly political way. Ironically, it was the multifacted imperial connection that made possible such "detachment" from the day-to-day work of colonization by subsidizing a "pure science" yet civic-minded institution to which ethnologists could retreat after their return from the field.

Up until this point I have concentrated on the pedagogic, scientific, and eventually socialist vocations of the museum. In retrospect it is clear that it also had another purpose: to provide legitimate scientific employment—in the continuing absence of university positions in anthropology in France—to the new generation of ethnologists Rivet and Mauss were training. It was Institut d'ethnologie students who staffed the museum in the 1930s, surviving on tiny fellowships or on their own funds, enthusiastically joined in a common cause. Under Rivet's mentoring they aspired to the newest scientific standards and thus not just to put French anthropology on a par with other countries, but to go a step further than their international rivals and seriously initiate the masses into the new science. After a century of fragmentation among anthropologists in France, an important institutional, intellectual, and cultural realignment appeared to be taking place—thanks not only to the liberal traditions of the Third Republic and the Popular Front, but also to the resources of the empire.[112]

This said, the retreat of most Musée de l'homme ethnologists from conscious involvement in colonial questions was partial at best and would prove of necessity to be short lived. It is worth noting here that none of the scholars who thought to turn to the colonial minister and administrators for support—Lévy-Bruhl, Mauss, Rivet, and Rivière—had themselves ever carried out substantial fieldwork in any part of the French empire. Their relationship to places such as Madagascar, Africa, and Southeast Asia was therefore very much a mediated one—experienced through the eyes of the press, the discourse of the civilizing mission, and correspondance or personal acquaintance with missionaries, administrators, and explorers. This inexperience, as well as the fact that in France no one yet was seriously contesting the legitimacy of colonial rule, helps to explain this older generation's apparent acceptance of empire. By choosing to send a good number of their students *sur le terrain colonial*, however, Rivet and Mauss were changing the professional rules of engagement significantly. Objects would now be associated with human faces, and modernization and its discontents might prove more tempting subjects of inquiry when devastating consequences were encountered directly. Trained to think sociologically and to respect basic human rights, the younger men and women who were drawn to the Institut d'ethnologie and the Musée de l'homme in the interwar years had every chance of posing—sooner rather than later—the hard questions about science, democracy, and empire that had eluded their "*chers maîtres*."

CONCLUSION

Was the foundation of the Musée de l'homme an example of democratic civil society in late republican France using the state to promote but also to impose a certain

[112] That the Muséum and Rivet's students were perceived as forming a "new school" of ethnology in France is evident in the fact that anthropologists outside the Musée de l'homme referred to them as the "École du Troca." See AMH, 2 AM 1 A 11, Jacques Soustelle à M. Roubault, 23 July 1938, no. 1207.

practice of tolerant citizenship premised on science? The answer depends on how one defines "democratic civil society" under a Popular Front government that was as accepting of scientific objectivity and truth as it was of the legitimacy of empire. Certainly the museum's socialist directors sought to communicate to all of France's citizenry the results of the most up-to-date science—results that proved the equality of all peoples and cultures, if not their equal degree of civilization. Only a nation that had accepted this truth could effectively combat racism and become humane colonialists. Democratic means for the most part were used to disseminate this progressive message. No one was coerced into the museum. The anthropology on offer was not the only one in town, so that individual citizens were free to make up their own minds about "the race question."

Such freedom of expression, however, was not a Popular Front invention. Indeed, I have explored the museum's prehistory at length to show, in part, that neither the institution nor its synthetic vision of man can be understood simply as a product of the socialist victory in 1936. Both grew out of a century of civic activism in the realm of science in France, during which time self-styled anthropologists argued incessantly over what constituted their discipline. While these personal, theoretical, and eventually institutional divisions prevented any early professionalization of anthropology in France, these scholars and the proliferation of organizations they spawned underscore just how vibrant and important associational life was to intellectual production under the entire Third Republic. And while civic activism in the realm of ideas does not, of course, always produce democratization of either public life or the state, in the case of the Musée de l'homme anthropologists, the correlation between exercising their rights as citizens and their humanist ethnology seems rather direct.

But it is also true that these same museum scientists believed that the best way to convince the public of the equality of all peoples was to instruct and explain, not consult those whose cultures were on display or those being educated. Does this mean the Musée de l'homme's display of tolerance unintentionally ended up silencing the very peoples that anthropologists were seeking to respect, by speaking for them through their objects? Undoubtedly. The latter were surrounded with labels, photographs, and panels of explanations and maps. Movies and sound recordings supplemented the written word or still lifes, but these, as has often been pointed out, were just another kind of object Western social science began collecting—with thus limited potential for bringing "objects to life." However admirable, the ideal of the masses—including the "colonial masses," or *indigènes*—"acting on the museum" was not seriously explored in 1938. But French educators in general behaved in the same imperial way towards all of their students, regardless of achievement, ability, rank, race, ethnicity, or gender. To point this out is not to condone the comparable practices at the Musée de l'homme, but to keep its shortcomings in historical perspective: within the realm of what was thinkable in the interwar decades, Rivet dared at least to assert that the science of man could be made accessible to all and to devote ten years of his career to popularizing ethnology in France.

Semi-imperial in its manner of asserting its scientific knowledge, the Musée de l'homme team was similarly so when it came to acquiring the bulk of its "ordinary" objects. Convinced of the possibility of scientific neutrality, they turned uncritically to the colonies for funds, artifacts, and legitimation. Such acceptance of empire placed the Musée de l'homme in the anomolous position of attempting to fight fascism with colonial forms of knowledge. The emphasis on collecting specimens untainted by European influence, for example, may have inadvertantly exoticized the very others that these scientists were hoping to humanize. Ethnologists did have a message of hope and tolerance to offer the world, but it was one directed at their own workers and fellow scientists rather than France's colonial subjects. Here again, however, we would be wise to think twice before criticizing too glibly. Curators of ethnographic museums today recognize the problems of past efforts of representation of other cultures, but have yet to resolve them.[113] To their credit, Rivet and his collaborators did not abandon the museum because of the difficulties inherent in synthesizing and educating their materials for a larger public. And there is every evidence that they accepted that the knowledge they were putting on display was provisional at best.

[113] For a recent debate in France on these issues, see Dominique Taffin, ed., *Du musée colonial au musée des cultures du monde* (Paris: Maisonneuve & Larose and Musée des Arts d'Afrique, d'Océanie, 2000).

Saving China Through Science:

The Science Society of China, Scientific Nationalism, and Civil Society in Republican China

*Zuoyue Wang**

ABSTRACT

The Science Society of China, the first comprehensive Chinese scientific association, was actually organized in 1914 by a group of Chinese students at Cornell University in the United States. Four years later, many members returned to China, where the association, until its dissolution by the Communists in the 1950s, played a crucial role in Chinese science and society, not the least by publishing the *Kexue* (Science) monthly. This paper presents a twofold thesis: first, the Science Society of China and the *Kexue* represent attempts by Chinese scientists to create a civil society and a public sphere in Republican China. Second, in contrast to the conventional Western model, the Science Society, even as it critiqued government actions in public, maintained intimate and complex connections with successive regimes in the Republican era. If professionalism drove the Science Society's rhetoric of autonomy, scientific nationalism, I argue, moderated the association's interactions with the state in practice.

INTRODUCTION

ON AUGUST 16, 1910, in Shanghai, Zhu Kezhen boarded the SS *China* to embark on a journey that would change his life. Less than a month before, he had passed a rigorous national examination and earned one of the seventy Boxer fellowships for studying science and engineering in America that year. Standing on the deck as the ship prepared to leave port, Zhu, still a diminutive figure at age 20, certainly recognized the significance of the moment as he and the other Boxer students bade farewell, if only temporarily, to a homeland on the eve of radical transformations.[1]

* Department of History, California State Polytechnic Univ., 3801 W. Temple Ave., Pomona, CA 91768; Zywang@csupomona.edu.

The author thanks Thomas Broman, Lynn Nyhart, Kathy Olesko, and anonymous referees for critical and helpful comments on earlier drafts of this paper and Zhang Li for assistance with materials. Unless otherwise noted, most Chinese names are rendered in *pinyin* with family names first and given name second, and translations from Chinese into English are my own.

[1] Xie Shijun, *Zhu Kezhen zhuan* (Biography of Zhu Kezhen) (Chongqing: Chongqing Press, 1993), pp. 30–7. See also Zhao Xinna and Huang Peiyun, eds., *Zhao Yuanren nianpu* (Chronological biography of Zhao Yuanren) (Beijing: Commercial Press, 1998), pp. 57–8. The *nianpu* is a Chinese literary genre especially valuable to historians because it usually reconstructs the subject's life based on letters and diary. This Zhao Yuanren *nianpu* is based on his papers deposited in the Bancroft Library, University of California, Berkeley.

 0369-7827/02/1701-0011$10.00

If there was one desire that united this group of Boxer students, it was a dream of saving China through science and technology. Convinced of the fundamental importance of farming for China, Zhu had decided to study agriculture at the University of Illinois.[2] Fellow student Hu Shi also would be studying agriculture, though he would be attending Cornell University. Accompanying him to Cornell were Hu Mingfu, a thin young man with training in business administration, and the shy, lanky Zhao Yuanren, gifted with a knack for acquiring local dialects. Both of them looked forward to learning physics and mathematics. Zhou Ren, who wanted to help "sharpen the tools for a strong [Chinese] nation," planned to major in mechanical engineering at Cornell.[3] As Zhu and his fellow Boxer scholars sailed for America on the SS *China*, they had no idea that once they returned to China, their faith in science and technology would be severely tested in successive waves of revolutions, wars, triumphs, and tragedies. They could not know that many of them would become prominent figures in modern Chinese history: Zhu Kezhen, for example, as a meteorologist, geographer, and science and education administrator, and Hu Shi, as a philosopher, intellectual, and diplomat. Neither could they realize that the informal ties they started to form with each other aboard the ship would blossom into the Science Society of China, an association destined to play a significant role in Chinese science, society, and politics during the first half of the twentieth century.

Founded by these and other Chinese students in the United States in 1914–1915, the Science Society became the largest and most influential general scientific organization in China after it moved its activities there in 1918, a position it would maintain until its dissolution in 1950. What happened to the Science Society of China and its members' dream of saving their country through science is the topic of this case study of science and civil society in modern China. The experiences of Zhu and several other leaders of the Science Society will be used to illustrate the collective aspirations and struggles of the first generation of modern Chinese scientists. The central argument of this paper has two parts. First, members of the Science Society sought to reshape China's destiny not only by making scientific contributions, but also by offering a significant, if embryonic, model of civil society and a liberal public sphere in Republican China. The Science Society rendered itself as an autonomous and voluntary association capable of critiquing state actions and public policies, especially through its journal *Kexue* (Science). Second, the Science Society represented a different kind of civil society from the conventional Western model: it was not an institution conceived in opposition to the state but rather one that maintained intimate and complex connections with successive governments, both national and local, throughout the Republican era. If professionalism—the scientists' pursuit of professionalization—drove the Science Society's rhetoric of autonomy, a sense of nationalism moderated their interactions with the state in practice. This tension between autonomy and dependence, rhetoric and reality, played itself out against a backdrop of extraordinarily turbulent social and political forces, including

[2] Zhu Kezhen zhuan bianjizu, *Zhu Kezhen zhuan* (A biography of Zhu Kezhen) (Beijing: Science Press, 1990), p. 11.

[3] For Hu Shi's description of his voyage, see Hu Shizhi (Hu Shi), "Huiyi Mingfu" (Recollections of Mingfu), *Kexue* 13 (1928): 827–34, on pp. 827–8. On Hu Mingfu, see Zhang Zugui, "Hu Mingfu," in *Zhongguo xiandai kexuejia zhuanji* (Biographies of contemporary Chinese scientists), ed. Lu Jiaxi, 6 vols. (Beijing: Science Press, 1991–1994) (hereafter cited as ZGXDKXJZJ), vol. 4, pp. 1–10. On Zhou Ren, see Zhou Peide and Dong Deming, "Zhou Ren," in ibid., vol. 5, pp. 783–91, on pp. 783–4.

Figure 1. *Boxer Fellows in 1910 just before their voyage to the United States to study science and technology. Reprinted from Yuen Ren Chao,* Yuen Ren Chao's Autobiography: First 30 Years, 1892–1921 *(Ithaca, N.Y.: Spoken Language Services, 1975).*

nationalism and wars in Republican China. In what follows, we will first take a look at the broad political background of modern China, then at the debate of civil society and science, before moving to examine the political and social roles of members of the Science Society of China, using the lives and careers of Zhu and several other society leaders as examples.

SCIENCE AND POLITICS IN MODERN CHINA

Back in 1910, when Zhu and other Boxer fellows hoped to use the science and technology they would learn abroad to serve and save China, they did not have a clear idea of who would be governing their beloved homeland in the future or how it would be governed. They certainly did not think highly of the present government, even though the imperial Qing regime busily asserted its authority over the students.[4] While still in Beijing following the July examination, the Boxer fellows were summoned to the Foreign Ministry for official "admonitions" before their trip overseas. (See Figure 1.) According to one report, they were told to carry a golden dragon flag of the Qing Empire with them at all times and never to join any revolution against

[4] Zhao Yuanren, for example, recalled that as early as 1908, when he was a high school student in Nanjing, he and his classmates looked forward to a revolution, "believing that the days of the Qing were numbered." They laughed, instead of cried, during the official mourning service for Emperor Guangxu and Empress Dowager Zixi in 1918, but nobody could tell the difference. Yuen Ren Chao (Zhao Yuanren in *pinyin*), *Yuen Ren Chao's Autobiography: First 30 Years, 1892–1921* (Ithaca, N.Y.: Spoken Language Services, 1975), p. 67.

the Qing. They also could not convert to foreign religions or marry foreign women.[5] The fearful and defensive Qing officials did not give Zhu and his colleagues much confidence in the Chinese state. Furthermore, the material and spiritual condition of Chinese society under the Qing greatly concerned them. For example, Zhu's family in rural Zhejiang, while jubilant about the honor of his being selected a Boxer fellow, could not afford to travel to Shanghai to see him off. When Zhu went to a barbershop in the port city to have his queue removed and hair cut before the voyage, the Chinese barber refused, fearing a capital punishment for disobeying the Qing rule about maintaining the queue. After arguing in vain that such measures were officially required before going abroad, Zhu had to find a Japanese barber to do the job.[6]

The introduction of modern science in China took place against a turbulent social and political backdrop. A survey of the modern Chinese political landscape might begin with the turn of the twentieth century, as Zhu and other members of the first generation of modern Chinese scientists went through their secondary education. The half century following it could be divided into four periods. From 1900 to 1911 the Qing government, headed by the ethnically Manchu imperial family in a predominantly Han Chinese nation, tried to implement political, social, and cultural reforms in the face of rising anti-Manchu sentiment at home and growing encroachment on Chinese national sovereignty from Western and Japanese powers from abroad. The reform measures proved to be too little and too late. In 1911–1912 the last emperor of China, the boy monarch Puyi, abdicated amid a republican revolution led by Sun Yat-sen's Revolutionary Alliance, thus opening the Republican era. The revolution did not, however, result in a unified, strong nation. Instead, regional warlords reigned in China under a nominal national government in Beijing from 1912 to 1927. It was not until 1927 that Sun's Nationalist Party, under the leadership of Jiang Jieshi (Chiang Kai-shek), established a true national government in Nanjing, where it remained until 1937 (thus the period came to be known as the Nanjing Decade). That year witnessed the start of the war of resistance against Japan and World War II (1937–1945), during which the Nationalist government moved the national capital to Chongqing in southwest China. The year the world war ended a civil war began between the Nationalists and the Communists under Mao Zedong. It raged until 1949 and resulted in Communist control of the mainland and the end of the Republican era (1911–1949).

Superimposed on this political-historical map were events and developments crucial to the experiences of first-generation scientists such as Zhu Kezhen. One incident that would have enormous impact on the history of modern science in China was the so-called Boxer uprising of 1900. An antiforeign movement eventually supported by the Qing government, the Boxer uprising not only resulted in the humiliating defeat of China by Western and Japanese forces, but also led to a treaty that specified a huge indemnity fund of $333 million (nearly twice the Qing's annual income), amortized until 1940, to be paid to the Western countries and Japan for their damages and loss of life during the incident. By the early 1900s, however, it was clear that the U.S. claim of $25 million was greatly exaggerated. The U.S. government then decided to return the surplus portion of the Boxer indemnity funds to China with the stipulation that they be used to send Chinese students to study

[5] Xie, *Zhu Kezhen* (cit. n. 1), p. 33.
[6] Ibid., pp. 36, 54–7. See also Zhao, *Yuen Ren Chao's Autobiography* (cit. n. 4), pp. 71–2.

science and technology in the United States. American policy makers expected that these students, when they returned to China, would extend American influence and help facilitate trade between the two countries.[7] The first batch of students was selected and sent to the United States in 1909. The next year Zhu and his fellow voyagers on the SS *China* were the second group to go. Over the next three decades, hundreds of other Boxer students followed in the footsteps of these pioneers. After they completed their studies in the United States and went back to China, these "returned students" did indeed, as American policy makers had envisioned, play a prominent role in Chinese science and education.[8] But ironically, aware of the origins of the Boxer funds, these scientists and engineers often harbored intensely nationalistic feelings and therefore were not always as pro-American in their political orientation as Washington had hoped.

The single most important cultural change in the Republican era was the May Fourth Movement of 1919. It started as a nationalist protest against Western and Japanese encroachment on Chinese sovereignty at the Versailles treaty negotiations but eventually evolved into a far-reaching intellectual revolution. The so-called New Culture Movement that was a part of the May Fourth Movement has often been termed the Chinese Renaissance or Chinese Enlightenment. Leaders of the movement, mostly literary figures but also a few scientists, called for the introduction of "Mr. Democracy" and "Mr. Science" into China to reform its traditional pattern of culture and politics.[9] It was against this backdrop of national crisis that Zhu's and other Chinese scientists' aspirations for science as a means to national salvation played out.

THE DEBATE OVER THE SEARCH FOR CIVIL SOCIETY IN CHINA

Amid the call for democracy and modernization in the first half of the twentieth century, did a civil society emerge? This question has been part of the general search for civil society in China that has attracted both excitement and controversy in recent decades. In the early and mid-1980s works on the possible existence and evolution of Chinese civil society began to appear, in an attempt to break away from the traditional view of Chinese society as essentially unchanging until the West came knocking on its door in the mid-nineteenth century.[10] Though it is thus clear that the search for civil society in China predates the student uprising in Tiananmen Square and the successful toppling of Communist rule in Eastern Europe in 1989, these tumultuous

[7] Michael H. Hunt, "The American Remission of the Boxer Indemnity: A Reappraisal," *J. Asian Stud.* 31 (May 1972): 539–59.

[8] For a study of the returned students before 1949, see Y. C. Wang, *Chinese Intellectuals and the West, 1872–1949* (Chapel Hill: Univ. of North Carolina Press, 1966). For the impact of returned students in China after 1949, see Li Peishan, "The Introduction of American Science and Technology to China before 1949 and Its Impact," in *United States and the Asia-Pacific Region in the Twentieth Century*, ed. Shi Xian-rong and Mei Ren-yi (Beijing: Modern Press, 1993), pp. 603–18.

[9] On the May Fourth Movement, see Chow Tse-tsung, *The May Fourth Movement: Intellectual Revolution in Modern China* (Cambridge, Mass.: Harvard Univ. Press, 1960); and Vera Schwartz, *The Chinese Enlightenment: Intellectuals and the Legacy of the May Fourth Movement of 1919* (Berkeley: Univ. of California Press, 1986).

[10] For a recent discussion of this historiographical change and its implications for the history of science in China, see H. Lyman Miller, *Science and Dissent in Post-Mao China: The Politics of Knowledge* (Seattle: Univ. of Washington Press, 1996), pp. 22–6.

events undoubtedly generated new energy for this search among Western scholars.[11] As is well known, the publication of Jürgen Habermas's 1962 work on *The Structural Transformation of the Public Sphere* in English translation that same eventful year further stimulated the interest of many scholars.[12]

Trying to explain the contours of modern Chinese history, several China scholars have advanced a twofold thesis on Chinese civil society. On the one hand, they have argued that an incipient civil society began to emerge in Ming (1368–1644) and Qing (1644–1911) China and, certainly by the early twentieth century, came to reshape the urban landscape as merchants, city residents, professionals, and intellectuals organized various institutions within the public sphere, including voluntary associations, newspapers, and periodicals. These new institutions played a mediating role between the state and society. On the other hand, these scholars have acknowledged that in the face of wars and revolutions and in the absence of a stable and tolerant state for much of the twentieth century, various efforts at civil society often failed to establish long-lasting institutional changes.[13]

Critics of the searches for civil society in China have expressed doubts about the degree of autonomy the various examples of public sphere and civil society enjoyed from state control.[14] They have also questioned whether such work is teleological in the sense that searchers are often imposing present-day concerns about social-political structure on historical settings. More radical challengers have even denied completely the validity of these searches. One critic, for example, accused those looking for civil society in Chinese history of being Eurocentric, of explicitly or implicitly assuming that the development of civil society was a necessary and desirable stage in social development in China just as in the West.[15] Indeed, Habermas himself has warned in his book against universalizing the concepts of civil society and public sphere. They cannot be "transferred, idealtypically generalized, to any number of historical situations that represent formally similar constellations."[16]

Interestingly, science and scientists have rarely figured in the debate on civil society in China. Is there value in searching for signs of civil society/public sphere among scientists in China? My view is that we can proceed with such investigations as long as we are aware of the inherent limitations and potential pitfalls. The concep-

[11] See, e.g., David Strand, "Protest in Beijing: Civil Society and Public Sphere in China," *Problems of Communism* 39(3) (1990): 1–19. See also Gu Xin, "A Civil Society and Public Sphere in Post-Mao China: An Overview of Western Publications," *China Information* 8(3) (1993–1994): 38–52.

[12] Jürgen Habermas, *The Structural Transformation of the Public Sphere*, 2d ed. (Cambridge, Mass.: MIT Press, 1991); originally published in 1989.

[13] These works include Mary Rankin, *Elite Activism and Political Transformation in China: Zhejiang Province, 1865–1911* (Stanford, Calif.: Stanford Univ. Press, 1986); William T. Rowe, *Hankow: Commerce and Society in a Chinese City, 1796–1889* (Stanford, Calif.: Stanford Univ. Press, 1984); idem, *Hankow: Conflict and Community in a Chinese City, 1796–1895* (Stanford, Calif.: Stanford Univ. Press, 1989); Marie-Claire Bergère, *The Golden Age of the Chinese Bourgeoisie, 1917–1937* (Cambridge: Cambridge Univ. Press, 1986); and David Strand, *Rickshaw Beijing: City People and Politics in the 1920s* (Berkeley: Univ. of California Press, 1989). Rowe provides an excellent review of this literature in "The Public Sphere in Modern China," *Modern China* 16(3) (1990): 309–29.

[14] See the papers in the special issue of *Modern China* 19(2) (1993), devoted to "'Public Sphere' and 'Civil Society' in China?," especially Frederic Wakeman Jr.'s "The Civil Society and Public Sphere Debate: Western Reflections on Chinese Political Culture," pp. 108–38; and Philip C. C. Huang's "'Public Sphere'/'Civil Society' in China? The Third Realm between State and Society," pp. 216–40.

[15] Adrian Chan, "In Search of a Civil Society in China," *Journal of Contemporary Asia* 27(2) (1997): 242–51.

[16] Habermas, *Structural Transformation* (cit. n. 12), p. xvii.

tual difficulties are not insurmountable as we take a closer look at them. First, on the question of exaggeration of autonomy, we should not presume a priori that, because of the weaknesses of some well-known cases, all cases will fail to measure up to the high bar of civil society and public sphere. The fact that few China historians, with or without a specialty in science, have examined the experiences of Chinese scientists in light of this debate on civil society perhaps reflects a continuing general indifference of historians of science to the civil society debate (and the general failure of China scholars to take science seriously). Second, on the question of teleology, one can only say that while we should not judge historical actors by our present standards, neither can we free ourselves entirely from bias in our historical research. Insights from our own lives and times are often useful in guiding such research as long as we and the reader are aware of the situation.

Finally, is the employment of civil society/public sphere in Chinese history intrinsically Eurocentric or ethnocentric? It is a complex issue because concepts such as "modernization," "freedom," "national distinctiveness," and "civilization" have certainly been used to justify various questionable political designs and ideologies. Yet we cannot be deterred from using these concepts because they could lead to abuses or because they are ideologically loaded. Even such commonly used terms as "science," "technology," "class," and "race" are, of course, culturally bound. If we deny the universality of "civil society" and "public sphere" a priori out of respect for the special local conditions of the Chinese society and culture, are we not falling into another form of ethnocentrism that says the Chinese are intrinsically unable to produce a civil society and liberal public sphere? As the China historian William T. Rowe explains: if one exempts China from Western liberal-democratic demands on the grounds of historical cultural differences, "we are justly suspected of orientalism: other, less 'civilized' societies cannot be expected to live up to the standards we [Westerners] set for ourselves."[17]

In many ways, the problems associated with the discussions of science and civil society in China parallel those engendered by Joseph Needham's famous question of why modern science did not arise in China, given its outstanding ancient technological achievements. For years historians have disputed, in the Needham debate, the meaning of key concepts such as science, modern science, theoretical science, technology, and national culture. Charges of teleology and ethnocentrism have also figured in the Needham discussion as some scholars have felt that the question is misplaced and China has been unfairly judged by what happened in modern Europe.[18]

Habermas's well-intentioned warning notwithstanding, I believe that it will be useful to apply the concepts of civil society/public sphere in historical situations other than seventeenth- and eighteenth-century Europe. Just as we use "science," "technology," "gender," and "class" in varied national, cultural, and historical contexts, we can use "civil society/public sphere" to study, perhaps most interestingly, how ideas and institutions "idealtypically generalized" are transformed and reconfigured when they cross those boundaries. As Thomas Broman points out in his

[17] William T. Rowe, "The Problem of 'Civil Society' in Late Imperial China," *Modern China* 19(2) (1993): 139–57, on p. 141.

[18] For a recent review in English of the debate on the scientific revolution in China, see Roger Hart, "On the Problem of Chinese Science," in *The Science Studies Reader*, ed. Mario Biagioli (New York: Routledge, 1999), pp. 189–201; Robert Finley, "China, the West, and World History in Joseph Needham's *Science and Civilisation in China*," *J. World Hist.* 11(2) (2000): 265–303.

study of Enlightenment science, the conceptual framework of civil society/public sphere can also serve as a great heuristic tool: "Its specific utility hinges on the way it helps our understanding of the 'public' for science in the period."[19]

In the case of science in Republican China, I believe the concepts of civil society and public sphere will help us bring out aspects of the interactions between science and civil society that other perspectives leave hidden, allowing us to better understand the political and social significance of science's professionalization and institutionalization in this setting. In one traditional narrative of professionalization, for example, the public's role is conspicuous for its disappearance: a field is "professionalized" when it bars amateurs. From this perspective, it is easy to interpret the formation of scientific organizations such as the Science Society of China as simply a step toward professional status. Attention to civil society, however, allows us to see much more clearly the extent to which the emergence of an autonomous scientific community in China actually depended on the presence of a civil society for funding and public support. The Science Society did indeed represent professionalization, but professionalization here was not just about the exclusion of amateurs: it was also about relative autonomy from the state; the freedom of members to order their own organization; the reliability of the state/government to act as a guarantor of civil society; and the readiness of amateurs to fight for, or at least to tolerate, the rights of experts to form their own associations. All of these conditions began to take shape in Republican China only to largely disappear during the Maoist era. In addition, the focus on Chinese scientists and civil society allows us to move beyond the confines of the traditional history of Chinese science and open a new frontier where we can examine Chinese science in the context of mainstream Chinese history and make cross-national comparisons. Recent scholarship seems to indicate that in contrast to the strict dichotomy between civil society and the state that existed in early modern western Europe and in modern Eastern Europe, the Chinese experiences are dominated by examples of civil society/public sphere institutions that involved much more interaction between state and society. As the China historian Philip C. C. Huang points out, "We need to employ instead a trinary conception, with a third space in between state and society, in which both participated."[20] Huang calls this intermediate space the "third realm" and explains its advantages:

> A value-neutral category, it would free us of the value-laden teleology of Habermas's bourgeois public sphere. It would also define more unequivocally than Habermas's public sphere a third space conceptually distinct from state and society. Such a conception would also prevent any tendency to reduce the third space to the realm of either the state or society. . . . We would see it as something with distinct characteristics and a logic of its own over and above the influences of state and society.[21]

Does the Science Society of China fit into this third realm model of a civil society/public sphere? In what follows I will try not only to demonstrate that the answer is a qualified yes, but also to go beyond the question to explore the dynamics behind the apparently more intense interaction between Chinese state and civil society insti-

[19] Thomas Broman, "The Habermasian Public Sphere and 'Science *in* the Enlightenment,'" *Hist. Sci.* 36 (1998): 123–49, on p. 124.
[20] Huang, "'Public Sphere'/'Civil Society' in China?" (cit. n. 14), p. 216.
[21] Ibid., p. 225.

tutions. Simply put, these dynamics hinged, in the case of the Science Society, on a tension between Chinese scientists' demand for professionalism as scientists and their equal, if not more strongly felt, desire to strengthen Chinese nationalism.

Before we examine the Science Society as a civil society/public sphere institution, a few more preliminary remarks on terminology are in order. The concepts of public sphere and civil society are notoriously ambiguous, in the context of both western and Chinese history. Indeed, as Rowe points out, whereas the Chinese political lexicon had long contained the word *gong*, which corresponds in general with "public," not until the late twentieth century, it seems, did a Chinese phrase for "civil society" appear.[22] Thus, instead of attempting precise definitions of these terms, I will follow Thomas Broman in viewing the "public sphere . . . as the cultural and political expression of the self-consciousness of members of civil society."[23] With the "third realm" modification in mind, we might use civil society to refer to public, nongovernmental institutions formed by private individuals, based on self-governance but not necessarily in opposition to the state. These could conceivably include chambers of commerce, voluntary associations of various kinds, and professional organizations such as the Science Society. The public sphere these institutions helped to create took the tangible forms of newspapers, periodicals, books, public meetings, and lecture series, among others.

Compared with "public sphere" and "civil society," "professionalism" and "nationalism" are more familiar terms but no more easily defined. In the context of this paper, I use "scientific professionalism" to refer to the ideals associated with science as a profession, such as freedom of conducting research without external interference, social respect for scientists' cognitive and professional authority, and internationalism. "Scientific nationalism" here is used to describe Chinese scientists' desire to create a strong, unified, and prosperous Chinese nation, free from foreign domination, based in part on the utilization of science and technology. In this sense, Chinese scientific nationalism is slightly different from, for example, the feeling of Japanese scientists who wanted their science to excel at the international level, or that of the German scientists who sought to use their superior science to redeem Germany's place in the world following World War I.[24]

PROFESSIONALISM, THE PUBLIC SPHERE, AND THE ORIGINS OF THE SCIENCE SOCIETY OF CHINA

The origin of the Science Society can be traced to a June 1914 gathering of about a dozen Chinese students at Cornell University, including several of those who had journeyed with Zhu on the SS *China* four years before. The tension in the world on the eve of World War I stirred nationalist sentiment among these Chinese students, who wanted to do something for their country. The group, mostly composed of science students, felt that what China lacked most was science: in all of China not a

[22] Rowe, "Problem of 'Civil Society,'" (cit. n. 17), p. 42.
[23] Broman, "Habermasian Public Sphere" (cit. n. 19), p. 125.
[24] On Japanese science, see James R. Bartholomew, *The Formation of Science in Japan* (New Haven, Conn.: Yale Univ. Press, 1989), especially p. 263. On the German case, see J. L. Heilbron, *The Dilemmas of an Upright Man: Max Planck as Spokesman for German Science* (Berkeley: Univ. of California Press, 1986); and Paul Forman, "Scientific Internationalism and the Weimar Physicists: The Ideology and Its Manipulation in Germany after World War I," *Isis* 64 (1973): 151–80.

single journal devoted itself to the field.[25] Subsequently, nine students signed a proposal establishing a science society (*kexue she*) with the main purpose of sponsoring a new journal in Chinese, titled *Kexue* (Science), to introduce science to their homeland.[26]

A closer examination of the background of the society's founders reveals remarkable historical connections between these students' scientific nationalism and the Republican revolution taking place inside China at the same time. Among the nine founders, seven were Boxer fellows who had come to Cornell in 1909–1911, including Hu Mingfu, Zhao Yuanren, Bing Zhi, and Zhou Ren. As Zhao Yuanren recalled later, few of the Boxer fellows were sympathetic to the Qing government and most were jubilant about its overthrow in 1911.[27] It is doubtful, however, that those seven would have initiated the idea of a science society or *Kexue* without the instigation of the two other signers of the proposal who were not Boxer fellows: Ren Hongjun and Yang Xingfo.

Born in 1886, Ren studied chemistry in Japan from 1908 to 1911, with the specific intent of learning how to make dynamite for the Republican revolution against the Qing. As an activist in Sun Yat-sen's Revolutionary Alliance in Japan, Ren returned to China in 1911 as soon as he heard the revolution had started. During the brief period (January to April 1912) when Sun Yat-sen was provincial president of the Republic of China, Ren served as one of his aides in Nanjing. After Sun turned over the presidency to the military strongman Yuan Shikai in April, as the result of a political compromise, Ren persuaded the new government to send him and a number of other young revolutionaries to study abroad. Ren had long wished to pursue studies in the West, as had Yang Xingfo, seven years Ren's junior, who had studied at the same middle school in Shanghai that Ren and Hu Shi had attended before becoming an aid for Sun in Nanjing.

While Ren and Yang waited for the bureaucracy to work out the details of their study abroad, they went to work for *Minyi Bao* (Public opinion press), a newspaper in Tianjin, a major city near Beijing. Ren became its editor-in-chief, and Yang its correspondent in Beijing. In their hands, the paper became a critical voice against Yuan Shikai's government in Beijing, and Yuan had it shut down for more than a month (it resumed publication just before Ren and Yang left for the United States in late 1912).[28] The experiences of working with Sun and at the *Minyi Bao* had a profound impact on the political outlooks of both men. They emerged from these

[25] Ren Hongjun, "Zhongguo kexueshe sheshi jieshu" (A brief history of the Science Society of China), *Zhongguo keji shiliao* (China historical materials of science and technology) (hereafter cited as ZGKJSL) 4(1) (1983): 2–13. Zhao Yuanren recorded in his diary of 10 June 1914 that he "in the evening went to an enthusiastic and serious meeting in H. Z. Zen's [Ren Hongjun] room to organize a Science Society for publishing a monthly." Zhao, *Yuen Ren Chao's Autobiography* (cit. n. 4), 79.

[26] Names of the nine founding members and the initial charter of the Science Society were recorded in Hu Shi's diary, which he reprinted in "Huiyi Mingfu," *Kexue* 13 (1928): 827–34, on pp. 829–30.

[27] See Zhao, *Yuen Ren Chao's Autobiography* (cit. n. 4), 79.

[28] Ren Hongjun, "Qianchen suoji (xia)" (Notes on my early life [pt. 2]), *Zhuanji wenxue* (Biographical literature) 26(3) (March 1975): 89–95. See also Tao Yinghui, "Ren Hongjun yu Zhongguo kexueshe" (Reng Hongjun and the Science Society of China), *Zhuanji wenxue* 42 (June 1974): 11–6; Xu Weimin, "Yang Xingfo: Zhongguo xiandai jiechu de kexue shiye zuzhizhe he shehui huodongjia" (Yang Xingfo: An outstanding organizer and social activist for the scientific enterprise in modern China), *Ziran bianzhengfa tongxun* (Journal of the dialectics of nature) (hereafter cited as ZRBZFTX) 12(5) (1990): 71–80.

experiences not only committed to a radical transformation of China but also convinced of the power of the media as a critical institution of the public sphere for effecting this process in the long term. Yet the fragility of Sun's political revolution and the corruption of Yuan's government also led them to believe that, at least for the immediate future, science, not politics, was the way to save China.[29] Ren and Yang decided to join Hu at Cornell, where Ren majored in chemistry and Yang in mechanical engineering. Although they were not required, like the Boxer fellows, to major in science and technology, they both did so with the knowledge that Sun, their leader, shared with them the dream of industrial nationalism based partly on the use and development of railroads. Given their revolutionary background, it is not surprising that Ren and Yang were leading instigators for the Science Society: the initial meeting on the matter was held in Ren's room, and Yang drafted the charter for the organization and *Kexue*.[30]

The goal of both the society and the journal was to advocate for science and promote the development of industry in China. When these students circulated their proposal for the establishment of the Science Society to Chinese students in other parts of the United States, in Europe, and at home in China, they received an enthusiastic response: more than seventy people joined the society during the first year of its existence. To meet the cost of running and printing the journal, the society was organized as a joint-stock company, with each member subscribing for one or more shares of stocks. Having a financial stake in the enterprise, the organizers thought, would give members of the society the extra incentive needed to ensure success. The first issue of *Kexue*, with articles written by Science Society members in the United States, appeared in Shanghai in January 1915.[31]

Nationalism and professionalism were incorporated in *Kexue*'s earliest public pronouncements about the new science society. Founders of the journal envisioned a dual purpose for it: as a means of scientific communication among members of the Science Society and as a way to popularize science among the Chinese literary public, with the ultimate goal of making China strong and respected in the international community. Thus in the journal's inaugural editorial in January 1915, most likely written by editor-in-chief Yang Xingfo, *Kexue* explained:

> All civilized countries have established scientific societies to promote learning. These societies in turn have sponsored periodicals to publish advances in scholarly research and inventions of new theories. Thus the academic periodicals in these countries are truly records of the rise of their scholarship and, in today's world, the means by which scholars communicate with each other. Because we are still at a stage of pursuing our

[29] Yang Cuihua, "Ren Hongjun yu Zhongguo jindai de kexue sixiang yu shiye" (Ren Hongjun and the scientific ideas and enterprise in modern China), *Zhongyang Yanjiuyuan jindaishi yanjiusuo jikan* (Contributions from the Modern History Institute of the Academia Sinica), no. 24, pt. 1 (June 1995): 297–324.

[30] On Ren Hongjun, see Fan Hongye, "Ren Hongjun: Zhongguo xiandai kexue siye de tuohuangzhe" (Ren Hongjun: A pioneer of the modern scientific enterprise in China), ZRBZFTX 15(3) (1993): 66–76. On Yang Xingfo, see Xu, "Yang Xingfo" (cit. n. 28). On the role of Sun Yat-sen's industrialization plan in Republican China, see William Kirby, "Engineering China: Birth of the Developmental State, 1928–1937," in *Becoming Chinese: Passages to Modernity and Beyond*, ed. Wen-Hsin Yeh (Berkeley and Los Angeles: Univ. of California Press, 2000), pp. 137–60.

[31] Ren Hongjun, "Waiguo kexueshe ji benshe zhi lishi" (A history of foreign scientific societies and our own society), *Kexue* 3 (1917): 2–18, on pp. 14–5.

studies, we have not been able to make many new discoveries or inventions, but we will try to convey what we have learned. . . . As our scholarship advances in the future, we hope to use this outlet to publish our new ideas and creative works.[32]

The editors went on to claim a prominent role for science in Chinese nationalism: "It is science, and only science, that will revive the forest of learning in China and provide the salvation of the masses!"[33] To accomplish this lofty goal, *Kexue* editors divided scholarship into "pursuing the truth" and "applications." The new journal would promote both but would exclude what its editors called "metaphysics" (*xuaixue*) and politics.[34] In this rhetoric on the place of science in general, and *Kexue* in particular, in China, the editors integrated scientific nationalism and professionalism by claiming that professional scientists, in conjunction with engineers and industrialists, uncorrupted by traditional Chinese learning or political powers, furnished the last hope of saving China from material and spiritual bankruptcy.

Scientific professionalism became a more explicit goal when the Science Society underwent a reorganization in 1915. Shortly after the appearance of the first issue of *Kexue* in Shanghai, a growing number of members recognized that, to realize their ambitious goals of promoting science and industry in China, merely publishing *Kexue* was not enough. Thus in October 1915, the Science Society, renamed the Science Society of China (adding *Zhongguo* or China before *kexueshe*), was formally reorganized as a comprehensive scientific society, the first in modern Chinese history. Newly elected president Ren Hongjun joined four others—Zhao Yuanren, Hu Mingfu, Bing Zhi, and Zhou Ren—to form the board of directors. (See Figure 2.) The funding mechanism changed from one based on stocks to one that relied on member dues as well as donations from individual and institutional members, universities, and government agencies. In addition to its board, the society, still headquartered in the United States, had several disciplinary divisions, an editorial department, a section devoted to translating scientific texts into Chinese, and another section planning for a Science Society library in China.[35] The society now sought, under a broader charter, to publicize science among the public, initiate a research tradition among scientists, write and translate scientific books, establish Chinese scientific terminology, hold scientific lectures to popularize scientific knowledge, and build libraries, museums, and research institutes in various disciplines in order to conduct scientific experiments and "to promote progress in scholarship, industry, and public-interest enterprises."[36]

The reorganized society held its first annual meeting in September 1916 at Phillips Academy in Andover, Massachusetts. The schedule included the election of officers, revisions of the society's charter, speeches and lectures, and a night of games involving mathematics and psychology, designed to promote friendship among members.[37] Out of a membership of 180, about 30 attended the two-day gathering. Chen Hengzhe, who studied European history at Vassar College, was the only female member

[32] "Liyan" (Rules), *Kexue* 1 (1915): 1.

[33] "Fa kan ci" (Inaugural notes), *Kexue* 1 (1915): 3–7, on p. 7.

[34] "Liyan" (cit. n. 32).

[35] Ren, "Waiguo" (cit. n. 31), pp. 16–7.

[36] Zhongguo kexueshe (Science Society of China), *Kexue tonglun* (Overview of science) (Shanghai: Science Society of China, 1934), pp. 463–4, quoted in Tao, "Ren Hongjun" (cit. n. 28), pp. 12–3.

[37] "Changnianhui jishi" (Record of annual meeting), *Kexue* 3 (1917): 69–88.

Figure 2. *The board of directors of the Science Society of China, October 1915, Ithaca, New York. Left to right: seated–Zhao Yuanren, Zhou Ren; standing–Bing Zhi, Ren Hongjun, and Hu Mingfu. (Reprinted from* Kexue *[Science] 13 [6] [June 1928].)*

at the meeting, perhaps reflecting the general dearth of women among Chinese students studying science and technology in the United States.[38] Interestingly enough, a sizable number of members, like Chen Hengzhe, specialized in the humanities and social sciences but presumably had an interest in natural science. This would be the case throughout the history of the society, indicating that the society's broad conception of science was akin to the German term *Wissenschaft*. In fact, some members started out in the natural sciences but eventually switched to other fields. Perhaps the most famous examples of this phenomenon were two of the Boxer fellows at Cornell, Hu Shi and Zhao Yuanren. As mentioned in the introduction, Hu Shi initially majored in agriculture but later switched to philosophy, eventually earning a Ph.D. in philosophy at Columbia University under John Dewey. Zhao Yuanren excelled in both physics and mathematics at Cornell, but later at Harvard he took a Ph.D. in philosophy and gradually turned his interest to linguistics and music.[39] Yet the two (Zhao Yuanren even more so than Hu Shi) would remain active in the Science Society and pen numerous articles for *Kexue*. In 1929 Hu Shi would write the lyrics and Zhao Yuanren compose the music for the society's anthem, a song that

[38] Chen later received a master's degree from the University of Chicago and became the first female professor at Beijing University in 1920. The same year she married Ren Hongjun. Chen Hengzhe, "Chen Hengzhe zizhuan" (Autobiography of Chen Hengzhe), *Zhuanji wenxue* 26(4) (1975): 83–4.

[39] On Hu Shi's experiences as a student in the United States, see his *Hu Shi liuxue riji* (Hu Shi's diary as a student abroad) (Taibei: Commercial Press, 1959). On Zhao Yuanren at Cornell and Harvard, see Zhao, *Nianpu* (cit. n. 1), pp. 57–93.

emphasized both the practical uses of natural knowledge and the joy of pursuing science.[40]

Beginning with its first annual meeting, the society conducted its own organization in a meticulously democratic fashion, with an elaborate system of elections, which were published in *Kexue* and presented as an example for the rest of Chinese society. This and subsequent society meetings also witnessed the emergence of what has been called a new arena of "democratic sociability" among these budding Chinese scientists and intellectuals.[41] In choosing to conduct its affairs in an open way, the society, according to Ren Hongjun, consciously rejected the traditional form of "study societies" in China, which were usually built around "a single master, a man of virtue and talent, who because of his great learning and lofty reputation, attracted swarms of students." The Science Society, Ren went on to claim proudly, followed the model of the "modern scholarly society," which "is formed by the mutual assent of specialists, similar in learning and knowledge, who want to improve themselves through discussion."[42]

Politically, although few members were revolutionaries like Ren Hongjun and Yang Xingfo, most were nationalists. They supported the revolution of 1911 and sought to use what they were learning to contribute to Chinese reconstruction. Typical of the Science Society members was Zhu Kezhen, who studied science and technology to make China strong and to reform its culture. The 1916 meeting marked the beginning of Zhu's active participation in the society's affairs. He found kindred spirits in both the Science Society members and the *Kexue* staff in particular. Although he was not among the society's nine original founders, he quickly became one of its leaders. By 1915 Zhu had already switched from agriculture to meteorology, pursuing a Ph.D. in that field at Harvard. (See Figure 3.) Summer trips to the American south not only brought him face to face with the reality of racial discrimination but also made him aware of the differences between Chinese and American agricultural operations. Meteorology appealed to him as a scientific field with potential applications in agriculture.[43] In 1915 several of the founders of the Science Society—including Zhao Yuanren, Hu Mingfu, and Ren Hongjun—joined Zhu at Harvard to pursue graduate studies. With these moves, the center of activities for *Kexue* and the Science Society shifted gradually from Cornell to Harvard. Zhu now became intimately involved in the running of *Kexue* and the reorganization of the society.[44]

[40] Hu Shi to Hu Jianzhong, 4 Jan. 1960, and attached "Ni zhongguo kexueshe shege" (Draft of the anthem for the Science Society of China), reprinted in Hu Shi, *Hu Shi xueji: Shuxin* (Selected works of Hu Shi: Letters) (Taibei: Wenxing shudian, 1966), pp. 175–7.

[41] For a discussion of "democratic sociability" in the European and Chinese contexts, see Rowe, "Problem of 'Civil Society,'" (cit. n. 17), p. 147 and sources identified in the article.

[42] Ren, "Waiguo" (cit. n. 31), as quoted and translated in James Reardon-Anderson, *The Study of Change: Chemistry in China, 1840–1949* (Cambridge: Cambridge Univ. Press, 1991), p. 98. See also Guo Zhengzhao, "Zhongguo kexueshe yu zhongguo jindai kexuehua yundong, 1914–1935" (Science Society of China and the movement of scientism in modern China, 1914–1935), *Zhongguo xiandaishi zhuanti yanjiu baogao* (Special research reports on modern Chinese history) 1 (1971): 233–81.

[43] At Harvard Zhu worked under Robert Ward and in 1918 received a Ph.D. with a dissertation proposing a new classification of typhoons in Asia. He recognized the subtle and formative influence of the university on him, including its emphasis on empirical rigor as reflected in its motto "Veritas." Bianjizu, *Zhu Kezhen zhuan* (cit. n. 2), pp. 11–2.

[44] Zhu and Zhao Yuanren were also among the handful of students who took courses in the history of science with George Sarton, the pioneer of the discipline in the United States. Zhu would later become a leading historian of ancient Chinese science and help promote Joseph Needham's well-known research in that field. On Zhu's taking courses with Sarton, see Xie, *Zhu Kezhen* (cit. n. 1),

Figure 3. *Zhu Kezhen, a long-time leader of the Science Society of China, in 1918, when he was finishing his Ph.D. in meteorology at Harvard. (Reprinted from Zhu Kezhen,* Zhu Kezhen riji [Diaries of Zhu Kezhen] *[Beijing: People's Press, 1984], vol. 1.*

At the Andover meeting in 1916, Zhu was elected to the society's seven-person board of directors. That same year he started writing for *Kexue*, becoming one of its most prolific contributors. Between 1916 and 1950 he would write fifty-two articles for the journal, on topics ranging from his research on raindrop levels in Chinese history, the formation of the West Lake in Hangzhou, and new classifications of typhoons, to popular pieces on wind, climate, weather, and geography. Despite the rules in the inaugural issue of *Kexue* proscribing politics, many of his articles also treated political and social issues as related to science and scientists.[45] He would be a long-time editor of the journal and would serve as president of the society from 1927 to 1930.[46]

The society's members had a chance to put their rhetoric of scientific professionalism into practice in 1918, when the Science Society of China moved its headquarters to Shanghai. For the first time in history, a scientific community began to take shape on Chinese soil. By then, many society members had finished their studies in the United States and returned to China, taking up leading scientific and engineering positions at universities, industrial firms, and governmental agencies. Zhu Kezhen,

---

pp. 57–8. On Zhao, see his *Yuen Ren Chao's Autobiography* (cit. n. 4), p. 84. On Zhu and Needham, see Zhu Kezhen's diary entry for Oct. 1944 in Zhu Kezhen, *Zhu Kezhen riji* (Diaries of Zhu Kezhen) (1936–1949), 2 vols. (Beijing: People's Press, 1984), vol. 2, pp. 787–92.

[45] See "Zhu Kezhen zhuzuo mulu" (Bibliography of Zhu Kezhen's publications), in *Zhu Kezhen wenji* (A collection of Zhu Kezhen's writings) (Beijing: Science Press, 1979), pp. 514–25.

[46] See "Jishi: Zhongguo kexueshe di shishanchi nianhui jishi" (Records of the thirteenth annual meeting of the Science Society of China in 1928), *Kexue* 13 (1928): 685–97.

for example, took a teaching position in geography and meteorology at Wuhan University in 1918; over the next nine years he would teach at universities in Nanjing and Tianjin as well. In the process he would help found some of modern China's first departments of geography and meteorology and train the first generation of scientists in these fields.[47] Zhou Ren, the mechanical engineer and another of the founders of the society, had already returned to China three years before Zhu, in 1915, after receiving a master's degree from Cornell. Over the years he would teach in Nanjing and work in industrial firms.[48] Like Zhu, Ren Hongjun and Yang Xingfo, the two revolutionaries, returned to China in 1918, Ren with a master's degree in chemistry from Columbia and Yang with a master's of business administration from Harvard. Not being research scientists, Ren and Yang initially pursued industrial projects for China but eventually would become organizers of science and education. In 1920 another society founding member, Bing Zhi, returned to China with a Ph.D. in biology from Cornell. In Nanjing, he founded one of China's first departments of biology at the Nanjing Advanced Normal College.[49]

Even as they engaged in disparate endeavors often under difficult conditions, these and other first-generation Chinese scientists and engineers found the Science Society of China their most valuable social support network and *Kexue* their most effective voice in articulating a place for science in the Chinese public sphere.[50] It was not a small matter to bring together a scientific community divided not only by disciplines and regions but also by different philosophies, a result of having attended different schools in the United States and Europe.[51] Through *Kexue* and other activities usually associated with the working of a Western scientific society, the Science Society helped, as James Reardon-Anderson points out, to both popularize and legitimize the study of science among Chinese at large and "to bind together and raise the spirits of the members themselves."[52]

Of course, the popularity of science, especially among young students, benefited greatly from the radical May Fourth Movement of 1919, which called for the introduction of "Mr. Science" and "Mr. Democracy" as the two pillars of modernization to reform and strengthen China. In turn, Science Society members and *Kexue* contributed to the New Culture Movement, which was a major part of the May Fourth agenda. *Kexue* promoted *New Youth*, the leading journal of the May Fourth New Culture Movement (edited by Chen Duxiu). Hu Shi, leader of the vernacularization of the Chinese language as part of a new culture, wrote for both journals.[53] *Kexue*

[47] Bianjizu, *Zhu Kezhen* (cit. n. 2), pp. 13–23.

[48] Zhou Peide and Dong Deming, "Zhou Ren," in ZGXDKXJZJ (cit. n. 3), vol. 5, pp. 783–91.

[49] Zai Qihui, "Bing Zhi," in ZGXDKXJZJ (cit. n. 3), vol. 1, pp. 458–68.

[50] On the difficulties faced by returned students in general and by Science Society members in particular, see Ren Hongjun's and Yang Xingfo's correspondence with Hu Shi in 1918 and 1919, published in *Hu Shi laiwang shuxin xuan* (Selected correspondence of Hu Shi), ed. Division on the History of the Republic of China, Institute on Modern History, Chinese Academy of Social Sciences, 3 vols. (Hong Kong: Zhonghua Shuju, 1983), vol. 1, especially Ren to Hu, 24 June 1918, pp. 15–6; Yang to Hu, 11 Dec. 1918, p. 22; 22 April 1919, pp. 39–40; and 31 July 1919, pp. 64–5.

[51] Norbert Wiener, MIT professor and founder of artificial intelligence who visited Qinghua University in Beijing in the 1930s, provided a vivid description of how many of the Chinese faculty carried the distinct styles of the country in which they were trained. Norbert Wiener, *I Am a Mathematician* (New York: Doubleday, 1956), p. 186.

[52] Reardon-Anderson, *The Study of Change* (cit. n. 42), p. 99.

[53] See Hu Shi, *Hu Shi koushu zizhuan* (Hui Shi's oral autobiography), trans. and ed. Tang Degang (Taibei: Zhuanji Wenxue Press, 1986). Hu credited the discussions with Ren Hongjun, Yang Xingfo,

was also the first journal in China to adopt horizontal typesetting and to use western-style punctuation, a move initially attacked by conservative Chinese critics. In their own defense, *Kexue* editors explained that the traditional, vertical typesetting would make it difficult to insert scientific formulas and the lack of punctuation would make difficult scientific reasoning even harder to follow. Scientific professionalism, in other words, mandated cultural changes.[54]

If criticism defined a central characteristic of the public sphere, Science Society leaders made ample use of *Kexue* and other media to attack traditional Chinese culture and beliefs as they established the authority of science and scientific method. Like the European philosophes in the eighteenth century, they sought to hold all ideas and practices to the test of free critical discussion.[55] If they appeared to advocate scientism by singling out science and the scientific method as the one best way of understanding the world, both natural and social, it was because they believed science—or what they called the "scientific spirit"—best exemplified free critical discussion. Thus in the inaugural notes, *Kexue* editors expressed their admiration for Galileo, who followed his natural curiosity to seek the truth and "fought bloody battles with religion for the freedom of thought." Looking at the Chinese political and cultural scene, the scientists lamented the desertion of scholarship and spiritual and material bankruptcy. "Although [we] closed our borders and tried self-reliance, it's still not enough [for China] to survive, especially in today's world. It is clear that to be a modern scholar one cannot just bury oneself in the old papers [meaning classical Chinese learning]." The solution was science, and *Kexue* was a way to spread it: "We hope to write [about what we have] learned every day in this journal and use it to stimulate truth-seeking minds and lead to ways for [scientific] applications."[56]

Science Society leaders' advocacy of science often went hand in hand with their social and cultural criticism. In a 1922 speech on "biology and women's education," Bing Zhi, for example, called for special attention to biology in women's education. For lack of biological knowledge, Bing argued, "people often misunderstood [natural] phenomena in their environment. Such misunderstanding led to superstitions and various associated harmful effects. If we reflect on the various bad habits of Chinese society today, such as the worship of dragon kings, tree gods, fox goddesses, and road ghosts . . . [we can see that] none of them was not caused by damned superstitions." He hoped that women would take the lead in learning biology and use it to combat such superstitions.[57] Likewise, Zhu Kezhen criticized government officials for resorting to "praying for rain and [the] banning of animal slaughter" as the solution to the problem of severe drought, calling such action a policy of "fooling the people." "We call our country a republic, thus everyone from the president at the top to the head of the county should be responsible to the people. [The best way to deal with] disastrous droughts or floods is to prepare for them before they come, by

---

psychologist Tang Yue, and literary scholar Mei Disheng, all fellow members of the Science Society, with launching him on the road of the literary revolution.

[54] See Liu Weimin, "Kexue zhazhi yu xin wenxue geming" (*Kexue* magazine and the new literary revolution), *Kexue* 49 (1997): 34–8.

[55] For a discussion of the role of criticism in the life of the public sphere during the Enlightenment, see Broman, "Habermasian Public Sphere" (cit. n. 19), pp. 129–31.

[56] "Fa kan ci" (cit. n. 33), on pp. 5–7.

[57] Bing Zhi, "Shengwuxue yu nüzi jiaoyu" (Biology and women's education), *Kexue* 7 (1922): 1175–80.

reforestation, by water conservancy, and by the establishment of a large number of meteorological stations."[58]

Of course, the Science Society and *Kexue* were not the only arenas for critical public discussions in Republican China, even for scientists. The emergence of an urban popular culture with its flourishing market for books, newspapers, and periodicals in late Qing and early Republican China, especially after the May Fourth Movement of 1919, provided ample opportunities for those scientists so inclined to engage in public debates over science, culture, and national politics. The Shanghai-based Commercial Press (Shangwu yinshuguan), which printed *Kexue* in its early years, also published other magazines and a large number of books related to science. Zhu's article on praying for rain, for example, appeared in Commercial's *Dongfang Zazhi* (Orient magazine), perhaps one of the most widely read periodicals at this time. During the 1920s, two key leaders of the Science Society, Ren Hongjun and Zhu Kezhen, actually worked in the Commercial Press, editing encyclopedia and science textbooks for the popular market.[59]

Regardless of the arenas they chose, leaders of the Science Society rarely missed an opportunity to advance the professional interest of science in the public sphere. During a famous debate over science and metaphysics in the early 1920s, for example, leading intellectuals of the May Fourth period, most of whom, on both sides, were members of the Science Society, exchanged polemics on whether science could govern a view of life (*renshengguan*). The argument started when the philosopher Zhang Junmai at Qinghua University in Beijing, returning from a tour in war-ravaged Europe, declared in a lecture in 1923 that science—objective and logical—could not govern a view of life that is subjective and intuitive. Ding Wenjiang, a British-trained geologist and later president of the Science Society, saw Zhang's challenge as one not only against scientism but also against scientific progress and Chinese modernization. Calling Zhang's beliefs "a metaphysical ghost," Ding responded that all phenomena, whether material or psychological, if they were "real," fell under science. If they could not be rationally analyzed, they were not "real." Others soon joined in the fray, including, on Ding's side, his close friends Ren Hongjun and Hu Shi.[60]

Although the debate was over philosophical and cognitive issues, the real stakes were professional and political. In his own contribution to the debate, Hu Shi argued that although World War I led some European intellectuals to question material progress and science, the situation in China was different:

> China at the present has not enjoyed the benefit from science, much less [suffered] the 'disasters' brought by science. Let us try to open our eyes and look around: the widespread divination altars and temples, the widespread magic prescriptions and ghost pho-

[58] Zhu Kezhen, "Lun qiyu jintu yu hanzai" (On praying for rain, banning of slaughtering, and drought), *Dongfang zazhi* (Orient magazine) 23(13) (1926), reprinted in Zhu, *Wenji* (cit. n. 45), pp. 90–9.

[59] For Ren Hongjun at Commercial, 1922–1923, see Zhao Huizhi, "Ren Hongjun nianpu (xu)" (Chronological biography of Ren Hongjun [continued]), ZGKJSL, 9(4) (1988): 37–47, on pp. 39–40. Zhu worked at the Commercial Press in 1925. See Bianjizu, *Zhu Kezhen zhuan* (cit. n. 2), p. 23.

[60] On the debate see D. W. Y. Kwok, *Scientism in Chinese Thought, 1900–1950* (New Haven, Conn.: Yale Univ. Press, 1965), especially chap. 6 ("'Science' Versus 'Metaphysics' in the Debate of 1923"); and Charlotte Furth, *Ting Wen-chiang: Science and China's New Culture* (Cambridge, Mass.: Harvard Univ. Press, 1970), chap. 5 ("Science and Metaphysics"). Ding, Ren, and Hu published their polemics in their own *Nongli* (Endeavor), a political weekly.

tography, such undeveloped transportation, and such undeveloped industry—how do we deserve to refuse science? . . . The Chinese view of life has never even encountered science face-to-face! At this time we are still troubled that science is not being promoted adequately, troubled that science education is not being developed, and troubled that the force of science is not enough to sweep away the evil spirit that spreads all over the country. Who could have expected famous scholars to come out to shout "European science has gone bankrupt," to put the blame of the cultural bankruptcy of Europe on science, to belittle science, to enumerate the crimes of scientists' view of life, and [to demand] that science not have any impact on view of life! How could people with faith in science not be worried about the current situation? How could [they] not come out and defend science in a loud and clear voice?[61]

This rhetorical defense of science in the public sphere was, however, only one part of the agenda of scientific professionalism. For Chinese science in general and the Science Society of China in particular to succeed, the rhetorical social capital had to be translated into real support in terms of financing and institutional building. This often involved negotiations with the government in the third realm.

SCIENTIFIC NATIONALISM, THE STATE, AND CIVIL SOCIETY

The creation of the Science Society and other disciplinary societies marked the introduction of what James Reardon-Anderson calls "the professional ideal" into the Chinese social order, an ideal that tended to have scientists aligned with society rather than the state, at least in rhetoric. As Reardon-Anderson explains:

The initiative [to organize scientific societies] came in all cases from outside the government, and in some from Chinese outside of China. The intention was to address the Chinese people directly, bypassing government. And the purpose was to foster an independent enterprise that would help remake China, irrespective of who ruled the country. Underlying these efforts was a conception, unstated and perhaps unconscious, of the scientific role: The scientist's purpose was to pursue knowledge, apply it in useful ways, and communicate it freely to others. His commitment was to an autonomous activity, separate from politics, yet serving in a disinterested way the public good.[62]

Among writings of Science Society leaders we find numerous arguments for distancing professional science from the government. In comparing the Science Society to the Royal Society of London, Ren Hongjun and his supporters saw themselves as forming a Republic of Letters in China, just as Robert Boyle and his supporters had in England. Like the Royal Society, the Science Society was a voluntary association of scientists for "self-cultivation" and "mutual assistance."[63] Zou Bingwen, another Boxer fellow and early leader of the Science Society, explicitly invoked the Royal Society in arguing, as early as 1914, that the Science Society should be independent of the government:

Although a science society [can]not avoid seeking assistance from the government for its operating funds, it should never let the government run it. This is because

[61] Hu Shi, "Kexue yu renshengguan xu" (Science and view of life preface), in *Kexue yu renshengguan* (Science and view of life), by Zhang Junmai et al. (1923; reprinted, Jinan: Shandong People's Press, 1997), pp. 12–3.
[62] Reardon-Anderson, *The Study of Change* (cit. n. 42), pp. 101–2.
[63] Ren Hongjun's phrases in *Kexue*, as quoted in Peter Buck, *American Science and Modern China, 1876–1936* (Cambridge: Cambridge Univ. Press, 1980), p. 120.

> administrators rarely excel at the promotion of scholarship. The Ministry of Education once established an Academic Council. But two years have passed and I have not seen anything coming out of it. Therefore I say, to build and maintain the Science Society we have to rely on our fellow scholars. The Royal Society of Britain is perhaps the oldest scientific society. Its establishment depended on the support of Boyle and Newton of the British scholarly world [and not the government].[64]

The fact that members of the Science Society sought to keep their distance from the government also reflected a radical departure from the traditional Confucian view of scholarship in service to the government/state. "It was legitimate and proper," as one student of the society noted, "for scholars to engage in scholarship and to see that scholarship as socially useful without feeling compelled to take part in government service."[65]

In a 1921 *Kexue* article titled "The Responsibilities of Geoscientists in Our Country," Zhu Kezhen echoed Zou's point by urging Chinese scientists to build up a nationalist science by relying on nonstate resources. He enumerated instances in recent Chinese history when a lack of geographic knowledge led rulers to make concessions of supposedly "valueless" territories to foreign powers, such as the Qing's transfer of Taiwan to Japan in the late 1890s. Not since the seventeenth century, he charged, had the government even tried to make a more accurate map of the country, leaving it to foreign powers, such as Germany and Japan, to survey and carve up the best coastal territories. He recorded the humiliating experience of finding out, on his trip back to China in 1918, that there were more books on Chinese geography by Japanese authors than by Chinese. To his dismay, he also discovered that the two meteorological stations on which the Chinese relied for the forecasting of destructive typhoons along the coast were founded and controlled by French missionaries in Shanghai and by the British government in Hong Kong. Zhu chided the Republican government for neglecting scientific developments:

> Today's government is concerned only with meeting the needs of the warlords. How can one expect it to spare resources and fund the development of meteorological stations and other institutions? To accomplish this, we have to rely on the whole society and the citizens. Every man shares the responsibility for the rise or fall of our country. After all, the government was not always the driving force behind geographic surveys in Europe, the United States, and Japan. [Even in our country] there are good examples of success where hard labors by individuals or cooperative efforts by scientific societies resulted in major books and projects [on Chinese geography].[66]

Though the scientists frequently shot such rhetorical arrows at the government, a close examination of their actual interactions with the state, especially during the

[64] Zou Bingwen, "Kexue yu kexueshe" (Science and science societies) (speech at the annual meeting of the Association of Chinese Students in the United States), *Liumei xuesheng jibao* (Chinese students in America quarterly) 2 (winter 1915): 4, as quoted in Qian Li, "Lun zhongguo kexueshe jianli de zongzi" (On the motivations for the establishment of the Science Society of China), Master's thesis, Beijing Univ., 1986, pp. 24–6.

[65] David Reynolds, "The Advancement of Knowledge and the Enrichment of Life: The Science Society of China and the Understanding of Science in the Early Republic, 1914–1930," Ph.D. diss., Univ. of Wisconsin–Madison,1986, p. 49.

[66] Zhu Kezhen, "Wuoguo dixuejia zhi zeren" (The responsibility of geoscientists in our country), *Kexue* 6 (1921), reprinted in Fan Hongye and Duan Yibing, eds., *Zhu Kezhen wen lu* (Essays of Zhu Kezhen) (Hangzhou: Zhejiang Culture and Arts Press, 1999), p. 8.

Nanjing Decade (1927–1937) under the Nationalists, tells a different story. Major portions of the support for the Science Society came from both national and local government. Government educational and research institutions often employed members of the society. Indeed, rather than bypassing the government, leaders of the Science Society actively sought support and sponsorship from it for their various endeavors. Even Zou and Zhu's critical comments above acknowledge the need for governmental support and, especially in Zhu's case, for an active role of government in scientific research, if not in the organization of the scientists themselves.

Scientific professionalism may have pushed scientists to seek autonomy from the government, but the practical needs of science and the scientists' sense of nationalism nevertheless created a powerful climate for collaboration in Republican China. Conducting scientific research in China was a process fraught with obstacles, paramount among them the lack of access to scientific literature and experimental facilities. Thus as a priority, the society sought to encourage its members to stay in research by creating local chapters in Shanghai, Nanjing, and Guangzhou and equipping them with science libraries.

Perhaps the most important step in transforming the society into a real research institution was the 1922 founding in Nanjing of its Biological Institute, complete with resident researchers and laboratories. This was the first private scientific research institution in China established and staffed by Chinese scientists. The zoologist Bing Zhi was its founding director, and the botanist Hu Xiansu, trained at the University of California, was his deputy.[67] With Bing and Hu turning out papers and monographs on Chinese fauna and flora from the laboratory, the society succeeded in its transition from discoursing on science to discoveries in science.

Why did the society choose to venture into biological research before any other scientific field? Expedience. As Ren Hongjun later explained, "In biological research, it is relatively easy to capitalize on local materials and the expenses for such efforts are also low."[68] In addition, there were several biologist members of the society on the faculty at the Southwestern University in Nanjing whose participation at the institute helped make it into China's premier center of biological research, especially in the taxonomy of Chinese flora and fauna.[69]

Where did the Science Society get the funding for all its endeavors? Perhaps nothing better demonstrates the society's nature as a third realm institution in Republican China than the juxtaposition of government, semigovernment, semipublic, and private sources for funding. In the earliest days when the society was headquartered in the United States, its leaders simply subsidized the expenses of publishing *Kexue* with savings from their own meager Boxer fellowships.[70] At the same time, Ren Hongjun and other leaders cultivated prominent figures in Chinese intellectual and

[67] Science Society of China, *Science Society of China: Its History, Organization, and Activities* (Shanghai: Science Press, 1931), pp. 1–3. See also Reynolds, "Advancement of Knowledge" (cit. n. 65). On Hu Xiansu, who fought against Hu Shi's vernacularization movement, see Shi Hu, "Hu Xiansu," in ZGXDKXJZJ (cit. n. 3), vol. 4, pp. 423–33; and Shen Weiwei, *Huimu xueheng pai: Wenhua baoshou zhuyi de xiandai mingyun* (Reexamining the *xueheng* school: The modern fate of cultural conservatism) (Beijing: People's Literature Press, 1999), chap. 3.

[68] Ren Hongjun, "Zhongguo kexueshe zhi guoqiu ji weilai" (The past and future of the Science Society of China), *Kexue* 8 (1923): 8.

[69] Yang, "Ren Hongjun" (cit. n. 29), pp. 312–3.

[70] Several members, including Zhao Yuanren, suffered malnutrition and became ill when they lived on soup and apple pies. See Zhao, *Yuen Ren Chao's Autobiography* (cit. n. 4), 79.

political circles who acted as patrons and sponsors of the association. The Science Society registered with the Ministry of Education in March 1916 and thereby gained the status of a legal organization (*faren tuanti*).[71] *Kexue*'s January 1918 special issue on the society's first annual meeting, in 1916, featured congratulatory messages from the president of Republican China, Li Yuanhong; former minister of agriculture and business and famous entrepreneur Zhang Jian (Jizhi); Beijing University president Cai Yuanpei; and Minister of Education Fan Yuanlian.[72]

To attract patrons without sacrificing its professional criteria, the society established six categories of membership: ordinary member, life member, junior member, honored member, honorary member, and supporting member. Ordinary membership was for those who conducted scientific research and engaged in scientific enterprises; life members were those ordinary members who contributed a one-time 100-yuan membership fee; junior members were ordinary members with "secondary standing." The three other categories were honorary in nature, usually reserved for patrons of the society.[73] In 1917 the Science Society awarded its first honored membership to Cai Yuanpei, an honorary membership to Zhang Jian, and supporting memberships to Fan Yuanlian and three others.[74] In 1922 the society reorganized its leadership structure so that in addition to an executive council, which consisted of the leaders of the society, there was a board of directors consisting of prominent patrons of the society. It included, among others, Cai Yuanpei, Zhang Jian, and Liang Qichao, perhaps the best-known Chinese liberal intellectual and reformer in late Qing and the early Republic era.[75]

The efforts at cultivating public and private patronage first paid off in 1918, when Cai Yuanpei made it possible for Beijing University to provide a 200-yuan monthly subsidy to the Science Society in return for the nominal service of helping Beijing University in buying and translating science books. As president of Beijing University, the most prestigious state-funded university, Cai was an influential leader of the May Fourth–era intellectuals, whom Ren Hongjun had cultivated from the beginning of the Science Society. The monthly subsidy, though not a large amount, was enough to prevent *Kexue* from going into bankruptcy and help it resume publication after an eight-month stoppage.[76] Encouraged by this development, Ren Hongjun and other leaders of the society asked Cai Yuanpei and Fan Yuanlian to sponsor a fund-raising drive in 1918 and 1919 to raise 50,000 yuan as the principal for a fund to support the society. Cai Yuanpei wrote an eloquent appeal to the Chinese public in which he called on both "the government of our country and those with resources

[71] Science Society, *Science Society of China* (cit. n. 67), Chinese section, p. 1.

[72] *Kexue* 3 (1918).

[73] Science Society, *Science Society of China* (cit. n. 67), p. 5.

[74] See Zhang Jian (not the same as the Zhang Jian in the text), "Cai Yuanpei yu zhongguo kexueshe" (Cai Yuanpei and the Science Society of China), *Shilin* (Forest of history) no. 2 (2000): 56–71, on p. 60.

[75] Science Society, *Science Society of China* (cit. n. 67), p. 3. On Liang Qichao, see Hao Chang, *Liang Ch'i-ch'ao* [Liang Qichao] *and Intellectual Transition in China, 1890–1907* (Cambridge, Mass.: Harvard Univ. Press, 1971); and Philip C. C. Huang, *Liang Ch'i-ch'ao and Modern Chinese Liberalism* (Seattle: Univ. of Washington Press, 1972).

[76] See Zhang, "Cai Yuanpei" (cit. n. 74), p. 61. On Ren Hongjun's appeal to Cai Yuanpei for support of the Science Society, see Ren Hongjun to Cai Yuanpei, Li Shizeng, and Jiang Jing Wei, 15 May 1915, and Cai Yuanpei and Li Shizeng to Ren Hongjun, June 1915, reprinted in Gao Pingshu, ed., *Cai Yuanpei lun kexue yu jishu* (Cai Yuanpei on science and technology) (Shijiazhuang: Hebei Science and Technology Press, 1985), pp. 33–4.

in our society" (meaning Chinese society in general, not the Science Society) to join together to raise more money than the Science Society was asking for. Only in this way "could one clear the shame that we Chinese are indifferent to science."[77]

This and other fund-raising drives, mainly from the private sector, met with mixed results. Even though Ren Hongjun devoted considerable time to the activity, by 1922 the society had reached less than half of its goal. A breakthrough came in 1922 when Zhang Jian, the entrepreneur who established a number of successful industrial operations in Nantong, donated 10,000 yuan to the society, which, along with other donations, allowed it to construct its long-planned Biological Institute in Nanjing.[78] That year the society met in Nantong, a coastal city north of Shanghai, for its seventh annual meeting, as a way both to thank Zhang Jian for his donation and to spotlight the success of Nantong as a model of local self-rule and of industrial development.[79] Further fund-raising efforts, however, proved difficult: there were plans for an institute on mathematics, physics, and chemistry in Shanghai that failed due to the lack of funding.

Although less publicized, government support proved crucial to the financial survival of the Science Society. In 1919 the Ministry of Finance approved the society's request to use some government buildings in Nanjing as offices, for a term of six years. In 1921 the Guangdong provincial government granted the society properties in Guangzhou for its local chapter's offices. In 1923 the society, with the help of its board of directors, gained a monthly subsidy of 2,000 yuan from the national funds (*guoku*) allocated through the Jiangsu provincial government. In 1927, after the establishment of the Nationalist government in Nanjing, the society applied for and received 400,000 yuan (a huge sum) from the Ministry of Finance, to be used as the principal for a foundation for the society. The new government also granted the Nanjing properties permanently to the society.[80]

In addition to the private, governmental, and semigovernmental (Beijing University) funding, there was one more source of funding, which we might term "semipublic." This was the China Foundation for the Promotion of Education and Culture, established in September 1924 with funds from a second batch of Boxer indemnity funds returned from the United States, totaling about $12.5 million. A joint China-U.S. board, appointed by the two governments but allowed to function on its own, was to distribute the funds to qualifying educational and cultural enterprises. In 1926 the China Foundation gave an annual grant of 15,000 yuan to the Science Society, plus a one-time gift of 5,000 yuan for the purchase of equipment by its Biological Institute. The annual grant continued at the same level until 1928; beginning in 1929, the foundation increased it to about 50,000 yuan, an arrangement that lasted until World War II.[81] Needless to say, the annual grant became the lifeline for the society, especially for its expanding Biological Institute, throughout those years.

[77] Cai Yuanpei, "Zhongguo kexueshe zhengji jijin qishi" (Announcement on fund-raising for the Science Society of China), 31 Dec. 1918, reprinted in Gao, *Cai Yuanpei* (cit. n. 76), p. 41.

[78] "Ben she shengwu yanjiusuo kaimu ji" (Report on the opening ceremony of the society's Biological Institute), *Kexue* 7 (1922): 846–8.

[79] "Zhongguo kexueshe di qi ci nianhui jishi" (Report on the seventh annual meeting of the Science Society of China), *Kexue* 7 (1922): 974–1013.

[80] Science Society, *Science Society of China* (cit. n. 67), pp. 3–4.

[81] Cao Yu, "Zhonghua jiaoyu wenhua jijinhui yu zhongguo xiandai kexue de zaoqi fazhan" (China Foundation for the Promotion of Education and Culture and the early development of modern science in China), ZRBZFTX 13(3) (1991): 33–41, data from p. 36.

Why did the China Foundation give so much funding to the Science Society? The society became one of the biggest beneficiaries of the foundation's largesse partly because of active lobbying by society leaders and patrons. In 1924 when the U.S. Congress first discussed the possibility of returning additional Boxer funds, and other countries were expected to follow suit, the Science Society leaders began to mobilize their network of social connections and capitalize on the society's position in the public media. On May 25, 1924, for example, Ren Hongjun wrote Hu Shi, then the center of a network of influential figures with ties to the Science Society, to ask him to use his considerable influence with American diplomats in Beijing to ensure that at least part of the returned Boxer funds would be used for scientific research:

> It is definite at the present that the rest of the American indemnity fund will be returned to China. Now that it is determined that the fund will be used for educational and cultural enterprises, our colleagues in the Science Society believe that we will be justified if we take advantage of the opportunity and propose to have part of the fund designated for supporting scientific enterprises (referring to scientific research enterprise in general, not just the Science Society). But we know very well that currently there are many who are working on getting this fund. If our Science Society joins in the competition, how should we go about doing it to be effective? Recently our colleagues in Nanjing and Shanghai have met countless times to talk about this matter. . . . Today, in Nanjing, the executive council of the Science Society met again to discuss this subject. We all agreed that we should not put off actions any longer and a resolution was passed that asked me to go to Beijing to talk it over with you.[82]

While Ren Hongjun tried to make the matter not one of purely institutional self-interest, Yang Xingfo was more direct in his letter to Hu Shi the next day. The Science Society was "rather eager to share the soup" of the returned funds and was even willing to pay for a trip by Hu Shi to the United States to lobby on its behalf.[83]

In this "Boxer funds for science" campaign, the Science Society not only utilized its informal social networks, but also cashed in its considerable social capital in the public sphere. In July 1924, the society widely publicized its "Science Society of China's Declaration on the Use of the Returned Boxer Indemnity" by publishing it in Chinese and English in *Kexue* and other media. Drafted by Ren Hongjun based on discussions with leaders of the society, it explained that the purpose of the declaration was to summarize public opinion from many quarters and to provide guidance for the government. It advocated the use of the returned Boxer funds from Britain and the United States for three specific causes: pure research (funding new and existing research institutes, subsidies for research at universities, and sending students abroad), research infrastructure and popularization of knowledge (libraries and museums), and international cultural exchange (endowed chairs on Chinese culture in British and U.S. universities, exchanges of professors, and the sending of British and American students for study in China). As to the management of the funds, the declaration insisted that first, the Chinese and the foreign government in question

[82] Ren Hongjun to Hu Shi, 25 May 1924, reprinted in *Laiwang shuxin xuan* (cit. n. 50), vol. 1, pp. 253–4.

[83] Ynag Xingfo to Hu Shi, 26 May 1924, reprinted in *Laiwang shuxin xuan* (cit. n. 50), vol. 1, p. 254.

should agree to set the principles for the uses of the fund based on the suggestions of prominent scholarly and educational organizations. Then they should jointly appoint to the governing board those "pure scholars and leaders of education and industry" who were "completely detached from any political or diplomatic relations." The board then should be given "complete freedom in execution within the bounds of the general agreement."[84]

By advocating a relatively autonomous board for the China Foundation, Ren Hongjun and other Science Society members walked a fine line between the desire to satisfy their professional interests and a strong sense of nationalism. Ideally, if China had a unified, democratically supported government, the funds should be turned over unconditionally to it for use in the best national interest. But China was under the rule of warlords. "In the current circumstance of anarchy and chaos, we should not hope that they [the United States] would 'unconditionally throw away' a fund of tens of millions," wrote Hu Shi. "Whom should they throw it away to? [If they] throw it to the government, we certainly would not be satisfied[.] [If they] throw it to the National Educational Association, or throw it to the Chinese Association for the Improvement of Education, would there be no disputes?"[85]

Nationalism nevertheless made an oblique but unmistakable appearance in the Science Society's declaration, which reminded the public that the funds were the "blood and sweat money of our people." The money should not be allowed to fall into the hands of manipulating politicians who claimed that they would pave roads and support education with the resultant revenues. "During this time of unstable political situation, when the citizens have long lost our power to monitor [the government] . . . how can we make sure that the incomes would go into supporting schools and not be misused for other purposes? Even a fool knows that it is hopeless."[86] In an editorial for *Kexue*, Zhu Kezhen amplified the point of the declaration. He reviewed the origins of the Boxer indemnity funds as the result of "foreign countries using their victor's brutal menace to force our people to pay this huge indemnity after the Boxer War." He implored the various intellectual organizations to unite and present a unified voice in demanding an active role in deciding the uses of the returned funds: "Since they are called returned funds, since they are supposed to be for Chinese educational and cultural enterprises, they should be Chinese funds used on Chinese enterprises, and we Chinese should be the owner of these funds and decide their uses."[87]

Ren Hongjun and his colleagues breathed a collective sigh of relief when the new China Foundation board adopted most of their proposals. The Chinese government even agreed to add Ding Wenjiang to the board, which was dominated by government and educational leaders, apparently as a response to Ren Hongjun's push to have "a real scholar" on it.[88] Ding Wenjiang also made it onto the board governing

[84] "Zhongguo kexueshe dui gengkuan yongtu zhi xueyan" (The Science Society of China's declaration on the use of the returned Boxer indemnity), *Kexue* 9 (1924): 868–71, on pp. 870–1.

[85] Hu Shi to Tao Xingzhi and Ling Bing, 25 April 1926, reprinted in *Laiwang shuxin xuan* (cit. n. 50), vol. 1, pp. 371–4, on p. 372.

[86] "Dui gengkuan yongtu zhi xueyan" (cit. n. 84), pp. 868–9.

[87] Zhu Kezhen, "Shelun: Gengzi peikuan yu jiaoyu wenhua shiyan" (Editorial: the Boxer Funds and educational and cultural enterprises), *Kexue* 9 (1924): 1015–9.

[88] Ren Hongjun to Hu Shi, 5 Sept. 1924, reprinted in *Laiwang shuxin xuan* (cit. n. 50), vol. 1, pp. 259–60. See other letters that Ren wrote to Hu, one undated in Sept. 1924, pp. 266–7; 6 Oct. 1924, pp. 267–8.

Figure 4. *Fourteenth annual meeting of the Science Society of China, August 1929, Beijing. (Reprinted from* Kexue, *14[3][March 1929]).*

the funds returned from Britain, as did Hu Shi.[89] Even better for the Science Society, Ren Hongjun himself joined the staff of the China Foundation in September 1925. Gradually moving up the ladder within the foundation, by January 1929 Ren Hongjun had become both a board member and a staff director, essentially running day-to-day operations.[90] With funding from the government and the China Foundation, the Science Society experienced steady growth in the 1920s. Membership grew from 35 in 1914 to 77 in 1915, from 180 in 1916 to 363 in 1918 when the society moved to China, expanding to 850 in 1927 when the Nationalist government began, and 1,005 by 1930.[91] Even after the establishment of many disciplinary scientific societies, such as the Chinese Physical Society, the Science Society remained the largest and most important general scientific society in China.

The highlight of the society's activities in any given year was usually the annual meeting. (See Figure 4.) Typically the society held these meetings in major cities (Hangzhou in 1919, Nanjing in 1920, Beijing in 1921, Nantong in 1922, Hangzhou in 1923, Nanjing in 1924, Beijing in 1925, Guangzhou in 1926, Shanghai in 1927, Suzhou in 1928, Beijing in 1929, and Qingdao in 1930). Typically, at the beginning of the meeting there would be welcoming speeches from local government officials and local organizations such as the chamber of commerce or a newspaper. Following the speeches the society's departments would report on the status of various programs and projects. There were sessions devoted to scientific and technological papers as well as popular lectures for the local audience. There would be numerous

[89] Ding explained to Hu Shi that he hoped to secure a grant from the British returned Boxer funds as the principal of a foundation for his Institute of Geological Survey in Beijing. Ding Wenjiang to Hu Shi, ca. May 1925, reprinted in *Laiwang shuxin xuan* (cit. n. 50), vol. 1, pp. 332–4.

[90] Zhao, "Ren Hongjun" (cit. n. 59), pp. 42–5.

[91] Science Society, *Science Society of China* (cit. n. 67), pp. 3–5.

banquets given by various local hosting organizations and by the society. Members had more opportunities for socializing on outings to local institutions or scenic spots. Perhaps most important, such excursions also brought Science Society members into contact with other parts of the emergent civil society in China in this period.[92]

Society members also reached out to layperson and scientist alike through its diverse publications. In addition to *Kexue*, the society published *Kexue Huabao* (Science pictorials), containing easy-to-read, well-illustrated articles on natural phenomena, as a way to popularize science, as well as *The Transactions of the Science Society of China*, which carried technical papers for the international scientific community. Beginning in 1928 the society also administered an annual student science-paper prize in memory of Gao Junwei, a young nutritional chemist and one of the few female members of the society, who had died of cancer that year. There were also an annual archaeological prize and a scholarship named after Madam Fan, a donor to the society, to be awarded to a student in the Biological Institute.[93] To fulfill both its own original goal of popularizing science and an "order of the central government," the society established the Bureau for Scientific Information in 1929 to answer science-related questions from the public. Questions and answers, by specialists among the members, were then printed in *Kexue*.[94] How effective such measures were in popularizing science is hard to determine, but there did seem to be a steady stream of questions from readers, some of whom used it as a way to give feedback to the editors.[95]

In addition to the Biological Institute in Nanjing, the society also operated a printing press, a scientific books and instruments company, and two science libraries, one in Nanjing, and one in Shanghai named after Hu Mingfu, who accidentally drowned in 1928. By 1927 the library in Nanjing had already amassed a sizable collection of materials, presumably related to science: 2,788 Chinese books, 10,572 books in western languages (mostly acquired through Sino-American exchanges), 3,087 issues of Chinese periodicals, and 20,493 issues of western ones.[96] The libraries were open to "students of science" (it is not clear whether nonmembers were permitted to check out books). As specialized journals for various scientific fields became available, *Kexue* shifted its editorial policy from publishing research papers to the popularization of science to satisfy an increasing public interest.[97] The Biological Institute expanded its research staff, scientific programs, and physical space devoted to laboratories, libraries, and exhibits. As the institute's research reputation rose, a whole generation of Chinese biologists sought to work there, and it became the "cradle" of modern biology in China.[98]

In view of their sense of scientific nationalism, their widespread yearning for a strong central government supportive of science, it is not surprising that the leaders

[92] See, e.g., "Zhongguo kexueshe di shierci nianhui jishi" (Report on the twelfth annual meeting of the Science Society of China), *Kexue* 12 (1927): 1616–54.

[93] See "Sheyou Gao Junwei nushi shilue" (Biographical sketch of member Ms. Gao Junwei), *Kexue* 13 (1928): 462–3; Science Society, *Science Society of China* (cit. n. 67), p. 30.

[94] "Zhongguo kexueshe fushe kexue zixunchu tonggao" (Announcement of the establishment of a Bureau for Scientific Information of the Science Society of China), *Kexue* 14 (1929): 759.

[95] See, e.g., "Kexue zixun" (Scientific information), *Kexue* 14 (1929): 1075–6.

[96] Science Society, *Science Society of China* (cit. n. 67), pp. 13–5.

[97] "Kexue jinhou zhi dongxiang" (The future orientation of *Science*), *Kexue* 19 (1935): 1–8.

[98] See Cao Yu, "Zhonghua jiaoyu" (cit. n. 81); Jia Sheng, "The Origins of the Science Society of China, 1914–1937," Ph.D. diss., Cornell Univ., 1995, chap. 5.

of the Science Society warmly embraced the establishment of the Academia Sinica in 1928, even if they harbored suspicions of the new Nationalist party state. Cai Yuanpei, one of the most persistent patrons of the society and an influential leader of the Nationalist party, used his leverage in the latter to press Chiang Kai-shek's new government to establish the Academia Sinica as a way to promote national science in China.[99] Cai modeled the Academia Sinica after the French and Soviet systems, establishing the academy as a centralized national scientific research institution with various research institutes, located mainly in Nanjing and Shanghai, where full-time staff researchers conducted research in different scientific fields.

Yang Xingfo, founding editor of *Kexue* and a key leader of the Science Society, became executive director of the Academia Sinica. After an unsuccessful career as an industrial accountant, Yang had returned to his revolutionary path in 1924 when he joined Sun Yat-sen's renewed revolutionary movement in Guangzhou. There Sun briefly led a rival national government against the Beijing regime controlled by the warlords. When the Nationalists succeeded in unifying the country in 1927, Yang answered Cai Yuanpei's call to assist in establishing the Academia Sinica.[100] Many members of the Science Society supported the new academy not only as a way to organize national scientific research, but also as a venue for representing China at international scientific conferences. The exclusion of China from the International Research Council, due to its lack of an official national scientific organization, had been a humiliation to many Chinese scientists, as had the discrimination against Chinese scientists at the 1926 Pan-Pacific Science Congress in Tokyo for the same reason.[101]

Given Cai Yuanpei's special relationship with the Science Society and its prestige, it is not surprising that many society leaders became heads of the new academy's research institutes: Zhu Kezhen became director of the Institute of Meteorology in Nanjing, Zhou Ren headed the Institute of Engineering in Shanghai, and Wang Jin, president of the Science Society 1930–1933, directed the Institute of Chemistry, also in Shanghai.[102] In 1933, when political enemies within the Nationalist party assassinated Yang Xingfo, Ding Wenjiang succeeded him as executive director of the Academia Sinica. After Ding left in the late 1930s, Ren Hongjun took over the post.[103]

Despite all these intimate connections with the government, the Science Society remained an independent organization. It seemed content to exist in the third realm, aligned with neither the state nor society. Certainly there were numerous problems, including the continuing conflict between the Nationalists and the Communists; the threat of foreign invasions; and the Nationalists' attempt to control science and edu-

[99] See Shiwei Chen, "Legitimizing the State: Politics and the Founding of Academia Sinica in 1927," *Papers on Chinese History* 6 (spring 1997): 23–41.

[100] Xu, "Yang Xingfo" (cit. n. 28), p. 79.

[101] See Zhang Yun, "Guoji xueshu yanjiu huiyi he Zhongguo kexue de fazhan" (The International Research Council and the development of science in China), and the attached note by Ren Hongjun, *Kexue* 11 (1926): 1391–1402; and Ren Hongjun, "Fan taipingyang xueshu huiyi de huigu" (Retrospective on the Pan-Pacific Science Congress), *Kexue* 12 (1927): 12. See also Xu Minghua, "Zhongyang yanjiuyuan yu Zhongguo kexue yanjiu de zhiduhua" (the Academia Sinica and the institutionalization of Chinese scientific research), *Zhongyang Yanjiuyuan jindaishi yanjiusuo jikan* (Contributions from the Modern History Institute of the Academia Sinica) no. 22, pt. 2 (June 1993): 233–9.

[102] See Zhang, "Cai Yuanpei" (cit. n. 74).

[103] Ibid; and Fan, "Ren Hongjun" (cit. n. 30).

cation and bend them toward military-industrial purposes as well as conservative cultural values. Yet by the 1930s scientists by and large felt that a happy medium between scientific professionalism and nationalism had been reached. Not only had the government embraced the goals—the development of modern science and technology—advocated by the scientists, but they themselves had opportunities to shape the course.[104] During the Nanjing Decade, Chinese science, under the leadership of these Science Society members and with the steady support of a relatively stable central government, made remarkable progress in all fields.[105]

The situation changed for the worse with the Japanese invasion of 1937. The Science Society of China survived the difficult years of the War of Resistance against Japan (1937–1945) but faced an uncertain fate when the Communists took over the mainland after a bloody civil war with the Nationalists that raged from 1945 to 1949. The corruption and political repression of the Nationalists had disillusioned many scientists, including members of the Science Society. Zhu Kezhen, who considered himself a liberal not unaware of the abuse of science in the Soviet Union under Stalin, was nevertheless hopeful that the Chinese Communists' focus, like his own, would be on rebuilding China. Thus in May 1949, when Nationalist leader Chiang Kai-shek sent for Zhu to retreat with him to Taiwan as the Communist forces advanced toward Shanghai, Zhu, then president of Zhejiang University, declined.

To avoid possible assassination by Nationalist agents, Zhu went into hiding at the Science Society of China in Shanghai. His first contact with the Communists occurred when he encountered the People's Liberation Army soldiers on the street of Shanghai the day after their takeover of the city. Their discipline, a stark contrast to the unruly and bullying behavior of the Nationalist soldiers he had known before, immediately impressed him. When Wu Youxun, a physicist and leader of the Science Society, came to discuss how the society should position itself in the new regime, Zhu responded with hope:

> I told [Wu] that in 1927, when the Nationalists launched the Northern Expedition [to defeat the warlords], the people rejoiced as much as they do today. But the Nationalists did not capitalize on the opportunity; they instead covered up embezzlements, failed to adhere to clear rules of rewards and punishments, and ended up being overthrown today. The people have welcomed the Liberation Army as they did clouds amid a severe draught. [I] hope that [the Communists] can work hard to the end and do not turn out to be as corrupt as the Nationalists. Science is extremely important to construction, and [I] hope the Communists will pay close attention to it.[106]

Perhaps it was the same hope for national reconstruction based on science that sent Ren Hongjun from Hong Kong to Shanghai to welcome the Communists. Like Zhu, he had refused to retreat with the Nationalists, perhaps because he believed that he could continue to lead his beloved Science Society into the new era.[107]

A rude awakening came to the scientists soon enough. On June 9, 1949, as Zhu Kezhen presided over the celebration of the twenty-first anniversary of the Academia Sinica in Shanghai, he heard two official speeches that would presage the dilemma

[104] For example, the geologist Wen Wenhao, a leader of the Science Society, was in charge of the powerful National Resource Commission in the 1930s. See Kirby, "Engineering China" (cit. n. 30).

[105] Reardon-Anderson, *Study of Change* (cit. n. 42), pp. 174–6.

[106] Zhu Kezhen diary entries for 25 and 26 May 1949, in Zhu, *Riji* (cit. n. 44), vol. 2, pp. 1255–6.

[107] Fan, "Ren Hongjun" (cit. n. 30), p. 75.

of science in the new regime. In his diary he noted that after he had given a talk on the history of the Academia Sinica, the mayor of Shanghai spoke:

> Mayor Chen Yi spoke for one hour, explaining the importance of theory to the [Chinese Communist] revolution. . . . He said that the Communists were humble and willing to hear [advice], that criticism should be penetrating . . . and that the essence of democracy was both that the minority followed the decision by the majority and that the majority should respect the opinions of the minority. What he said was very reasonable. Next Feng Ding from the [Communist party's] Department of Propaganda spoke of Marxism and Leninism as the highest principles of all the theories in the world[. He] said that the subjective opinions of the proletariat were more objective than the objective opinions of the bourgeois. What he said was really hard to understand.[108]

This was the first, but certainly not the last, time of thought reform, or ideological indoctrination, that Zhu and other scientists had to endure in the Mao era, despite Zhu's being appointed vice president of the new Chinese Academy of Sciences, in late 1949, by the Communists.

As for Ren and Zhu's Science Society, it fared no better than any of the other institutions of civil society that had grown up in the Republican period. Under pressure to nationalize and collectivize all enterprises and activities, Ren reluctantly turned over the assets of the Science Society piece by piece to the government. In 1949 the Science Society was pressured to dissolve when the new, official All-China Confederation of Special Societies in Natural Sciences was established.[109] Within the Communist party, the Science Society was viewed as politically untrustworthy, particularly because of its leaders' extensive former connections with the Nationalist government.[110] Zhu Kezhen, now actively involved in central scientific organization and planning as vice president of the Chinese Academy of Sciences, recognized the futility of sustaining the Science Society. Thus Zhu recorded in his diary on January 22, 1952:

> This morning I wrote to Ren Shuyong [scholarly name for Ren Hongjun] because the Science Society of China sent me a letter last November asking for member registration. . . . But when the All-China Confederation of Special Societies in Natural Sciences was established in September two years ago at the Conference of the Deputies of Scientific Associations, it did not include comprehensive scientific groups [as its members]. The implied hope was that, to avoid rivalry, there should not be another comprehensive scientific organization. This time the call for reregistration by the Science Society will unavoidably be viewed as making a statement of dissent. . . . The proposed new charter of the society listed an item on enterprises, including the running of institutes, publication of scientific journals—such extravagant self-promotion . . . is indeed wrongheaded because the government is just beginning to consolidate scientific journals.[111]

[108] Zhu Kezhen diary entry for 9 June 1949, in Zhu, *Riji* (cit. n. 44), vol. 2, pp. 1260–1.

[109] Fan, "Ren Hongjun" (cit. n. 30), p. 75.

[110] Fan Hongye et al., "Huang Zongzhen fangtan lu" (An interview with Huang Zongzhen), ZGKJSL 21(4) (Dec. 2000): 316–23, especially p. 321.

[111] See also Zhu Kezhen's diary entries on affairs of the Science Society in Zhu Kezhen, *Zhu Kezhen riji* (Diaries of Zhu Kezhen), vols. 3–5, 1950–1974 (Beijing: Science Press, 1989–1990): 5 Aug. 1950, pp. 82–3; 15 Aug. 1950, p. 85; 21 Aug. 1950, pp. 87–8; 25 and 26 Aug. 1950, p. 89; 12 Oct. 1950, p. 110; 19 Dec. 1950, p. 128; 23 Dec. 1950, p. 133; 27 Dec. 1950, pp. 134–5.

Gradually and reluctantly, Ren gave up. In 1951 *Kexue* was combined with, or rather absorbed by, the confederation's *Ziran Kexue* (Natural sciences).[112] In 1953 Ren Hongjun presided over the relinquishing of the *Kexue Huabao* to the official Shanghai Association for the Popularization of Science. The next year he saw to it that all the materials and staff of the Science Society's Biological Institute were transferred to the Chinese Academy of Sciences. In 1956 he turned over the society's library, printing press, and instrumentation company in Shanghai to the government.[113] During the brief liberalization period in 1957, Ren Hongjun revived *Kexue*, but it did not last for long. By 1959 he finally had turned over every asset of the Science Society to the government, and *Kexue* had stopped publication. Ren Hongjun died in 1961 not long after he wrote a short history of the Science Society.[114]

CONCLUSIONS

Peter Buck, in his provocative *American Science and Modern China* (1980), sees the Science Society of China as a voluntary association modeled partly after the burgeoning chambers of commerce in southeastern China, where most Science Society members originated, and partly after western scientific associations. But rather than viewing it as a seed for civil society in China, Buck casts architects of the Science Society of China as an elite, like the gentry leaders of the chambers of commerce, and as superficial and imperfect conveyors of American science detached from the Chinese reality. American science was exported to China via the Science Society and the Rockefeller-funded institutions as a way to transform Chinese society and politics. But it failed because science cannot be separated from its social context. "In China, where American science achieved a considerable measure of autonomy from Chinese society and politics, its freedom only ensured its irrelevance," according to Buck.[115] Thus Buck, like the Communist science policy makers in the 1950s, views the Science Society as irrelevant to Chinese society.

But it may be premature to call the Science Society a failure, especially in light of the fact that the society and its ideals inspired many of the second generation of Chinese scientists. Tang Youqi, a leading biochemist at Beijing University who received his Ph.D. from the California Institute of Technology and worked with Linus Pauling in the 1940s, vividly remembers reading Ren Hongjun's speeches in *Kexue* as a youth and being inspired to pursue a career in science.[116] The late Hua Luogeng saw his first scientific publication appear in *Kexue* in 1929 when he was only a school teacher without a university education. The paper led to his "discovery" by several prominent Chinese mathematicians at Qinghua University and eventually to his becoming a world-renowned mathematician.[117]

Tang and Hua were not the only young scientists who benefited from the Science Society. Through it, the first generation of modern Chinese scientists, such as Ren

[112] On the cessation of *Kexue* in 1951 and its brief reappearance in 1957–1960, see Xu Weimin, "Kexue zazhi de liangdu tingkan yu fukan" (The two-time cessations and resumptions of *Science* magazine), ZRBZFTX 14(3) (June 1992): 24–9.

[113] Fan, "Ren Hongjun" (cit. n. 30), p. 75.

[114] Ibid., p. 76.

[115] Buck, *American Science* (cit. n. 63), pp. 120–1, on p. 236.

[116] Tang Youqi, interview by author, Claremont, Calif., Nov. 2000.

[117] Wang Yuan, *Hua Luogeng* (Beijing: Kaiming Press, 1994), pp. 29–31. See also Stephen Salaff, "A Biography of Hua Lo-keng [Luogeng]," *Isis* 63 (1972): 143–83.

Hongjun and Zhu Kezhen, recruited thousands of scientists and engineers to continue their quest of saving China through science. Most of the leaders in twentieth-century Chinese biology went through Bing Zhi's and Hu Xiansu's Biological Institute. As there can be no question about the importance of the Science Society for Chinese science, to question the relevance of the society is to question the relevance of science and related technology in modern China. Of course, depending on one's perspective, there may well be disagreement about the relevance and importance of science and technology to modern Chinese society. But even without singling out the crucial role of the Chinese atomic bomb in the 1960s in raising the nation's international stature and healing wounded national pride, one can point to the formation of a scientific and educational infrastructure in the era of the Science Society as proof that the efforts of Ren, Zhu, and their fellow Boxer students helped to strengthen, if not save, China at critical junctures in its turbulent recent history.

Its remarkable achievements in promoting science and technology in China aside, was the Science Society a failure as a model of civil society? After all, most institutions of civil society disappeared during the Mao era, especially during the Cultural Revolution (1966–1976). Yet, as one examines the patterns of science-state interaction during the Mao years, one may find that there was more continuity than first appears. The personal networks formed in the Republic era did not suddenly vanish under the Communist rule, and scientific leaders such as Zhu Kezhen capitalized on these networks to continue to promote scientific research and industrial development. Zhu even continued his critique of party-state policy during the dangerous Cultural Revolution years, not in the pages of *Kexue*, of course, but in his private diaries. In a way, when the public sphere was squeezed out in an authoritarian regime, civil society found expression in private space. Zhu used his diaries not only to express his skepticism and criticism of Maoist excesses, but also to help former associates and students clear themselves of the Red Guards' charges.[118]

In the post-Mao decades of the 1980s and 1990s, there was such a remarkable revival of civil society institutions and a liberal public sphere that scholars such as Merle Goldman speak of a "restarting" of Chinese history.[119] As the market-oriented economic reform radically reshaped the science-state relationship, there began to emerge various private scientific organizations, especially ones related to environmental protection. But limits to critical public discussions remain. Though *Kexue* resumed publication in 1985, the Science Society has not been revived, and *Kexue* is no longer an independent journal sponsored by a private scientific organization. It is now published by the government-run Shanghai Science and Technology Press. In 1995 Jiang Zemin, the general secretary of the Communist Party, pronounced "Spread Science and Uplift National Strength" as the journal's new mandate.[120] Although the new *Kexue* still echoes the scientific nationalism of the old Science Society of China, it is a far cry from the critical voice it used to be more than half a century ago when Zhu and other Boxer Fellows sought to save China through engaging science in the third realm between the state and civil society.

[118] See Zhu, *Riji* (cit. n. 111), vol. 5 (covering the Cultural Revolution years of 1966–1974).
[119] Merle Goldman, "Restarting Chinese History," *Amer. Hist. Rev.* 105(1) (2000) :153–64.
[120] See the journal's official Web site: www.kexuemag.com.

Scientists and the Problem of the Public in Cold War America, 1945–1960

*By Jessica Wang**

ABSTRACT

In 1927, the philosopher John Dewey asked whether public political authority could survive and prosper in an age of experts. This essay takes Dewey's question as the central problem of science and civil society in twentieth-century America, and examines the place of scientists in public life during the early cold war years. Using the atomic scientists' movement as a case study, the author argues that scientists experimented briefly but vigorously with public engagement, only to succumb to cold war political repression and its circumscription of civil society during the late 1940s and early 1950s. Yet, even as scientists became victims of totalitarian political forces, science itself became feared as a potentially totalitarian locus of power in cold war America.

INTRODUCTION

> "To the village square we must carry the facts of atomic energy. From there must come America's voice."
>
> Albert Einstein, 1946[1]

> "Democracy is not government by majority but government by discussion."
>
> Kathleen Lonsdale, 1958[2]

In *The Public and Its Problems* (1927), the philosopher John Dewey identified the essential question of twentieth-century American politics: Could the democratic

* Department of History, University of California, Los Angeles, P.O. Box 951473, Los Angeles, CA 90095-1473; jwang@history.ucla.edu.

I thank participants at the April 2000 conference on "Science and Civil Society," especially Thomas H. Broman, Lynn K. Nyhart, and Ronald L. Numbers; my friends and colleagues in the Los Angeles Social History Research Seminar; and Kathy Olesko, Ellen W. Schrecker, and an anonymous reviewer for their comments and criticisms in response to this article at various stages of its development. At a crucial moment, Amy Crumpton supplied valuable information about the early history of the AAAS. Last but not least, I am grateful to the undergraduate and graduate students in my fall 2000 seminar, "Political and Social Theory: Hobbes to Habermas," for their vigorous discussions about the problem of civil society and its implications.

[1] Albert Einstein, "The Real Problem is in the Hearts of Men," interview by Michael Amrine, *New York Times Magazine*, 23 June 1946, p. 44.

[2] Kathleen Lonsdale, "Scientists and the People," *Bulletin of the Atomic Scientists* 14 (Sept. 1958): 245.

 0369-7827/02/1701-0012$10.00

ideal of an informed and active citizenry engaged in reasoned debate and decision making survive the rise of an increasingly technological and bureaucratic society? Writing in a context of growing disillusionment among American intellectuals about the public's capacity for rational decision making, Dewey readily acknowledged the atomizing effects of the machine age and the complexity of modern society, both of which, he believed, had eroded the public's ability to form reasoned opinion. As he observed, "The Great Society created by steam and electricity may be a society, but it is no community."[3] But where other intellectuals despaired and suggested that Americans abandon an obsolete faith in the public in favor of objective, scientific, expert-directed decision making, Dewey maintained a sincere and deeply held hope in the possibilities of public life. His vision encompassed more than a narrow, technocratic rationality; it sought to meld reason with a sense of profound moral commitment as a way to reinvigorate American democracy.[4]

Rather than blame the public for its apathy as Walter Lippmann and other prominent intellectuals had done, Dewey indicted a political and economic system that removed people from meaningful forms of civic engagement and left them with a sense of powerlessness.[5] A bewildered and atomized public, he argued, could not be abandoned to a "glorification of 'pure' science."[6] Expert rule without public participation inevitably deteriorated into "an oligarchy managed in the interests of the few." Instead, experts needed to express their expertness not through "framing and executing policies," but through "discovering and making known the facts upon which the former depend."[7] They needed to become a part of associational life and actively encourage and participate in the forms of communicative action Dewey believed were critical for overcoming the atomization caused by modern society and restoring the public to its proper place within American politics. Only then might "the Great Community" flourish as a lively realm of political discussion and bold experimentation in the face of the massive social problems of industrial society.[8]

Although Dewey was not writing specifically about natural scientists, his thoughts neatly encapsulate one of the central problems of American science in the twentieth century. Historians of American science have often spoken of the conflict between science and democracy, expertise and public rule, but they have not fully understood

[3] John Dewey, *The Public and Its Problems* (New York: Henry Holt & Company, 1927; Athens: Ohio Univ. Press, 1991), p. 98. On early twentieth-century ambivalence about the public, see Edward A. Purcell, *The Crisis of Democratic Theory: Scientific Naturalism and the Problem of Value* (Lexington: Univ. Press of Kentucky, 1973), especially chap. 6; and Christopher Lasch, *The True and Only Heaven: Progress and Its Critics* (New York and London: W. W. Norton, 1991), pp. 340–68. Walter Lippmann's *The Phantom Public* (New York: Harcourt, Brace, 1925) provided the most articulate and widely read indictment of the public in the 1920s, and Dewey's *The Public and Its Problems* was in part a reply to Lippmann.

[4] See James T. Kloppenberg, "Pragmatism: An Old Name for Some New Ways of Thinking?" *J. Amer. Hist.* 83 (June 1996): 100–38.

[5] Dewey observed, "Instead of individuals who in the privacy of their consciousness make choices which are carried into effect by personal volition, there are citizens who have the blessed opportunity to vote for a ticket of men mostly unknown to them. . . . There are those who speak as if ability to choose between two tickets were a high exercise of individual freedom. But it is hardly the kind of liberty contemplated by the authors of the individualistic doctrine." Dewey, *The Public* (cit. n. 3), p. 120.

[6] Ibid., p. 175.

[7] Ibid., p. 208.

[8] Ibid., chap. 5. Although Dewey criticized Rousseau's "general will" and other basic conceptual categories of political theory as overly vague and rooted in imaginary universals rather than real experience, his own concept of "the Great Community" seemed just as amorphous.

the issue as fundamentally one of civil society and the problem of the public. Throughout the past hundred years, American political thought has revolved around a debate over active civic engagement and whether or not it is truly necessary for the functioning of democratic political institutions. The question proved a critical one for American scientists as they became experts with substantial public visibility and political power, and as they struggled to understand and develop their political roles in the midst of the profound domestic and international crises of the mid-twentieth century.[9]

This essay considers the place of scientists in American public life and the dynamic between scientists, the public, and the state during the early years of the cold war. The scientific profession had relatively little experience with public civic engagement before the 1940s. Professional organizations became centers of institutionalized associational life for American scientists in the mid-nineteenth century, but in the decades that followed, they showed only minimal promise as conduits to the public. After World War II, however, scientists stretched the political possibilities of professional modes of association as they never had before. In the aftermath of Hiroshima and Nagasaki, the Manhattan Project generation of scientists explored the political possibilities offered by voluntary association and the entry of physicists into public life. But this time of creative political experimentation was also a period of growing political constraints, as the binding of science to the state, the consolidation of wartime ties between science and the military, the realignment of research toward cold war priorities, and the political repression of scientists also came to define the postwar era for American science. Scientists' public ventures soon clashed with the political priorities of the cold war state, and the state responded with totalitarian modes of action that terminated the Manhattan Project generation's lively effort to bring scientists into public life. Yet, the growing political and cultural authority of science in the nuclear age meant that science itself came to be feared as a potentially totalitarian force. Simultaneously dominant and dominated, science and scientists occupied an uneasy, ambivalent cultural space during the cold war.

FROM PROFESSIONALIZATION TO POLITICS: THE ATOMIC SCIENTISTS' MOVEMENT AND THE EMERGENCE OF A SCIENCE-PUBLIC RELATIONSHIP

As recent work on the history of science in early modern Europe has indicated, the mechanisms for the production and dissemination of knowledge about the natural world were part and parcel of the new institutions for public discussion and debate that Jürgen Habermas has defined as the public sphere. Defining a realm of publicly debated knowledge did not necessarily imply a commitment to public political life, however. Gradually, the professionalization of science exerted new forms of authority over knowledge that postulated strictly limited notions of science as a public activity. In the United States, the professional organizations of science that emerged in the mid-nineteenth century claimed public purposes for themselves, but those

[9] Silvan S. Schweber has argued for the influence of Dewey on American physicists in the 1930s and suggested that the American theoretical physics community embodied Dewey's democratic ideals. It is intriguing to think that there might thus be a direct connection between Dewey's conception of the public and scientists' political activities after World War II. See Schweber, *In the Shadow of the Bomb: Oppenheimer, Bethe, and the Moral Responsibility of the Scientist* (Princeton, N.J.: Princeton Univ. Press, 2000), pp. 8–9.

public roles remained ancillary to their professional objectives. For example, the American Association for the Advancement of Science (AAAS), established in 1848, and the National Academy of Sciences, chartered in 1863, paid homage to the public dimension of science in their founding statements but left no doubt that their main objectives lay in the promotion of science as a professional activity, not in creating forums for public discussion of pressing political matters related to science.[10] Decades later, in the 1940s, their respective senses of professional mission still hampered AAAS's and the academy's tentative efforts to confront the politics of anticommunism with their own explicitly public and political responses.[11]

By the turn of the century, professional organizations increasingly performed public functions, but their estrangement from the Deweyan public persisted, and they acted as special-interest groups, not associations striving to encourage public reflection and debate. For example, there is no question about the enormous public impact of the American Medical Association, which exercised overwhelming influence in the passage of the licensing laws and other legislation that ensured professionalized, scientific medicine's dominance over medical practice in the United States by the early twentieth century.[12] But although regular physicians undoubtedly felt the weight of public purpose as they campaigned against what they argued was "quackery," the AMA existed to serve its members and its profession, not the public. While it assumed an identity of interest between its professional aspirations and the needs of the public, and it participated in the political efforts that medical professionals saw as necessary to serve both, the stimulation of public discussion was never more than a side effect of the professionalization mission. This distinction might seem like hairsplitting, but it is critical to understanding the political potential of the public as an inclusive realm for the lively exchange of ideas and criticism, as opposed to the interest-group model of politics, which restricts knowledgeable debate and influence to specialized communities of interest and excludes the larger public as a politically significant entity.[13] According to the interest-group theory, which came to dominate the political thought of American liberals by the 1950s, the active engagement of the public in political life is neither necessary nor even desirable; the public merely

[10] AAAS stated its "Objects and Rules" as follows: "By periodical and migratory meetings, to promote intercourse between those who are cultivating science in different parts of the United States; to give a stronger and more general impulse, and a more systematic direction to scientific research in our country; and to procure for the labours of scientific men, increased facilities and a wider usefulness." *Proceedings of the American Association for the Advancement of Science, First Meeting, Held at Philadelphia, Sept. 1848* (Philadelphia: John C. Clarke, 1849), p. 8. The desire to build a "more general impulse" behind the scientific enterprise and serve "a wider usefulness" indicated a sense of public purpose in the founding of the AAAS, but not one any broader than the promotion of the scientific profession. The 1863 Act of Incorporation of the National Academy of Sciences indicated the public role of science in its clause that specified the academy's responsibility to assist the federal government with its scientific expertise when called upon to do so, but as a quasi-governmental organization, the academy embodied public officialdom, not the public itself.

[11] On this point, see Jessica Wang, *American Science in an Age of Anxiety* (Chapel Hill and London: Univ. of North Carolina Press, 1999), pp. 184–204, 215, 242–50.

[12] Paul Starr, *The Social Transformation of American Medicine* (New York: Basic Books, 1982), pt. 1, chap. 3.

[13] For example, Habermas's assessment of the decline of the political public sphere in the twentieth century depends upon such a distinction. See Jürgen Habermas, *The Structural Transformation of the Public Sphere: An Inquiry into a Category of Bourgeois Society*, trans. Thomas Burger (Cambridge, Mass., and London: MIT Press, 1989), pts. 5, 6; originally published in German as *Strukturwandel der Öffentlichkeit* (Darmstadt and Neuwied: Hermann Luchterhand Verlag, 1962).

registers approval or disapproval.[14] The contemporary advocates of civil society, by contrast, view the invigoration of public life and spirited public participation in politics—in short, civic life—as essential to the health of the polity.

Although professional organizations as they emerged in the second half of the nineteenth century did not themselves constitute institutions that upheld the public's political role, they did provide an entryway for science into more inclusive forms of public life. During the early twentieth century, American engineers, for example, began to develop a new consciousness about their identities that encompassed more than just professional practice. The Progressive Era movement toward social responsibility in engineering reflected a growing sense that the social upheaval associated with modern industrial society required those most responsible for the machine age to look beyond their profession's immediate needs. Although such discussions did not move far beyond questions of professional ethics, they did begin to push at the boundary between the professional and the public.[15] Some early social scientists also looked to maintain the tradition of social and political activism that defined the origins of their profession, but professionalization applied strong pressures for objective detachment that ultimately shifted active political engagement to the edge of the social sciences.[16]

The coming of the Great Depression prompted a new wave of activism that expanded to include the natural sciences. The Depression-era science and society movement of the AAAS and the newly formed American Association of Scientific Workers debated the role of scientific and technological advances in the economic breakdown of the 1930s, the special role scientists could play in ensuring the beneficent application of scientific knowledge, and the general need for scientists to exert a sense of responsibility over the social effects of science and technology. The radical American Association of Scientific Workers went further, arguing that scientists had to become involved in pressing political conflicts outside science and join the era's larger fights against racism and fascism. Both organizations sought to use professional identity as a base from which to build a public identity for science.[17]

The 1930s activism of the AAAS and the American Association of Scientific Workers produced a broad array of discussions, public statements, fund-raising efforts, and public gatherings dealing with a wide variety of political issues. At the same time, professionalized forms of action circumscribed the ventures of both organizations. Neither stretched the limits of professional modes of operation too far. Even the political efforts of the more activist American Association of Scientific Workers remained confined largely within scientific, academic, and intellectual circles.[18] Nonetheless, scientists achieved a new level of political involvement and public engagement during the Depression decade.

[14] For a critique of twentieth-century American liberalism and the rise of a political model that denied the significance of civic participation, see Lasch, *The True and Only Heaven* (cit. n. 3), chaps. 8, 10.

[15] Edwin T. Layton Jr., *Revolt of the Engineers: Social Responsibility and the American Engineering Profession* (Cleveland: Press of Case Western Reserve Univ., 1971).

[16] See Mary O. Furner, *Advocacy and Objectivity: A Crisis in the Professionalization of American Social Science, 1865–1905* (Lexington: Univ. of Kentucky Press, 1975).

[17] Peter J. Kuznick, *Beyond the Laboratory: Scientists as Political Activists in 1930s America* (Chicago: Univ. of Chicago Press, 1987).

[18] Wang, *American Science* (cit. n. 11), p. 6.

It was the post–World War II atomic scientists' movement, however, that provided the most startling twentieth-century example of public engagement and political action by scientists. The atomic bombings of Hiroshima and Nagasaki in August 1945 provided a grim counterpoint to the elation with which Manhattan Project scientists had celebrated the Trinity test a month earlier. Even as the war ended, scientists began to imagine the terrifying possibilities of the next great war, one they believed would be fought with nuclear weapons and might conclude with the complete obliteration of humanity.

Manhattan Project scientists responded to these sobering realizations with a program of political organization and launched what became known as the atomic scientists' movement. Immediately after the bombings of Hiroshima and Nagasaki, they began a series of discussions at the various Manhattan Project sites around the country about the international ramifications of the atomic bomb and the steps necessary to avoid future nuclear warfare. They soon reached a consensus that the inevitable loss of the American nuclear monopoly as other countries directed their research efforts toward acquiring nuclear capability, combined with the lack of any conceivable defense against nuclear weapons, meant that only a sound system of international control of atomic energy could ensure world safety.

The atomic scientists quickly created formal organizations: the Atomic Scientists of Chicago, the Association of Oak Ridge Scientists (later the Association of Oak Ridge Engineers and Scientists), the Association of Los Alamos Scientists, and other scientists' associations around the country. The individual local associations soon established themselves as part of a national-level organization, the Federation of Atomic Scientists, later renamed the Federation of American Scientists (FAS). Younger working scientists composed the core of these organizations and did much of the legwork that kept them going on a day-to-day basis. A November 1945 photograph of four physicists—H. H. Goldsmith of the Atomic Scientists of Chicago, Irving Kaplan of the Association of Manhattan Project Scientists (centered in New York City), Lyle B. Borst of the Association of Oak Ridge Scientists, and W. M. Woodward of the Association of Los Alamos Scientists—young (all in their 30s, except 29-year-old Woodward), earnest, mostly bespectacled, and slightly awkward in suits and ties as they announced the formation of the Federation of Atomic Scientists, speaks volumes about the ordinary scientists who imbued the scientists' movement with its energetic, improvisational, and inventive political character. As one of the founding editors of the *Bulletin of the Atomic Scientists*, Goldsmith became an especially vigorous advocate of nuclear arms control until his premature death in an accident in 1949. The leaders of the atomic scientists included William A. Higinbotham, a group leader in electronics in Los Alamos who was in his mid-thirties at the end of the war, and Katherine Way, a Chicago Metallurgical Laboratory physicist who had compiled data on nuclear cross sections (the so-called Kay Way tables), contributed to reactor design, and conducted theoretical work on the decay of fission products during her work for the Manhattan Project. Way directed publicity for the Atomic Scientists of Chicago during the fight against military control of atomic energy in the fall of 1945, and she coedited *One World or None*, a best-selling collection of scientists' essays about the nature of the nuclear threat, in 1946. Higinbotham helped to man the FAS's Washington, D.C., office during the immediate post–World War II period, and he played a major role in setting policy and coordinating the federation's activities during the late 1940s and early 1950s. Over the following decades, as he pursued a career in physics at Brookhaven National Laboratory, he remained devoted to nuclear arms control as the other half of his life's work.[19]

Some of the older, more established physicists also played key roles in the postwar politics of atomic energy. For example, Harold C. Urey, a Nobel Prize–winning chemist who had spent the war at the Manhattan Project's Substitute Alloy Materials Laboratory at Columbia University, played a major role as a publicist in the year and a half following the end of the war. Urey's long-standing devotion to progressive left causes provided the atomic scientists' movement with some continuity going back to the 1930s. One of the most widely quoted of the atomic scientists, he spent much of his time talking to reporters, giving speeches, and testifying before Congress, as well as writing widely published essays about the dangers of the nuclear age.[20] The prominent theoretical physicist Edward U. Condon was also one of the more active elder statesmen in the scientists' movement. Condon had helped set up the radar project at MIT's Radiation Laboratory before becoming assistant director of Los Alamos in 1943; disastrous personality clashes and sharp disagreements with General Leslie R. Groves over security policy and living conditions forced Condon's resignation after only six weeks. He spent the remainder of the war as a part-time consultant to the uranium separation project at Berkeley. After the war, he took the lead in helping the atomic scientists lobby for civilian control of atomic energy, and he gained official status in Washington when he became technical advisor to the Senate Special Committee on Atomic Energy and director of the National Bureau of Standards in November 1945.[21]

In forming the FAS and its local chapters, these and other scientists created an organizational entity that melded professional identity with political purpose.[22] For the next eighteen months, they carried on a lively and imaginative campaign for civilian and international control of atomic energy. They achieved a partial success with the August 1946 passage of the McMahon Act, which placed atomic energy and the development of nuclear weapons under the aegis of a civilian agency, the Atomic Energy Commission, rather than the military. International control, however, went down to defeat in the United Nations early in 1947.

At its peak, the atomic scientists' movement represented an integration of scientists into public life beyond anything that had ever happened previously in the United States, and it demonstrated the potential of scientists to enrich and enliven public discourse at every level of politics. During the fight for civilian and international

[19] For the photograph of the young physicists who announced the formation of the Federation of Atomic Scientists, see ibid., p. 19. On Higinbotham, see Ronald Sullivan, "William A. Higinbotham, p. 84; Helped Build First Atomic Bomb," *New York Times*, 15 Nov. 1994, D29; and Alice Kimball Smith, *A Peril and a Hope: The Scientists' Movement in America, 1945–1947* (Chicago: Univ. of Chicago Press, 1965), pp. 151, 246. On Way, see Murray J. Martin, Norwood B. Gove, and Ruth M. Gove et al., "Katherine Way," in *Women in Chemistry and Physics: A Biobibliographic Sourcebook*, ed. Louise S. Grinstein, Rose K. Rose, and Maria H. Rafailovich (Westport, Conn.: Greenwood Press, 1993), pp. 572–80.

[20] On Urey's politics, as well as his turn toward cold war liberalism by 1948, see Wang, *American Science* (cit. n. 11), pp. 55–8.

[21] Jessica Wang, "Science, Security, and the Cold War: The Case of E. U. Condon," *Isis* 83 (June 1992): 238–69, especially pp. 241–4.

[22] The FAS was not composed exclusively of scientists, but the vast majority of its members were scientists or engineers. Local chapters required that two-thirds of their membership possess at least a bachelor's degree in a science or engineering field and that they be employed in an area related to their scientific or engineering expertise. Smith, *A Peril and a Hope* (cit. n. 19), p. 237.

control of atomic energy, the scientists initiated a wide range of public appeals and flooded the media with interviews, articles, and radio addresses in which they discussed the proposals before Congress and the United Nations, as well as the general implications of atomic energy. They also established a presence in Washington, D.C., where they learned how to rub shoulders with congressmen and senators, and hobnob in the elite social circles of the nation's capital. They learned the techniques of political lobbying as well, testifying before congressional committees and assaulting federal lawmakers with letters from scientists and other citizens in support of their cause.

More important, the atomic scientists did not limit themselves to high politics. They believed fervently in the need to build ties with the public and provide ordinary people with the knowledge to make intelligent decisions about the nuclear age. In a June 1946 article in the *New York Times Magazine*, Albert Einstein invoked the metaphor of the "village square" to describe the scientists' task. Even without advanced scientific expertise, Einstein noted, anyone, "if told a few facts," could aspire to informed decision making and "understand that this bomb and the danger of war is a very real thing, and not something far away." Moreover, he emphasized, the nuclear age "directly concerns every person in the civilized world," and difficult decisions could not be delegated to the political elite. Instead, choices about survival hinged "ultimately on decisions made in the village square."[23]

In taking to the village square, scientists went far beyond political elites and the distanced organs of the media. They spoke at Rotary Clubs, Kiwanis Clubs, science clubs, meetings of local business groups, schools, churches, synagogues, local meeting halls, and other public places. They organized a speakers' bureau, prepared information packets, and distributed filmstrips for classroom use. The Federation of American Scientists also reached out to the labor movement, educational organizations, religious groups, and professional associations for cooperation and assistance in appealing to the public. Working with representatives from about fifty such organizations, the scientists helped to form the National Committee on Atomic Information (NCAI) as an umbrella organization that would reach the public and "promote the widest possible understanding of the facts and implications of developments in the field of atomic energy." Given the size of its member organizations, the National Committee on Atomic Information had a potential audience of ten million people.[24]

The pages of *Atomic Information*, the main publication of the NCAI, open a window onto the dynamic partnerships between scientists and the general public that developed in locales around the country. One issue told the story of "Mrs. W., housewife," who "attended a Washington church conference, and heard a young atomic scientist tell of the profound social and political implications of atomic energy." She then returned home to Pittsburgh determined to energize her community. First, she

[23] Einstein, "Hearts of Men" (cit. n. 1), p. 44. Amrine, credited as the interviewer for this article, was a journalist who devoted himself to working as the scientists' movement's professional publicist in the FAS's early years. In all likelihood, he wrote the piece in the *New York Times Magazine*, and Einstein merely approved it for publication. Amrine was also the ghostwriter of Urey's famous article "I'm a Frightened Man," published in the 5 Jan. 1946 issue of *Collier's*.

[24] On the public activities of the scientists' movement, see Smith, *A Peril and a Hope* (cit. n. 19), chap. 10, especially pp. 290–3, 299–302, 324–5; and Paul Boyer, *By the Bomb's Early Light: American Thought and Culture at the Dawn of the Atomic Age* (New York: Pantheon Books, 1985), chap. 5. The statement of the NCAI's purpose comes from "The Federation of Atomic Scientists: National Committee on Atomic Information," *Bulletin of the Atomic Scientists* 1 (24 Dec. 1945): 5.

arranged an invitation for Oak Ridge scientist Paul Henshaw to speak at a local church. The gathering, "Pittsburgh's first public meeting on atomic energy," attracted an audience of six hundred. Mrs. W. reported triumphantly, "Sixty questions were asked from the floor! A list of these questions was forwarded to Senator McMahon. . . . Key community leaders had been invited to the first public meeting. Fired with new enthusiasm, they met with the guest scientist the next morning and planned a second, larger meeting." Six weeks later, the next meeting, sponsored by local religious organizations, women's groups, fraternal associations, as well as business and labor interests, attracted two thousand people to hear Harold Urey and discuss civilian control of atomic energy.[25]

Similar gatherings took place in countless cities and towns throughout the country in the first half of 1946. The atomic scientists and citizens in Kansas organized a series of "Atomic Age Conferences" in eight communities around the state.[26] In Philadelphia, some 21,000 locals participated in a series of meetings about atomic energy.[27] Bostonians put together a citywide rally.[28] Young people got into the act. High school students in Oak Ridge formed the Youth Council on the Atomic Crisis and pledged to be informed and politically active.[29] Six high school students in Washington, D.C., organized a radio broadcast in which they discussed the ramifications of civilian and international control of atomic energy, including governmental versus industrial control, the question of whether or not the bomb could be successfully outlawed, and the relative merits of international control within the United Nations versus a future world government.[30] In Nebraska, students in a church group organized a reenactment of hearings before the Special Senate Committee on Atomic Energy for an expected audience of three hundred.[31] Over a period of just a few months, the NCAI reported receiving thousands of letters asking for advice about how to mobilize their locales. Many of these correspondents were already actively generating heated discussions about the nuclear age in their communities and encouraging them to make their opinions known to Washington.[32]

One should be cautious about taking the reports from *Atomic Information* purely at face value. The NCAI existed to promote political action, and *Atomic Information* was hardly an unbiased source in its claims about the successes of local organizing. Nonetheless, one must take seriously the extent to which the atomic scientists' movement directly engaged the public and helped make it an active force in the early postwar debate over atomic energy. The nearly 71,000 letters, telegrams, and

[25] "Pittsburgh Did It—Your City Can, Too!" *Atomic Information*, 4 March 1946, p. 3.

[26] "State-Wide Meetings in Kansas to Study Atom," *Atomic Information*, 25 March 1946, p. 3.

[27] "News Round-Up," *Atomic Information*, 22 April 1946, p. 7.

[28] "National Organizations Reach State and Local Groups with Atomic Information," *Atomic Information*, 17 June 1946, p. 2.

[29] "Oak Ridge Scientists' Kids Start a New Lobby," *Atomic Information*, 25 March 1946, p. 3. Although the headline identified these students as children of scientists, the text of the article identified them as "the sons and daughters of the atomic bomb workers in the 'city without a past.'" It is important to remember that scientists alone did not build the bomb: the Manhattan Project employed a total of half a million workers during the war. In many respects, Oak Ridge was a working-class city. See Russell Olwell, "Help Wanted for Secret City: Recruiting Workers for the Manhattan Project at Oak Ridge, Tennessee, 1942–1946," *Tennessee Hist. Quart.* 18 (spring 1999): 52–69.

[30] "Student Radio Forum Shows High Listener Interest Plus Educational Value," *Atomic Information*, 8 April 1946, p. 8.

[31] "National Organizations Reach State and Local Groups with Atomic Information," *Atomic Information*, 17 June 1946, p. 2.

[32] "How Can I Help to Prevent an Atomic War?" *Atomic Information*, 22 April 1946, pp. 1, 3.

petitions in favor of civilian control of atomic energy that the Special Senate Committee on Atomic Energy received in the spring of 1946 did not arise spontaneously from mass opinion. They were the product of the atomic scientists' creative and effective entry into public life, which generated just the kind of communicative action that Dewey had called for nearly two decades earlier.[33]

The scientists' movement thus broke through the professionalized modes of activity that had characterized other scientific organizations and experimented with wholehearted participation in and promotion of public life. The atomic scientists sometimes worried that they had gone too far in abandoning the canons of scientific objectivity and feared that their political activism might undermine the credibility of science. As scientists, what right did they have to speak about political issues? They surmounted this difficulty by appealing to their dual identity as scientists and citizens. For example, in a November 1945 letter to the *New York Times*, several members of the Association of Oak Ridge Scientists responded to an earlier editorial that had sharply censured scientists for claiming political authority beyond their expertise. They replied that their identity as citizens made their actions appropriate: "Since we are, because of our daily activities and obligations, average citizens, we feel that we have a normal understanding of political and social affairs. On these matters, we have not intended to speak as experts." Their knowledge as scientists, however, did provide them with "one distinct advantage over other citizens with respect to the present problem. We have known of atomic power and its possibilities for years. We are familiar with the potentialities of its future development for both peace and war." This special awareness about the atomic bomb and its implications obligated scientists to speak out: "Since we are best equipped with this knowledge, we have assumed the responsibility of aiding in the education of those who are not aware of the revolutionary nature of atomic power."[34] According to this conception of the scientist's political role, scientists' specialized knowledge granted no special privileges, but neither did scientists lose their responsibilities as citizens in taking up the obligations of their profession. With this emphasis on citizenship, they defined themselves as part of a larger public, within which they participated as equals, but to which they offered their expertise for the purposes of information, consideration, and criticism.

Scientists and the public were not the only arbiters of the cold war relationship between science and civil society, however. The state, as the source of organized coercive power in society, played a critical role in limiting the political potential of scientists in public life. With the rise of the cold war, the atomic scientists' exploration of the public realm did not last long. The postwar resurgence of anticommunist ideology doomed serious discussion about arms control and other subjects that lay outside the boundaries of cold war orthodoxy. During the late 1940s, political repression forced the Federation of American Scientists to turn inward and abandon the

[33] "70 Thousand Letters Back the McMahon Bill," *Bulletin of the Atomic Scientists*, 15 April 1946, p. 6. As the historian Paul Boyer has noted, postwar appeals for civic engagement over atomic policy "were not merely routine calls for more public discussion on a matter of current interest." Rather, "[T]hey had about them an urgency that is difficult to recreate after the passage of forty years." Boyer, *By the Bomb's Early Light* (cit. n. 24), p. 30. The public's vigorous response to those calls is also difficult to re-create, but it is no less real for that difficulty.

[34] Warren H. Burgus, Raymond P. Edwards, Howard Gest, Lawrence E. Glendinen, Charles W. Stanley, and R. Williams Jr. of the Association of Oak Ridge Scientists, letter in "Topics of the Times," *New York Times*, 3 Nov. 1945, p. 14.

public-oriented politics that had constituted its most original contribution to the politics of American science.

SCIENTISTS AND COLD WAR ANTICOMMUNISM: LOYALTY TESTS, SURVEILLANCE, AND THE DESTRUCTION OF PUBLIC LIFE

In one way or another, most contemporary discussions of civil society focus on the institutions and values necessary to ensure freedom, open political discussion, and the assertion of public authority over governmental actions. "Transparency" has become a favorite term to denote the openness of the state to public oversight, and the willingness of the state to submit itself to public scrutiny has become an indicator of the health of civil society as the public realm of association between the individual and the state. What this emphasis on freedom and openness deemphasizes, however, is the traditional function of coercive power as a prerogative of the state. Coercion serves to maintain civil society in so far as it guarantees the rule of law and enforces the civility of relationships within both private and public life. That same coercive power, however, threatens civil society when it is put to the private purposes of the individual public officials who make up the state, or when it exerts a level of surveillance and control that curtails the political functions of the public as a social space for discussion and debate. Hence the relationship between state and society is always an uneasy one, for the state simultaneously guarantees the vitality of civil society and threatens its eradication.

During the postwar decade, anticommunism severely curtailed the political dimensions of public life. In some respects, the cold war political order came to mirror the totalitarianism that it claimed to oppose. In order to appease Republican critics and establish a domestic political consensus over the containment of the Soviet Union abroad, the Truman administration launched an enormous expansion of the internal security state within U.S. borders. Transparency and openness retreated, as official secrecy, once restricted to wartime, became an integral part of government during times of peace. The security clearance system, largely ad hoc during the war, became permanently institutionalized. The establishment in 1947 of the federal loyalty program, which required loyalty clearance for all federal employees, further increased the power of the Federal Bureau of Investigation (FBI) to monitor and probe the lives of ordinary citizens. Congressional and state-level sedition committees expanded their purview, and by the 1950s, liberals and radicals in the universities, public school systems, labor unions, civil rights organizations, and other institutions could fear attention and harassment from the House Committee on Un-American Activities (HUAC), the Senate Internal Security Subcommittee (SISS), or any number of state-level committees. Within a decade after the end of the war, anticommunism encompassed virtually all areas of American life, and anyone who tried to launch a meaningful challenge to the prevailing political order risked charges of disloyalty, loss of employment, and constant surveillance.[35]

[35] On the postwar institutionalization and expansion of the internal security apparatus, see Daniel Patrick Moynihan, *Secrecy: The American Experience* (New Haven, Conn., and London: Yale Univ. Press, 1998). Elsewhere I have described in greater detail the top-down political process by which the cold war political consensus was established. See Jessica Wang, "American Science in an Age of Anxiety: Scientists, Civil Liberties, and the Cold War, 1945–1950," Ph.D. diss., Massachusetts Institute of Technology, 1995, pp. 27–54.

Cold war political repression curtailed American scientists' political activities through two means: loyalty and security investigations of individuals, and official surveillance of organizations. In the case of the former, the state confronted civil society within the intimate sphere of selfhood by requiring loyalty and security tests for a wide range of individual scientists. Scientists employed by the government had to submit to investigations under the federal loyalty program, while scientists involved in classified research required security clearances. Increasingly, by the 1950s, scientists outside government employ also faced loyalty investigations even when their research lacked any national security implications. Applicants for government grants or passports confronted the possibility of being denied funds or the right to travel on the basis of loyalty tests, while politically active academic scientists risked invasive inquiries from HUAC, the SISS, and state sedition committees.

Loyalty and security investigations inevitably encroached upon the public life that took place within the intimate sphere of thought and sentiment, as well as the web of social relationships that formed a part of open political engagement. In the absence of evidence of overt acts that disqualified an individual for clearance, the loyalty-security system inevitably determined employee loyalty by making judgments about a person's character, motivations, beliefs, and associations. Under both the federal loyalty program and the security clearance system, the vague standard of "sympathetic association" with any organization or group of persons considered totalitarian, fascist, communist, or subversive could form the basis for a finding of disloyalty and loss of employment. Procedural safeguards, such as the right of the accused to a hearing, provided for due process, but they did not prevent loyalty inquiries from becoming explorations into personality and character rather than determinations about the truth or falsehood of specific charges. As Walter Gellhorn pointed out in his 1950 study of science and the loyalty-security system, "What is really being appraised in a personnel security case is not any particular question of fact but is, in a word, a man."[36] The Atomic Energy Commission itself could be quite explicit about this aspect of security clearance investigations. As the AEC commissioners deliberated over the political associations of Frank P. Graham, president of the Oak Ridge Institute for Nuclear Studies, they noted that "it must be recognized that *it is the man himself* [italics mine] the Commission is actually concerned with."[37] In essence, the state claimed the right to enter the mind of the individual in order to make a determination of loyalty. The interior realm of personhood was thus invested with a political significance that moved the normal life of the individual outside the domain of the private and into the public arena. Under such conditions, the intimate sphere of thought and rumination about public issues became part of a public realm embattled by the intrusive intervention of the state.

The case of a young physicist named Robert H. Vought dramatically illustrates how the loyalty-security system sought to investigate the subjective realm of thought, feeling, and character. In the fall of 1946, Vought, a newly minted Ph.D., accepted a position at General Electric (GE) and applied for a security clearance to work on contract projects for the Atomic Energy Commission. His application was rejected. After a two-year ordeal, during which he left GE for academe, he finally received a statement of charges and was granted a hearing. Most of the charges

[36] Walter Gellhorn, *Security, Loyalty, and Science* (Ithaca, N.Y.: Cornell Univ. Press, 1950), p. 90.
[37] Quoted in ibid., p. 91.

pertained to Vought's past membership in a Philadelphia area civil rights organization that the FBI considered communist infiltrated. The AEC accused Vought of membership in an organization that followed the communist party line, association with alleged communist members of the civil rights group, and presence at a meeting with a communist speaker. Vought's hearing, however, focused not on the charges and what he did or whom he knew, but on *who he was*.

Essentially, Vought had to prove the propriety of his political beliefs and associations through a demonstration of his character. Hence, he began his appearance before the local AEC security board with a summary of his life history to show that he led "a normal life, normal American life, with no indications of instability or insecurity." In order to indicate his good character, reliability, and loyalty, he also presented the board with an array of affidavits from physicists, friends, his college advisor, leaders of the Boy Scouts from his childhood, and even his supervisor back at the dining halls in college, and he brought seven character witnesses to testify before the board in person. Years later, when asked about the means he had used to defend himself, he answered, "I suppose I brought some of that [the affidavits] in because, what else do I have to offer, other than denying what they have accused me of. . . . What else did I have? Except I'm a good guy?"[38]

Vought's experience points to the ways in which loyalty-security investigations turned ordinary activities of personal and political life into objects of suspicion. Ironically, the most illuminating insights into the general workings of the internal security state come from the 1950s theorists of totalitarianism, despite their commitment to anticommunism, which led them to insist their ideas applied only to the most authoritarian regimes.[39] For example, in 1952, Jacob Talmon described the patterns of thought, speech, action, and inaction that the totalitarian state viewed as evidence of subversion. He observed:

> To have remained silent on some past and half-forgotten occasion, where one should have spoken; to have spoken where it was better to hold one's peace; to have shown apathy where eagerness was called for, and enthusiasm where diffidence was necessary; to have consorted with somebody whom a patriot should have shunned; avoided one who deserved to be befriended; not to have shown a virtuous disposition, or not to have led a life of virtue—such and other 'sins' came to be counted as capital offences, classifying the sinners as members of that immense chain of treason.[40]

Talmon wrote this passage as part of a larger argument about the supposed origins of totalitarianism in the French Revolution, but the statement also captures the patterns of thought and action that characterized cold war anticommunism in the United States. Failure to show enthusiasm for cold war foreign policy; open advocacy of U.S.-Soviet accommodation, arms control, greater U.S.-Soviet cooperation in science, civil rights for African Americans, and noncentrist labor politics; insufficient

[38] Robert H. Vought, interview by the author, tape recording, 9 Sept. 1993. Tape and transcribed notes in possession of the author. Four months after the hearing, the AEC finally notified Vought that his "employment on work of a classified nature would not have endangered the common defense and security." John E. Gingrich to Robert H. Vought, 3. Feb. 1949, personal possession of Robert A. Vought. For a fuller account of the Vought case, see Wang, *American Science* (cit. n. 11), chap. 2, pp. 102–14.

[39] This point will be expanded upon in the next section of this essay.

[40] J. L. Talmon, *The Origins of Totalitarian Democracy* (London: Secker & Warburg, 1952), p. 129. Talmon's book was one of the classic cold war works on totalitarianism.

expression of antipathy toward communism or the Soviet Union; vigorously stated opposition to loyalty investigations; display of interest in Marxism or other radical political ideas; association with communist or radical family members, friends, or acquaintances; attendance or participation in meetings of left-wing organizations or gatherings where supposed communists were present; refusal to name and denounce friends or acquaintances—all such manners of political action or inaction could find their way into an FBI report, departmental loyalty-security investigation, or congressional hearing and form the basis for questioning an individual's loyalty to the United States and the cold war political consensus.

While loyalty-security investigations encroached upon the intimate sphere of the self and only indirectly impeded the formation of bonds of mutuality in civil society, state-sponsored surveillance of political organizations exercised a more direct curtailment of the associational life of the public. In the early cold war years, FBI surveillance soon wrought irreparable damage upon the new mixture of scientific expertise, political association, and public action pioneered by the atomic scientists. The FBI's monitoring of the Federation of American Scientists moved the politics of science into a dangerous netherworld of informing, which distorted relationships of trust between individuals and violated organizational integrity by injecting the state's presence into the everyday functions of the public created by the scientists' movement.

As the atomic scientists gained ground on the American political landscape in the fall of 1945 and winter of 1946, the FBI began to pay close attention to the Federation of American Scientists. The bureau assiduously gathered information on every major FAS chapter, particularly the Washington, northern California, New York, and Cambridge affiliates. FBI agents watched scientists enter and leave gatherings in public places and private homes, took down license plates of cars parked at meetings, noted names and physical descriptions of persons they saw, and even attended meetings themselves, where they gathered literature and monitored discussions. The bureau also employed wiretaps to eavesdrop on discussions within the Northern California Association of Scientists, the Association of New York Scientists, and the Association of Cambridge Scientists.[41]

Most of the FBI's intelligence on the atomic scientists' movement came not from its own surveillance activities, however, but from confidential informants who provided information about every FAS chapter. Informants could be anyone—a concerned citizen such as the taxi driver who thought documents left behind in his cab might involve atomic espionage, or a conservative congressional staffer who reported to the FBI after attending an FAS meeting. Most of the FBI's informants

[41] See, e.g., San Francisco Field Office, Report, "Northern California Association of Scientists," 6 March 1946, FBI HQ 100–344452-3X2, vol. 1, FBI File on the Federation of American Scientists (hereafter cited as FBI-FAS), Federal Bureau of Investigation, Washington, D.C., pp. 3–4, 9–13; San Francisco Field Office, Report, "Northern California Association of Scientists," 12 June 1946, FBI-FAS, FBI HQ 100–344452, vol. 5, p. 3, for evidence of physical surveillance. On the wiretapping of the northern California association's phone, see Memorandum, SAC, San Francisco, to the Director [J. Edgar Hoover], 12 Sept. 1947; teletype, Kimball to the Director, 2 Oct. 1947; and Hoover to Communications Section, undated (or date deleted), all in FBI-FAS, FBI HQ 100–344452, vol. 14. On the use of technical surveillance to monitor discussions between New York and Boston area FAS members on a possible merger between the FAS and the American Association of Scientific Workers, see Memorandum, SAC, Boston, to the Director, 5 Dec. 1946, FBI-FAS, FBI HQ 100–344452-35, vol. 6.

about the FAS, though, came from within the atomic scientists' movement and involved either members of the FAS itself or of the National Committee on Atomic Information. These informants delivered literature to the bureau, including internal documents not meant for distribution, conveyed their thoughts and opinions about individual FAS members' characters and political proclivities, and reported in detail about discussions at various meetings and the general activities of the federation. Some informants developed long-term relationships with the FBI, reporting regularly to the bureau and, in a few clear cases, using information provided by the FBI or other security-oriented agencies to sway internal personnel decisions within the FAS.[42]

The FBI's use of informers to monitor the scientists' movement constituted an invasion of public life that spread the atomizing forces of fear and distrust throughout the federation. Informants understood that their actions breached a trust, and when they arranged to meet with FBI agents, they emphasized their need to protect their anonymity and avoid exposure. The federation's leaders also knew that the scientists' movement was under surveillance and that anything said within the scientists' movement might become known to the FBI. As FAS chairman William A. Higinbotham wrote in a worried memo in July 1946, "There is no question in my mind but that the FBI is watching us closely. If they ever get anything at all on us they and the Un-American Affairs Committee will go to town."[43] Higinbotham responded by taking steps to isolate the federation from suspect political contacts and excise any hint of communist influence in the FAS's public statements.

The federation also assisted in the firing of Daniel Melcher, director of the National Committee on Atomic Information, because anticommunists within the FAS and the NCAI believed Melcher was following the communist line. Throughout the three-week internal battle that culminated in Melcher's ouster, informers provided the FBI with a blow-by-blow account. Two representatives from the FAS, almost certainly Higinbotham and FAS treasurer Joseph H. Rush, also visited Attorney General Tom C. Clark and inquired about the possibility of obtaining information about suspected communists within the federation. An informant relayed the substance of the conversation to the FBI: "[Name deleted] reported to the Attorney General that he was fearful that the communists were obtaining positions in his organization for the purpose of taking over. Mr. Clark is reported to have stated if [names deleted] would give him the names of the suspects he would have them checked out and turn the information over to the Federation of American Scientists for their confidential use."[44] The FBI, of course, would have been the source of any such information. After all but one of Melcher's staff members resigned following

[42] I have discussed the nature of FBI surveillance and its effects on the atomic scientists' movement in detail in Wang, *American Science* (cit. n. 11), chap. 2, pp. 58–77.

[43] William A. Higinbotham to the Administrative Committee, 18 July 1946, Box 120, Folder, "FAS July-Dec. 1946," J. Robert Oppenheimer papers, Manuscripts Division, Library of Congress, Washington, D.C.

[44] Washington Field Office, Report, "National Committee on Atomic Information; National Committee for Civilian Control of Atomic Energy," 17 Sept. 1946, FBI-FAS, FBI HQ 100–344452, vol. 4, p. 27. See also [name deleted], Special Agent to Guy Hottel, SAC, Washington Field Office, 31 July 1946, Washington Field Office File on the Federation of American Scientists, FBI-FAS, FBI WFO 65–4736-30, sec. 2. Higinbotham and Rush's identities are revealed in a follow-up note Higinbotham sent to the attorney general. William A. Higinbotham to Tom C. Clark, 4 Aug. 1946, Box 1, folder 7, Joseph H. Rush papers, Special Collections, Regenstein Library, Univ. of Chicago.

his dismissal, however, Higinbotham decided that the FAS could do without the Justice Department's assistance, at least for the time being. Higinbotham informed Clark that the NCAI could now "start with a clean slate," and the federation could manage to monitor itself. "I feel that we have a pretty good idea whom to watch among the scientists," he wrote, "and at this time it does not seem to us that any of them are in a position where they can do any harm."[45] State surveillance, he implied, would not be necessary if the scientists could watch themselves.

The effects of FBI monitoring on the scientists' movement were subtle, but no less devastating for the nuanced character of surveillance. Other than limited technical surveillance, the FBI files on the Federation of American Scientists reveal little in the way of the blatant abuses of power—wiretapping, break-ins, mail intercepts, infiltration, counterintelligence, active harassment—that most scholars have emphasized in their studies of the bureau under J. Edgar Hoover's leadership. But the simple act of state-sponsored surveillance, by itself and decoupled from more actively repressive measures, represented a damaging intrusion of the state into the realm of intimate, face-to-face social relationships essential to open political discussion and action. Combined with internecine struggles, surveillance severely weakened the federation. The National Committee on Atomic Information never recovered from the purge of Daniel Melcher. A year later it closed its books, and the FAS lost its best resource for reaching the public and tapping a national political base. FBI monitoring also fractured and destabilized the federation as the chapters most affected by factionalism and informing deteriorated. Several major FAS chapters, including the Association of New York Scientists, the Association of Cambridge Scientists, and the Association of Los Alamos Scientists, folded completely. In April 1950, satisfied that the scientists' movement posed no threat to U.S. national security, the FBI shut down its active investigation of the federation. The FAS lived on, and it still operates as a political watchdog group today, but it has never recovered its original sense of spontaneity and creativity in politics or its ability to appeal to a mass political base. Scientists' postwar experiment with what were, for them, new forms of participation in public political life thus came to a premature end.

SURVEILLANCE AND THE TOTALITARIAN STATE

Modern democratic theory has tended to downplay or ignore the role of coercion and police power as functions of the state, emphasizing instead questions about pluralism, participation, and rights. Totalitarian theory, by contrast, in postulating social existence in a setting where the state is all and civil society nonexistent, reaches more directly toward a sociological understanding of surveillance and its implications for public life. Most accounts of cold war anticommunism and the harmful effects of state-sponsored surveillance have stressed the violation of individual rights, the constriction of political discourse, and the general curtailment of political freedoms. Such damage to the political realm is serious enough on its own terms, but at another level, surveillance digs deeper into the heart of social existence and defines a social system based on distrust, anomie, and atomization. Totalitarian theory speaks precisely to this realm of alienated human relationships.

No single person provided more evocative descriptions of totalitarianism's effects

[45] Higinbotham to Clark, 4 Aug. 1946 (cit. n. 44).

than Hannah Arendt. In the 1950s liberal anticommunist intellectuals in the United States seized upon *The Origins of Totalitarianism* (1951) and wholeheartedly adopted Arendt's sweeping analysis of the character of totalitarian regimes and the historical circumstances that had brought them to power.[46] The book's place in cold war thought has led scholars critical of anticommunism to dismiss much of Arendt's work on totalitarianism as cold war hyperbole. Rather than disregarding her findings outright, however, it is more interesting to use them to illuminate the patterns of behavior associated with political repression in a broader variety of contexts.

Arendt herself would not have endorsed such an approach; she insisted that one had to differentiate between totalitarian and authoritarian regimes, that totalitarianism's distinctiveness lay in the systematic and widespread use of terror, and that Nazi Germany and the Soviet Union constituted the only true totalitarian states in human history. Her representation of the totalitarian state's claim to the inner life of the individual and its distortion of social relationships between persons, however, provides extremely compelling material with which to understand surveillance in the United States. Moreover, recent formulations of totalitarian theory have abandoned terror as the defining characteristic of the totalitarian state and now recognize softer forms of "civilized violence"—heavily restricted access to information from the outside world, "harassment rather than physical terror," the abandonment of "big confessions" through show trials in favor of "petty confessions" still capable of eliciting "total submission of the individual to the lie," self-censorship rather than official censorship—as sufficient to subdue the populace and maintain control.[47] Indeed, from the perspective of the Eastern European intellectuals who first articulated post-totalitarian theory, Soviet-sponsored hegemony served only as an extreme case of a dangerous trend present throughout modern societies.[48] In light of these recent theoretical trends, it seems reasonable to try to take Arendt's insights and use them to shed light on the American experience during the cold war.

Arendt identified state intrusion into the interior life of the individual and the disruption of person-to-person relationships as among the central characteristics of the totalitarian state. The effort to "hunt secret thoughts" and target the inner realm of personhood for terror distinguished totalitarianism from ordinary despotism.[49] The essential unknowability of the human mind, however, made the attempt to penetrate the inner life of the individual fruitless and reinforced suspicion and distrust as the basis of social relationships in totalitarian society. As Arendt observed:

[46] On *The Origins of Totalitarianism* and its impact on American intellectuals, see Richard H. Pells, *The Liberal Mind in a Conservative Age: American Intellectuals in the 1940s & 1950s* (New York: Harper & Row 1985), pp. 83–96.

[47] Jacques Rupnik, "Totalitarianism Revisited," in *Civil Society and the State: New European Perspectives*, ed. John Keane (London and New York: Verso, 1988), pp. 273, 275–7. The phrase "civilized violence" comes from Milan Simecka, while Karel Bartosek developed the distinction between "big confessions" and "petty confessions" and argued that the latter still produced "submission to the lie." On the meaning of post-totalitarianism, see also Abbott Gleason, *Totalitarianism: The Inner History of the Cold War* (New York and Oxford: Oxford Univ. Press, 1995), chap. 9.

[48] For example, as Abbott Gleason has noted in the case of Vaclav Havel, "Havel connected mass society, the loss of coherent social identities and structures, and the emergence of totalitarianism, contending that the post-totalitarian society 'is only an inflated caricature of modern life in general.'" Gleason, *Totalitarianism*, p. 186. See also Vaclav Havel, "Anti-Political Politics," in *Civil Society and the State*, 381–98. (Both cit. n. 47.)

[49] Hannah Arendt, *The Origins of Totalitarianism*, new ed. with added prefaces (New York: Harcourt Brace Jovanovich, 1973), p. 422.

> Simply because of their capacity to think, human beings are suspects by definition, and this suspicion cannot be diverted by exemplary behavior, for the human capacity to think is also a capacity to change one's mind. Since, moreover, it is impossible ever to know beyond doubt another man's heart . . . suspicion can no longer be allayed if neither a community of values nor the predictabilities of self-interest exist as social (as distinguished from merely psychological) realities. Mutual suspicion, therefore, permeates all social relations in totalitarian countries and creates an all-pervasive atmosphere even outside the special purview of the secret police.[50]

Under such conditions of widespread suspicion, surveillance soon became ever present. First, the policing arm of the state sought to penetrate and dominate the intimate realm of face-to-face relationships through complete and total surveillance. As Arendt wrote:

> The police dreams that one look at the gigantic map on the office wall should suffice at any given moment to establish who is related to whom and in what degree of intimacy; and, theoretically, this dream is not unrealizable although its technical execution is bound to be somewhat difficult. . . . [T]he Russian secret police has come uncomfortably close to this ideal of totalitarian rule. The police has secret dossiers about each inhabitant of the vast country, carefully listing the many relationships that exist between people, from chance acquaintances to genuine friendship to family relations.[51]

The police, however, constituted only a part of the state's surveillance apparatus. As Arendt made clear elsewhere in her analysis, informers vastly augmented the power of the police and further constricted civil society:

> In totalitarian regimes provocation, once only the specialty of the secret agent, becomes a method of dealing with his neighbor which everybody, willingly or unwillingly, is forced to follow. Everyone, in a way, is the *agent provocateur* of everyone else; for obviously everybody will call himself an *agent provocateur* if ever an ordinary friendly exchange of 'dangerous thoughts' (or what in the meantime have become dangerous thoughts) should come to the attention of the authorities. . . . In a system of ubiquitous spying, where everybody may be a police agent and each individual feels himself under constant surveillance; under circumstances moreover, where careers are extremely insecure and where the most spectacular ascents and falls have become everyday occurrences, every word becomes equivocal and subject to retrospective "interpretation."[52]

With these words, Arendt described a social system of distrust that distorted face-to-face relationships by turning all persons into potential informers who brought others under the gaze of the state. The invisibility of the system, which hid informers under the cloak of anonymity, further expanded the state's power to generate fear and restrict spontaneity by instilling a sense of being watched among all citizens, whether or not they were actually under surveillance.[53]

[50] Ibid., p. 430.
[51] Ibid., p. 434.
[52] Ibid., pp. 430–1.
[53] In the 1970s Michel Foucault similarly emphasized the importance of invisibility to the operation of the police power of the state. "In order to be exercised," he observed, "this power had to be given the instrument of permanent, exhaustive, omnipresent surveillance, capable of making all visible, as long as it could itself remain invisible." Panopticism disciplined the individual by creating the impression of complete and total surveillance, which forced persons to act as if they were being watched, whether or not surveillance was actually taking place. Michel Foucault, *Discipline and Punish: The Birth of the Prison*, trans. Alan Sheridan (New York: Pantheon Books, 1978; originally published as *Surveiller et Punir: Naissance de la Prison*, Paris: Editions Gallimard, 1975), pt. 3, chap. 3, on p. 214.

In the U.S. context, loyalty investigations constituted the kind of effort to "hunt secret thoughts" to which Arendt referred. The FBI's zeal for compiling information and sorting out the bonds of kinship, acquaintance, friendship, and common purpose that connected diverse individuals and organizations matched Arendt's description of the police state's "dream" of total surveillance. Scientists in particular came within view of the FBI's watchful eye because of the postwar connection between scientific research and national security. The result, as the mathematician Oswald Veblen observed at a meeting of the National Academy of Sciences in 1948, was that "We are now living to a very large extent under a police state. Nearly every man in this room has a dossier in the FBI to which he has no access, and which presumably includes all kinds of gossip which he probably knows nothing about."[54] Veblen's comment pointed to the simultaneous invisibility and omnipresence of surveillance that cast a gloomy shadow over postwar science. The pernicious effects of informing within the FAS and, presumably, elsewhere in the scientific community, added an additional layer of meaning to Veblen's observation. In accordance with Arendt's vivid description of how totalitarian pressures made everyone "the *agent provocateur* of everyone else," scientists took part in the process of official surveillance and consequently foreshortened their own participation in the politics of the public realm. Open political discussion and a vibrant public life depend upon trust and openness, both of which surveillance inevitably undermines.

Arendt cautioned against the drawing of false analogies between totalitarianism and other forms of repressive social organization. Post-totalitarian theory, however, provides the rationale for taking the discussion of totalitarianism beyond the Nazi and Soviet contexts. Cold war political repression in the United States did not depend upon violence and terror, but it did employ pressure and harassment through loyalty tests and surveillance. These soft forms of repression did more than simply ensure passive compliance; they also encouraged acceptance of official ideology, or what Vaclav Havel has called "living within a lie."[55] For example, scientists denied loyalty or security clearance resorted to demonstrating their awareness of the threat of communism rather than protesting the workings of the loyalty-security system. Whether or not they believed in the politics of anticommunism, they had to operate within its confines if they wanted to appeal an adverse decision.[56] Loyalty oaths provided another means of enforcing anticommunist discipline. In material terms, a reluctant signature was a small price to pay for the right to a fellowship or job, but in an abstract sense, it constituted a significant subordination of the individual by the state. As the mathematician Marshall Stone wrote in 1949, when explaining his objection to a congressionally mandated loyalty oath and noncommunist affidavit as a condition for AEC fellowship recipients, even if the oath were nothing more than

[54] Transcript of NAS business session, 27 April 1948, p. 48, Organization 1948: NAS: Meetings: Annual: Business Sessions: Transcript, NAS Archives, National Academy of Sciences, Washington, DC.

[55] As Havel observed in 1980, "One need not believe all these mystifications [of official ideology], but one must behave as if one did, or at least put up with them tacitly, or get along with those who use them. But this means living within a lie. One is not required to believe the lie; it is enough to accept life with it and within it. In so doing one confirms the system, gives it meaning, creates it . . . and merged with it." Quoted in Rupnik, "Totalitarianism Revisited," p. 271. See also Gleason, *Totalitarianism*, pp. 183–8 on Havel's understanding of post-totalitarianism. (Both cit. n. 47.)

[56] For specific examples, see my discussion of Eugene Rabinowitch and Robert H. Vought in Wang, *American Science* (cit. n. 11), chap. 3, especially pp. 116–7.

"a kind of empty ritual," it involved a level of coercion that would produce "a certain sense of humiliation and a corresponding feeling of resentment" in the signer. For Stone, the oath was not a mere gesture made for the sake of political necessity; it was an offense against "the political principles on which our nation is founded." Stone rejected the expedient course of living the lie and instead resigned from the AEC's Postdoctoral Fellowship Board in the Physical Sciences.[57] At the greatest extreme, informing in order to protect one's organization or oneself constituted the most serious betrayal of trust, both of oneself and of others. But, as the example of the loyalty oath indicates, lesser forms of post-totalitarian coercion posed just as serious a threat to the integrity of the individual and the health of public life.

SCIENCE AND TECHNOLOGY AS TOTALITARIANISM IN COLD WAR AMERICA

Totalitarian patterns of state action thus curtailed the entry of scientists into public life during the postwar years. But the problems of the nuclear age assured physicists and scientists in related fields a prominent place in national-level policymaking nonetheless. As a result, scientists' political roles were relegated to the province of the elite, where they could exercise influence in high-level policymaking circles as long as they did not deviate too widely from cold war orthodoxy.[58] The ironic consequence was that having been denied by totalitarian political pressures a vigorous presence in public life, scientists themselves came to be perceived as a potentially totalitarian locus of power. By the mid-1950s and early 1960s, an increasing number of American intellectuals and other observers, up to and including the president, expressed growing fears about the power of both science and scientists, and their potential to disrupt democratic forms of governance.

Within the American scientific community, faith in expertise coexisted uneasily with concerns about preserving democratic rule during the early cold war years. The elite scientist administrators, most notably Vannevar Bush and J. Robert Oppenheimer, paid little mind to the potentially undemocratic nature of expert-based decision making. Bush believed strongly that, in order to make effective decisions, scientists and other experts needed to be granted a free hand and shielded from what he viewed as external political interference. Oppenheimer also preferred the policymaking elite to mass-based politics. As Silvan S. Schweber has noted, Oppenheimer "came to believe that if a few wise people at the top had the right ideas, they could effect change. In both politics and physics, he was a member of that elite and perhaps assumed that therefore things would be fine."[59]

[57] Marshall H. Stone to Detlev W. Bronk, May 31, 1949, NAS: Fellowships 1949: AEC-NRC Fellowship Boards: Security Clearance: NAS: Council Statements: First: General, NAS Archives.

[58] As the sociologist Edward A. Shils noted in 1957: "It is a paradox that the past decade during which the roar of anti-intellectualism and of distrust for scientists was louder than it has ever been in America, was also the decade of the greatly enhanced influence of scientists within public bodies and of a moderate but nonetheless unprecedented effectiveness of scientists outside the government seeking to influence opinion and policy." Shils, "Freedom and Influence: Observations on the Scientists' Movement in the United States," *Bulletin of the Atomic Scientists* 13 (Jan. 1957): 15.

[59] Schweber, *In the Shadow of the Bomb* (cit. n. 9), p. 23. On Bush's politics, see Nathan Reingold, "Vannevar Bush's New Deal for Research: Or the Triumph of the Old Order," *Hist. Stud. Phys. Biol. Sci.* 17(2) (1987): 299–344. One should note that faith in an elite was not solely the province of government insiders. Michael Bess has taken Leo Szilard, the consummate outsider, to task for "his elitist and overly rationalized view of politics." Bess, *Realism, Utopia, and the Mushroom Cloud:*

By contrast, those in the postwar atomic scientists' movement devoted considerable attention to the question of their place in public life and the potential conflict between science and democracy. Nor were such concerns limited to the atomic scientists and their campaign to energize the public. During the debate over the National Science Foundation, for example, liberal and leftist scientists also strove to establish the means to ensure public control over science, and they insisted that science policy could not be left purely in the hands of scientific elites. In response to Bush's fears of political interference in science, they argued that politics was a part of modern scientific life, and scientists needed to keep expert authority subordinate to democratic will.[60]

Heated discussions over the relationship between scientists, the public, and democratic governance cooled by the 1950s, as anticommunist political pressures forced scientists to concentrate on defending freedom in science itself. For example, attacks on scientists' loyalty, abuses of the security clearance system, and restrictions on travel rights dominated the *Bulletin of the Atomic Scientists* throughout the first half of the decade. Scientists did continue to write about social responsibility in science during these years, but they focused more on questions of ethics and individual choice than on political participation and public life. The waning of McCarthyism might have revived discussions, but scientists preferred to direct their attention to the renewed prospects for serious arms control negotiations. In 1956, when Argonne physicist David R. Inglis paid heed to "the democratic role of the public" that had been "utterly lost" in the newly resurrected policy debate over arms control, his voice was a relatively rare exception to the general lack of rumination about the Deweyan public in the *Bulletin*'s pages.[61]

Meanwhile, from outside the scientific community, critiques of science and the totalitarian potential of technological society became increasingly apparent. Such concerns grew out of a long-standing critical tradition. Although the first half of twentieth-century history is sometimes portrayed as a time of optimistic faith in the power of scientific and technological progress, such generalizations overlook the ambivalence about technological society that developed long before the 1960s. In European intellectual circles in the 1920s, the Frankfurt School had already begun to associate modern technological society with totalitarianism and the stifling of the individual. On the other side of the Atlantic, John Dewey also viewed the machine age and the rise of organizational society as responsible for a widespread sense of isolation and helplessness that left the public in disarray. In the 1930s Americans debated the extent to which technological unemployment was responsible for the deprivations of the Great Depression. Then, after the end of the World War II, the atomic bomb came to symbolize the destructive capability of science and the inability of humankind to control the technological power unleashed by human creativity.

Four Activist Intellectuals and Their Strategies for Peace, 1945–1989 (Chicago and London: Univ. of Chicago Press, 1993), chap. 2, on p. 84.

[60] I have examined this point at length in "Liberals, the Progressive Left, and the Political Economy of Postwar American Science: The National Science Foundation Debate Revisited," *Hist. Stud. Phys. Biol. Sci.* 26(1) (1995): 139–66, especially pp. 151–6.

[61] David Inglis, "Armament Decision in a Democracy," *Bulletin of the Atomic Scientists* 12 (Sept. 1956): 284. See also Eugene Rabinowitch, "The Role of Science and Scientists in Public Life," *Bulletin of the Atomic Scientists* 13 (March 1957): 80–1; and Lonsdale, "Scientists and the People" (cit. n. 2), pp. 242–5.

Postwar affluence raised a new set of concerns. Echoing the fears of the Frankfurt School, sociologists such as David Riesman, William H. Whyte Jr., and C. Wright Mills worried about the stagnating effect of bureaucracy on white-collar workers and the loss of individual initiative and creativity that seemed to plague modern industrial society. Mills, like Habermas several years later, believed that the rise of mass society and the concentration of political power in the hands of elites had played a major role in replacing an engaged public with an alienated and powerless mass.[62]

Concerns about the dehumanizing character of mass society led to more direct critiques of science. Within the Frankfurt School, the rise of Nazism combined with general trends in the nature of modern society led Max Horkheimer and Theodor W. Adorno to condemn the extent to which the Enlightenment vision of scientific truth had become one of instrumentalism and domination. "Enlightenment," they contended, "is totalitarian."[63] A few years later, C. Wright Mills's attack on bureaucratic society identified scientists as a part of the "power elite" that had wrested political authority from the public.[64] But this kind of criticism of science did not come only from leftist intellectuals such as Mills or the neo-Marxist tradition of Frankfurt School sociology. By the early 1960s concerns about the totalitarian potential of science emanated from the political mainstream. Two particularly prominent figures eloquently articulated such fears: C. P. Snow and Dwight D. Eisenhower.[65]

Snow's 1959 Rede Lecture at Cambridge on "The Two Cultures and the Scientific Revolution" made the British writer and physicist, if not exactly a household name, at least an haute cultural personage. Even if one could not identify Snow himself, the phrase "two cultures" carried a certain cachet and resonance throughout the 1960s. The idea neatly crystallized the troubling sense that the distance between scientific possibility and humanistic understanding was all too dangerous in an age threatened with nuclear destruction.[66]

In the "two cultures" lecture, Snow proposed the adoption of a scientific mindset as a way out of the dilemma of the two cultures. Science, he argued, promised possibility, material abundance, and optimism about the human condition, all of which offered an escape from the tragedies of politics.[67] Had he stopped there, Snow might be remembered solely as a booster of the ideology of progress. But his 1960 Godkin Lectures on "Science and Government," delivered at Harvard University, advanced a much darker portrayal of science and the perils of expert rule in the modern world.

Snow's Godkin Lectures sounded a warning against secrecy in science and the pernicious implications of allowing critical decisions to be made by an isolated elite.

[62] See C. Wright Mills, *The Power Elite* (Oxford and London: Oxford Univ. Press, 1956), pp. 303–4.

[63] Max Horkheimer and Theodor W. Adorno, "The Concept of Enlightenment," in *Dialectic of Enlightenment* (New York: Continuum, 1999; originally published as *Dialektik der Aufklärung: Philosophische Fragmente*, Amsterdam: Querido, 1947), p. 6.

[64] Mills, *Power Elite* (cit. n. 62), pp. 216–8.

[65] I have discussed Snow and Eisenhower at greater length in "Merton's Shadow: Perspectives on Science and Democracy since 1940," *Hist. Stud. Phys. Biol. Sci.* 30(1) (1999): 279–306, especially pp. 285–9.

[66] The gap between the sciences and the humanities was actually only one of three problematic divisions identified by Snow. The other two were the gap between industrialized and nonindustrialized nations, and the cold war cleavage between the United States and the Soviet Union. C. P. Snow, *The Two Cultures and the Scientific Revolution* (Cambridge: Cambridge Univ. Press, 1959), pp. 4, 39.

[67] Ibid., pp. 5–6, 11, 38–44.

He opened with a grim depiction of cold war political realities: "One of the most bizarre features of any advanced industrial society in our time is that the cardinal choices have to be made by a handful of men in secret, and, at least in legal form, by men who cannot have a first-hand knowledge of what those choices depend upon or what their results may be." The most basic decisions—"those which determine in the crudest sense whether we live or die"—had been delegated to a handful of persons, and public oversight over those decisions ranged from minimal to nonexistent. In such a situation, references to "phrases like 'the free world,' or 'the freedom of science'" rang hollow. The greater threat to freedom in the United States and Western Europe, Snow implied, came not from without (in the form of Soviet totalitarianism) but from within, in the concentration of power among political and scientific elites.[68]

World War II and the cold war, Snow indicated, had allowed secrecy and science to become an all too insidious combination. Here Snow seemed to echo the warnings of the Frankfurt School in his attack on an instrumentalist worldview that tried to reduce difficult political choices to technical problems. The arms race, he argued, was the product of "a typical piece of gadgeteers' thinking" that "has done the West more harm than any other kind of thinking." Secrecy further distorted policymaking by propagating "an unbalancing sense of power" that led scientists and policymakers to lose a proper sense of perspective. The result, he concluded darkly, was a form of lunacy: "It takes a very strong head to keep secrets for years, and not go slightly mad. It isn't wise to be advised by anyone slightly mad."[69]

Months later, in his "Farewell Address," outgoing president Dwight D. Eisenhower provided an even starker statement of the totalitarian potential of science and scientists. Even as the country battled Soviet totalitarianism abroad, he observed, it faced a potentially more serious threat at home. The military-industrial complex and the perpetual mobilization of science posed a new kind of danger to American democracy. Eisenhower did not mince words: "The total influence—economic, political, even spiritual—is felt in every city, every state house, every office of the Federal Government. . . . Our toil, resources and livelihood are all involved; so is the very structure of our society." The military-industrial complex, he continued, portended a "disastrous rise of misplaced power." Not only did it threaten to dominate American institutions, but it carried the menacing potential to invade every area of public and private life, including the innermost recesses of the human spirit. Eisenhower further cautioned that scientists themselves might come to constitute a source of power independent of public control. As he declared, in one of the most famous passages of the speech, "Public policy could itself become the captive of a scientific-technological elite."[70]

One expects such a clarion call from a leftist intellectual such as a C. Wright Mills, but not from a conservative, Republican, U.S. president. But Eisenhower's meditation about the dangers of the military-industrial complex and the cold war mobilization of science reflected more widely held anxieties and itself touched a chord. Throughout the early 1960s, no major discussion of the relationship between

[68] C. P. Snow, *Science and Government* (Cambridge, Mass.: Harvard Univ. Press, 1961), pp. 1–2.
[69] Ibid., pp. 70, 72–3.
[70] Dwight D. Eisenhower, "Farewell Address," *Vital Speeches of the Day* 27, 1 Feb. 1961, pp. 228, 229.

science and society could take place without referring to Snow or Eisenhower. Studies of science policy, ranging from the early work of the political scientist Robert Gilpin in *American Scientists and Nuclear Weapons Policy* (1962) to the political theorist Don K. Price's masterpiece, *The Scientific Estate* (1965), to the dark vision of scientific dominance and "the tyranny of technology" in physicist-writer Ralph E. Lapp's *The New Priesthood* (1965), all attempted in one way or another to grapple with the new political power of scientists and the threat science and technology posed to public rule.[71] Stanley Kubrick's brilliant 1964 film, *Dr. Strangelove*, also captured for the popular mind the "cardinal choices" of the cold war, the insidious power of science and scientists within the military-industrial complex that Eisenhower had warned of, and the insanity that Snow had identified as a part of government secrecy, all pushed to their utmost extreme.

These cold war anxieties about science and scientists, however, ultimately had less to do with science itself than with the nature of governance in a modern, bureaucratic society. Domestic politics during the early cold war did not simply revolve around short-term struggles of right versus left, or free speech versus repression. Cold war political culture reflected a larger conflict over who had the right to make politics. Scientists' postwar political experiences took place within a historical context of deep ambivalence, and even hostility, toward the public as an active political force. The old Tocquevillian ideal had already been on the defensive for decades as interest-group pluralism steadily supplanted voluntary modes of association in the first decades of the twentieth century, and, in an increasingly complex, bureaucratic society, expertise supplanted citizenship as the main criterion for active political participation.[72] The 1930s signaled some new possibilities for the public, in the form of a nascent and vibrant civil rights movement in the South, as well as renewed efforts to bridge the culture of expertise and the public as an alternative to fascism. Scientists launched modest efforts in this realm through the AAAS and the American Association of Scientific Workers, while the big technology projects of the New Deal also attempted to find ways to reconcile technological society with values of localism and popular government.[73] During the cold war, however, liberal anticommunists, distrustful of what they perceived as an irrational, easily manipulated, and even authoritarian public, reasserted a preference for expert-directed, interest-group

[71] Robert Gilpin, *American Scientists and Nuclear Weapons Policy* (Princeton, N.J.: Princeton Univ. Press, 1962); Don K. Price, *The Scientific Estate* (Oxford: Oxford Univ. Press, 1965); and Ralph E. Lapp, *The New Priesthood: The Scientific Elite and the Uses of Power* (New York: Harper & Row, 1965).

[72] For critiques of interest-group liberalism, see, e.g., Christopher L. Tomlins, *The State and the Unions: Labor Relations, Law, and the Organized Labor Movement in America, 1880–1960* (Cambridge: Cambridge Univ. Press, 1985), and Lasch, *The True and Only Heaven* (cit. n. 3).

[73] On the New Deal and civil rights, see Patricia Sullivan, *Days of Hope: Race and Democracy in the New Deal Era* (Chapel Hill and London: Univ. of North Carolina Press, 1996). Purcell, *The Crisis of Democratic Theory* (cit. n. 3), pt. 3, details American intellectuals' renewed dedication to the democratic ideal during the 1930s. On scientists in the 1930s, see Kuznick, *Beyond the Laboratory* (cit. n. 17). New Deal programs such as the Greenbelt Town Program, the Tennessee Valley Authority, and the Rural Electrification Administration all claimed to be undertaking large-scale, government-sponsored development for the sake of upholding the values of localism and extending popular democracy. See, e.g., David E. Lilienthal, *TVA: Democracy on the March* (New York and London: Harper & Brothers, 1944). Recent scholarship has tended to be critical of such pretensions and has emphasized the antidemocratic, elitist tendencies of the New Deal. At the same time, certain aspects of the New Deal arguably represented genuine efforts to reconcile expertise and democracy, even if such efforts were not entirely free from self-contradiction or successful.

politics over mass political movements.[74] Within this context, the debate over the totalitarian potential of science implied a larger question: Which was more totalitarian, the experts or the masses?

The Deweyan problem of the eclipse of public authority in an age of experts has yet to reach a final resolution. For a time in the 1960s, the revival of the civil rights movement and the emergence of the New Left heralded new life and creativity in the public and its politics. Scientists also showed signs of renewing their sense of public civic engagement through vigorous protests against the Vietnam War and the military uses of science.[75] In general, however, scientists' experiments with public life have proven somewhat abortive. On the whole, professional pressures and interests have historically militated against the expansion of the public dimension of science. Moreover, as the experience of the atomic scientists' movement during the cold war illustrates, efforts to build a presence for science within the realm of the public do not take place within a political vacuum. A reinvigorated and newly assertive public risks a backlash from the state and entrenched forms of power. The tendency of science itself to represent concentrated power further complicates its potential role as a force to revitalize public life.

The problem of science and civil society is thus not merely an academic one. The history of science in the cold war suggests both the creative possibilities of science as a part of public life, as well as the extent to which science itself has dampened the political potential of the public. In the 1920s Walter Lippmann and John Dewey clashed over whether the rise of experts in an age of high science and technology had made the public obsolete or if the public remained more necessary than ever. The basic dilemma they identified remains just as relevant three-quarters of a century later.

[74] Pells, *The Liberal Mind in a Conservative Age* (cit. n. 46), chap. 3, especially p. 130; and Lasch, *The True and Only Heaven* (cit. n. 3), chap. 10, especially pp. 460–5 on Seymour Martin Lipset's idea of "working-class authoritarianism."

[75] Stuart W. Leslie, *The Cold War and American Science: The Military-Industrial-Academic Complex at MIT and Stanford* (New York: Columbia Univ. Press, 1993), chap. 9.

COMMENTARY

The 'Creative Possibilities of Science' in Civil Society and Public Life: A Commentary

*By Celia Applegate**

ABSTRACT

The commentary categorizes some of the ways the authors have depicted the relationship of science to civil society and considers some of the parallels between these analyses of civil society and those in comparable fields (art and music especially)—fields that also are attempting to broaden their perspective to the place of their subjects in society as a whole.

IN AN EFFORT to establish a new source of authority for truth at a time when old sources of authority seemed no longer compelling, the American scientist and progenitor of pragmatism, Charles Sanders Peirce (1839–1914), famously wrote that truth is what reasonable and competent people eventually agree upon: "the opinion which is fated to be ultimately agreed to by all who investigate it is what we mean by truth and the object represented by this opinion is real."[1] The contributors to this volume have reached such a consensual truth about the role of science in the public life of the past several centuries and have placed science just about where Peirce (himself a representative of the public cultures of science under discussion here) might have expected to find it—among the wellsprings of change and creativity in collective life, the reception and influence of which depend on the circumstances of politics, profession, and power. These authors have put together a new narrative about the place of science in the modern world, a narrative in which scientists are not heroic pursuers of truth or instruments of state power, not merely or only avatars of professionalized self-interest or remote purveyors of socially improving expertise. They appear instead as participants in the broader social and political movements of their times, players with important, even constitutive roles in the debates of public life.

These authors also illuminate the ways in which the circumstances of politics, institutions, and interests—many of them of the scientists' own making—have shaped this participation in civil society and public life, though often offstage from

* Department of History, University of Rochester, 461 Rush Rhees Library, Rochester, NY 14627-0055; capg@mail.rochester.edu.

[1] A witty and unexpected use of Peirce's pragmatism may be found in Franklin Foer's defense of "Conventional Wisdom: Why What Everyone Thinks Is Usually Right," *The New Republic*, 19 March 2001, pp. 23–5.

the activities that take center stage in these articles. While the scientists appearing in these pages were all in their own contexts committed to the significance of the public sphere as a space for scientific activism and engagement in rational-critical debate, the articles illuminate the extent to which the communicative space of the public sphere remained as much an ideal as a reality. The limitations on its vitality and even existence that troubled Habermas in his original discussion of its origins and fate (and others, such as gender limitations, that did not) are very much in play in the settings considered here. As Jessica Wang writes in the concluding—and in many ways the darkest—article in this volume, the examination of scientists in civil society "suggests the creative possibilities of science as part of public life, as well as the extent to which science itself has dampened the political potential of the public." Over the past centuries, the scientific project has provided the strongest possible case for the social benefits of free and rational debate, but it has also contributed to the isolation of knowledge within either private or exclusively state-serving communities of experts. Moreover, science's very engagement with the ideology of civil society has led, as Thomas Broman points out in his introduction, to an entanglement with interests that are far from universal and truth seeking.

We can approach these problems first by comparing the kinds of stories that the authors in this volume tell about the "creative possibilities of science as part of public life" and second by looking at the limitations to both the civil place of science and the existence of a vital public sphere that these articles reveal. The authors tell essentially three kinds of stories about the creative possibilities of science for public life, sometimes linking two or even three of these stories.[2] The first narrative concerns scientific knowledge as a source of guidance for those who would devise or describe an ideal form for collective life. As Harold Cook tells us, this quasi-public use of science relied on a simple, persistent, and at the same time misleading metaphor (as all metaphors are) embodied in the term "body politic." If political, which is to say collective, life operates analogously to the human body, then what scientists know about the human body must obviously have implications for how we live together. This essentially banal observation has, as Cook writes, informed the speculations of a whole raft of political thinkers and scientists turned political thinkers. The power of this analogy between bodies human and bodies politic "has remained with us," Cook argues, and subsequent articles in this volume bear him out. John Carson's cast of Enlightenment thinkers, in Europe and America, puzzled over whether differences in the "virtues and talents" of people derived from nature or from experience and reached in vain into the well of scientific knowledge to try to resolve the "ambiguities surrounding 'talents.'" Andreas Daum's energetic investigators and Lynn Nyhart's teachers of biology remained indebted to what Daum calls Alexander Humboldt's "cosmic view of nature," within the capacious framework of which all manner of conclusions about how people ought to live together could be formulated. And Elizabeth Hachten's Russian scientists, Theodore Porter's utopian statistician, Zuoyue Wang's students returning from abroad to invigorate modern science in China, and Jessica Wang's atomic scientists had in common, despite indulging in quite different

[2] I have tended, along with several of the authors, to treat civil society and the public sphere as "virtually synonymous," or at least as interchangeable terms, a tendency in historical writing as a whole that Jessica Wang remarks upon in a particularly insightful footnote. This is simply a means of placing discussions of science within the process of rational-critical debate that defined the public sphere for Habermas.

degrees of optimism and self-aggrandizement, the view that their scientific knowledge represented a resource for the improvement of society, not just physically and technologically, but politically and morally as well.

It is worth noting that this view of science's public role, as an essential part of the knowledge that animates and improves society as a whole, assured science an institutional presence and a degree of state and private sponsorship in nineteenth- and twentieth-century societies. Such an assessment of the crucial place of scientific knowledge in an enlightened society lay behind the establishment of scientific societies, new universities, and other institutions dedicated to the promotion of science for the good of society. Perhaps even more interesting, other branches of knowledge followed suit, often claiming an analogy to science's recognized value to justify their own claims to recognition and funding. Within the capacious concept of *Bildung* in the German case or *civilisation* in the French one, soon such people as art historians and music theorists were themselves claiming the need for university departments, massive new art museums, concert halls, and conservatories—indeed, in the case of the musically adept, they routinely claimed to possess "scientific knowledge."[3] To be sure, music and art had been part of European culture, and the objects of considerable private and princely expenditures, for centuries. But in the wake of the spread of scientific knowledge, art and music, too, began to come forward as autonomous entities, not merely decorations of a noble life or appendages to religious ceremony. As such, they took on a social and morally improving purpose in the eyes of their enthusiasts. An art museum in the nineteenth century, as James Sheehan has shown, was not just a place where the general public could look at art, but "a setting in which visitors can comprehend the connections between art, truth, and morality."[4] Perhaps the clearest expression of this redefinition of art in the wake of the scientific revolution and enlightenment came in Friedrich Schiller's *On the Aesthetic Education of Man*, published in 1795 in the form of letters to his patron, the duke of Augustenburg. For Schiller, reason alone—which for the sake of argument we can understand here more concretely as the *application* of reason to the understanding of human beings and society—was insufficient to solve contemporary problems of social fragmentation and cultural disharmony. The free play of humans in creating and experiencing art alone held the possibility of bringing reason and sensuality, needs and choices into harmony, and thus creating a truly free society.[5] In the context of this volume, the point to emphasize about Schiller is the extent to which his whole argument assumed a general agreement among the educated people to whom he addressed himself that a rational, scientific approach provided the most promising means of dealing with the problems of social life and morality. The new narrative

[3] See, e.g., the musical pedagogue, historian, and theorist Adolf Bernhard Marx, *The Music of the Nineteenth Century and Its Culture*, trans. C. Natalia MacFarren (London: Robert Cocks & Co., 1855), p. 8. On art and *Bildung*, see James J. Sheehan, *Museums in the German Art World from the End of the Old Regime to the Rise of Modernism* (New York: Oxford Univ. Press, 2000). On music and *Bildung*, see David Gramit, *Cultivating Music* (Berkeley: Univ. of California Press, 2000); and Celia Applegate, "How German Is It? Nationalism and the Idea of Serious Music in the Early Nineteenth Century," *19th Century Music* 21 (1998): 274–96.

[4] Sheehan, *Museums* (cit. n. 3), p. 3.

[5] Friedrich Schiller, *On the Aesthetic Education of Man, in a Series of Letters*, ed. and trans. Elizabeth M. Wilkinson and L. A. Willoughby (Oxford: Clarendon Press, 1982); see also Josef Chytry, *The Aesthetic State: A Quest in Modern German Thought* (Berkeley: Univ. of California Press, 1898), pp. 70–105.

about science and civil society put forth by these articles holds out to cultural and political historians the promise of linking science to art, philosophy, music, and so on through an understanding of how all these forms of knowledge emerged in particular contexts and processes of institution building.

But as the very notion of an autonomous art or music or scientific knowledge implies, science's place in the emergence of civil society is not exhausted by the story of its "usefulness." The second story these articles tell understands civil society less as the recipient or beneficiary of scientific knowledge, in one form or another, than as a work in progress to which scientists, among others, have contributed through their inquiries and public activity. To use Broman's definition, civil society functions in these articles not as an empirical reality, but "as an ideology that has had an important impact on history and has interacted in significant ways with science." Or perhaps we could describe civil society slightly differently and regard it, like Ernest Renan's nation, as an ideal image of collective life that must be constantly re-created, renewed, and relegitimated to exist. Our nineteenth-century musical informant, A. B. Marx, had just such a view of nation, which for him really was the same thing as civil society: the expression of freedom and human creativity—and something that would fade away if not constantly invigorated by the efforts of its "conscious parts." "It is as necessary in art as in every other concern of life," he wrote in 1854, "that those who invent, and those who examine, improve, and apply, should go hand in hand with brotherly love; that every one should unite his own interests with those of others, and find his gain and reward in the gain of all." Turning to a scientific metaphor, Marx claimed that "each of us is merely a link of the electric chain through which flashes the spark that is to kindle a light among the people; no one forming a link of this chain is to isolate himself from the rest, no one is to neglect his portion of the work; every one must be ready to receive, that he may be able to give, and freely to dispense whatever he may have gathered or matured."[6]

Such sentiments would not have been in any way strange to a number of the scientists and scientific promoters described in these articles. Daum and Hachten, for instance, analyze developments in Germany and Russia in which men who were neither state servants nor (in Daum's article) recognized professionals did not just work in the public interest (story number one) but acted publicly as well. In their writing, communicating, and organizing, they thought of themselves as sustaining a foundering civil society. The eighteenth-century *Ladies' Diary* that Shelley Costa examines gains its significance for its work in constituting a public sphere that was just emerging at the time, as well as ensuring that mathematical puzzlers (including women, but only in a limited way) would be part of it. Zuoyue Wang defends the use of the concept of civil society in China because we need some term to identify what Hachten calls the "space where the interactions between individuals with divergent agendas can be mediated through civic institutions, shared values, and common social goals." That space, Wang argues, was precisely what returning Boxer students opened up by founding science societies across China and disseminating scientific knowledge through journals. In our second narrative, then, the "creative possibilities of science" must include its crucial, possibly pioneering service in establishing a public sphere, particularly in societies late to industrialize and ruled by states jealous

[6] Marx, *Music of the Nineteenth Century* (cit. n. 3), pp. 2–3.

of action outside their spheres of control. The implication one can draw from Wang's and Hachten's research is that science in the modern world has the potential to slip past the barriers states erect to public debate and discussion, by virtue of its promise of technological progress. If this is true, then we are looking at quite a different phenomenon of civil society than, for instance, the German singing or gymnastic associations of the early nineteenth century, which served as cover for liberal nationalists, ultimately more interested in promoting their political cause in the face of censorship and political oppression than in actually singing or exercising together. If indeed scientific societies could create pockets of free exchange in otherwise repressive societies, then historians need to investigate more fully, as Broman suggests, whether such activity could actually "undermine the legitimacy of political authority" to the end of political liberalization, and whether such undermining had more than connections of convenience to the science itself.

That in turn suggests the third story these articles tell about the civil implications of science—the story of science as a marker of national or urban vitality, a story that has less to do with the specifics of scientific knowledge or the participation of scientists in some putative public sphere than with science's privileged status within civil societies and states. The scientists of Russia and China, depicted in Hachten's and Zuoyue Wang's articles, seem to have been at least as motivated by thoughts of what they were doing for their nation's international reputation as by their desire to pursue specific scientific agendas. In the articles of Alice Conklin and Glenn Penny, this kind of competitive consciousness becomes dominant. Ethnographers in both France and Germany—driven, in Penny's phrase, by "honor, image, and prestige"—fought for ethnographic museums as public manifestations of scientific prowess. To be sure, these ethnographers wanted as much of the public to visit their museums as possible. Yet as Conklin acknowledges and Penny argues outright, the ethnographers' desire to enlighten the broader public did not reflect any longing for engagement in a Habermasian rational-critical dialogue with visitors, but a simpler need to impress and show off, to one's own city, to the nation, and to the world. In this, scientists were clearly part of a much broader nineteenth-century trend to make the ownership and display of knowledge the ultimate marker of both urban and national strength, whether through art museums, universities, or world-renowned orchestras.

In fact, in these latter two articles, one begins to understand some of the difficulties encountered by scientists and other knowledge brokers in trying to realize their vision of a society made better through the pursuit of learning. The actual public often proved elusive as an active participant in its own enlightenment. Despite the fact that the founders of the Museum of Man held favorable views toward "elite-masses collaboration," they seem to have had few ideas about how to achieve this beyond keeping the museum open on some evenings and providing a considerable amount of explanatory text. Likewise, Alfred Lichtwark, the first professional director of Hamburg's Art Museum, pursued an ambitious program of cultural promotion, production, and consumption, but often expressed despair at not having reached the people he most wanted to influence.[7] And after a decade of trying to wean the Berlin public from its demoralizing attachment to intellectually vapid entertainment music,

[7] Jennifer Jenkins, "Provincial Modernity: Culture, Politics, and Local Identity in Hamburg," Ph.D. diss., Univ. of Michigan, 1997.

A. B. Marx wondered whether the changes he anticipated through an improvement of artistic life would come in his lifetime.[8] We seem to confront in each of these cases, disparate in time and place, a central tension in the very idea of creating a civil society through the free exchange of knowledge—a tension that has less to do with the actual content of the knowledge, be it scientific, philosophical, or musical, than with the sociology of how we learn. For although the scientists and other teachers wanted to create settings that would be open to everyone, they almost invariably came up against the structural inequalities in the world surrounding them, whether those be the exhaustion of the worker or the inability of people to imagine the relevance of debate to life. The result of inequalities that cannot be rectified by extended opening hours is a kind of passivity imputed to the public, present only to admire the magnificent display techniques of cutting-edge ethnographers or to learn their scientific, musical, and artistic lessons without being changed by them. Civil society as an ideal of free and open exchange remains just that, an ideal, perhaps a guiding one but never quite realizable.

Every article in this collection highlights the extent to which the idea of scientific work proceeding freely, democratically, and effectively to the benefit of society as a whole looked rather different—and far more interesting—in the effort at realization than in the theory. The range of problems that scientists encountered in such efforts to constitute a public life of science is a familiar one, characteristic also of the problems encountered in political, social, and cultural life more generally. For Habermas and the Frankfurt school in general, the greatest imposition on the communicative space of the public sphere was the force of the market. Fear of commercialization as the ultimate enemy of *Bildung* or self-development haunted German educators of all stripes throughout the nineteenth and twentieth centuries, even as the capacity of the market to work in concert with knowledge production proved itself in many ways. For instance, the consumer desires that fueled travel and tourism supported arts both high and low: the literature on nineteenth-century spas attests to the centrality of musical performance to the spa experience; the literature on travel speaks similarly to the importance of tourists to new art, historical, and scientific museums. Certainly science, too, as a cultural field among others, proved accessible to the marketplace and hence vulnerable to the kinds of tensions, often quite productive ones, that arose between the market's peremptory measures of value and any others. As Penny argues, "Major decisions governing large acquisitions or changes" in the ethnographic museum in Hamburg "were consistently based on their importance for the city's image," while in Leipzig, a museum with chronically "tenuous" finances had to pay constant attention to the popular appeal of its displays. Ethnologists, like other scientists, had to learn to "sell their sciences" to the proper authorities and, like David Lodge's latest protagonist, become media intellectuals and "masters of the scientific sound-bite."[9]

Penny, for his part, sees this process of self-promotion as essentially productive, and in any case not especially corrupting of the scientific endeavor as a whole. But not every tension between an ideal and its realization can be regarded so sanguinely.

[8] A. B. Marx, "Abschied des Redakteurs," *Berliner Allgemeine Musikalische Zeitung* 7(52) (24 Dec. 1830): 414–6.

[9] David Lodge, *Thinks . . .* (New York: Viking, 2001).

First, in their sometimes self-appointed, sometimes popularly requested, and sometimes state-supported roles as experts on the problems that beset societies, scientists do not necessarily provide the best answers or even the best means to an answer. Therefore, the habit of expecting them to provide answers—a habit the acquisition of which so many of these articles illuminate—carries all-too-familiar hazards. As both Cook and Carson explore in different ways, the efforts of philosophers and politicians to solve such questions as how political authority should be exercised or who should rule, by reference to the contemporary body of scientific truth, produced debate, certainly, but no resolution that either author is willing to label beneficial. In the seventeenth and eighteenth centuries, one could still argue that such "continual renegotiation" (as Carson characterizes the debate about the importance of promoting excellence or equality) nevertheless sustained a desirable alertness to the greater social good among citizens, precisely the sort of effect a public sphere ought to make possible. But with the rise of the debate-stifling expert, alongside a growing consensus that scientific knowledge has become too complex for the layperson, the habit of turning to science for answers begins to carry significant costs. Moreover, as a number of these authors note, deference to science as a source of guidance for public life does not always produce merely a lively exchange of opinion; sometimes it produces policy. Reading Porter's account of Karl Pearson, one can only be grateful that he did not acquire the kind of influence in public life to which he aspired. Nor, to put it mildly, can one be completely sanguine about the biologistic paradigms of Nyhart's teachers, even though in their particular context they embodied the democratic promises of decentralization, localism, and experientially based knowledge. The words of the well-read, intelligent, and publicly minded Oliver Wendell Holmes expressing the opinion of the U.S. Supreme Court on the compulsory sterilization of Carrie Bell—"Three generations of imbeciles are enough"—haunt our understanding of science in the public sphere.[10] These articles suggest the importance of investigating why and how public institutions and social formations adopt some kinds of science and not others, as well as the processes by which public understanding of science changes, moves forward, transforms. What, to put it otherwise, does a paradigm shift look like in the context of civil society?

That in turn suggests a further question that lies behind the study of science and civil society: What precisely is the relationship of expertise and professionalization to the idea of civil society? From one perspective, the rise of the free professions—law, journalism, medicine, some kinds of teaching—formed the very backbone of political and social liberalization in much of Europe. Just as academic tenure was meant to ensure freedom of speech and conscience, so too could it be observed historically that the security of professional norms and protections allowed men (mostly) to stand outside the state and question its purposes, while at the same time proposing remedies to social ills. The more a group consolidated such professional norms, developed specialized knowledge, and organized nationally and even internationally, the more freedom of thought and action might be expected to develop. But, of course, we know things did not entirely work out that way. Studies of professional organizations in Germany show the extent to which the state found common

[10] Henry Friedlander, *The Origins of Nazi Genocide from Euthanasia to the Final Solution* (Chapel Hill: Univ. of North Carolina Press, 1995), pp. 8–9.

purpose with them, resulting in a trade-off by which the freedom to develop the profession as such was bought at the cost of speaking more generally in public life.[11] The whole phenomenon of the expert professional created in its wake that somewhat hapless person of public discourse, the layperson. To be sure, lay status can be an effective rhetorical platform from which to engage in public debate, as both Nyhart and Daum demonstrate. But the dichotomy of expert/layperson can also threaten the flourishing of public debate, by the propagation of knowledge that only extensive study can make intelligible, by the proliferation of specialized language, by the intrusion of considerations of career and promotion into research and study. The list is a familiar one, and implicates us all as we talk to each other, about each other, and for each other. The associational life of an idealized civil society provides a congenial atmosphere for clubbiness, and professional agendas are not always compatible with the civic ones from which they may once have come. For instance, a fine line existed between the promotion of the scientific profession and the engagement with a general social good in the Russian case Hachten examines and the Chinese case Zuoyue Wang examines. Conklin's treatment of French ethnography tells a story about the struggle for recognition, funding, and resources, which brought in its wake imperial attachments inimical to the egalitarian hopes of the ethnographers and to their capacity to act freely and politically in the public sphere. There is strong evidence here, to paraphrase Habermas, for a structural transformation of the public sphere out of existence, such that rational debate itself reflects not the exercise of judgment but the calculation of professional self-interest and the appeal to power. The invocation of democracy, equality, and liberty among museum promoters rings false, and the possibility that the museum might work to combat racial theories remained rhetorical, as long as a colonial state lay behind its very existence.

The capacity of state power to overwhelm the associational, nonhierarchical, and intermediary spirit of civil society has manifested itself unevenly across the groups and gatherings of social life. But it has had more fateful consequences for scientists than most other participants in the public world, perhaps because of science's uniquely privileged place among the different kinds of knowledge in modern life. Important though most Germans believed music was to their national life, no German government intervened to control musical activities and associations until 1933, and even then the actual practice of musical "coordination" contained so many inconsistencies and areas of inattention that very little changed for non-Jewish musicians and musical amateurs.[12] The same lack of interest cannot be found for any branch of scientific inquiry in the Third Reich, or for many before then, as numerous studies of physics, mathematics, biology, psychology, and even cancer research have overwhelmingly shown.[13] Nor was Nazi Germany a unique case, even if an extreme one. The paradox of science at once dominating public debate and being dominated

[11] See, e.g., Hannes Siegrist, ed., *Bürgerliche Berufe* (Göttingen: Vandenhoeck & Ruprecht, 1988).

[12] On this complicated question, see especially Pamela Potter, *Most German of the Arts: Musicology and Society from the Weimar Republic to the End of Hitler's Reich* (New Haven, Conn.: Yale Univ. Press, 1998); Michael Kater, *The Twisted Muse: Musicians and the Music in the Third Reich* (New York: Oxford Univ. Press, 1997).

[13] For a sampling, see Robert Proctor, *The Nazi War on Cancer* (Princeton, N.J.: Princeton Univ. Press, 1999); Götz Aly, *Cleansing the Fatherland: Nazi Medicine and Racial Hygiene* (Baltimore: Johns Hopkins Univ. Press, 1994); Thomas Powers, *Heisenberg's War: The Secret History of the German Bomb* (New York: Da Capo, 1993); Geoffrey Cocks, *Psychotherapy in the Third Reich: The Göring Institute*, 2d ed. (New Brunswick: Transaction Publishers, 1997); Sanford Segal, "Mathemat-

by state power informs Jessica Wang's classically tragic account of atomic scientists' effort to act as citizens *and* scientists and their fatal entanglement in the repressive political atmosphere of cold war America. For Wang, the central problem in the relationship of science to civil society in the twentieth century was that of the public's ability to maintain control in the face of science's claim to power, a problem exacerbated by the need of political power for scientific knowledge. As Broman points out, at some level scientists simply could not be both scientists and citizens, both providers of special knowledge and speakers of universal truths.

Moreover, the thicket of interests, causes, and platforms from which and for which scientists might speak has only become denser since the 1950s. If we are to understand what we need from science in public life and under what conditions scientists themselves can speak most effectively, then studies of the sort represented in this volume are invaluable. The evidence from contemporary debates about global warming or genetic research presents us with a self-evident need for scientific understanding in the public realm. But in these cases and others, we remain mired in controversies over the politics of scientific claims, unable to assess just what these politics might amount to and how they might shape knowledge, in part because of a very real lack of historical perspective on how science has operated publicly in the past. Much, in other words, remains to be studied about how the intermediary groups of scientists in civil society came into being, what sustained them financially, and how important specialized knowledge as such was and is to their broader legitimacy. But quite apart from contemporary relevance, the study of associational life in general has proven its worth as a means of breaking down simple narrative lines of historical development and complicating monolithic explanations for historical change. A nineteenth-century Germany inhabited by people other than Prussian martinets and ineffectual, emigrating democrats was recovered thanks in part to a renewed emphasis on the parties, clubs, journals, institutes, and movements that flourished with and without state assistance, attention, and aggravation. But long before revisionists took on the German *Sonderweg*, generations of social historians in Great Britain, starting with the Hammonds in the 1920s, had challenged the complacency of ameliorative Whig history in large part by finding a level of civil society in which (to paraphrase David Cannadine) more happened, more dramatically than was once thought.[14] The very notion of an industrial revolution, still so central to our understanding of the modern world, came about in part because Barbara Hammond discovered extensive evidence of civil disturbances in the British Home Office records from the late eighteenth and early nineteenth centuries.[15] From there it was only a few simple steps to E. P. Thompson's *Making of the English Working Class*, E. J. Hobsbawm's *Captain Swing*, and other classic studies of civil action and consciousness. Insofar as the history of science wishes to remain free of all unidirectional narratives and participate in the writing of general history, then the consideration of civil society, with all the complexity, contradiction, and general mess it brings to our understanding of the past, can only liven things up for all of us.

ics and German Politics: The National Socialist Experience," *Historia Mathematica* 13 (1986): 118–35.

[14] David Cannadine, "British History: Past, Present—and Future?" *Past Present* 116 (1987): 183.

[15] Stewart Weaver, *The Hammonds: A Marriage in History* (Stanford, Calif.: Stanford Univ. Press, 1997).

Notes on Contributors

Celia Applegate is Associate Professor of History at the University of Rochester, where she teaches modern German and European History. She received her Ph.D. from Stanford University in 1987 and has published *A Nation of Provincials: The German Idea of Heimat* (Berkeley: University of California Press, 1990) as well as numerous articles on German regionalism and German music. She edited with Pamela Potter the volume on *Music and German National Identity* (Chicago: University of Chicago Press, 2002).

Thomas H. Broman is Associate Professor of History of Science and History of Medicine at the University of Wisconsin–Madison. His 1997 book, *The Transformation of Academic Medicine in Germany, 1750–1820*, was published by Cambridge University Press, and he is currently writing a book on the evolution of the public sphere and the periodical press in 18th-century central Europe.

John Carson is Assistant Professor in the Department of History at the University of Michigan. His publications include "Minding Matter/Mattering Mind: Knowledge and the Subject in Nineteenth-Century Psychology," *Studies in the History and Philosophy of the Biological and Biomedical Sciences* 30 (1999): 345–376, and "Army Alpha, Army Brass, and the Search for Army Intelligence," *Isis* 84 (1993): 278–309. He is currently working on a book manuscript tentatively titled *Making Intelligence Matter: Cultural Constructions of Human Difference, 1750–1940* for Princeton University Press.

Alice L. Conklin is Associate Professor at the University of Rochester, where she teaches modern European history, the history of colonialism and imperialism, and women's history. She is the author of *A Mission to Civilize: The Republican Idea of Empire in France and West Africa (1890–1930)* (1997), which won the Berkshire Prize, and of several articles on French colonialism. She has also written on the challenges of teaching colonial history, and coedited a reader on modern European imperialism for classroom use. She is currently at work on a book, tentatively titled *In the Museum of Man: Ethnographic Liberalism in France, 1920–1945.*

Harold J. Cook, Professor and Director, Wellcome Trust Centre for the History of Medicine at the University College London, continues to investigate subjects related to early modern English medicine, but now gives most of his energy to medicine and natural history in the Dutch Golden Age, including the activities of the Dutch East India Company, in an attempt to reassess the relationships between the beginnings of a worldwide trading system and a worldwide exchange of information about nature. Among his publications are *Trials of an Ordinary Doctor: Joannes Groenevelt in Seventeenth-Century London* (Baltimore: Johns Hopkins University Press, 1994), and *The Decline of the Old Medical Regime in Stuart London* (Ithaca, N.Y.: Cornell University Press, 1986).

Shelley Costa is currently a writer and independent scholar living in Cincinnati. She earned a Ph.D. from Cornell University's Department of Science and Technology Studies in 2000.

Andreas W. Daum is a John F. Kennedy Memorial Fellow at the Minda de Gunzburg Center for European Studies at Harvard University. From 1990 to 1996, he taught as an assistant professor of modern history at the University of Munich (Germany), and he was a Research Fellow at the German Historical Institute in Washington, D.C. from 1996 to 2001. Daum has worked on the history of popular science, Alexander von Humboldt, and German-American relations during the cold war. He is the author of *Wissenschaftspopularisierung im 19. Jahrhundert* (Munich: Oldenbourg, 1998; paperback forthcoming).

Elizabeth A. Hachten is Associate Professor of History at the University of Wisconsin–Whitewater. She works on the history of bacteriology and public health in late Imperial and early Soviet periods of Russian history, and is now completing *Disciplining Disease: Science and Practice of Public Health in Russia, 1860–1930.*

Lynn K. Nyhart is Associate Professor and Chair of the Department of the History of Science at the University of Wisconsin–Madison. She is the author of *Biology Takes Form: Animal Morphology and the German Universities, 1800–1900* (1995) and is currently working on a book project titled *Civic Zoology and the Rise of the Environmental Perspective in Germany, 1848–1914.*

Theodore Porter is Professor of History and member of the program in history of science and medicine at the University of California, Los Angeles. His books include *The Rise of Statistical Thinking, 1820–1900* (1986) and *Trust in*

Numbers: The Pursuit of Objectivity in Science and Public Life (1995). He is also a coauthor of *The Empire of Chance: How Probability Changed Science and Everyday Life* (1995). Most recently he has coedited, with Dorothy Ross, *The Cambridge History of Science, Volume 7: Modern Social Sciences* (2002). His paper in this volume is part of a larger project on the wide-ranging researches and passions of Karl Pearson's early career, and how Pearson put them together into a statistical program of scientific reformation that defined his life's work from about 1892.

H. Glenn Penny is Assistant Professor of Modern European History at the University of Missouri–Kansas City. He has published articles on ethnographic and history museums, edited a volume on the history of German anthropology, and just completed a book manuscript titled *Museum Chaos: Cities, Collecting, Displays, and the Ethnographic Project in Germany, 1868–1914*.

Jessica Wang is Associate Professor of History at the University of California, Los Angeles and the author of *American Science in an Age of Anxiety: Scientists, Anticommunism, and the Cold War* (1999). She is currently working on a project on science, technology, and New Deal political thought.

Zuoyue Wang is Assistant Professor of History at California State Polytechnic University, Pomona. His research interests include science and politics in China and the United States in the twentieth century. He is at work on a history of the U.S. President's Science Advisory Committee during the cold war and on a study, with Benjamin Zulueta, of Asian-American scientists and engineers.

Index

SUGGESTIONS FOR CONTRIBUTORS TO OSIRIS

OSIRIS is devoted to thematic issues, often conceived and compiled by guest editors.

1. Manuscripts should be **typewritten** or processed on a **letter-quality** printer and **double-spaced** throughout, including quotations and notes, on paper of standard size or weight. Margins should be wider than usual to allow space for instructions to the typesetter. The right-hand margin should be left ragged (not justified) to maintain even spacing and readability.

2. Bibliographic information should be given in **footnotes** (not parenthetically in the text), typed separately from the main body of the manuscript, **double-** or even **triple-spaced,** numbered consecutively throughout the article, and keyed to reference numbers typed above the line in the text.

a. References to **books** should include author's full name; complete title of the book, underlined (italics); place of publication and publisher's name for books published after 1900; date of publication, including the original date when a reprint is being cited; page numbers cited. *Example:*

[1]Joseph Needham, *Science and Civilisation in China,* 5 vols., vol. I: *Introductory Orientations* (Cambridge: Cambridge Univ. Press, 1954), p. 7.

b. References to articles in **periodicals** should include author's name; title of article, in quotes; title of periodical, underlined; year; volume number, Arabic; number of issue if pagination requires it; page numbers of article; number of particular page cited. Journal titles are spelled out in full on first citation and abbreviated subsequently. *Example:*

[2]John C. Greene. "Reflections on the Progress of Darwin Studies," *Journal of the History of Biology,* 1975, *8*:243–272, on p. 270; and Dov Ospovat, "God and Natural Selection: The Darwinian Idea of Design," *J. Hist. Biol.* 13(2) (1980):169–174, on p. 171.

c. When first citing a reference, please give the title in full. For succeeding citations, please use an abbreviated version of the title with the author's last name. *Example:*

[3]Greene, "Reflections" (cit. n. 2), p. 250.

3. Please mark clearly for the typesetter all unusual alphabets, special characters, mathematics, and chemical formulae, and include all diacritical marks.

4. A small number of **figures** may be used to illustrate an article. Line drawings should be directly reproducible; glossy prints should be furnished for all halftone illustrations.

5. Manuscripts should be submitted to OSIRIS with the understanding that upon publication **copyright** will be transferred to the History of Science Society. That understanding precludes OSIRIS from considering material that has been submitted or accepted for publication elsewhere.

OSIRIS (SSN 0369-7827) is published once a year.

Subscriptions are $50.50 (hardcover) and $33.00 (paperback).

Address subscriptions, single issue orders, claims for missing issues, and advertising inquiries to *Osiris,* The University of Chicago Press, Journals Division, P.O. Box 37005, Chicago, Illinois 60637.

Postmaster: Send address changes to *Osiris,* The University of Chicago Press, Journals Division, P.O. Box 37005, Chicago, Illinois 60637.

OSIRIS is indexed in major scientific and historical indexing services, including *Biological Abstracts, Current Contexts, Historical Abstracts,* and *America: History and Life.*

Hardcover edition, ISBN 0-226-07371-8
Paperback edition, ISBN 0-226-07372-6

Osiris

A RESEARCH JOURNAL DEVOTED TO THE HISTORY OF SCIENCE AND ITS CULTURAL INFLUENCES

EDITORIAL OFFICE
BMW CENTER FOR GERMAN & EUROPEAN STUDIES
SUITE 501 ICC
GEORGETOWN UNIVERSITY
WASHINGTON, D.C. 20057-1022 USA
osiris@georgetown.edu